Creativity and the Performing Artist

Explorations in Creativity Research

Series Editor

James C. Kaufman

Creativity and the Performing Artist

Behind the Mask

Paula Thomson and S. Victoria Jaque
California State University, Northridge, CA, United States

ELSEVIER

ACADEMIC PRESS
An imprint of Elsevier
elsevier.com

Academic Press is an imprint of Elsevier
125 London Wall, London EC2Y 5AS, United Kingdom
525 B Street, Suite 1800, San Diego, CA 92101-4495, United States
50 Hampshire Street, 5th Floor, Cambridge, MA 02139, United States
The Boulevard, Langford Lane, Kidlington, Oxford OX5 1GB, United Kingdom

Notices

Knowledge and best practice in this field are constantly changing. As new research and experience broaden our understanding, changes in research methods, professional practices, or medical treatment may become necessary.

Practitioners and researchers must always rely on their own experience and knowledge in evaluating and using any information, methods, compounds, or experiments described herein. In using such information or methods they should be mindful of their own safety and the safety of others, including parties for whom they have a professional responsibility.

To the fullest extent of the law, neither the Publisher nor the authors, contributors, or editors, assume any liability for any injury and/or damage to persons or property as a matter of products liability, negligence or otherwise, or from any use or operation of any methods, products, instructions, or ideas contained in the material herein.

British Library Cataloguing-in-Publication Data
A catalogue record for this book is available from the British Library

Library of Congress Cataloging-in-Publication Data
A catalog record for this book is available from the Library of Congress

ISBN: 978-0-12-804051-5

Publisher: Nikki Levy
Acquisition Editor: Nikki Levy
Editorial Project Manager: Barbara Makinster
Production Project Manager: Lisa Jones
Designer: Matthew Limbert

Typeset by MPS Limited, Chennai, India

Contents

Section V Professional Reality of a Performing Artist

Foreword

Before I fell in love with creativity research—before I fell in love with psychology—I fell in love with theater. Some of my favorite childhood memories include seeing *Into the Woods* in its original pre-Broadway San Diego tryout on the heels of discovering *Sweeney Todd* via a local college production (and, soon, the video with George Hearn). From *The Fix* to *Arcadia* to *Weird Romance* to *House of Blue Leaves* to *Cabaret*, I have had a special place in my heart for the magic that happens on stage.

Much of the core creativity research and theory is rooted in the arts, but writing and visual art have been more traditional subjects. Yet studies of acting, dancing, and other performance-based arts have continued to grow, and it is my delight to introduce *Creativity and the Performing Artist: Behind the Mask* as the latest volume to the *Explorations in Creativity Research* series from Academic Press. Paula Thomson and Victoria S. Jacque have produced a comprehensive, meaningful work that is readable, grounded in the literature, and of great interest to scholars and performers alike. I hope that you will enjoy this book as much as I have!

If you are new to the series, there are two previous authored books, *Creativity in Engineering: Novel Solutions to Complex Problems*, written by David Cropley, and *Domain Specificity of Creativity* by John Baer. In addition, there are three edited volumes: *Animal Creativity and Innovation* (edited by Allison Kaufman and myself), *Video Games and Creativity* (edited by Garo Green and myself), and the forthcoming *The Creative Self: How Our Beliefs, Self-Efficacy, Mindset, and Identity Impact Our Creativity* (edited by Maciej Karwowski and myself).

There are more exciting books slated—and if you are a fellow creativity scholar, do feel free to drop us a line and propose a volume!

James C. Kaufman
Neag School of Education
University of Connecticut

Preface

If we shadows have offended,
Think but this, and all is mended,
That you have but slumber'd here
While these visions did appear.

A Midsummer Night's Dream, *Act V, Scene II, Lines 54–57*
Shakespeare, 1554/1974

In writing this book, the authors reveal the creative nature of performing artists. The authors add to the field of creativity research by focusing directly on the performing arts. They address the imbalance in current creative arts studies. This book explores the creative process and the creative person; in the performing arts field, investigating the creative product is fundamentally the study of the person. Performers are both the creative person and the creative product. Their primary goal is to express art with an audience witnessing it (American Heritage Dictionary, 2011). Simply stated, performers are the communicators in a public performance. Expression is manifested via performers' bodies; they produce sound and movement for artistic expression, although they may interact with other objects such as musical instruments, props, or sets. Anyone interested in the performing arts, including the performers themselves, will find this book relevant; they will gain insight into the individuals who dedicate their lives to sharing artistic expression with their audience.

Performers are regarded as interpreters (actors, comedians, dancers, magicians, circus artists, musicians, singers) (Kogan, 2002; Thomson, Keehn, & Gumpel, 2009); however, many of the interpreters are also generators (composer, writer, choreographer) or directors (stage directors, conductors, film directors, directors of photography). Essentially, performing arts is a collaborative art form that involves performers, generators, design teams, managers, and the audience; collectively they give the performance life. The product is always dependent on the expressivity of the interpreters/generators and the receptivity of the audience; it is an interactive and intersecting experience that involves person, process, and product.

Beyond the intersecting relationships among key figures working in the performing arts, performers themselves add a unique perspective. They do more than interpret a script, score, or choreography. Performers add their own personal statement and vision. The collective performance product has its own aesthetic that exists beyond each of the single factors (Osipovich, 2006). It has temporal immediacy and human actions and reactions. For example, for audience members who attend performances, the most important aspect of their theater-going experience is being engaged at an emotional level (involvement and empathy), followed by their cognitive responses (complexity of the work and thought-provoking impulses) (Boerner & Jobst, 2013). Personal engagement with the performance is much stronger than influences such as the physical environment of the theater and the amenities associated with it (Boerner, Jobst, & Wieman, 2010). The authors appreciate the powerful engagement of the audience; they consciously include all who influence and shape the aesthetic appeal and expressivity of performers.

Investigating the performing arts requires awareness that complex interacting variables coexist at all times (Sawyer, 2006a). Throughout this book, the authors address this complexity. They provide an interdisciplinary examination that encompasses sociopolitical, aesthetic, development, biology, genetic, cognitive, emotion, personality, temperament, economic, talent assessment, psychopathology, and performance-related health issues. Currently, empirical research studies primarily investigate Western performing artists. This uneven distribution is reflected in this book; however, the authors include studies from other regions of the world whenever possible. There is also an imbalance in studies that investigate specific performing art domains; for example, musicians followed by dancers are the most frequently studied, with minimal research investigating singers and actors.

This book is organized into five sections with chapters included that elucidate these larger topics. In the first section of this book, Creativity Theory and Performing Artists, major integrative models are outlined. A brief description and definition of the major domains that exist within the performing arts are followed by an elaboration on the domains and talents of performers. Lastly, the experience of creativity and the imaginative engagement that operates within performing artists is discussed in Section I. The second section, Biological Manifestations of Creativity in Performing Artists, offers some exploration into the neurobiology and physiology of creativity and flow. Because of the complexity of neurobiology and physiology, the chapters in the second section offer a synthesis of information that is directly pertinent to creativity in the performing arts. The third section, Development of the Performing Artist, explores the developmental trajectory of performing artists, including early attachment, parenting, play theories, personality, motivation, and training. The fourth section, Emotion: Regulation, Resilience, and Pathology, examines self-regulation (i.e., emotion regulation, resilience), mood, unresolved mourning, PTSD, and psychopathology (i.e., anxiety, mood disorders, eating disorders). The fifth section, Professional Reality of a Performing Artist, outlines the realities of a performing career, including injury, rehabilitation, health concerns, and burnout.

The primary focus of this book is to see behind the performance mask, to reveal the unique qualities of creativity that reside within creative interpreters. Performers are familiar with the spotlight on stage, but in research investigations, they are seldom included as examples of creative artists. It is now time to place them in the spotlight of creativity research. The authors shine a spotlight on these talented, disciplined, and courageous artists. Their journey from childhood to elite performer is revealed, along with myriad challenges and achievements inherent in a performing arts career. Unlike Puck's soothing suggestion to the audience at the end of *A Midsummer Night's Dream*, we move performers out of the shadows and offer visions that reveal a story rife with contradictions and complexity.

Author Biographies

Paula Thomson, PsyD, is professor in the Department of Kinesiology, California State University, Northridge (CSUN). She is a licensed clinical psychologist and works in private practice in California. She is codirector of the Performance Psychophysiology Laboratory at CSUN, adjunct faculty at Pacifica Graduate Institute, OperaWorks faculty, and professor emeritus at York University's Departments of Theatre and Graduate Studies (Canada). She was a professional dancer and continues work as choreographer and movement coach in dance, theater, and opera. Past professional choreographic company work includes Canadian Opera Company, Canadian Stage Company, Stratford Shakespearean Festival, Northern Lights Dance Theatre, Ballet Jorgen, and UCLA On the Edge of Chaos. In 2013, she was named one of the top 20 female professors in California.

S. Victoria Jaque earned her PhD in Exercise Science from the University of Southern California, and is professor and graduate coordinator in both the Department of Kinesiology and Assisted Technology Studies and Human Services at California State University, Northridge (CSUN). She is codirector of the Performance Psychophysiology Laboratory at CSUN. An exercise physiologist, she has conducted research in a variety of areas, including the influence of physical activity on the development of peak bone mass in humans and rodents. Her current research focuses on developing a better understanding of how the autonomic nervous system responds to physiologic and psychological stressors in performing artists. She is an active researcher in laboratory and field settings (concert halls, studios).

Acknowledgments

Thank you is extended to the following individuals who offered their expertise and support: Ann Baltz; Maurice Godin; Dr. Danielle Jarvis; Dr. James Kaufman; Dr. Donna Krasnow; Amy Pell; California State University, Northridge, CA, USA; York University, Canada; and OperaWorks.

Section I

Creativity Theory and Performing Artists

Understanding creativity in the performing arts

1

On those nights, the words were for me alone. They came up unbidden from my heart. They spilled over my tongue and spilled out my mouth. And because of them, I, who was nothing and nobody, was a prince of Denmark, a maid of Verona, a queen of Egypt. I was a sour misanthrope, a beetling hypocrite, a conjurer's daughter, a mad and murderous king.

Jennifer Donnelly, Revolution

Definitions and expert evaluations

Creativity is commonly defined as something that requires novelty (or originality) and usefulness (or effectiveness), features that describe a creative product (Sternberg, 2001). Although both features are essential, research has demonstrated that novelty is the largest predictor of creativity and indirectly influences usefulness (Diedrich, Benedek, Jauk, & Neubauer, 2015). This bipartite definition emerged as a formal explanation by Barron (1955) and Stein (1953) and it continues to be cited in most works investigating creativity (Runco & Jaeger, 2012). When looking at creativity in the Western tradition of the performing arts, an even earlier description exists. In *Poetics*, Aristotle (c.335 BCE/1958) claimed that poetry (which included music, dance, and drama) is imitation and that this imitation must transcend the mundane replication of an object or event and "startle" the audience. He outlined the elements of creativity long before the term came into existence, claiming that the production of art imitates life (character, actions, emotions) through melody and rhythm. For Aristotle, the usefulness of imitation was that it revealed humanity, whether accurately or as the artist wished the audience to perceive it. Imitation was considered art if it followed the basic principles of dramatic form and if the audience recognized that which was expressed.

Aristotle alluded to two other factors contained in the traditional definition of creativity. When something is regarded as novel or original the product must have a sense of authenticity. Plato, on the other hand, suggested that the artistic product must contain the inspiration of divine power. According to Plato (c.338 BCE/1976), artists are able to create authentic and meaningful works because the divine speaks through them. Artistic works are based on an inherent truth contained within imitation and inspiration (Box 1.1).

Creativity and the Performing Artist. DOI: http://dx.doi.org/10.1016/B978-0-12-804051-5.00001-9

Box 1.1 Definition of Creativity

Creativity is commonly defined as something that requires novelty (or originality) and usefulness (or effectiveness), features that describe a creative product. Authenticity and aesthetic appeal are now added to the two-criterion definition.

In the performing arts, not only must the generator (composer, choreographer, playwright) provide a sense of inspired authenticity, but also so must the performers who are expressing the work of art. Further, within the performing arts, a keen sense of aesthetic appeal must be incorporated into the novelty and effectiveness of the works performed. Aesthetic appeal may in fact be the major value within a piece of music, a dance work, or a theatrical production. Although aesthetic appreciation in the performing arts may be bound by time and place, many performance works transcend time and place in the audiences' lived experiences. This capacity to transcend time and place reflects the novelty and effectiveness of eminent works of art. Ultimately, a sense of meaningfulness is elicited during a transcendent aesthetic experience. It is highly rewarding, as evidenced by subjective reports and brain imaging that demonstrate activation of the reward centers in the brain (Martindale, 2007).

Kharkhurin (2014) suggested that by including authenticity and esthetics to novelty and usefulness (effectiveness) the two-criterion bipartite theory would become a four-criterion construct of creativity, and that this construct would capture what is considered creative within the arts. Certainly this proposal comes closer to how the creative qualities within the performing arts are perceived. Weisberg (2015) suggested that the subjective determination of usefulness or effectiveness is a judgment about the value of the creative work. This variable may change over time because what is viewed as novel in one generation may be dismissed as old fashioned and boring by a different generation. This is evidenced by each new generation's aesthetic bias regarding popular music. For example, the Beatles were once regarded as one of the greatest popular bands ever; however, the current generation may view them as simplistic and irrelevant (Gibson & Mumford, 2013; Kirschenbaum, 1998). By including intentionality to novelty, Weisberg believed that we could contextualize the creative product's novelty based on the intention at the time the work was produced. This would counter the subjective valuing that changes over time. Weisberg recommended that the criterion of value (useful, effective) be replaced by intentionality. If this was practiced, and the other two criteria suggested by Kharkhurin (2014) were included, the four-criterion model would consist of novelty, intentionality, authenticity, and esthetics.

Expert analyses regarding how artists guide the audiences' perceptions are important in the performing arts. Aristotle can be regarded as one of the earliest critics, and his expertise, although challenged by other experts such as Plato, was given serious consideration (c.335 BCE/1958). This practice of criticism has persisted throughout the production of Western performing arts. Philosophers and historians were originally regarded as the primary experts (Aristotle, c.335 BCE/1958), and later the formal field of professional criticism emerged. Critics are deemed to be experts and they determine whether a product is novel and effective; fundamentally, they discern if it is creative or not. In research today this assessment process is termed the consensual assessment technique (Amabile, 1996; Kaufman & Baer, 2012; Kaufman, Cropley, Baer, Reiter-Palmon, & Sinnett, 2013). Consensual assessment is considered the gold standard to evaluate whether a product is creative, in part because being recognized as an expert means that he or she has attained recognition within a specific domain. Experts have completed their training and are actively working and contributing at a level that is deemed elite (Amabile, 1996). These experts can then evaluate a product independently and determine if it is novel and effective. In the performing arts, expert evaluators are often the critics who review new productions. Their reviews are shared in multiple media sources such as newspapers, radio, television, or the Internet. Experts from domain-specific organizations that represent artists determine who should be recognized with awards. The most notable is the Oscar award, but each domain has its respective awards. The practice of experts judging a creative product implies that there is consensus among the panel of judges (Simonton, 2004). This consensus can be seen in the overlap of awards given by different organizations to the same artists or productions. An example is a pattern of awards given to the same actor or film by organizations, such as the Golden Globe, Screen Actors Guild Awards, and the Academy Awards (Oscars). This practice of expert assessment has existed for centuries in the performing arts, including in research conducted on creativity in the performing arts (Simonton, 2004).

Often the evaluation of performers is conducted to determine whether the performers have the capacity to effectively express the creative product. Artistic directors identify and engage generators and interpreters to create new productions, and educators follow the practice of auditioning, then accepting, and finally training performers. These educators commonly identify an individual as talented and trainable. For those recognized as talented and who enter specific performance domains, years of training follow. Research suggests that it takes approximately 10 years of deliberate domain-specific training to become an expert; then these trained performers audition once again to move into a professional performance group (Ericsson, Roring, & Nandagopal, 2007; Gardner, 1993; Simonton, 1999). Only when this is achieved do performers begin to establish their place as professionals; however, it may take even more years before they are considered experts within their specific domain.

What is unique to the performing arts is that a consensual assessment of the creative product must also include an assessment of the performers and the generators of these artistic domains. Performers (person) are integral to the product. They are both the creative product that is to be assessed and the person who creates and gives life to the generators' vision. To fully appreciate the complexity of the performing arts, both the production and the performer must be assessed to determine whether the performance is novel, effective/intentional, authentic, and aesthetic (Fig. 1.1).

Figure 1.1 "Rehearsal: A View Through the Mirror." Dancer: Jessica McCann. Photographer: Lee Choo.

Integrative models

Investigating creativity in the performing arts must include the three traditional dimensions studied in creativity research: person, process, and product. Assessing these three overarching dimensions has provided substantial information (Ma, 2009). A fourth dimension that has frequently been examined, and is relevant to performing arts, is studying "environmental" influences on creativity (Rhodes, 1961/1987; Runco, 2004). Key variables offered within an environment (family, community, cultural values, sociopolitical views, financial resources, optimal work settings, and opportunities) enhance creativity. These include challenge, intellectual stimulation, positive collegial exchanges, material resources, support, and a perceived sense of autonomy (Hunter, Bedell, & Mumford, 2007). The environmental climate has a powerful influence on the success of the created product, as evidenced in all rehearsals and performances.

The four dimensions (person, process, product, and environment) can be explored independently; however, the collaborative nature of the performing arts indicates that more complex and integrative models may add to our knowledge. Attempts to evaluate complex psychological variables within the individual are important, such as examining the interactive effectiveness of creativity training, determining the environmental variables that promote or deter creativity, and investigating the relationship of cognitive and personality factors (i.e., intelligence, imagination, and openness to experience) (Ma, 2009). Further, there are a set of variables and a sequence of actions that converge during the making of each performance (Lubart, 2001; Sawyer, 2006a). This multidimensional concept of creativity, although often appearing unwieldy and unfocused, shares similarities with "fuzzy" concepts, and like "fuzzy" theories, researchers strive to grapple with this complexity (Fryer, 2012).

During the first wave of research that investigated creativity, the focus was primarily on the individual creator. Personality studies examined who were the eminent figures and what made them uniquely creative. Unfortunately, during this first wave of investigation, few researchers studied performing artists, in particular, because the focus was primarily on eminent figures (generators) (Sawyer, 2006a). This wave of research led to a second wave that expanded the attention beyond the individual creator to the social and cultural dimensions of creativity. This shift led to a focus on how creativity is distributed across people and environments. Larger systems theories were developed to demonstrate how the individual was influenced by the surrounding field of others working in a specific domain, (1) the uniqueness of the domain itself, (2) the body of works existing in the domain, and (3) general knowledge acquired to work within the domain (Csikszentmihalyi, 1996, 1999). During this second wave, understanding the role of collaboration and group dynamics in creativity also expanded (Farrell, 2001). This second wave has introduced complex theories to examine the nature of creativity—a practice that is particularly important when studying the performing arts.

It is undeniable that the performing arts are creative, although they are often regarded as the poor cousin to the fine arts such as literary writing, visual art, and musical composition. Now that a more diverse approach to investigate creativity exists, the complex factors operating and intersecting within the performing arts may finally be understood and appreciated. Certainly, the works of Mozart or Bach would not survive without the great musicians who recreate and reinspire this music. And Shakespeare, although appreciated when read, springs to life when actors inhabit the characters and resuscitate Shakespearean verse. Studying the performing arts necessitates that creativity is examined at the intersection of a complex array of variables (Batey, 2012). Examining some of these second wave models may help identify the creative properties inherent in the performing arts. The next section will briefly examine these theories, and how they relate to the performing arts.

Heuristic assessment: three-dimensional and geneplore models

A heuristic basis of assessing creative potential enables researchers to examine the strategies that underlie creativity (Vessey & Mumford, 2012). By focusing on the underlying heuristics, common themes may emerge that operate during creativity in the performing arts. One such heuristic model outlines a three-dimensional matrix (Batey, 2012). In this model, the level of creativity to be assessed is determined by a decision to focus on the individual, the group, the organization or the culture. These four focal points can be explored independently and then interactively. The second heuristic dimension examines the synergistic effects of the four traditional dimensions (facets) of person, process, product, or environment. Lastly, the third dimension explores the measurement designs to investigate creativity. It is proposed in this matrix model that researchers should analyze the interactive effects on these three dimensions (level, facet, and measurement). This heuristic practice, when applied to the performing arts, may address the layers of variables that coexist.

The geneplore model describes two main creative processing phases, a generative phase and an exploratory phase (Sternberg, 2005). In the generative phase, memory retrieval, mental synthesis, and transformation facilitate the emergence of novel ideas or associations. Conscious and unconscious processes operate during this phase and they are equally important during the exploratory analysis of the ideas or associations (Finke, 1996). The importance of this model is that it resembles the creative practice in the performing arts. It is readily applied to the generators of the performing arts, and to a lesser degree, the interpreters. Playwrights, composers, and choreographers generate new works that are then explored by both the generators and the interpreters. Both invent ways to express these performance works (Bourgeois-Bougrine et al., 2014; Rocamora, 2008). Although this two-phase model is generally applied to sole creators (Hagood & Kahlich, 2007), the collaborative nature of the performing arts necessitates that this process is distributed among several artists. It is a model that many researchers employ when investigating the performing arts (Stevens, Malloch, McKechnie, & Steven, 2003; Stevens & McKechnie, 2005).

Componential model of creativity

Amabile (1996) proposed a componential model of creativity to include the confluence of multiple variables. In this model, Amabile included three dominant components: domain-relevant skills, creativity-relevant skills, and task motivations. Within each component, specific factors are needed and they are dependent on other factors to be realized. For example, within the domain-relevant skills, individuals must have knowledge, technical skills, and special talents within the domain and these variables are contingent on innate cognitive, motor, and perceptual abilities in combination with formal and informal education. Creative-relevant skills include appropriate cognitive styles (divergent thinking, remembering accurately, perceiving novelty), implicit or explicit knowledge regarding idea generation, and effective work styles. These skill sets are dependent on past training and experience in generating new ideas coupled with personality traits that support creative exploration. Within task motivation, individuals must perceive that they are initiating the motivation to undertake the task and that the task is worth pursuing. Task motivation is dependent on intrinsic motivation, extrinsic constraints and rewards, and an ability to minimize the constraints.

According to the component model of creativity, these three skill domains must be operationalized during the process of creating a new work. This process includes identifying a problem or task, preparing to solve the problem, generating response possibilities, evaluating the possibilities, and ultimately selecting responses to complete the task. When applying this to the creation of a new performance, the generator may engage in this process alone during the initial formation stage; however, the performers and the other creative interpreters will participate in the final creation of the product. All who are involved require domain-relevant skills, creativity-relevant skills, and task motivation to create the product. A confluence of these skills will inform the creative collaboration involved in realizing the performance. What is unique about this model is that it has an inherent set of feedback loops (Amabile, 1996) that support the recursive process of collaborative creation in the performing arts. This model has received robust empirical study and has proven to be reliable and valid; further, it has stimulated other formulations of complex models to examine creativity (Weisberg, 2006).

The Darwinian evolutionary theory of creativity

In the Darwinian evolutionary theory, Simonton (199b) described the essential interactions of multiple factors such as genetics, child rearing, age, birth order, intelligence, personality, mental illness, productivity, and contemporary social and cultural conditions. These factors all converge to support the inheritance of creative eminence or genius (Martindale, 2007; Simonton, 1999b). If an individual possesses several of these factors, the possibility of reaching eminent creativity is then dependent on three conditions: first, generation of solutions to solve novel problems; second, selection of a solution or solutions; and third, preservation and reproduction of these solutions in a creative product (Simonton, 1999a; Weisberg, 2006). These conditions usually operate without knowledge of the end product. Like evolutionary theory, chance mutations of the species shape survival and this holds true for creative thought as well. For creative thought, a broad knowledge base supports loose associations that are available for subsequent permutations. According to Simonton, idea generation is more intuitive and less analytical; creative thought is shaped by chance, either by externally derived random interactions and serendipitous moments or by unconscious mental processes that emerge during the creative process (Weisberg, 2006). Applying Simonton's evolutionary theory to the performing arts offers a unique perspective. It can readily be argued that the interpreters of the performing arts draw from external and internal sources and that chance moments during a rehearsal process frequently give shape to a performance or a production. This is most evident in improvisational performances by jazz musicians, contact improvisational dancers, freestyle urban street dancers, and by actors who perform improvisational theater (Sawyer, 2000; Fig. 1.2).

Investment and propulsion theory

According to the investment theory by Sternberg and Lubart (1991), creative people are willing and able to buy low and invest high in the realm of ideas. Creative individuals persist despite adversity and eventually their creative product is realized and recognized. This process is certainly true in the performing arts, although

Figure 1.2 "Jazz ensemble." Performers: Adam Hersh (piano), Daniel Massey (bass), Kirk Portuguez (drums), Jordan Leicht (tenor saxophone). Photographer: Lee Choo.

buying low is a debatable criterion. The generators of new works, and the performers who give these works life, spend years training and preparing, even though the actual work of producing a new work may take place in a relatively short span of time. According to the investment theory, six distinct but interrelated resources must converge. They include intellectual abilities, knowledge, thinking styles, personality, motivation, and environment (Sternberg, 2012). These resources can be divided into five person-centered and one context-centered resources (Zhang & Sternberg, 2011). Ultimately, creativity involves all these resources and they operate as a system (Csikszentmihalyi, 1999). What is unique about the investment theory is that it anticipates that the initial presentation of a new creative product will be soundly rejected. For example, Stravinsky's *Rite of Spring* provoked riots in the theater when it was first performed. Bizet's *Carmen* was ridiculed and the opera was immediately closed. Years later they were resurrected and hailed as masterpieces. From the investment theory, these examples reflect the common pattern of society's initial rejection. The presentation of a novel work often means that the performers and generators of these artistic works must gradually influence society by repeatedly performing them. Over time these works are then perceived as valuable and important; however, the performing artists have usually moved on to explore other novel works. This is the repeated pattern of buying low and selling high in the investment theory (Sternberg, 2012).

Within the investment theory, Sternberg and colleagues suggested that creative individuals make different decisions based on how to express their creativity and how they will invest in their creative resources. They are propelled to move their ideas forward, whether through replication or major redirections (Sternberg, 2006). This propulsion theory can be observed in the performing arts. Major works that exist in the traditional repertoire of theater, music, and dance companies are repeatedly reexplored and performed in different settings and with different performers.

It could be argued that performing artists are propelled to reinvestigate plays, scores, and dances. The propulsion theory outlines multiple styles of creative thinking that are engaged to evoke change. These include redefinitions of the current status of the field, incremental forward direction, redirection of the field either toward a new focus or a return to a previous focus (a variation on reinitiation), or a synthesis of several fields into one (Sternberg, 2005).

In the performing arts these multiple styles of creativity are readily identified; performing artists consciously or unconsciously implement them. For example, redefining the field of music happened when there was a shift to the 12-tone compositional technique orchestrated by Arnold Schoenberg or when the minimalist movement took place in the 1970s. Composers such as Philip Glass, John Riley, Steve Reich, and John Adams all propelled this minimalist redirection. Incremental forward movement is evident in the increased range of technical skills that are now expected in today's ballet dancers. A redirection is evident in the continuous remounting of past musical theater works, whether on Broadway or at regional theaters around the world. New choreography, music, and staging give the past work a more current appeal, reflecting the increased skill level of the performers and the expectations of the audience. A synthesis of several fields is evidenced in genres such as in musical theater, physical theater, and dance theater. These art forms expect performers to be able to dance, sing, act, and often play music. This propulsion model readily identifies the essential nature of the performing arts; performing artists are highly motivated (propelled) to reinvestigate and redefine their art forms.

Distributed creativity theory

Distributed creativity theory (Sawyer & DeZutter, 2009) describes how creative cognitive processes are distributed across groups. The focus of this research approach is not to examine the results of groups engaging in activities such as during brainstorming activities, but rather what is the interaction between individuals within a group who are creating the new work. Cognitive scientists have studied distributed cognition by conducting interactional analyses that are qualitative and observational. This same approach has been applied in studies investigating distributed creativity. Examining videotaped collaborations over time reveals emergent patterns of distributed creativity (Sawyer & DeZutter, 2009). This form of analysis is ideal when a new performance work is created through a series of improvisations. Improvisational theater companies create exclusively with this approach; however, improvisational explorations are common practices when constructing theater of witness performances or devised theater productions. Jazz musicians actively explore new musical interactions among each other when performing together, and musical bands incorporate improvisation when creating new songs. Pure collaborative emergence of creative performances transpires when there is no set endpoint, each performer's actions are contingent on other group member's actions, and these actions take place in the present moment; however, there are always underlying structures that guide purely improvisational works. For example, jazz performers may rely on familiar musical "licks" or actors may lean on loose everyday plot development. These provide a safety net when performing. Distributed creativity among performers is also contingent on group safety. To some degree individualism needs to be suppressed in order to sensitively respond to group needs, and through this effort the group establishes its own unique identity (MacDonald & Wilson, 2005). The experience of improvising in front of an audience is complex. It emerges on a continuum from completely novel discoveries to small variations on established musical or language (semiotic) structures (Sawyer, 2000), and through this process a group identity emerges that separates the sounds of one jazz band from another and one improvisational theater group from another (MacDonald & Wilson, 2005).

The final criterion for distributed creativity is that all participants are equal contributors in the process. This last element is important, especially if the performance work is completely improvisational and is presented to an audience in real time. What makes improvisation a powerful source of distributed creativity is that, like all complex systems, emergent properties manifest that are far greater than any one individual element (Lewis & Granic, 2000; Sawyer, 2000). Emergent phenomena are unpredictable and hard to explain or anticipate based on the group's component parts (Sawyer, 2006b). Emergence is readily recognized when participating in or observing improvisational theater or jazz ensembles (Montuori, 2003). The unpredictable direction of an improvisation may lead performers into modes of expression that they never anticipated. When a group is deeply connected and a new idea emerges the group may experience a flow-like state, something that is usually reserved for individual experiences (Csikszentmihalyi, 1990). The entire group senses flow (a positive feeling of effortless performance) and each member of the group has equally joined in the experience. The nature of group flow is also an emergent phenomenon; however, like all emergent properties, it is difficult to predict a flow experience before the performance takes place (Sawyer, 2006b).

Social network theory and small world groups

Unlike improvisational groups that are equally distributed, most performing arts experiences are not equally distributed; in fact, performing arts are typically hierarchically organized. They are tightly constrained by the repertory to be performed by orchestras, theater, or dance companies and by traditional hierarchical structures (Montuori, 2003). There is usually one identified leader who determines the final product that will be prepared and performed in front of an audience. Once there is an identified leader, then a pure collaborative emergence is not possible; however, group flow can still occur despite the hierarchical arrangement. For example, group flow can be experienced among performers who are performing scores, plays, or dances and a sense of improvisation can emerge, especially when someone in the group makes a mistake. The group can adapt to momentary mistakes and return to the set performance dictated by the script, score, or choreography. Group flow and a sense of spontaneous adaptation demonstrate the collaborative nature of performing groups.

Many performing arts groups operate like small world groups (Uzzi & Spiro, 2005). In small world groups, the network of artists tends to be densely clustered with individuals who have direct or indirect knowledge of each other and of the skills that they possess (Schilling, 2005). This clustering ensures the maintenance of a theatrical convention. This can be witnessed in the deep traditions of Broadway musical theater productions, the sound and identity of a symphony orchestra, or the style of a ballet company. Innovation is guaranteed when these networks also include outside artists who are not part of the small world network. Introducing an outside artist into the small world group may inspire the group and provoke a sense of invention, but it is a curvilinear pattern because too many outside artists will not maintain enough convention to ensure cohesiveness in the performance genre (Uzzi & Spiro, 2005). Artists who continually participate within the small world grouping give it stability; however, a threshold effect necessitates that new individuals are introduced into the small world network. This unique paradigm is very relevant to the majority of performing arts domains. Densely clustered networks operate whether at the local level of small fringe performances or internationally among world-class artists who perform as guest artists in local companies.

Although slightly different than the distributed creativity found in improvisational groups, small world groups engage in creation at the intersection between members of the group (Sawyer & DeZutter, 2009). In general, within performing arts groups the small world network ties are considered weak, in that the members of the group don't have to socialize nor are they dependent on each other for daily emotional and interpersonal security (Perry-Smith & Shalley, 2003). The weak ties guarantee shared skill sets and shared performance experiences but there is enough separation to enhance novelty among the performers. These weak network ties facilitate creativity, as does performing on the periphery rather than occupying a central position in the organization. When performing artists are continually forced to occupy a central role, or if the small world network is comprised of strong ties (dependent on each other for personal security), the small world network will begin to constrain creativity (Baer, 2010a; Perry-Smith & Shalley, 2003). Once again a curvilinear balance of relational ties and periphery-central positioning is optimal to support creativity in social networks that operate within small world groupings.

Systems model

The systems model (Csikszentmihalyi, 1996, 1999) is an integration of cultural, social, and psychological processes. Because the phenomenon of creativity is complex, a systems approach essentially captures the dynamics of intrapsychic processes interacting with intersubjective fields (Csikszentmihalyi, 1999). In this three-part model, individuals, referred to as person (part one), create in an environment that is both cultural and symbolic, and is referred to as the domain (part two). The social aspects within the domain are referred to as the field (part three). The system is comprised of an interaction between individuals, domains, and fields (Csikszentmihalyi, 1999). The implicit and explicit "rules" of these three components must be internalized to enable creative fluidity. This internalization provides knowledge regarding what is viewed as relevant, novel, and appropriate. Having this knowledge enables creative individuals to function and communities to nurture creative individuals. According to Csikszentmihalyi, creativity cannot be recognized unless it operates within a system of cultural rules and is supported by social acceptance.

This systems model captures the essence of the performing arts. The environment that supports the performing arts recognizes the cultural and symbolic role they play. The social needs of the community are enhanced when they experience the performances, whether in a formal concert hall, a nightclub, or outside on the streets. The artists are nurtured when they are accepted and encouraged to perform. Of note, the social and cultural contexts and the individual performer all integrate within the performing arts, perhaps due to the shared social and cultural values held by the individual artists and the audiences who attend the performances (Glaveanu, 2010). This model assesses the performer embedded within a complex system; however, the same principle can be applied to assess creativity within small world groups and distributed organizations; they are social creative phenomena that resemble an individual creator working within a complex system (Watson, 2007). Socially constructed entities, such as Dreamworks or Pixar, can be assessed as creative organizations in the same way the systems model describes individual creativity (Fig. 1.3).

Figure 1.3 "Intensity." Performers: Kiah Childress, John Ghera, Tiffany Grant. Photographer: Lee Choo.

Complexity theory and creativity classification system

Complexity theory suggests that random chaos is also self-organizing; the integration of elements that are perceived to be chaotic actually have emergent properties that reorder themselves (Lewis & Granic, 2000). Research studies have demonstrated that creative individuals are comfortable with complexity and can tolerate more chaotic ambiguity (Feist, 1998). Creativity can be regarded as a constant effort to organize, disorganize, and reorganize, a hallmark of complex self-organizing systems (Montuori, 2003). The creative classification system places the individual at the center of a chaotic living system and it is the creative individual who is attempting to reorganize the system and the self. Whether creating something new or revamping something already in existence, the goal is to make the system more effective (Kirschenbaum, 1998). In this system, the attributes of the individual (flexible thinker, imagination, curiosity, willingness to engage in tasks, spontaneous, and self-motivated) interact during the creative process in a feedback loop. Through multitasking, insight, seeking stimulation, persistence, and high standards to assess the process and the product, a creative work emerges.

This theory is supported by research that investigated whether collectivism versus individual orientations within a group enhanced creativity. What was found was that individuals within a group were more creative than the collective group (Goncalo & Staw, 2006). This study sheds light on group dynamics in the performing arts. Interpreters and generators operate as individuals who work to create a shared performance work. Even when they appear more divisive and confrontational, this group dynamic can foster greater ingenuity. Kirschenbaum (1998) believes that a careful assessment of these factors will help identify creative individuals and the environmental factors that can support or restrict their creative abilities. This assessment approach can identify individuals with unique talents within a specific domain. By identifying, and addressing, the needs of these creative individuals, it is hoped that order and newness can be enhanced in societies and cultures. Within the performing arts, the need to identify creative individuals and to situate them within a supportive environment guarantees the success of the performing arts. Training programs, performance organizations, and theatrical producers all contribute directly or indirectly to creativity in the performing arts.

Summary

- Definitions and expert evaluations: definitions of creativity; bipartite and four-component criteria (novelty, effectiveness/intentionality, authenticity, esthetics); experts, critics, and the consensual assessment technique.
- Integrative models: first wave of psychology focused on the individual and the second wave focused on social–cultural dimensions.
- Heuristic assessment: three-dimensional matrix model (level, facet, measurement) and geneplore model (generative and exploratory phases).

- Componential model: domain-relevant skills, creativity-relevant skills, and task motivation.
- Darwinian evolutionary theory: internal, external, and chance variables that interact.
- Investment and propulsion theory: six resources (five person-centered and one context-centered); propulsion and styles of creative thinking (redefine, reinitiate, redirect, synthesize, incremental movement forward or backward).
- Distributed creativity theory: creative cognitive processes distributed across groups (interactions, intersections, improvisation, group flow).
- Social network and small world group theories: social networks and hierarchies, small world groups (innovation and convention, weak–strong ties, peripheral-central position, constraints).
- Systems model: intrapsychic and intersubjective fields; individual, social, and cultural values and how they interact and provide support.
- Complexity theory and creativity classification system: self-organizing emergent properties (organize, disorganize, reorganize); the individual as the central organizing figure within chaos; collectivism–individual orientation within groups.

Concluding remarks

The integrative models discussed in this chapter provide guidelines on how to comprehend the interacting and intersecting variables that exist within the performing arts. Ultimately, each model attempts to describe the interrelated factors that make the performing arts creative. The interpreters and generators are dependent on each other to create a product, and the interpreters are also the product that is presented. The people/products engage in a creative process while constructing the performance, and the audience who views the performance completes this process. They share the creative lived experience of the performance on an emotional and cognitive level (Boerner & Jobst, 2013). What makes the performing arts so unique is that participating in the experience of preparing a performance and then performing it in front of an audience amplifies a sense of identity formation, a process that can potentially transpire during aesthetic experiences that shape society and culture (Sullivan & McCarthy, 2009). Audiences and performers can collectively join in a group experience that may transcend to a flow experience. Perceptions can potentially be altered for all involved.

Domains in the performing arts

2

Any actor who tells you that they have become the people they play, unless they're clearly diagnosed as a schizophrenic, is bullshitting you.
Gary Oldman

Introduction: domain-general and domain-specific creativity

Actors, dancers, and opera singers have very different talents, training practices, and skills. Performers and audiences readily recognize their domain-specific talents. For example, the skills of a cello player are different from those of an opera singer, ballet dancer, or jazz musician. However, they also share similar skills relevant to performance, such as expressing works resonant with aesthetic and emotional meaning. Research has addressed a few of these unique differences and similarities but many questions still remain. Some of the questions include: Is creativity domain-specific or domain-general in the performing arts? Are performing artists (interpreters) creative? Preliminary answers to these questions reveal a lack of consensus among researchers. Some researchers claim that creative individuals in one field manifest creativity in other areas (domain-general); whereas, others argue that individuals have unique creative talents that manifest as "islands of creativity" (domain-specific) (Baer, 2010b; Baer, 2015; Silvia, Kaufman, & Pretz, 2009). And other researchers argue that creativity is both domain-specific and domain-general (Sternberg, 2005; Box 2.1).

To adequately examine some of these questions, researchers explored how individuals manage similar creative tasks. They then determined whether the task itself influenced individual creative problem solving (Baer, 2015; Reiter-Palmon, Illes, Buboltz, Cross, & Nimps, 2009). Findings in the Reiter-Palmon et al. study (Reiter-Palmon et al., 2009) demonstrated that creativity was contingent on the task, including the degree of task complexity; however, the level of involvement of the participants also influenced creative problem solving. In a different study, musicians, like engineers, endorsed similar descriptive adjectives to explain their creativity—a finding that suggested creativity was domain-general (Charyton & Snelbecker, 2007). These studies support the domain-general theory of creativity; whereas, other studies have identified domain-specific characteristics. Through a latent class analysis, domain-specific group characteristics were differentiated (Silvia et al., 2009). Performing artists were distinguished as a separate group; they were unlike visual artists and a noncreative group. In other studies, musicians' traits were classified as domain-specific; unfortunately, the other performing arts groups were not included in these investigations (Hass, 2014; Woodward & Sikes, 2015).

Creativity and the Performing Artist. DOI: http://dx.doi.org/10.1016/B978-0-12-804051-5.00002-0

Box 2.1 Definition of Domain-specific and Domain-general

Domain-specific creativity examines the particular elements that are unique for a specific field. For example, a violinist has very different technical demands when compared to a percussionist, ballet dancer, singer, or actor. These technical requirements influence creative demands on the individual.

Domain-general creativity includes all creative demands that cross all fields. A scientist, composer, painter, actor all engage in generating original ideas and evaluating them to determine if the ideas have merit.

Interpreters, generators, and intermediaries

As a group, performing artists are viewed as the interpreters (actors, dancers, singers, musicians) of the performing arts. They are often described as the craft workers who merely give expression to the generators' creative works (playwrights, choreographers, composers) (Sawyer, 2006a); in short, they are regarded as less creative compared to the generators (Kogan, 2002). Despite this assumption, those working within the performing arts are well aware that the interpreters engage equally in the creative process, in fact, performing artists self-define as creative (Copland, 1957). Aaron Copland, an eminent 20th century composer, stated unequivocally that the performing arts must be reinterpreted in order to live. The interpreters, although serving the vision of the composer, playwright, or choreographer, have their own unique experiences that are creatively revealed when performing (Copland, 1957). Copland believed that the audience appreciated and recognized the special relationship that exists between the generator and interpreter.

It is often assumed that the generators are free to creatively explore and produce new works, whereas the interpreters are constrained by their task. This assumption has been contradicted in study findings. Gluck, Ernst, and Unger (2002) demonstrated that free artists (visual artists) and constrained artists (architects, designers) similarly self-define creativity. But there were also differences between the two groups. The free artists claimed, "creative people in their field should be imaginative, talented, risk-taking, unconventional and intellectual ... and hard work(ing)" (p. 64). Constrained artists endorsed "assertiveness, cultural competence and intrinsic motivation and ability to problem solve" (Gluck et al., 2002, p. 64). In a study that examined interpreters and generators within the performing arts, there were no significant differences in fantasy proneness between the two groups. The interpreters actually scored significantly higher on imaginative absorption compared to the generators (Thomson et al., 2009). These findings reinforce the reality that performing artists (interpreters) are creative.

Although both performers and generators work creatively to give life to the performing arts, intermediaries are also essential participants. They help ensure that performances are produced; they directly guarantee that the production reaches the

Figure 2.1 "Callings." Performers: Neil Harvey, Paula Thomson. Photographer: Cylla von Tiedemann.

market place (Kushner, 2003). These intermediaries include artist representation (managers, talent agents, casting agents), presenting organizations (producers, publicists, stage technicians, front-of-house management, ticketing, and box office personnel), policy makers, and artist unions. If the performance is to be recorded then recording studios, sound engineers, film editors, and other technicians will be employed in an effort to deliver the performing arts product to the audience and the market place. These intermediaries help balance supply and demand between performing artists and their audiences. Their creative talents, skills, and dedication are vital to the health and growth of the performing arts industry (Fig. 2.1).

Performing arts domains/fields

For the purpose of this book, performing artists are differentiated from entertainers, a term that describes someone who wants to amuse or please an audience, such as comedians, street performers, or circus performers (American Heritage Dictionary, 2011). This differentiation is subtle because most performers, even if they are not defined as entertainers, desire to "please" their audience. Entertainers, such as standup comedians or magicians, may interact more directly with their audience; however, both performing artists and entertainers engage their audience. What is unique to the performing arts is the reality that the performers and audience gather in a specific venue at a specific time to share the experience of the performance. Performing arts exist in a wide variety of venues, including high-art concert halls, nightclubs, and small black box theaters. Digital media has expanded this experience. Performing artists still perform but it may not be to a live audience, and yet audiences do view their performances, whether it is in a movie theater or in their home.

In 2013, the National Endowment for the Arts (NEA) reported that arts and culture, as a whole, accounted for 3.2% of the United States current-dollar gross domestic product (GDP), higher than travel and tourism. This translated to $504 billion. This figure demonstrates the significant contribution to the US economy. The NEA identifies five separate art and culture fields (domains): dance, opera, music, media arts (film, television, radio, Internet), and theater/musical theater. Within each of these five fields there are a myriad of subcategories. These subcategories may operate as mainstream branches within the field or exist on the fringe. The fields will be discussed in alphabetical order to avoid any implied bias that one field is more valuable or important than the other fields (Tables 2.1–2.3).

Table 2.1 Audience attendance in 2012

	Percentage	Number (millions)
Music		
Jazz	8.1	53.7
Classical	8.8	53.1
Opera	2.1	10.0
Latin	5.1	28.8
Outdoor festivals	20.8	
Plays		
Musical	15.2	72.5
Nonmusical	8.3	42.1
Dance		
Ballet	2.7	9.6
Other	5.6	27.3
Other nonspecified	11.6	
Free nonspecified	15.9	

Source: From NEA, 2015 Report #58.

Table 2.2 Performers engaged in performing arts in 2012

	Percentage	Number (millions)
Music	5.0	11,749,677
Singing (all forms)	8.7	20,444,438
Acting	1.4	3,289,910
Dance	1.3	3,054,916
Social dance	31.6	74,257,960
Artistic films or videos	2.8	6,579,819

Source: From NEA, 2015 Report #58.

Table 2.3 **Adult participation in classes**

	2012 (%)	Lifetime (%)	18–24 year olds in lifetime (%)
Music	2.0	35.6	42.9
Acting and theater	0.3	7.1	10.6
Dance	1.8	16.7	17.3
Photography or film	1.2	9.4	9.8

Source: From NEA, 2015 Report #58.

Dance

The NEA describes dance as "encyclopedic in scope and international in its aesthetic traditions" (http://arts.gov/artistic-fields/dance#sthash.UaZorP3z.dpuf). According to Curtis Sachs (1937/1965), "[t]he dance is the mother of the arts. Music and poetry exist in time; painting and architecture in space. But dance lives at once in time and space" (p. 3). The mode of communication is fundamentally the human body in motion. Traditionally, dance represented cosmic harmony, existing in the present moment and organized within temporal rhythmic structures and spatial dimensions. From the earliest beginnings, dance and music were integrally linked in purpose and intention and this integration has remained a common practice throughout dance history, and across all dance genres. Not surprising, this coalition of music and dance has neurobiological roots; they are deeply interconnected in the brain (Hagen & Bryant, 2003). Originally, dance and acting were also aligned; they were both rooted in action. When spoken text and song began to dominate performance, the two art forms separated (Thomson, 2011). The overarching intention of dance is aesthetic movement (Christensen & Calvo-Merino, 2013; Thomson, 2011).

Dance transcends time and culture, as evidenced in today's myriad dance categories and subcategories. Each category and subcategory can be distinguished by different skill sets and aesthetic appeal. According to the NEA (2013), the major dance categories include (1) *ballroom* (e.g., waltz, quickstep, swing, jitterbug, Charleston, tango, samba, mambo, salsa, rhumba, cha cha), (2) *club* (freestyle, boogie-woogie, cakewalk, Chicago stepping, disco, country two step, line dances, Texas swing, Texas slide, voguing, grinding), (3) *concert* (ballet, contemporary, jazz, modern, Bharata Natyam, dance theater, physical theater), (4) *contact improvisation*, (5) *folk* (country dance), (6) *historical* (medieval, Renaissance, Baroque), (7) *liturgical* (praise dance groups, Shaker), (8) *street* (e.g., break, hip hop, krumping, electric boogaloo, jerkin', locking, popping, turfing, lyrical), (9) *tap*, and (10) *traditional* (e.g., African, Native, Chinese, Bangra, belly, Flamenco). The authors are aware that all dance forms can be performed with an audience in attendance; however, for the purpose of this book, the primary focus will be directed toward concert dance (Box 2.2).

Dance, as a performing art, has evolved over centuries, as has the relationship between choreographers (generators) and dancers (interpreters). Throughout dance

Box 2.2 Dance Categories (NEA, 2013)

1. ***Ballroom***	Waltz, quickstep, swing, jitterbug, Charleston, tango, samba, mambo, salsa, rhumba, cha cha
2. ***Club***	Freestyle, boogie-woogie, cakewalk, Chicago stepping, disco, country two step, line dances, Texas swing, Texas slide, voguing, grinding
3. ***Concert***	Ballet, contemporary, jazz, modern, Bharata Natyam, dance theater, physical theater
4. ***Contact improvisation***	Derived from somatic movement practices
5. ***Folk***	Country dances
6. ***Historical***	Medieval, Renaissance, Baroque
7. ***Liturgical***	Praise dance groups, Shaker
8. ***Street***	Break, hip hop, krumping, electric boogaloo, jerkin', locking, popping, turfing, lyrical
9. ***Tap***	Rhythmic dance derived from stepping dance form
10. ***Traditional***	African, Native, Chinese, Bangra, belly, Flamenco

history, choreographers have created dances based on the style and tradition of their time. For example, during the Baroque period a formal and rigid sequence of dances were developed and performed for royalty (Cohen, 1974). Both the choreographers and composers were constrained by the rules imposed by the court. In the Late Baroque period, dance and song became less formally ordered and structured; consequently the choreographers had more freedom to express dance ideas (Cohen, 1974). Gradually, the emergence of a unique dance form, ballet, appeared. Ballet soon separated from the theatrical spectacles that dominated the Late Baroque period; it was recognized as its own artistic domain. With this separation, choreographers were able to explore multiple stylistic variations. For example, during the Romantic era, ballets focused on a desire to transcend human mortality. The ballerina moved from soft ballet slippers to pointe shoes to portray the sylph-like quality of the romantic ballets. In the Classical period, full-evening story ballets were created and composers joined in the collaboration of these new productions (Cohen, 1974). *Swan Lake, Cinderella, Sleeping Beauty*, and *The Nutcracker* were all created via collaboration between the choreographer and composer.

Unlike today's acceptance that choreographers are the creative generators, in the late 1800s and early 1900s, choreographers were regarded as master interpreters. This notion changed with Isadora Duncan; she initiated a choreographic rebellion in the late Romantic period. During this period, the pioneers of modern dance rejected the rigid codification of dance, and the assumption that choreographers were master interpreters (Cohen, 1974). Individual choreographers, who were also dancers in their own right, pushed the boundaries of tradition as they struggled to portray the unconscious realm of the human psyche. They also joined the movement of abstract art; choreographers explored abstract spatial design and worked with music that resided outside the traditional dance lexicon (Hagood & Kahlich, 2007). Choreographers in the 20th century were officially considered the creative generators of dance; they were no longer viewed as master interpreters.

Today, choreographers typically respond to external sensations and ideas that are then translated into kinesthetic movement ideas, although choreographers may also begin exploring through an internal kinesthetic sensation that gives rise to choreographic ideas (Hagood & Kahlich, 2007). For example, the early modern dance pioneers rejected all formal dance steps and sought a new vocabulary that was more relevant to their time. Street choreographers followed this same path 70 years later. Because dance movement is considered an expressive language in its own right, it is not uncommon that generators also employ preexisting dance vocabularies to express new dance ideas (Blasing, Tenenbaum, & Schack, 2009). This is evident when choreographers draw from ballet vocabulary; however, the movement is altered to meet specific aesthetic goals. George Balanchine is a prime example of this choreographic approach (van Delinder, 2005). Many generators successfully work within codified dance traditions such as tap, ballet, ballroom, and jazz; these choreographers do not feel constrained by the existing dance vocabulary (Martin, Ric, & Hristovski, 2015).

Dancers are the interpreters, although some dancers are also generators. Martha Graham, both a consummate choreographer and dancer, stated:

> *I am a dancer. I believe that we learn by practice. Whether it means to learn to dance by practicing or to learn to live by practicing living, the principles are the same. In each it is the performance of a dedicated precise set of acts, physical or intellectual, from which comes shape of achievement, a sense of one's being, a satisfaction of spirit. One becomes in some area an athlete of God.*
>
> *(1991, p. 3)*

Regardless of dance category and subcategory, dancers must undergo a specific training regime to be proficient enough to perform. Each dance category requires different training and skill sets. To be accomplished at professional levels most dancers must identify a specific dance form and train in that style for years, although dancers may elect to train in multiple styles. Trained dancers, regardless of style, learn to think and remember in movement actions. Dancers are able to embody and express a language that is universally understood; this embodied expression resonates deeply in the conscious and unconscious processes of both the performer and the audience (Dijkstra, Kaschak, & Zwaan, 2007).

Dancers, such as ballerinas and urban street dancers, are acutely aware of body image. Their bodies are their instruments, and like singers who are preoccupied by their vocal health (Sandgren, 2002, 2009), dancers are painfully aware of their body image (Heiland, Murray, & Edley, 2008), including the stereotyped images of dancers within their fields (Clabaugh & Morling, 2004). Dancers work in group settings. Unlike singers or musicians, who tend to practice alone, dancers must attend daily classes together, and rehearse and perform in groups. They interact in a similar way as other small world groupings. Sadly dancers' careers tend to be shorter than any of the other performing artists, and when their career ends they must seek

Figure 2.2 "Classical Dancer." Performer: Tharini Shanmugarajah. Photographer: Lee Choo.

other career paths. This transition often reflects domain-general talents, because successful dancers tend to be equally successful in their new careers (www.career-transition.org; Fig. 2.2).

Audiences are able to appreciate dance, in part, because of innate abilities to perceive movement and sense another's movement as if it occurred within their own body. Strong dancers attract audience attention primarily because their training promotes powerfully coordinated expressive movement. Audiences respond to a corporeal focused attention, including nonmusical sounds such as the dancers' breath rhythms and the sounds of bodily contact with the floor and other dancers (Reason et al., 2016). Regardless of gender-neutral or gender-oriented dance performances, audiences are also more aware of gender differences during performance, including heightened attraction to gender-oriented movement (Weege, Lange, & Fink, 2012). Audience members from all cultures readily respond to dance; our brains are designed to recognize and understand movement, as well as the emotional and aesthetic appeal inherent in dance (Dijkstra et al., 2007).

Dance has a popular appeal, as evidenced by the 2012 survey conducted by the NEA (2015). There were 36.9 million adult American audience members who attended a dance performance in a public concert hall. Three million adult dance performers were employed and performed in 2012. Nonprofessional adult Americans also studied dance, with approximately 16.7% of the population taking at least one dance class in 2012. This number could easily increase given the recent research findings that demonstrated the neuroprotective benefits of dancing in the aged. Superior sensory, motor, and cognitive performances were evident in elderly individuals who danced regularly compared to elderly individuals who did not dance regularly (Kattenstroth, Kolankowska, Kalisch, & Dinse, 2010). Participating in dance also enhanced cognitive flexibility and quality of life in the elderly (Coubard, Duretz, Lefebvre, Lapalus, & Ferrufino, 2011). These NEA survey statistics and research findings indicate that audience members and performers welcome the aesthetic experience of dance (Christensen & Calvo-Merino, 2013).

Dance is a coalition of music and dance (Hagen & Bryant, 2003). The immediacy of dance makes it a powerful medium; it can unite performers and their audience through an interpersonal connection facilitated by our brains' sensory and motor systems. Dance is an art form that has existed since the beginning of time; regardless of historical and cultural changes, human expression through movement prevails. Dance is the mother of all the arts (Sachs, 1937/1965).

Media arts

According to the NEA (2015), media arts include screen-based projects presented via film, television, radio, audio, video, the Internet, interactive and mobile technologies, video games, transmedia storytelling, and satellite, as well as media-related printed books, catalogs, and journals (http://arts.gov/artistic-fields/media). Performing artists engage in all but the printed media. The performers are usually actors; however, musicians, dancers, and opera singers can participate in any of these media arts. The generators in media arts include screenwriters who write for films, and writers who create for television and radio. Unique to television, show runners are both the creators (writers) and the producers of the show. In film, producers/directors often generate the idea for a film; however, many other people join the creative team, including screenwriters and casting directors, who select the talent (actors) that will perform in the film.

The creation of film (motion pictures) usually involves several stages: development, preproduction, production, postproduction, and distribution. The "talent" includes the actors, designers, and the crew. In small budget films, it may be one person who produces, directs, acts, films and edits the motion picture; however, most films require at least a few skilled individuals. Actors, set and costume designers, camera operators, sound technicians, editors, and producers who finance and seek distribution options are all integral participants in filmmaking (Nelmes, 2004). Television and radio productions share many features with the film industry. Regardless of the media format, many talent-specific individuals share their expertise; a media product is always a collaborative enterprise.

The success of media arts may be directly related to financial gain, although many great artistic media products fail to garner financial success (Simonton, 2005). This pattern is true for American films, but it is not accurate for the French film industry, where artistic merit is always financially rewarded (Hadida, 2013). Other variables that influence success include the desires of the audience. All film genres (horror, thrillers, romantic comedy, comedy, science fiction, action, adventure, drama, melodrama, film noir, documentaries, docudramas) directly appeal to unique audience tastes, with many audience members drawn to several or all genres. Importantly, audience selection of a specific genre influences a positive response to the film and influences its success rating (Hilscher, Cupchik, & Leonard, 2008).

In television, a similar array of genres exists, although they are more immediately embedded within a social and artistic context. The creative novelty and success of television shows is directly related to the times and situations that support them. A major example of this is Norman Lear's creation of *All in the Family*. It was first influenced by the British series, *Till Death Do Us Part*. *All in the Family*, a series that is regarded as one of the greatest television shows of all time, is a clear example of a creative generator seizing the influences of his time and reflecting these social constructs in a comedy genre that was edgy and provocative (Rich & Weisberg, 2004). Another genre that has dominated television is reality TV (although reality shows also occur in theatrical settings). Real people are cast as the "actors" to perform their own stories. This practice is intended to remove the distancing between actors and their audience. Reality TV focuses on audience identification with the real people in the shows (Upton, 2011). Docudramas and documiniseries are another form of reality programming. In these shows, actors play real people in real situations that have been reported in recent news stories. Unlike reality media, Theater of Witness performances hope to address truth and reconciliation for the performers and the audience members alike (Malpede, 1999). This media form attempts to educate audiences about particular human atrocities, while promoting healing and awareness. All these performance formats blur boundaries between drama and reality in an effort to increase audience awareness of the human condition endemic in modern societies (Paget, 2007).

Film and television editing, regardless of genre, profoundly influences audience perceptions. The editors can direct audience attention, often without the audience's conscious awareness. Editing style has shifted drastically since the early years of cinema and television. Editing is direct and strategic. Rapid cuts, rather than fades or dissolves, heighten emotional intensity and direct audience attention. Television and cinema editing capitalize on viewers' brief moments of attentional distraction. Today's editors match the cognitive visual processes that operate in our brains (Shimamura, Cohn-Sheehy, & Shimamura, 2014); e.g., rapid cuts actually replicate neurological visual-movement patterns (Cutting, DeLong, & Brunick, 2011), and match-action edits with an overlap technique resemble our brains' ability to process and perceive visual and movement stimuli simultaneously.

Producing film and television also requires actors to work very differently than in theatrical settings. They must accommodate the demands and schedules inherent in media arts. In cinema, the scenes are often shot out of chronological order. The actors must also film a scene multiple times before moving on to another scene (Richards, 1992). On set, actors not only collaborate with other actors and directors, but they also interact with the director of photography and the camera itself. In fact, the camera is always a scene partner that is "invited" into the interior unconscious processes of the actor (Richards, 1992). Unlike theater actors who have total control of their performance, from the opening scene until the final bow, the cinema actor has minimal control. Often scenes are excluded in the final edited version of the film. And the viewing of the film occurs months or years after the actual filming process. This temporal dissonance can be disorienting for actors, even though the audience receives the film as a new theatrical experience. The audience is largely unaware of this temporal disconnect, except on rare occasions when a lead actor has died before the release of the film.

Film music is another powerful ingredient in media arts, particularly film scores (Simonton, 2007a); the music directly influences and augments audience emotional responses (Hoeckner, Wyatt, Decety, & Nusbaum, 2011). Cinema composers are excellent examples of creators who have constraints placed on them (honoring character and plot development, length of time for a scene, or a transition from scenes, etc.). In a study by Simonton (2007b), film composers' careers were similar to composers in general; both the constrained creators (film composers) and the free creators (composers) produced music that was personally rooted, meaningful, and creative.

Media art is a complex and highly successful industry. Approximately 47% of the American population has witnessed a media product and 71% witnessed performing arts productions on the Internet (NEA, 2015). In 2012, 59.4% of adults in the Unites States attended a film, whereas only 30.4% attended a sporting event (professional or amateur). There were 6.6 million performers (2.8% of the US adult population) who participated in film and video in 2012 and 9.4% of the adult population studied filmmaking in this same year. These statistics demonstrate the popularity of media arts. It is perhaps one of the most collaborative of all the performing arts.

Music

Music is a diverse field with many different categories. Whether it is consciously or unconsciously recognized, to hear a piece of music involves at least a generator (composer) and an interpreter (musician or singer), and also perhaps a sound engineer who can digitally transmit the music. These three distinct groupings of musical expertise are required to create a musical product that is heard by audiences; however, these three talents can be found within a single individual or multiple individuals.

The NEA (2015) supports a wide range of musical genres, educational programs, performance ensembles, and organizations. It recognizes the incredible musical diversity that exists within traditional and nontraditional genres. Music genres are organized into shared traditions or sets of conventions. The three large categories are as follows: (1) *art music*, (2) *popular music*, and (3) *traditional music*. The first category, *art music*, is derived from "classical" musical forms; however, art music predates the formal historical Classical period. Music created from the medieval era until the present can be classified as art music or "classical" music. Some of the many subcategories include chamber music, chant, choral, early music, medieval, High Classical, Romantic, minimalism, and orchestral music. Opera is often considered art music; however, it is also regarded as its own artistic field. The second category of music,

Box 2.3 Music Categories (NEA, 2015)

1. *Art music*	Chamber music, chant, choral, early music, medieval, High Classical, Romantic, minimalism, and orchestral music. Opera is often considered art music; however, it is also regarded as its own artistic field
2. *Popular music*	Jazz, operetta, alternative, pop, funk, punk, rock, heavy metal, grunge, blues, country, hip hop, anime, house, techno, honky tonk, swing, R & B, soul, new age
3. *Traditional music*	Lullabies, bluegrass, gospel, Latin, world, reggae, holiday, folk

popular music, is easily accessible to the general public. Jazz and operetta are sometimes considered art music; however, they are usually classified as popular music. Subcategories within popular music also include alternative, pop, funk, punk, rock, heavy metal, grunge, blues, country, hip hop, anime, house, techno, honky tonk, swing, R & B, soul, and new age. The third category, *traditional music*, includes lullabies, bluegrass, gospel, Latin, world, reggae, holiday, and folk (Box 2.3).

Music can be delivered to an audience through multiple media such as in large concert halls, nightclubs, salons, or in the street. It can also be recorded in studios, including on sound stages for film or television. Original artists who created the work can perform the music or "cover" artists can perform the original work and add their own personal ideas (Uzzi & Spiro, 2005). When music is recorded on sound stages, the sound engineers are also regarded as creative interpreters. They influence the identity and quality of sound based on how they mix and edit the final product.

Music, as one of the five performing arts, has generators (composers) and interpreters (instrumentalists, singers, conductors). Like dancers who are also choreographers, instrumentalists and singers can generate and interpret musical works. Musicians can perform in large groups (orchestras, choral singers), small groups (chamber, bands), or as soloists. They also rely on conductors, both in rehearsals and during public performances. Conductors are the master interpreters of a composer's work, but they also add their own personal interpretation to the composition. Great conductors are recognized by the quality of sound that is produced by the orchestra, and their musical aesthetic that is understood and appreciated by an informed audience (Copland, 1957). A conductor has a unique set of skills, including mastery of one or more musical instruments, aptitude in organization and leadership, and musical imagination. They motivate and inspire members of the orchestra through a coordinated integration of sensory and motor cues that are primarily gestural in nature (Van Weelden, 2002; Zaza, Charles, & Muszynski, 1998). These demands are heightened during public performances, a reality that can increase physical and psychological strain on the conductor (Jankovic & Ashoori, 2008; Jaque, Karamanukyan, & Thomson, 2015).

Music always resides within a social–cultural climate, and the creation of new music reflects these changes (McIntyre, 2008). For example, popular music is considered one of the most emotionally evocative musical forms (Schellenberg

& von Scheve, 2012). It is argued that popular music is the "mirror for society" (Schellenberg & von Scheve, 2012, p. 196). Examining the trajectory of popular music in the United States reveals intriguing sociocultural shifts. Currently, there is an increasing trend toward songwriters creating songs that evoke sadness or mixed emotions (Vuoskoski & Eerola, 2012). This same pattern was found in the songs written by the Beatles between the years 1963 and 1970 (Petrie, Pennebaker, & Sivertsen, 2008). McIntyre (2008) suggests that songwriters are directly influenced by both their knowledge of the field and the social organization they reside within. This interactive systems model may explain the rising popularity of sad or mixed emotion songs, as well as a rising popularity of female recording artists. It is speculated that the increase in successful female singers may be related to gender equality, or perhaps to the current rampant objectification of female artists (Vuoskoski & Eerola, 2012). Not surprisingly, a similar pattern was observed in listener preferences. They tend to prefer music that is mood-congruent to their own interpersonal situations (DeMarco, Taylor, & Friedman, 2015). Listeners also experience more aesthetic emotional experiences such as "chills" when they are feeling happy or sad but not worried and anxious (Nusbaum et al., 2014). The combination of the emotional state of the listener and the capacity of music to elicit emotions increases musical appreciation and emotional communication. Songwriters and their listeners influence and reinforce current trends in popular music.

Unlike most forms of popular music that are more universally situated, jazz is deeply rooted within the American identity (Tietze, 2006, 2008). Interestingly, jazz musicians are one of the most frequently studied groups, in part because researchers are intrigued by the overt creativity that is manifested when jazz players improvise; they are simultaneously generators and interpreters (Sawyer, 2006a). In fact, public opinion often identifies jazz musicians as the most creative group (Sawyer, 2006a; Fig. 2.3).

Figure 2.3 "Jazz musicians." Performers: Jordan Leicht (tenor saxophone), Daniel Massey (bass). Photographer: Lee Choo.

The popularity of music is hard to estimate. The sound of music permeates our daily life, whether it is playing in the background while riding an elevator, watching an advertisement that is digitally displayed, or receiving a message on our mobile device. According to the NEA (2015), over 60 million people attended a music event in 2012, and 33% of all adult Americans have taken music lessons in their lifetime. When sampling adults between the ages of 18 and 24, an even higher proportion (43%) took a music lesson. These are certainly underestimated figures, especially if we include the daily listening of music that is not surveyed. Almost everyone hears music. We are directly and indirectly influenced by music on a daily basis.

Opera

According to the NEA (2015), opera is a multifaceted art form that comprises music and text to move a story or dramatic concept forward. The origin of the word *opera* is derived from the Latin *opus*, meaning *work*. Opera has a rich history that grew from the early Greek plays and the mystery plays of the Middle Ages. The first official opera, *Dafne*, was created in Florence, Italy in 1597 (Jacobs & Sadie, 1985). This form of music quickly spread across Italy, and then, throughout Europe. Famous composers were drawn to the complex musical demands inherent in opera. Powerful stories, usually based on mythology, the Bible, or political and social events, are given life through an integration of singers and the orchestra (Jacobs & Sadie, 1985).

Performing opera requires a diverse range of interpreters, although a single composer and librettist are usually responsible for generating the scored and scripted work. Opera singers are musicians who portray characters and further plot development. Directors (musical and stage), designers, and, often, choreographers, participate in developing the opera production. This art form integrates highly trained musicians, singers, and directors, in part, because opera repertory, regardless if it is one of the traditional works or a newly composed modern work, demands precise theatricality and musicality, including the coordination of singers with instrumentalists. To work professionally as an opera singer requires years of specific and deliberate training to produce the sound expected in operatic performances. The pressure to produce this sound imposes intense demands on the singers, whether they are performing as soloists or as members of an opera chorus. Audiences and other performing artists tend to be highly critical of opera singers. Sadly, singers report constant exposure to negative criticism and this pressure adds to their fears of vocal strain or vocal fatigue (Sandgren, 2009).

Opera singers experience unique pressures compared to other performing artists. Their voice is their musical instrument; however, they must also embody characters in a manner similar to actors. Singers have a heightened performance load in that they must integrate cognitive, sensory, and sensorimotor mechanisms that underlie musical timing and performance (Corlu, Maes, Muller, Kochman, & Leman, 2015).

Part of an opera singer's training is to learn to manage these demands without the debilitating effects of performance anxiety and negative perfectionistic evaluations (Kenny, Davis, & Oates, 2004; Sandgren, 2002, 2009). As discussed earlier, the influence of the environment on creativity is even more apparent in opera (Runco, 2004). Singers are vulnerable to harmful substances that are often endemic in theater settings. Singers often complain about dry or hot theaters and that these conditions negatively affect their voices. These complaints are often related to airborne toxins that cause harm to singers' respiratory systems (Richter et al., 2002). Other environmental factors also influence the creative conditions of opera singers; however, one of the most important of these is support. Singers maintain their performance levels through ongoing vocal coaching and expert guidance from vocal teachers. They work closely with conductors and stage directors to build a performance that is both musically and dramatically sound. Generally, singers are aware of other singers and have a keen sense of international standards (Sandgren, 2002). Working as a singer in opera resembles a small world group. The group dynamic can enhance singer success but it can also quickly destroy the career of a singer (Fig. 2.4).

Opera may be the most constrained of all the performing arts. To perform traditional and modern classical opera repertoire requires the convergence of multiple disciplines (music, theater, and often dance) and a complex array of skills. Opera singers must adhere to the dictates of the musical score, the dramatic intention of the

Figure 2.4 "Singing duet." Performers: Alicia Spilleas, Emily Works. Photographer: Lee Choo.

libretto, and the actions of the theatrical performance. To appreciate this art form, the audience must be educated and prepared. Opera requires singers to have incredible natural talent and years of disciplined and deliberate practice. It is an expensive training for the performers. To produce opera, it is equally expensive because it requires highly skilled musicians, conductors, directors, designers, and producers. The attendance at opera (2.1% of the American population) is similar to ballet performances (2.7% of the American population). Both art forms surveyed in 2012 are smaller compared to the attendance at other performing arts (NEA, 2015). In 2012, 10 million people attended an opera performance (NEA, 2015), although a greater number of adult Americans sang professionally in a performance (8.7%) (NEA, 2015).

Theater

The NEA (2015) differentiates theater (nonmusical and musical) from media arts (film, television, radio, Internet). This delineation is useful when gathering statistical data on audience attendance and participation (practice or training); however, it is not as helpful when examining actors as performing artists because they straddle the fields of theater and media. Because of the NEA separation of theater from media arts, and for the purposes of this section, the discussion will focus primarily on live theater (nonmusical and musical).

Drama is generally described as a play or writing that tells a story. It is performed in one of the four media: cinema, radio, television, or theater (Merriam Webster Dictionary). The Internet is another popular medium to view drama, including small "webisodes" or amateur creations that are uploaded onto YouTube. *Drama* is derived from a Greek word that means *action* or *to act*, hence the terms *actor* and *acting* to designate the person and process (Thomson & Godin, 2011). The two masks from Greek theater, comedy and tragedy, designate the primary dramatic modes. These primary modes were further divided into extreme forms of farce and melodrama, respectively (Brockett & Hildy, 2007). The dramatic arts are equivalent to theater arts, and as such these words are used interchangeably. This performing art field requires unique skill acquisition that sets actors apart from the other performing artists, although the primary process of preparing a production to be viewed by an audience is similar across the fields.

Playwrights who write scripts are the generators in this domain. Generators can also take existing writings, whether literary or news stories, and devise new scripts to be performed as "devised" theatrical events. Devised theater can also be created from collaborative explorations that are then shaped and developed into theatrical productions (Oddey, 1993). Other collectively scripted works come from Theater of Witness works. In this form of theater, the participants create dramatic works to "testify" their traumatic experiences. The dramatic scripts are built from testimonial explorations (Caruth, 1996; Culbertson, 1995; Felman & Laub, 1992; Malpede, 1996, 1999). This practice follows Aristotle's (c.335 BCE/1958) claim that the dramatic purpose of tragedy was catharsis for the individual and city-state.

The theater generators (playwrights) often have previous acting training and many also direct their plays; they are deeply informed about the theatrical tradition, including how to transform a play into a great production and how to guide the actors who will perform the works. Edward Albee is an ideal example; he was a prolific playwright who was trained as an actor and worked as a director (Rocamora, 2008). The interpreters are the actors and the directors. Acting is an ancient practice; in fact, in early Greek drama the moment that the actor, Thespis, stepped out of the chorus and stated, "I am Dionysius," a theatrical revolution was born. He had the courage to embody another (Thomson & Godin, 2011). By pretending to be someone else, actors portray characters and give life to scripts; performing in real time with an audience present enables direct communication. Unlike actors, the director has a direct relationship with actors but less immediate connection with the audience. Directing is a position that emerged in the 19th century (Cole & Krich Chinoy, 1976). Originally termed the *metteur en scene* (master of the scene), directors originally controlled the appearance of the stage, but as theater changed from spectacle to a more realistic portrayal of the human condition, the directors assumed greater responsibility (Cole & Krich Chinoy, 1976). With this change, actor training became paramount; actors were required to express the vision of both the director and the playwright (Hodge, 2000; Fig. 2.5).

Actors must memorize vast amounts of text, invest the text with intentions while staying true to the character being portrayed, and engage in the emotional resonance that is evoked by a combination of the actions of the play and the interaction with the other actors on stage. According to Noice and Noice (2002), the creative process for the actors involves two distinct stages. First, the actors cull the script for underlying intentions of the characters, and second, they actively experience the cognitive–emotive–motor processes inherent in all human interactions. Actors are strongly motivated to participate in the theatrical production, in part, because the experience is stimulating and exciting (Martin & Cutler, 2002). For actors,

Figure 2.5 "Theatrical portrayal." Performers: California State University, Northridge (CSUN) Performers. Photographer: Lee Choo.

performing is often a highly creative experience. They can readily merge action with awareness and skills with challenge; consequently, during performance they enter a flow state that is often transmitted to the audience (Csikszentmihalyi, 1990; Robb & Davies, 2015). During memorable performances, the actors and audience share in a highly engaging, pleasurable, and transformational experience.

The larger domain of theater also includes comedians; they are regarded as their own subcategory. According to Greengross and colleagues (2012), stand-up comedians should be regarded as a distinct group of performing artists, although many actors who are gifted in comedic performance may share similar features. Comedians are both generators of their stand-up comedy routine and the interpreters of this work. Like actors, comedians often present a stage persona that is distinct from their true personality traits (Greengross & Miller, 2009). They also interact with audiences through nonverbal and verbal communication. Not surprisingly, comedians tend to score higher on verbal intelligence and humor production (both positive and negative) compared to noncomedian college students (Greengross et al., 2012). Comedians strategically employ affiliative humor to capture the attention of their audience. To support a career in this field, they must also sustain a rapport with club owners, booking agents, and other comedians.

Musical theater is another popular form of theater; in fact, 72.5 million people attended a musical theater performance in 2012 (NEA, 2015). This form of theater requires a core group of diverse artists to develop a production, including a composer, lyricist, librettist, choreographer, directors (stage and music), and producer(s) who manage the financial costs and investments. This art form is a collaborative engagement and reflects both distributed creativity and small world groups. According to Uzzi and Spiro (2005), musical theater shows originate when at least one artist develops the initial material for the production and then other artists are recruited to develop it further. Broadway shows, the most prominent showcase for musical theater, gradually develop with a specific team of artists, along with the original cast of performing artists who participated in the early rehearsal and performance phases. A Broadway show is usually premiered outside of New York; further edits and restaging transpires based on audience responses. When it finally is performed on Broadway it has usually undergone many changes. Despite all these modifications, a Broadway show may never progress past preview performances, although many productions run for years (Uzzi & Spiro, 2005).

Musical theater emotionally engages audiences and can even influence audience beliefs and change audience attitudes (Heide, Porter, & Saito, 2012). In Heide and colleagues' study, the results challenge a popular view that musical theater is lightweight and unimportant. Telling a story that includes song and dance may entertain an audience but it also contains the key elements of a creative work: novelty, usefulness/intention, authenticity, and aesthetic appeal. Key to the Heide et al. study is the empirical evidence that audiences' attitudes can be shifted as a result of viewing musical theater. Certainly *West Side Story, Cabaret, South Pacific, A Chorus Line*, and *Godspell* are a few of many musical theater productions that challenged and changed social beliefs and assumptions.

Despite the fact that theater has many subcategories and genres, the consistent appeal of theater is its immediate contact with an audience. Actors (or comedians) tell stories through embodied actions that are both verbal and nonverbal, sung and spoken. The fact that Thespis dared to break from tradition, stepping away from the chorus and publicly stating to the audience that he was Dionysus, was a major paradigm shift in performing arts history. Not only did he claim to be another person, he actually claimed to be God. He gave life to another, and through his embodiment of Dionysus, he was able to ask his audience to willingly suspend their disbelief and join him in this imaginative journey (Thomson & Godin, 2011). Storytelling was no longer a reporting of events; it was now an enactment of these events. Theater can engage the performers and audience members alike. All are vicariously drawn into the situations and circumstances of the characters portrayed in the theatrical production (Hetzler, 2008).

Not surprising, theater attracts many participants and audience members. According to the NEA (2015), in 2012, approximately 115 million adult Americans attended a theatrical production and almost 1% of Americans participated as a performing artists. Even more studied acting, with 7.1% of Americans taking at least one acting class in 2012. The NEA (2012) also reported that students who were exposed to theater arts education had greater academic achievement and future career success (Catterall, Dumais, & Hampden-Thompson, 2012). Given these statistical reports, it is apparent that millions of people derive great personal benefit from participating in theater arts or attending as an audience member.

Performing artists working outside the five primary NEA fields

Circus arts

Circus, although not included as one of the five performing arts by the NEA, is often considered a performing art. It is identified as an institutionalized cultural medium with its own skilled physical performances and visual style (Arrighi, 2015). Circus arts are present in all cultures, each with its unique colorful history and each transforming to meet the needs of the time (Loring, 2007). Within this art form, a network of company performers and producers creates circus performances. There is a unique distribution of power and agency among the participants, a feature that promotes resilience and adaptability. The internationally recognized company Cirque du Soleil is probably the best current example of circus art as performance art; it is a company that employs countless numbers of skilled performing artists (Rantisi & Leslie, 2015). This company has successfully transformed traditional views of circus, providing audiences around the world rich opportunities to see highly talented performers.

Unlike the other performing arts, circus's primary focus is spectacle. The audience is not asked to identify with a character, situation or mood; rather, they are asked to observe and appreciate the powerful, almost superhuman, talents of the performers (Carmeli, 2001). One of the hallmarks of circus life, including the survival as an organization, is resilience. Circuses adapt and respond to unpredictable economic and political states, in part, because they must diversify and adapt to audience desires. Performers also diversify to meet the needs of the circus organizations (Loring, 2007). Circus organizations are constructed similar to small world groups, although there is more transience among the performers, in part, because of events such as career-ending injuries (Loring, 2007). Circus organizations are like nomadic tribes that require loyalty to the tribe and yet flexibility to adapt to the shifting environments that support the tribe (Loring, 2007).

The emergence of Cirque du Soleil resembles this pattern of resilience. A few talented street performers situated in Montreal, Canada built an international empire. Although Nouveau Cirque in France preceded and influenced the creators of Cirque (Cruz, 2014), the trajectory of Cirque was informed by the added influential factors of place-specific support (multicultural environment of Montreal, Quebec) and path-dependent relationships (original creators' friendship, talents, and shared vision). The creation of this organization reveals how distributed creativity and the system in which the group leaders were embedded facilitated this international phenomenon (Leslie & Rantisi, 2011). Today, most circus performances integrate spectacle with art. Artistic composition and thematic structure are integrated into the productions. These are the key ingredients that were introduced by Nouveau Cirque and they are the factors that revitalized circus tradition and contribute to circus survival today (Loring, 2007).

Performance art

The NEA recognizes and supports performance art work, but it is not included as one of the five cultural fields identified by the NEA (2015). Performance art is a nontraditional performing art that draws from the other art forms such as acting, dance, theater, visual arts, film, music, and poetry (Merriam Webster Dictionary). Typically performance art work is focused on political or social themes and reflects the artistic statement of an individual artist. Performance art can be performed by soloists or small groups and can be presented in galleries, theaters, or outside venues. These works often share features with site-specific performance art works; however, site-specific works may feature only dancers or actors. The intention of performance art is to explore boundaries between art disciplines and to seek new forms of expression. According to the NEA (2015), it is an effort to fuse or transform artistic disciplines.

Curiously, the very nature of performance art is to break the mold of institutionalization and to seek nonart audiences and yet it is now recognized and supported in a similar fashion as other established art domains (Wheeler, 2003). Performance art has also been described as an ethnographic-based performance approach. In fact, it is proposed that this art form become a qualitative research tool, one that is performative in nature and includes both the performer and the audience as investigative subjects (Bagley, 2008). Performance art exists on the margins of the performing arts; it endures and even influences the established performing arts domains.

Summary

- Introduction: definition and studies on domain-specific and domain-general criteria; National Endowment for the Arts (NEA) statistics for 2012; interpreters as creative artists; intermediaries who bring the performance product to the market.
- Dance: origins and definition; categories and subcategories; choreographers; dancers; audience appreciation and neurobiological influences; NEA statistics for dance; positive effects of dance.
- Media arts: categories and generators of media arts; stages of development in film; determining success (financial and social); editing and the influence on audience perceptions; actors; film music and emotional appeal; NEA statistics.
- Music: categories; NEA statistics; composers, conductors, and musicians; effects of music on the audience; popular music and current trends; mood and music; systems model and music; jazz as distributed creativity.
- Opera: definition and origins; opera as a small world network; complex demands of opera productions; singer demands and pressures; constraints and NEA statistics.
- Theater: definition and categories; NEA statistics; generators; actors and directors; creative skills needed in acting; comedians; musical theater (definition and description, influence on emotions and values, audience appeal).
- Circus arts: definition; resilience; Cirque du Soleil.
- Performance art: definition; intention of performance art.

Concluding remarks

Are the performing arts fields domain-specific or domain-general in terms of creativity? This question was raised at the beginning of this chapter. Empirical studies still need to be conducted to answer this question; however, based on experiential knowledge, there are many domain-general variables that operate within all the fields of the performing arts. One domain-general variable is that all fields rely on a collaborative process between the generators, the performers, and the artistic team of supporters such as directors, conductors, designers, producers, and technicians. To bring a performance to the public is also similar across domains. Each draws upon the raw material contained in a script, score, or choreography. The interpreters analyze and memorize this material (although musicians may not need to memorize the score). The preparation phase occurs during the rehearsal process, and during rehearsals the interpreters, individually and collectively, identify and deliberately practice specific actions that will give expression and meaning to the work. The product is then performed to an audience and during this public performance the interpreters are prepared to adjust to any unexpected events that may arise during a live performance, such as props or costume pieces falling on the stage, performers missing or late for an entrance, sudden illness or injuries that require last minute replacements, or technical problems such as sound or lighting system failures. They work as a collaborative small world group to maintain a coherent performance product while simultaneously accommodating the needs of the group.

Another domain-general variable is that all the performing arts are based on movement and action. The interpreters manifest a corporeal creativity, a felt-body experience that operates within a specific performance environment and within a specific set of prescribed rules (Julmi & Scherm, 2015). The performing arts' unions establish policies and procedures that support a safe working environment and the professional performers and producers adhere to these guidelines. This practice helps ensure a stable and consistent working environment, one that allows the performers to productively and collectively analyze, clarify, and express the performance work. Further, performers constantly adapt to changing environments and audience expectations. Surprisingly, domain-specific creativity may only be evident in the specific talents and skills required to work within dance, music, or acting.

Performing arts are an integral part of our social and economic culture. In 2012, according to the NEA (2015), 59.4% of the adult American population attended at least one performing arts event compared to 30.4% adult Americans who attended an amateur or professional sports event. As stated earlier, the performing arts contribute substantially to the GDP. Many Americans studied one or several of the performing arts fields in childhood, and as adults, many continue to do so.

The person: talent and performing artists

3

People have asked me why I chose to be a dancer. I did not choose. I was chosen to be a dancer, and with that, you live all your life.
Martha Graham (1991, p. 5)

Talent or giftedness

Talent is defined as "a person or group of people with a special ability to do something well; innate ability, aptitude, or faculty; above average ability" (Merriam-Webster). Differentiating between "giftedness" and "talent" has offered many challenges to researchers and educators (Morelock, 1996). Today, the distinction between the two terms, although subtle, helps clarify the terminology. Giftedness describes the rapid and asynchronous development in childhood and is fundamentally related to IQ scores. Gifted children have an atypical trajectory of development that can provoke added stress and requires educational programs that address their unique developmental needs. Most researchers view giftedness as involving high intellectual or technical aptitude, including heightened mental energy that contributes to the giftedness (Mayer, Perkins, Caruso, & Salovey, 2001). Talent refers to multileveled potentials in domain-specific fields (performing arts, sport, technology, etc.). Talented individuals have the will to work hard and they have the possibility to advance to a superior level in their domain (Rasmussen & Rasmussen, 2015). Talent can be fostered first through identification of the talent followed by environmental support. It resides at the juncture between domain-specific abilities and an individual striving for mastery within the domain (Box 3.1).

Disambiguating the two terms is important when placing children into programs geared for giftedness or domain-specific talents; however, many individuals can be both gifted and talented. Multiple studies have included samples of individuals who are regarded as highly gifted intellectually and also demonstrate profound talents in domain-specific performing arts fields (Mayer et al., 2001). Subotnik, Olszewski-Kubillius, and Worrell (2011) suggested that an inclusive definition should be standardized. In their definition of giftedness they included performers and generators who are distributed at the upper end of domain-specific talents. They indicated that giftedness is developmental, with children demonstrating potential and adults realizing their potential. Including a developmental trajectory into the definition of giftedness incorporates the critical psychosocial variables that influence talent development. Following the suggestion of Subotnik and colleagues, giftedness and

Creativity and the Performing Artist. DOI: http://dx.doi.org/10.1016/B978-0-12-804051-5.00003-2

Box 3.1 Talent and Giftedness Definitions

Talent	Gifted
Multileveled potential in domain-specific fields, high intrinsic motivation to realize superior level of talent	High intellectual aptitude, rapid and atypical developmental trajectory

talent are overlapping concepts. They are ultimately social and cultural constructs (Morelock, 1996). For example, despite every effort to create objective measurements, many gifted and talented individuals remain unnoticed during their early years of development simply because opportunities to demonstrate talent were not provided (Gaztambide-Fernandez, Saifer, & Desai, 2013).

Pyramid of talent development model and three-ring model

Piirto (1998) stated that talent was a natural proclivity but in order to link talent with creativity it must be developed. In her pyramid of talent development model, she identifies a sequence of variables that influence and are also part of talent. At the foundation of the pyramid is genetic inheritance. The next layer is personality attributes, and above this layer is cognitive competence. Climbing higher on the pyramid is the layer that Piirto terms the "thorn." It is sensing a "call" to pursue a talent. If this call is answered then a higher level of talent development must be realized and that is the deliberate and specific training of the talent. At the apex of the pyramid is environmental support that facilitates the manifestation and recognition of talent. This pyramid model is a comprehensive description regarding how talent is nurtured, including the nurturing of multitalented individuals (Piirto, 1995). Piirto emphasizes that creativity is possible in everyone and that "those who are more creative have learned to be so" (1998, p. 42), a statement that amplifies the significance of environmental support (Fig. 3.1).

In 1972, the United States Department of Education mandated the inclusion of the arts into the definition of gifted and talented students. This policy ensured that artistic aptitude was equal to intelligence and should be assessed independently (Oreck, Owen, & Baum, 2003). Despite this policy decision, most schools still do not assess creativity and artistic abilities, and most parents do not question this omission. Rather than evaluating all students, the general practice is to select and assess only students that have already been deemed talented. Performing arts students are traditionally identified based on talent demonstration, e.g., selecting dance talent is contingent on body type, coordination, musicality, and personality; whereas, selecting an opera singer is based on quality of voice, musical aptitude, and personality. These subjective assessments are the norm in the performing arts and follow the practice of consensual assessment, but they tend to reinforce cultural stereotypes of the idealized

Figure 3.1 "Reaching outward." Performer: Sarah Wood. Photographer: Lee Choo.

dancer, musician, or actor (Oreck et al., 2003). The most problematic feature of this assessment process is its lack of predictability. Many young prodigies burn out and never enter a professional career (Winner, 2000). Those who are recognized as talented, even prodigies, are identified primarily because of parental attention and external support. This basic level of support is required to nurture talent. Children who are raised in unsupportive or underprivileged environments may never be identified despite possessing superior talents (Tannenbaum, 1992).

Aware of the difficulties in identifying talent, Renzulli (1978) developed a three-ring conception of giftedness that included above average ability, creativity, and task commitment. According to Renzulli, the interaction of these three variables determined talent. Results from an assessment instrument that was developed based on Renzulli's three-ring conception of giftedness indicated that no single dominant skill or characteristic could distinguish talented individuals, rather it was a combination of the three factors (Oreck et al., 2003). To identify talented individuals based on Renzulli's model, Oreck et al. created an evidenced-based assessment protocol. This protocol requires a group of experts to assess the students and come to a consensual decision, and it requires the students to be exposed to different performing arts fields multiple times prior to assessment. This repeated exposure to authentic arts-based experiences actually facilitated students' ability to declare a preference for a particular art form such as increased interest and desire to learn to sing, dance, play a musical instrument, or act. In every instance the student preferences matched their talent aptitude. This assessment protocol countered the traditional subjective stereotyped appraisals and enabled a more culturally and socioeconomically diverse group of students to be identified and selected into gifted educational programs. Because expert assessors and teachers were involved during the assessment process, they were also able to gather valuable information about the strengths, interests, and learning styles of every child selected to gifted–talented programs (Oreck et al., 2003).

The more traditional creative testing methods are far less discriminating of talent, in part, because they do not capture the complexity of creativity as described in Renzulli's three-ring model (Lemons, 2011). Traditional acceptance practices are frequently deeply flawed. Economically, socially, and culturally advantaged students are more likely to be accepted simply because they have previously received training and been exposed to the art form (Gaztambide-Fernandez et al., 2013). They are already more skilled and educated about the art form, even before being assessed for talent. Once accepted into special art schools, the students, faculty, and administration are dedicated to fostering hard work and discipline, an environment that reinforces an internalized sense of talent and competence. And if students who are accepted into these programs lack financial support, especially during high school and college years, they are often forced to leave an exclusive arts program. For these students, the demands of part-time employment run counter to the mandated requirements in conservatory programs. The programs expect students to spend long hours dedicated to rehearsing and performing while maintaining academic standards required to graduate (Gaztambide-Fernandez et al., 2013). (The obvious importance of environmental support and its relationship to talent will be examined further in the chapters on attachment and development.)

Whether a performing artist is selected into an elite arts education program or not, one of the hallmarks of their ultimate success is task commitment, the third ring of Renzulli's model. Task commitment within individual performing artists is consistent across all the performing arts fields. Performers practice regularly and usually it is a self-motivated initiative to practice. They experience a unique "call" to explore their field, often in early childhood, although this call may come later if they have not been exposed to the performing arts field that matches their talent (Piirto, 1998). Identifying talent in children may be related to natural aptitude, but even more consistent is their almost obsessive desire to practice and learn. In all of the performing arts fields, successful adult performers state that they practiced regularly, often willingly missing social and school functions so that they could attend lessons, daily practice, and rehearsals (Piirto, 1998). It is common knowledge that self-educated visual artists can achieve recognition and acceptance, but few performing artists can reach professional levels without training from experts. The underprivileged performers who do succeed find mentors despite adversity; they seek alternative ways to explore their "calling." This persistence and task commitment is evident during the years that the artist performs, and for many, even long after they retire. Twyla Tharp referred to dance and creativity as a habit. She stated, "dancers are totally governed by ritual. ... What makes it a ritual is that they do it (daily dance class) without questioning the need" (Tharp, 2003, p. 16). A young reporter interviewed Pablo Casals, who is regarded as one of the greatest cellists of all time. The reporter queried, "Mr. Casals, you are 95 and the greatest cellist that ever lived. Why do you still practice six hours a day?" And Mr. Casals answered, "Because I think I'm making progress" (Graham, 2007).

Multiple intelligences

Performing artists, even more than the generators who create the works to be performed, have body intelligence and emotional expressivity (Piirto, 1998). Howard

Figure 3.2 "Vocalise." Performers: Ann Baltz, Nan Cui, Lilia Kibarska. Photographer: Lee Choo.

Gardner (1993) believed that talent and creativity operate within a domain-specific field. Even if there are domain-general creative traits, individuals who train in one specific performance field manifest unique talents that are common to other individuals working within the same performance field. For example, dancers' physical talents are expressed very differently from the physical talents of musicians. But both performing artists require speed, agility, and stamina even though they also possess very different talents and specific physical skill sets. Gardner identified seven different forms of intelligence, and many of these forms are evident in performing artists. For example, actors have linguistic, spatial, bodily–kinesthetic, interpersonal, and intrapersonal intelligences. Musicians have musical, bodily–kinesthetic, and often logical–mathematical intelligences, and dancers share many intelligences found within actors and musicians, especially strong kinesthetic and visual expressivity (Alter, 1984). Certainly, an individual performer, regardless of field, will demonstrate strengths differently but the essential advantage of defining intelligence as multiple possibilities is that it supports the identification and ultimately the nurturing of talented individuals who may not be recognized as gifted based on IQ scores (Gardner, 1993; Piirto, 1995, 1998; Fig. 3.2).

Prior to Gardner's concept of multiple intelligences, general intelligence was viewed as a single phenomenon. The distinction between general intelligence and multiple intelligences follows the pattern of domain-general and domain-specific creativity. Gardner (1983) described the following seven domain-specific intelligences:

1. Musical–Rhythmic: Individuals with this form of intelligence are sensitive to sounds, rhythms, tones, and music. Musicians, opera singers, and musical theater performers have

substantial literacy in music, including a deep understanding and recognition of rhythm, pitch, meter, tone, melody, or timbre. Dancers also have substantial musical intelligence, even if it is expressed in a more bodily–kinesthetic way.

2. Visual–Spatial: An aptitude for spatial judgment and the ability to visualize with the mind's eye are hallmarks of this form of intelligence. Spatial ability is included in general intelligence testing, although visual–spatial intelligence as described by Gardner captures the creative manipulation of space that is common practice in the arts. With the exception of musical performances, all the performing arts require this form of intelligence in order to stage or film performances.
3. Verbal–Linguistic: Like visual–spatial intelligence, verbal–linguistic intelligence is evaluated in general intelligence testing. This form of intelligence is amplified in individuals who display a facility with words and languages. Verbal–linguistic intelligence is critical in most of the performing arts, especially in opera, theater, and film. Performing artists in these fields have a great aptitude to memorize words and tell stories.
4. Logical–Mathematical: This form of intelligence features logic, abstraction, reasoning, numbers, and critical thinking. Logical reasoning is part of general intelligence. Many performing artists demonstrate a strong aptitude for logical reasoning; e.g., choreographers and composers frequently engage this type of intelligence to create new works, and the performers quickly learn to decipher this creative expression.
5. Bodily–Kinesthetic: The control of one's bodily motions and the capacity to handle objects skillfully is the hallmark of this intelligence. Inherent in this form of intelligence is an innate sense of timing, coordination, and goal-directed actions. Every performing artist has heightened bodily–kinesthetic intelligence. In fact, this form of intelligence could be regarded as one of the domain-general factors in all performing artists.
6. Interpersonal: Most performing artists, regardless of personality traits such as introversion or extraversion, have high interpersonal intelligence. This form of intelligence is characterized by heightened sensitivity to others' moods, feelings, motivations, and needs. Individuals with interpersonal intelligence are able to work as part of a group, a variable that is critical within all the performing arts.
7. Intrapersonal: This form of intelligence includes introspective and self-reflective capacities. Performing artists spend long hours engaging in this form of intelligence, especially a deep understanding about the strengths and weaknesses of the self. This depth of introspection enables performers to emotionally express the works they perform. Actors are perhaps the most highly talented in this form of intelligence but all the performing arts fields require self-reflection.

Later Gardner added naturalistic, existential, and moral intelligence to his earlier seven intelligences. He believed that each individual possessed a unique pattern and blend of intelligences. In *Creating Minds* (1993) these multiple intelligences were identified in the case studies of unambiguous creative individuals. The influence of these multiple intelligences also informs imaginative exploration. The complexity provided by these interacting forms of intelligences is manifested in all the great performing arts works. The creators manipulate their craft to give an aesthetic and authentic appeal, the performing artists give expression to these works, and the audience implicitly or explicitly recognizes the complexity of these multiple intelligences. All are components within generative cognition (Policastro & Gardner, 1999) and the deliberate practice of performing artists. They are understood best when referenced to a cultural context and to educational background

(Matare, 2009). Talent, culture, education, and multiple intelligences all influence the unique attributes of performing artists.

Theory of positive disintegration

Dabrowski suggested that developmental potential is characterized by "three main features: domain-specific abilities or talents, the five forms of overexcitability (OE), and a strong autonomous drive to achieve individuality" (Mendaglio & Tillier, 2006, p. 69). Once again, drive, or task commitment, is identified as a key feature in those who manifest talent. Dabrowski also clearly stated that ability is domain-specific, a statement that is shared by Baer (2015) and Gardner (1993). The third feature, the five OE dimensions, are regarded as variables that give talent its power (Piirto & Fraas, 2012). In fact, gifted and talented individuals can often be identified via their unique OE profiles (Chavez-Eakle, Eakle, & Cruz-Fuentes, 2012; Thomson & Jaque, 2016a).

In Dabrowski's theory of positive disintegration he suggested that talented and gifted individuals mature through a series of adaptive disintegrations and reintegrations. During the development of talented individuals this recursive process leads to a sense of autonomy. They have a multilevel awareness of their temperament and the environment in which they work (Mendaglio & Tillier, 2006). When talented individuals are educated in domain-specific programs this awareness is further promoted and amplified (Gaztambide-Fernandez et al., 2013). The capacity to disintegrate and reintegrate resembles the emergent processes described in complexity theory. In both theories, creativity and human development are fundamentally achieved through a process of organizing, disorganizing, and reorganizing events (Lewis & Granic, 2000; Montuori, 2003). Dabrowski suggested that this process of positive disintegration facilitated developmental potential in the talented (Mendaglio & Tillier, 2006).

The theory of positive disintegration is a five-stage hierarchical process, one that Dabrowski observed in talented individuals (Ackerman, 2009). The first level, primary integration, involves psychological integration, harmony, and low levels of inner conflict. Level two, unilevel disintegration, is marked by brief and often intense crises that evoke disintegration. These events may propel the development of personal growth or the individual can return back and conform to the social norms of the group. This juncture is critical in development since a person may either progress or regress. In this second stage, there is a growing awareness of multiple value systems. A sense of stability results from trust in authority figures, and for performing artists this may extend to teachers and mentors. If performing artists are not guided toward their talents, then this second stage may be fraught with instability. The talented individual oscillates between either following the values of his family and cohort or pursuing his talents in the performing arts. During stage two, talented individuals are able to sense the call to pursue their domain-specific interests.

A spontaneous beginning marks level three, referred to as spontaneous multilevel disintegration (Ackerman, 2009). Talented individuals have a clear sense of their calling. They have a multilevel awareness of the lower social norms and the higher imagined ideals and alternative choices. During this developmental process, the individual is asked to imagine an ideal life that is not necessarily endorsed by the social norms. For performing artists, this ideal life is easily imagined because they are usually interacting with other talented artists who share their passion. During this third phase, the disintegration is marked by a more existential crisis of self. There is a dynamic internal struggle that manifests as self-dissatisfaction and a sense of inferiority. Feelings of shame and guilt are more pronounced, primarily because talented individuals fully understand what is required to achieve their goals. According to Dabrowski, the suffering is no longer regarded as a psychological illness because the individual is engaged in active choices with an awareness of building autonomy and integration (Ackerman, 2009). In this third phase, many performing artists grapple with the uncertainty of a professional life and whether their talent and hard work will culminate in a viable career. This third phase is often painful for performing artists; however, for professional performers, their call to pursue their passion inevitably pushes them forward. Martha Graham, like so many performing artists, clearly described how she built a life around dance, despite all odds (1991).

In level four, the person must consciously choose to follow their higher ideals, and this decision is supported by a combination of temperament and environmental factors. Level four is referred to as the organized (directed) multilevel disintegration. It is more conscious and less spontaneous; talented individuals assume full responsibility for their choices during this stage (Ackerman, 2009; Piirto, 1998). It is a time of self-reflection, self-evaluation, and growing self-control. Not only is education focused on mastering domain-specific skills but also the intrapersonal and interpersonal demands of a career in the performing arts are addressed. A growing sense of autonomy emerges; however, this process may necessitate seeking professional counseling or psychotherapy (Ackerman, 2009), or more direct mentoring from those who have already attained the fifth level. Often during this stage performing artists will seek new teachers or move to larger professional companies.

The fifth level, secondary integration, is one of integrated harmony. The person can now manage crises with multilevel options that are supported by the complex ideals that shape behavior and thought. This level reveals the capacity to positively disintegrate and reintegrate. Operating at this level allows talented individuals increased tolerance to manage ambiguity, emotional intensity, and uncertainty. Northrop Frye (1963) argued that educating the imagination allowed creative people to simultaneously detach and engage, a process that "stretches us vertically to the heights and depths of what the human psyche can conceive" (p. 42). Because individuals have deliberately practiced and mastered their skills, including fully educating their imaginative skills, inner conflict is reduced. The talented individual has answered the "call" and is actively engaging in creative work. Achieving this fifth level doesn't mean that there are no struggles; rather the individual is more adaptable. Increased awareness and an "educated imagination" make the external conflicts less troublesome. Like Piirto's concept of the apex in her pyramid of talent model, working effectively and productively within the talent-specific domain is possible because the lower layers are fully integrated and support the individual. At the fifth level, according to Dabrowksi, the potential development of talent is fully realized (Ackerman, 2009).

The five overexcitabilities

Dabrowski also described talented individuals as sensitive, intense, and overly excitable (Mendaglio & Tillier, 2006). He believed that overexcitabilities, innate dispositional traits of talented individuals, must be present and that the talented individual must learn how to regulate them. Talent is facilitated by overexcitabilities (Ackerman, 2009; Mendaglio & Tillier, 2006; Piirto & Fraas, 2012). Martha Graham (www.goodreads.com/quotes/tag/dance) implicitly understood these overexcitabilities as evidenced in her statement, "No artist is pleased. [There is] no satisfaction whatever at any time. There is only a queer divine dissatisfaction, a blessed unrest that keeps us marching and makes us more alive than the others."

Piechowski operationalized OE into five dimensions (Ackerman, 2009; Falk, Lind, Miller, Piechowski, & Silverman, 1999; Mendaglio & Tillier, 2006; Piirto & Fraas, 2012):

1. Psychomotor OE: surplus energy that is expressed as movement, insatiable drive, restlessness, or emotional tension and nervousness. It is sometimes confused with attention deficit disorder; however, talented individuals with high psychomotor excitability are most at home when allowed to express through psychomotor activities.
2. Sensual OE: enhanced differentiation of sensory experiences and seeking outlets for inner tensions. Like all the OEs, sensual OE can be expressed positively through heightened aesthetic explorations or negatively through overeating, buying sprees, or sexual overindulgences.
3. Imaginational OE: thought creations that manifest as frequent daydreaming, intensely engaging fantasy, vivid imagery, and rich associations. High imaginational OE is marked by low tolerance for boredom and a strong desire to engage in imaginative play.
4. Intellectual OE: intensified and accelerated activity of the mind that is differentiated from intelligence. This OE defines behaviors that reveal a love of problem solving, or strong desires to explore or discover new ideas. It can also manifest as a love of thinking about thinking, introspection, and reflection.
5. Emotional OE: extremes of feeling and affective memory that intensify somatic expressions, and promote strong attachment bonds. The emotional OE can augment positive expressions of compassion and responsibility as well as the negative feelings of insecurity, shyness, anxiety, and shame. This OE can intensify the existential awareness of death and increase suicidal thoughts and depression.

In talented performers one or several OEs may be present in greater abundance; however, some talented individuals may not display any of the five OEs (Alias, Rahman, Majid, & Yassin, 2013). The OEs can be expressed positively and negatively and they influence how talented individuals learn. The expression of the OEs can make peers feel uncomfortable, which is why performing artists and gifted students thrive in programs that are dedicated to nurturing their domain-specific talents. In these programs, other students who share their passions surround them. The OEs also provoke internal and external conflicts, a process that is critical to realize the fifth level of integration. But OEs can also increase difficulty interacting with others, even while they are integrating their personal talents (Mika, 2005; Wellisch & Brown, 2013). High-OE individuals are extremely intense and sensitive, and as a result, they are often excluded or alienated; labeled as a nerd, strange, or difficult; or worse, misdiagnosed with a psychological disorder (Alias et al., 2013). Despite potential positive talent development provided by the OEs, when talented individuals experience repeated negative interactions, a shame-based sense of self may form and compromise optimal talent development (Wellisch & Brown, 2013; Fig. 3.3).

There have been numerous studies attempting to identify an OE profile that discriminates individuals with and without talent. Alias et al. (2013) identified four profiles that characterized gifted students: Profile 1—high imagination, Profile 2—all five OEs high, Profile 3—high psychomotor, and Profile 4—all OE dimensions low. This study demonstrated that intellectually gifted students were not homogeneous in their OE characteristics. Other studies have suggested that the three major OEs—intellectual, imaginational, and emotional—are typically higher in intellectually gifted students (Mendaglio & Tillier, 2006), although Piirto and Fraas (2012) did not find any significant differences between gifted and nongifted students. In studies on creative samples, highly creative individuals scored higher on sensual, intellectual, and imaginational OEs (Chavez-Eakle et al., 2012); whereas, athletes scored higher on psychomotor OE (Chang & Kuo, 2009). When comparing athletes to opera singers and dancers a slightly different profile was identified (Thomson & Jaque, 2016a). All the OEs were higher in dancers and opera singers, with the exception of psychomotor OE, which was similar between dancers and athletes. Both the singers and dancers scored significantly higher than the athletes on sensual, imaginational, and emotional OEs. Not surprising, the dancers had higher

Figure 3.3 "Psychomotor talent." Performer: Jacob "Kujo" Lyons. Photographer: Christopher Haggard.

psychomotor OE scores compared to the opera singers, in part because they rely primarily on movement for artistic expression. The high scores on imaginational OE may be related to either an innate disposition (Goldstein & Winner, 2009) or the imaginational dimension may have been fostered through training. As well, a career in the performing arts necessitates that performers communicate emotionally with their audience. Singers and dancers need to fully embody emotional expression in a sensually visceral manner. The high emotional OE in dancers and singers suggests that these performing artists had higher emotional intelligence, a factor that may be related to emotional giftedness (Mayer et al., 2001). The findings in the Thomson and Jaque study (2016a) were similar to earlier study results, suggesting that this pattern—high imagination, emotional, sensual, and psychomotor OE—may be unique to performing artists (Piechowski & Cunningham, 1985; Piechowski, Silverman, Cunningham, & Falk, 1982).

Given the mixed OE findings to date, it is difficult to state conclusively that creative and talented individuals manifest a distinct pattern of OEs. Understanding the influence of the OEs on performing artists may be helpful to teachers who train and support performers in their domain-specific fields. Further investigation needs to occur to understand Dabrowski's model of talent development. The relationship between OE and psychopathology needs to be further investigated as well. Perhaps a high OE profile increases a preexisting vulnerability to psychopathology. In multiple studies on performing artists, a subgroup of performing artists has been identified with clinical levels of psychopathology (Kenny, 2005; Kenny et al., 2004; Neihart, 1999; Thomson & Jaque, 2015b); they also had higher OE dispositions (Thomson & Jaque, 2016b).

The mixed OE findings make this concept troublesome to implement as a means to identify talented individuals. Perhaps this concept is better served when applied to the training practices of talent-specific educational programs (Alias et al., 2013), especially considering that in several studies the performing artist group profiles were significantly different compared to the athlete group profiles (Piechowski & Cunningham, 1985; Piechowski et al., 1982; Thomson & Jaque, 2016a). A simplistic model to identify talent may be fundamentally problematic given the complex nature of performing artists.

Talent-development mega-model

Subotnik et al. (2011) addressed the complexity of understanding talent in their talent-development mega-model (TDMM). Like Piirto's pyramid of talent development model and Renzulli's three-ring model, TDMM is an integrative description of talent. It explicitly states that the intended outcome of talent is to be realized at an eminent level. This model differentiates creative performers (athletes, musicians, dancers, singers, actors) from producers (composers, choreographers, writers, scientists/scholars/academics). Both performers and producers have high-level knowledge in their domain and they are skilled, including psychosocial skills to manage success and adversity. Performers and producers (generators) are usually evaluated differently in their early and later development. For example it is easier to assess and predict performer talent; however, it is much more difficult to identify the success of producers.

The TDMM model includes variations on when talent is fully expressed; e.g., some domains such as dancing usually culminate in early adulthood; whereas, others can extend throughout the lifespan. The developmental trajectory of realizing full potential is contingent on the development of psychobiological factors such as muscle mass, coordination, abstract cognition, and emotional expressivity and regulation. The maturation of talent is also contingent on developmental opportunities such as appropriate training, interactions with other individuals who are deemed talented in the same domain-specific field, and external recognition to promote further development. For example, young dancers must receive early training that will prepare them for a future career as a professional dancer, they must dance with other talented dancers to kinesthetically experience the competitive rigors of this field, and they need to be observed by key figures who can promote and help facilitate placement into a professional company. Not only must these opportunities be available but also talented individuals must seize the opportunities (Fig. 3.4).

Talent recognition changes throughout the lifespan. To continue to engage fully in domain-specific fields, talented individuals must develop substantial skills (domain-specific skills and psychosocial skills) that require ongoing deliberate practice and are supported by strong intrinsic motivation (task commitment). External support from family, peers, mentors, community, and the general public enable individuals to aspire to realize their talents. Subotnik et al. (2011) argue that recognizing and supporting talent is a benefit to the individual and to society. Their TDMM addresses the complexity of identifying and promoting talent. It is an inclusive model that gathers the wisdom of previous models into one overarching model.

Figure 3.4 "Talent revealed." Performers: Darrell Hardaway, Jesse Ricaldi, Christian Vidaure. Photographer: Lee Choo.

Summary

- Talent and giftedness: definitions and overlapping concepts.
- Pyramid of talent development model and three-ring model: inclusion of arts into a gifted definition and gifted programming; consensual assessment of talent.
- Multiple intelligences: musical–rhythmic, visual–spatial, verbal–linguistic, logical–mathematical, bodily–kinesthetic, interpersonal, intrapersonal.
- Theory of positive disintegration: talent, ability, and drive; stages of talent development.
- Five overexcitabilities: psychomotor, sensual, imaginational, intellectual, emotional.
- Talent-development mega-model: intended outcome of talent is to be realized at an eminent level.

Concluding remarks

Identifying and nurturing talent is important for several reasons. When individuals with natural abilities are afforded opportunities to develop their talent, and when they realize their optimal potential, society benefits along with the individual (Subotnik et al., 2011). In all theories discussed in this chapter, realizing talent involves opportunity, support, and commitment. Identifying talent is often limited by stereotypes; it is always a cultural and social construct (Gaztambide-Fernandez et al., 2013).

Talent in the performing arts may be easy to recognize because it is strongly related to specific art forms and aesthetic values. Children who demonstrate sensitivity to musical rhythms and pitches are often those who imagine playing the piano, flute, or violin. Spontaneously jumping, turning, and replicating physical positions, especially when music is played, often indicate that these children want to become dancers. Many children sing quietly on their own; they enjoy any opportunity to learn new songs. They want to sing and they imagine a career as a singer. Young actors often spend time imitating others and role-playing popular characters. They are the children who create plays in their family backyards or in the neighborhood parks. Children may express these interests even if they initially present as seemingly ordinary children. When given educational opportunities to learn about these domain-specific skills, they often show an even stronger indication of talent. They gain confidence because they are receiving recognition for their interests, and if motivated, they will demonstrate task commitment; simply stated, they will practice, practice, practice. These opportunities, coupled with a willingness to respond to the "call" to seriously pursue their talent, all inform the definition of talent.

"Talent development is the partnership between the talented and the field" (Piirto, 1998, p. 337). To achieve eminence in the performing arts, innate talent characteristics must be present. Unfortunately, practice without sufficient talent will not lead to expertise, nor will talent without practice be sufficient (Ruthsatz, Detterman, Griscom, & Cirullo, 2008). Stella Adler, one of the greatest acting teachers, stated in her book, *The Art of Acting*, "The thing that makes you say, 'I want to do something'—that is the beginning of talent" (www.goodreads.com/quotes/tag/acting). In the words of Alan Rickman, "Talent is an accident of genes, and a responsibility" (www.goodreads.com/quotes/tag/acting).

Creative experiences

4

Creating dance is the thing I know best. It is how I recognize myself. Even in the worst of times, such habits sustain, protect, and, in the most unlikely way, lift us up. I cannot think of a more compelling reason to foster the creative habit.

Tharp (2003, p. 243)

Introduction

In psychology, the creative process has generally been described as a four-stage process (preparation, incubation, illumination, and verification) (Bourgeois-Bougrine et al., 2014; Wallas, 1926). Because of the nature of the performing arts, the performers actually describe a pattern that differs from this four-stage model, suggesting that the creative process may be more domain-specific (Bourgeois-Bougrine et al., 2014; Sternberg, 2005). This chapter will outline common creative processes that manifest within performers, in part, because they have received less attention compared to the generators of the performing arts (choreographers, composers, and writers). Unlike the first three chapters, this chapter will focus on the actual process of the performers. As interpreters they embody the essence of being human and not simply what human beings do; just like performers we are all essentially interpretive beings (Crusius, 1991). Performing artists model our fundamental interpretive nature when they perform. This human dimension is revealed through the power of artistic aesthetic performances.

In the first chapter, creativity in the performing arts was defined as a complex integrative model. This theme of complexity persisted when examining the domain-general and domain-specific nature of the performing arts and when identifying talent within the performing artists. In this chapter, the specific process of performers is discussed and once again it is comprised of a complex array of factors.

Interpreters and the creative process

In the four-C model of creativity, a graded level of creativity is defined (Kaufman & Beghetto, 2013). Professional performing artists are considered pro-C creative individuals; they have achieved expert-level creativity. Although generators in the performing arts are often identified as Big-C creators, a term that identifies genius or eminent status, some interpreters are also described as Big-C level. Big-C generators and interpreters reach star status and receive all the domain-specific awards,

Creativity and the Performing Artist. DOI: http://dx.doi.org/10.1016/B978-0-12-804051-5.00004-4

such as the Kennedy Center Awards. The little-c, everyday creativity, plays a potential role for performers during the countless hours of training and practice. They often have brief moments of little-c creativity such as new ways to transport their instruments or novel ways to prepare pointe shoes. Lastly, mini-c creativity, which is the subjective self-discoveries that are regarded as meaningful and novel to the individual, may also occur, perhaps more frequently during early training; however, these mini-c moments can emerge at any point during a rehearsal or a performance once professional status has been reached. Although still needing to be empirically tested, anecdotal reports from performing artists suggest that they have experienced all four levels of creativity. Perhaps the sheer enjoyment of experiencing creativity, regardless of level, helped propel them toward pursuing a career in the performing arts.

To be an accomplished interpreter of the performing arts, a prodigious mastery of technical skill is essential, along with an acute personal awareness of aesthetic appeal. Interpreters must acknowledge the embedded personality of the generator and the context in which the work was created. For example, when performing Mozart's opera, *The Marriage of Figaro*, all involved must respect the actual score and libretto that was produced in 1786 in Vienna. It was a controversial opera that challenged the cultural mores of the time. Over the last 200-plus years, performers of this opera consciously or unconsciously sense the presence of Mozart's personality and the era in which he lived. On some level, the audience who views a performance of this opera is also aware of Mozart's personality and the cultural sensibilities of his time. Concurrently, the audience is aware of the personalities of the performers who are interpreting the opera. The conductor, stage director, designers, musicians, and singers bring their own personalities and lived histories into the performance. This complex layering of realities gives the work a power and life that resonate between the performers and their audience. As Aaron Copland (1957) stated, "For composition is, after all, an organism. It is a living, not a static thing. That is why it is capable of being seen in a different light and from different angles by various interpreters or even by the same interpreter at different times.... The relationship of the performer to the composition that he is recreating is therefore a delicate one" (p. 160).

Interpreters face a formidable task. They must fundamentally engage as a partner in a dyadic relationship. During the creation of the production and while performing it, performers must join with the text, score or choreography. They establish an interdependent couple relationship. The performer must remain faithful and continually infuse life into the text and the performance (Pavis, 2008). They must be faithful to the text or score while working at the interface between life and the "shadow" of life. Audiences are aware that they are watching real performers in the act of expression. Both experience the ephemeral nature of the performing arts, one that attempts to capture the fragility of life. Artaud described this process as the theater and its double; the performance is a "shadow of reality" that looms just outside of reach of the performer and the audience (Artaud, 1958).

As stated in earlier chapters, it is commonly believed that composers, choreographers, and writers are creating via self-exploration and self-realization; whereas, performers participate in the ritual act of performance (Nagy, 2015). Although this may be accurate, performers are also actively engaged in self-exploration and self-realization and generators value the ritual act or habit of creating (Copland, 1957; Tharp, 2003). Many consider performers constrained by the limits of scores or scripts but these constraints can actually enhance creativity (Haught, 2015). Performers are trained to extract meaning from the works to be performed and they infuse their sensibilities into the expression of these works. The seeming constraints of the works are also doorways for further self-exploration.

Performers have been likened to athletes who openly compete to win through physical prowess, but unlike athletes, performers' physicality is the vehicle to expressively communicate the performance work (Press & Warburton, 2007). Martha Graham described a dancer as an athlete of God (1991) and Artaud (1958) stated, "The actor is an athlete of the heart" (p. 133). In an effort to be an athlete of God and the heart, performers must rely on motor control; this control operates as a complex neurobiological inhibitory feedback system. Studies have demonstrated that the creative process functions during cortical and subcortical disinhibition, a process evidenced during improvisation and creative thinking; however, neurobiological inhibitory motor constraints are necessary to produce greater motor originality and creativity (Scibinetti, Tocci, & Pesce, 2011). These neurobiological constraints during motor creativity facilitate the unique moments of magic when performers produce a sound, complete a dance phrase, or act in a scene, moments that leave the audience and the performers aesthetically fulfilled (Csikszentmihalyi, 1996; Scibinetti et al., 2011). Unlike novices, accomplished performers cultivate cognitive structures that enhance motor control. Prescribed movement, whether it is the fingering action to play the piano, the vocal musculature to shape sound in spoken and singing text, or the execution of an arabesque or assemble in ballet, resides in long-term memory as perceptual-cognitive structures (Blasing, Tenenbaum, & Schack, 2009). These mental structures are the seeds from which motor creativity springs.

A predilection for reflective rumination in performers also enhances these moments (Jones, Roy, & Verkuilen, 2014). Unfortunately, performers are vulnerable to the brooding rumination inherent in clinical depression, but when it operates as reflective rumination it supports creative endeavors (Jones et al., 2014; Verhaeghen, Khan, & Joorman, 2005). Although these studies examined musicians, the other performing artists demonstrate a similar ruminative mindset. It helps them master technique and it helps them analyze and interpret a score, text, or choreography. Whether rumination is brooding or reflective it does indicate that the individual/performer is self-absorbed. Absorption in performing artists has been demonstrated in multiple studies (Thomson, Keehn, & Gumpel, 2009; Trapnell & Campbell, 1999).

Rehearsal and the creative process

The actual preparation for a performance involves a structure that has persisted for centuries. Performers prepare for rehearsals, attend rehearsals, independently work extra hours outside rehearsal, and return for more rehearsals, which finally culminate in a performance. And during the performance season, more rehearsal and independent work continues. In France, rehearsals are actually called *répétition*, literally meaning repetition. Rehearsals are prescribed times and places to facilitate the creative process. In rehearsal, creative cognition in the performing arts takes the form of creative realism; a process that involves both structured ordered thinking and spontaneous divergent imagination (Finke, 1996). Structured creativity implies that the interpreters seek connections that are inherent within the scores, scripts, or choreography. The performers must also imaginatively and freely explore the creative implications within the work in order to create a performance that is insightful, novel, spontaneous, and authentic (Finke, 1996; Nemiro, 1997).

Building from Dewey's stage model (1910), creative processing includes perceiving a difficulty, defining the problem, exploring solutions, elaborating on them, and then testing the validity of the solution. During rehearsals the interpreters begin with a clearly defined problem and a perception of the difficulty. The conductors, directors, or choreographers usually guide the exploration of solutions, and the validity of the solution is tested when the work, or a section of the work, is "run." When creatively exploring, all interpreters draw upon a complex tapestry of memories (Damasio, 1999), including (1) autobiographical memory that informs the interpretation, (2) remembering previous rehearsals when musical landmarks, staging or choreography were set, (3) the actual memorization of scores, scripts, or dance sequences, and (4) the procedural memory acquired from years of training and practice. Performing artists are acutely aware that assuming specific body postures will facilitate the retrieval of autobiographical memories, whether for the purposes of enhancing the emotional expression of the work or to simply remember the staging (Dijkstra, Kaschak & Zwaan, 2007; Kogan, 2002). These memory systems inform the structured thinking during rehearsal; the spontaneous freedom is facilitated by playful interactions, unexpected discoveries or random happenstances. Spontaneous moments of exploration may or may not remain in the final product but they become shared experiences that influence future explorations among the performers.

In all the performing arts, rehearsals require embodied exploration (Kogan, 2002). For musical theater performers they must physically explore the three elements of dance, script, and score. They are asked to retain information from choreographers, stage directors, and musical directors. In dance, the cognitive process is primarily an embodied engagement of time, space, motion, and physical expression; it is a nonverbal and verbal language that is shared between dancers and the choreographer. In music, the cognitive process is musical language that operates in time and motion. All performers continually and simultaneously engage declarative and procedural memories. Declarative memory is usually encoded as verbal symbols, but at an expert level, even the most nonverbal procedural art forms such as dance are processed as a declarative language (Stevens & McKechnie, 2005). Ultimately, these two dual systems are part of what could be termed motor "cognition" or embodied cognition (Kristensen, 2004). Talented performing artists excel in this form of cognition, and with it they also gain an enhanced sense of self and other (Jackson & Decety, 2004; Leube et al., 2003). They increase crossmodal self-recognition (Platek, Thomson, & Gallup, 2004) and an embodied integration of linguistic meaning and communication (Bernardis & Gentilucci, 2006; Gibbs, 2003). These by-products of embodied cognition are critical performance enhancers.

Over time, they develop an embodied knowing, a way of perceiving oneself from the inside out. The performer is aware of feelings, movement, and intentions; this embodied knowing has also been described as the "thinking body-mind" (Chappell, 2007; Stinson, 2004). They also develop an aesthetic understanding that is based on both cognitive and physical feedback. For performing artists, aesthetic knowledge is grounded in embodied knowledge; it is a form of felt understanding or embodied cognition that guides creative appraisals during all deliberate practicing, rehearsals, and performances (Chappell, 2007; Dijkstra et al., 2007; Fig. 4.1).

Embodied cognition also influences the emotional tenor of creative expression. It is a felt-affective process (Damasio, 1999). Over time, professional performing artists learn to filter out felt-affective processes that are unrelated to the performance work, even though they may be grappling with intense emotional turmoil outside of the rehearsal hall or theater. They must paradoxically employ their embodied knowing and feelings and yet compartmentalize that which is not appropriate to the work (Lussier-Ley & Durand-Bush, 2009). All performers struggle to realize the perfect integration of appropriate embodied cognition with aesthetic expressivity. This gap is sometimes large and in moments of optimal performance (flow) the gap is almost indistinguishable (Csikszentmihalyi, 1990).

Figure 4.1 "Directing." Director: Geoff Hyland. Photographer: Jean O'Sullivan.

Performing a character

The specific demands of portraying a character influence the creative process. Actors, musical theater artists, opera singers, and dancers must inhabit the character they are to portray. They usually follow a two-phase process to achieve this task (Noice & Noice, 2002). First, they prepare via an intense analytic exploration of the script, and/or score, and/or choreography. Most actors and musical theater performers subdivide the script into units of action and smaller beats (goal-directed actions). Through this process they determine the intentions or objectives of the character and they discover who the character is based on the situations in which the character is placed and what the other characters say about the character (Thomson & Godin, 2011). For opera singers, they also prepare with a similar textual analysis but they must add an analysis of the music to glean insight about the emotional states-of-being of the character. Determining if the music is written in a major or minor key, if it is an allegro or adagio, and even which instruments are playing while they are singing all add insight about character motivation, intention, and emotional state. Dancers follow a similar process as well; they must analyze the music, the specific dance steps to be performed, and the spatial configuration of these steps. In dance, the use of pathways, levels, and planes can lend rich meaning regarding the character they are to portray. Working with the script, score, or choreography in this manner facilitates a more efficient memorization and retention process, one that enlivens the text/score/choreography and moves it beyond simple reciting, singing, or doing steps. This process begins before formal rehearsals begin and continues during the rehearsal phase. Stanislavski (1924/1963) stated, "If our preparatory work is right, the results will take care of themselves" (p. 25).

In the second phase of preparation, the active experiencing phase, performers actually do what the character is doing (Noice & Noice, 2002). They experience the character in their own emotive, cognitive, motor, and physiological systems (Lussier-Ley & Durand-Bush, 2009). They also gain deeper understanding about why a particular word, note, or step is used by the writer, composer, or choreographer. During this phase, memory retention is deepened, along with emotional intensity and involvement. For some performers this phase is also marked by self-identity boundary blurring (Burgoyne, Poulin, & Rearden, 1999) and perceived personality changes (Hannah, Domino, Hanson, & Hannah, 1994). In this active experiencing phase, for some performers, the character or the character's journey may be exhausting or distressing, a situation that can compromise the creative process. Although the actual expressive experience can enhance emotional regulation (Goldstein, Tamir, & Winner, 2013) this is not always the case for performers (Thomson & Jaque, 2012a, 2012b). A poor or hostile working environment can hamper this phase, whether due to poor direction, cast dissention, or being marginalized as a performer (Nemiro, 1997). The social context, both inside and outside the rehearsal hall, can also influence character development and either increase or decrease performer identity tensions (Beech, Gilmore, Cochrane, & Greig, 2012; Fig. 4.2).

According to Stanislavski, the three basic drivers behind creating a role include (1) "mind" (for analysis and understanding), (2) "will" (for control and physical

Figure 4.2 "Embodied characters." Performers: Elissa Brock, Jocelyn Brooke. Photographer: Lee Choo.

preparedness), and (3) "feeling" (fostering passionate and zestful relationships with the characters being created) (Hodge, 2000; Stanislavski, 1924/1963). These three drivers are evident in all the performing arts that portray characters. Hamlet proclaimed, "Suit the action to the word, the word to the action" (Shakespeare, 1600/1987). The merging of action and word (step) is an embodied creative process that requires intense preparation and active experimentation. And when these variables effectively converge, the audience experiences "the theater and its double" (Artaud, 1958).

Performing noncharacter works

Not all performances are character driven. Concert music, whether classical or popular, and dance, whether stage or street, are based on compositional forms and structures. They may be mood-driven or intellectual-abstract rooted. They require different performance intentions but the preparation is often quite similar to that of character-driven works. The capacity to memorize and retain texts, scores, or choreography is at the heart of the early preparation phase. Studies have demonstrated that memory is enhanced by "subject-performed tasks"; literally the performer must perform the work (Noice & Noice, 2002). The performers must deliberately practice motor tasks, whether it is playing an instrument or dancing an abstract work.

When dancers prepare to perform a non-story-driven or character-driven dance, during rehearsals they must carefully watch the choreographer's (or coach's) movement and reproduce it, and yet incorporate their own unique physicality and phrasing. It is a process of looking, thinking (cognitive and motor), and then moving (Lille, 2010). For musicians, it is a process of listening, thinking (cognitive and motor), and then moving (Piirto, 1998). Musicians have cultivated a refined ability to differentiate sounds within an orchestra, ensemble or band, and to modulate, correct, or amplify these sounds (Fig. 4.3).

Both the preparatory work and the active experiencing phase of creating involve doing the action. Phrasing in music and dance, defining dynamics, replicating them,

Figure 4.3 "Language of music." Performer: David Aks. Photographer: Lee Choo.

and incorporating an aesthetic expressiveness, even if it is nonemotional, activates cognitive, motor, and physiological responses in the performer. Again, the social context inside and outside the rehearsal and performance hall positively or negatively affects the creative process and the performance experience.

Experiencing the creative process

How do performers experience creativity? Nelson and Rawlings (2009) conducted a series of qualitative phenomenological studies to investigate the experience of creativity. They were able to identify two major dimensions and multiple variables contained within these dimensions. The experiential dimension describes the ways creativity is experienced; whereas, the existential dimension refers to the meaning derived from the creative experience (Nelson & Rawlings, 2009). The experiential variables include:

1. Distinct Experience: The creative process is identified as something different than everyday experiences. There is an awareness of a loss of self with a sense that boundaries are breaking down. Greater emotional intensity is experienced coupled with a heightened sense of emotional stability. Individuals express a feeling that a force

beyond them propels them, and many sense a heightened spiritual dimension. Distinct creative experiences are marked by heightened awareness, and increased technical and expressive abilities.
2. Anxiety: During the creative experience some individuals feel more anxious and vulnerable. These feelings are often more intense after the process but some also feel it during the process.
3. Absorption: Experiences of deep absorption in the creative activity and a sense of inspiration are accompanied with heightened states of discovery, receptivity, and openness to the direction of the creative work. Usually in this experience there are marked oscillations between absorption and the distancing process of critical awareness.
4. Power/Pleasure: A sense of control, power, and pleasure are experienced, including a heightened capacity for analytic mental processes.
5. Clarity/Preparation: The creative experience is infused with a feeling of certainty and clarity. This feeling informs the direction, meaning, and purpose of the creative work. When clarity and preparation operate concurrently, individuals are able to cultivate appropriate moods necessary for the creative work regardless of how distressing or joyful.

In a recent study by Thomson and Jaque (2016a), dancers experienced heightened distinct experiences, absorption, and power/pleasure compared to a group of athletes. There were no differences between the two groups for anxiety or clarity/preparation. This study supports the notion that dancers not only experience performance as creative, they are also creative artists in their own right.

Although the majority of the experiential variables discriminated dancers from athletes, the existential dimension and its variables were even stronger discriminators (Thomson & Jaque, 2016a). For the dancers, performing was extremely meaningful and existentially creative. The existential variables include (Nelson & Rawlings, 2009):

1. Transformation: This existential variable describes how the creative process enables a deeper engagement with the self and the world. The transformation occurs through an ability to broaden awareness, which increases confidence in the subjective experience of creating. It is a powerful force that has a healing quality.
2. Centrality: This existential variable captures the need to engage in artistic activities and the centrality that creativity plays in the life of the individual. The creative process provides meaning and purpose, so much so that going without creative experiences will increase anxiety. It is almost an addictive need to create; however, it also offers a sense of healing and a deeper connection to spirituality. The primary motivation is one of experiencing joy and self-discovery.
3. Beyond the Personal: Existential meaning is related to an interpersonal concern with the context of the artwork. The creative process transcends the personal; it is a freedom to explore artistic possibilities.

The experiential and existential dimensions and their variables are valuable ways to quantify the creative experience in performing artists. It is well documented that artists are more open to experience, including more paranormal experiences (Ayers, Beaton, & Hunt, 1999), and a transcendent sense of belonging to something larger than the self (Bernard, 2009), findings that transcend cultures (Fleith, Rodrigues, Viana, & Cerqueira, 2000). This openness also suggests a willingness to creatively explore, including allowing more wide-ranging associations to enter the creative experience. Although all these experiential variables are associated with latent inhibition and schizotypy features, it must be remembered that there is an inverted U pattern in artist samples. This pattern explains the notion that creative individuals have features of schizotypy but not necessarily in the range of psychotic or affective disorders (Nelson & Rawlings, 2010). Despite this inverted U association with psychopathology and creativity, identifying experiential and existential dimensions helps clarify common behaviors inherent in artists, especially the existential experiences that promote joy, meaning, and purpose. These intrinsic creative experiences are rewarding and they certainly outweigh any extrinsic monetary rewards for performing artists (Lacaille, Koestner, & Gaudreau, 2007; Fig. 4.4).

The experiential and existential dimensions are closely related to optimal flow experiences (Nelson & Rawlings, 2010). Like Nelson and Rawlings (2009), Csikszentmihalyi and colleagues (1990, 1996) began their work with qualitative phenomenological studies. Through countless interviews they noticed that creative individuals repeatedly described the same nine variables. These variables were first identified as factors that contributed to a sense of optimal performance (also described as flow states or performing in the "zone"); when creating, eminent artists reported these factors. For performing artists, flow states occur when performing in front of an audience and during rehearsals. The creative process is one of embodied motor engagement, a feature that is inherent in all flow (performance and creative) experiences. When looking exclusively at the creative experience through the lens of flow theory, it is easy to recognize the conditions that amplify creative flow states for performing artists. Certainly, during the actual performance, performers have vast opportunities to achieve moments of optimal flow; they also have ample opportunity to feel the joy of discovery and the depth of meaning that is elicited during the preparation of a work to be performed.

Figure 4.4 "Communal world." Performers: Allen Chung & COSMIC. Photographer: Lee Choo.

The nine flow elements are as follows (Csikszentmihalyi, 1996; Jackson & Eklund, 2004; Kawabata & Mallett, 2011):

1. Clear Goals: In flow conditions, the performer always knows what is expected; e.g., musicians and dancers know the performance order of musical works or dances. Actors, opera singers, and musical theater performers know the scene breakdown of the entire show. All expect that this order will be followed. The goal is to perform the work and the audience is clearly aware of this goal.
2. Immediate and Unambiguous Feedback: Because the performing arts are an embodied, kinesthetic experience, the performers immediately "know" if they have played or sung the right note, recited the right verse, or executed the right step. This feedback informs the performer if all is going well or not.
3. Challenge–Skill Balance: In flow states, whether during a performance or during the preparation for a performance, the ability to meet the technical challenges with adequate skills is essential. If the challenge is low then performers are easily bored and if the challenge is too high then anxiety is increased. An imbalance between challenge and skill can compromise flow experiences but when there is a balance between these two variables, flow states are more likely to occur.
4. Merging of Action and Awareness: Performers have cultivated an acute sense of embodied cognition that enhances their awareness and their ability to coordinate all actions with heightened attention to detail. They value this skill and derive pleasure from it.
5. Concentration and Ability to Exclude Distractions: Performers are able to focus on the performance tasks when the challenge and skill levels are balanced; however, if they are bored then they may easily allow their thoughts to wander, especially during long performance seasons when the same work is performed repeatedly. Part of the training is to invest every moment with a here and now presence so that concentration can be maintained (Stanislavski, 1924/1963).
6. Sense of Control: When a performer has a feeling of control, flow states are enhanced. The performers are not preoccupied with worries of failure, especially because they have gained sufficient skills to manage the task of performance.
7. Loss of Self-Consciousness: During performance, the performers are usually not preoccupied by what others think about them. If the critical internal or external judgments increase then flow is unlikely to transpire.
8. Transformation of Time: A sense of clock time may be altered, with some moments experienced as prolonged moments of pleasure while other moments disappear rapidly. After a performance many performers are surprised that hours have passed.
9. Autotelic Experience: This condition literally captures a sense that the performance activity was an end in itself. The joy of performing a work for an audience is the essential goal. It is meaning-making and gratifying. This variable resembles the earlier discussion of existential gains achieved during the creative process.

These nine conditions have been investigated in countless studies, with robust results indicating the positive nature of flow. For example, in performing artists, greater flow was associated with decreased anxiety (Thomson & Jaque, 2011–2012), blunted dissociative processing (Thomson & Jaque, 2012c), diminished self-doubt (de Manzano, Theorell, Harmat, & Ullen, 2010), and more perceived playfulness (Tan & Chou, 2011). Flow increased in relationship to greater intrinsic motivation (Martin & Cutler, 2002) and subjective emotional wellbeing (Fritz & Avsec, 2007). Features identified within talented individuals such as novelty seeking, persistence, and self-transcendence are positively related to flow, suggesting that personality traits and genetic predispositions may augment flow-like experiences (Mosing et al., 2012; Teng, 2011; Ullen et al., 2012). The capacity for goal directedness was also related to higher autotelic personality traits (Busch, Hofer, Chasiotis, & Campos, 2013). These findings support the commonly accepted reality that performing artists love to perform; they are able to experience flow states at an intense level (Thomson & Jaque, 2016a) (Box 4.1).

In all performing arts settings, environmental and relational variables can enhance or compromise experiences of creativity and flow states (Hefferon & Ollis, 2006). For example, when the physical conditions in rehearsal halls and theaters are adequate, performers can meet the challenges of the work. These include good floor surfaces (sprung floors, marlee surfaces), acoustic conditions (dry, dead, resonant), temperature (heating and air conditioning, drafts), and lighting (side lights, follow spot, ambient). Costuming and makeup can enhance character portrayal and performance aesthetics but they can also constrain movement or undermine character development. Preperformance rituals to support optimal performance can be compromised (insufficient time, space, or privacy) and relationships with others can be either supportive or destructive (receiving corrections in a timely or inopportune and inappropriate manner, competitive rivalry that is bullying or devaluing). One of the most important influences on performers' flow is the emotional contagion between them and their teachers or mentors. Throughout a career in the performing arts, performers engage with teachers, coaches, and mentors. There is a powerful reciprocal relationship that influences flow states of both (Bakker, 2005). Perhaps flow contagion can affect an audience as well.

Box 4.1 Relationship Between Creative Process Factors and Flow Factors

Creative process factors	Flow factors
Distinct experience	Merging action and awareness
Anxiety	Immediate and unambiguous feedback
Absorption	Concentration on the task at hand
Power/pleasure	Sense of control
Clarity/preparation	Challenge–skill balance
Transformation	Transformation of time
Centrality	Loss of self-consciousness
Beyond the personal	Autotelic experience

Summary

- Introduction
- Interpreters and the creative process: four-C model (Big-C creator, Pro-C professional, little-c everyday creativity, and mini-c subjective creativity); complex requirements to become an accomplished interpreter; text–performer relationship; ritual and self-exploration; motor constraints and creativity; rumination (reflecting or brooding).
- Rehearsal and the creative process: structured and spontaneous processes; creative realism; problem solving and memory; embodied knowing; embodied cognition.
- Performing a character: two-phase (preparation and active experiencing); creating a role (mind, will, feeling).
- Performing noncharacter works: preparatory work and active experiencing phrasing; defining dynamics; incorporating aesthetic expressiveness.
- Experiencing the creative process: experiential dimension (distinct experience, anxiety, absorption, power/pleasure, clarity/preparation); existential dimension (transformation, centrality, beyond the personal); flow (clear goals, unambiguous feedback, challenge–skill balance, merging actions and awareness, concentration and ability to exclude distraction, sense of control, loss of self-consciousness, transformation of time, autotelic experiences).

Concluding remarks

Experiencing the creative process is fundamentally positive. Even when struggles plague the process, performers willingly continue to grapple with the problems in an effort to realize a performance that is aesthetically compelling and meaningful for the audience and for them. Performers have unique talents and years of training that facilitate this process. In the words of actor Tom Hiddleston, "Actors in any capacity, artists of any stripe, are inspired by their curiosity, by their desire to explore all quarters of life, in light and in dark, and reflect what they find in their work. Artists instinctively want to reflect humanity, their own and each other's, in all its intermittent virtue and vitality, frailty and fallibility" (www.goodreads.com/quotes/tag/acting).

The creative process is a rich opportunity for performers. They have the opportunity to derive meaning and purpose that can be expressed in their careers and their lives. Performers attempt to realize that which looms just outside of our grasp. This effort propels them to actively explore themselves and the work they are to perform. In great performances, the audience and the performers are acutely aware that they share in the creation of an illusion, a shadow of reality, and yet, it is also an experience that is vibrantly alive and real. As Artaud claimed, the creative process in the performing arts is one that produces "the theater and its double" (1958).

Imagination and fantasy

5

Creative ability belongs to the sphere of reality as much as the realm of fantasy.

Mary Wigman (1966, p. 12)

Imagination

In the words of legendary opera singer, Maria Callas, "An opera begins long before the curtain goes up and ends long after it has come down. It starts in my imagination, it becomes my life, and it stays part of my life long after I've left the opera house" (www.goodreads.com/quotes/tag/opera). This statement captures the reality for most performing artists. Accessing the imagination allows performers to dream of entering a career in the performing arts. It is engaged during rehearsals and performances and it enriches the long hours of daily practice. It can easily be considered the bedrock of the performing arts.

Imagination is defined as "the act or power of forming a mental image of something not present to the senses or never before wholly perceived in reality" (Merriam Webster Dictionary). It is concerned with new insights, whether it is a brief moment of clarity or a prolonged exploration. Imagination is intentional and is employed in response to a particular context (Coleridge, 1817/1976; Policastro & Gardner, 2005). It can vary in intensity and valence and always draws upon and integrates knowledge and experience. Engaging the imagination means that a coherent representation will be formed over time. It is the representation of that which does not yet exist (Kristensen, 2004) and it requires a willingness to initiate an exploration into the imagination (Lin, Chang, & Liang, 2015). It can also be employed to reengage previously formed images or practiced actions. When rekindling images or experiences, performers can discover emergent patterns that were formerly unrecognized and yet are embedded within the image (Finke, 1996), or they can crystalize and then transform them so that they can operate in different settings or contexts (Lin et al., 2015). For example, a singer can imagine a musical phrase, hear the texture of the sound, and discover nuances that were previously unnoticed, and through this exploration, the singer can consciously alter the performance. It is a process of active imagination; the performer literally interacts with the flow of internally generated images, gains insight about these images, and allows them to inform the creative exploration (Johnson, 1986).

On some level, all performing artists generate new images and reinvestigate earlier images. As performers, their imaginative explorations are embodied kinesthetic imaginings (Lussier-Ley & Durand-Bush, 2009). Optimally, they also integrate aesthetic, sensory, and emotional meaning into the image. The artists are able to

Creativity and the Performing Artist. DOI: http://dx.doi.org/10.1016/B978-0-12-804051-5.00005-6

actively interact with these images to alter, amplify, or further explore new possibilities (Johnson, 1986). Performing artists can conjure an array of diverse images, although field-specific images, and imaginings are typical. For example, musicians have auditory images, literally, musical ideas; they are not intellectual or literary ideas but completely rooted within musical themes (Copland, 1957). All performers and generators, during the creative process, engage their imagination in these field-specific patterns.

Performing artists embody the "lives" or emotional reality of the performance work; it is simultaneously separate from them and shared by them. Many actors report that, from an early age, they created imaginary worlds (Goldstein, 2009). They gathered other children and "produced" theatrical events, either performances that took place in the backyard or park, or as digital performances that were shared on the Internet. Dancers remember creating dances and performing them in their bedrooms or living rooms, and musicians and singers offered spontaneous concerts for their friends and families (Piirto, 1998). Goldstein and Winner (2009) investigated professional actors and found that actors were drawn to fictional characters and role-playing as children. Individuals who created elaborate fictional worlds during childhood were likely to continue this practice in adulthood, such as writing stories that were performed on stage or film (Taylor, Mottweiler, Naylor, & Levernier, 2015). Childhood constructions of virtual imaginative realms or embodied real-time theater-making may be the training ground for persistence, absorption, discovery, and synthesis, all features of creativity (Root-Bernstein & Root-Bernstein, 2006). In fact, studies that examined creative adults reported that they engaged extensively in "worldplay" (creation of alternate lives) as children; their extent of involvement was not different from that reported by adult performing artists. As such, childhood worldplay activities may indicate a predisposition toward creative behavior, in general, and not specifically a predisposition for the performing arts (Goldstein & Winner, 2009; Root-Bernstein & Root-Bernstein, 2006).

Fantasy and fantasy proneness

Fantasy is defined as "something that is produced by the imagination: an idea about doing something that is far removed from normal reality" (Merriam Webster Dictionary). Fantasy is believed to be a more subjective personal fabrication that is employed as a means of escape from the pressures or mundane existence of daily life. It serves as an adaptive function to handle conflicts, boredom, or wishes. Unlike imagination, which generates potentially creative ideas, fantasy tends to generate illusions (Policastro & Gardner, 2005). In fantasy, individuals passively observe the internally generated fabrications, regardless if these fantasies repeat themselves in loops of worry or escape images (Johnson, 1986).

Generally, as terms, imagination and fantasy are used interchangeably (Sternberg, 1999). To differentiate them, it helps to determine the intention or purpose of the image. Imagination is employed when seeking new ideas within a specific context; whereas, fantasy offers an internal escape. Although they have different purposes, both require a deep state of absorption (Levin & Fireman, 2001–02; Lynn & Rhue, 1988; Rauschenberger & Lynn, 2002–03). Absorption states are marked by a loss of internal and external awareness; an intense concentration is narrowly directed toward a specific activity (Garrido & Schubert, 2011). It is a state of heightened imaginative or fantasy involvement (Cardena, Lynn, & Krippner, 2000; Lynn & Rhue, 1988), something that performing artists commonly experience. The capacity to become deeply absorbed is one of the factors operating during deliberate practice sessions and the active exploration phase during rehearsals and performances. Intense absorption states have also been associated with increased enjoyment of music, including music that evokes negative emotional moods, perhaps because absorption states increase tolerance for emotional arousal (Garrido & Schubert, 2011).

Fantasy, including absorption states, are everyday occurrences; however, approximately 4% of the population are considered "fantasy-prone," a term that describes a set of characteristics such as spending long periods of time absorbed in vivid daydreams (Wilson & Barber, 1983). According to the original study by Wilson and Barber (1983) high fantasy-prone individuals share common traits and childhood backgrounds. They report an ability to fantasize imagined or real events with all sensory modalities fully engaged; the intensity of their fantasy is so real that many fantasy-prone individuals have difficulty distinguishing reality from daydreams. For example, some experienced physical symptoms such as high fevers, nausea, and severe pain when they witnessed or imagined others with an illness, and many have experienced a false pregnancy, which included amenorrhea, breast changes, morning sickness, abdominal enlargement, and imagined fetal movement. Other characteristics include heightened frequency and intensity of paranormal experiences such as telepathy and precognition, out-of-body experiences, and religious visions. In childhood, fantasy-prone individuals report spending long periods of time alone, happily exploring their fantasy worlds. Some described protracted adventures with fairies, angels, elves, or leprechauns; some had imaginary companions; and all had elaborate relationships with their dolls and stuffed animals. They believed these imaginary creatures were real and as they got older they learned to hide their beliefs with others but continued to engage with these imaginary figures into adulthood. Factors that lead to fantasy-prone traits and extensive fantasy involvement included encouragement from adults to fantasize, isolation or loneliness, a need to escape from troubled environments, or significant early stressor(s) (i.e., severe illness, death of family member; Fig. 5.1).

Fantasy proneness and absorption traits have consistently been linked to creativity (Lack, Kumar, & Arevalo, 2003). These individuals are able to produce more complex and elaborate narratives (Marinelli et al., 2012) and are better storytellers (Merckelbach, 2004). Fantasy proneness has also been linked to higher scores on IQ testing, suggesting that fantasy proneness may be a feature of talent and giftedness (Dunn, Corn, & Morelock, 2004). Fantasy proneness shares many features with the imaginational overexcitability dimension discussed earlier (Piechowski, Silverman, Cunningham, & Falk, 1982). They are both associated with talent and giftedness. Performing artists manifest a heightened capacity to engage in vivid

Figure 5.1 "Fantasy reality." Performer: Yvonne Ng. Photographer: Cylla von Tiedemann.

imaginings; in fact, fantasy proneness, and imaginational overexcitability were higher in dancers and opera singers compared to elite athletes (Thomson & Jaque, 2016b). Fantasy proneness, and the ability to become deeply absorbed, is also associated with paranormal beliefs and out-of-body experiences (Gow, Lang, & Chant, 2004; Irwin, 1990). They enhance abilities to perceive apparitions, a quality that is much different than the internal visions of the imagination. Seeing ghosts or hearing apparitional voices may be positive symptoms of psychotic illnesses, but they have also been reported by well-balanced individuals, ranging from shamans and mystical leaders to everyday people who report seeing a deceased loved one or receiving spiritual guidance from the departed (Parra, 2006).

Based on previous research, two factors are identified as antecedents for high fantasy proneness. The first is that fantasy proneness shares features with absorption and schizotypy; hence it is considered a personality trait (Merckelbach & van de Ven, 2001; Sanchez-Bernardos & Avia, 2004, 2006). The second is that fantasy proneness is a coping strategy to manage distressing situations. It is mobilized in early childhood and persists into adulthood as a coping strategy (Rhue & Lynn, 1987; Somer, 2001; Wilson & Barber, 1983). In all studies, only a proportion of the high fantasy-prone sample had childhood adversity, which indicates that it may be a personality trait, much like schizotypy, and under stress it is mobilized as a protective defense structure (Lynn & Rhue, 1988; Sanchez-Bernardos & Avia, 2006). Because these studies on fantasy proneness describe a convergence of developmental and personality factors, it might be more advantageous if it were regarded as a multidimensional construct (Sanchez-Bernardos, Lloreda, Avia, & Bragado-Alvarez, 2015), one that may be dimensionalized as normative or psychopathological (Klinger, Henning, & Janssen, 2009). This suggestion is consistent with findings demonstrating that the majority of high fantasy-prone individuals are well adjusted (Dunn et al., 2004).

In a minority of individuals, fantasy proneness is debilitating. High fantasy proneness and intense frequent daydreams are related to sleep disorders (Giesbrecht & Merckelbach, 2006). Nightmares and nightmare distress are associated with higher levels of fantasy proneness and psychological absorption, and they increase when fantasy and absorption are both concurrently present at high levels. Exhaustion and psychopathology intensify when these sleep disorders persist (Levin & Fireman, 2001–02). Some also suffer increased anxiety and dissociative disorders (Giesbrecht & Merckelbach, 2006; Rauschenberger & Lynn, 2002–03). Compulsive fantasy, a feature that may be present in high fantasy-prone individuals, is a form of addiction, one in which the individual is not able to stop the fantasies (Somer, 2001, 2002). They are physically and emotionally exhausted by the prolonged periods of intense kinesthetic and sensory activation during their fantasies, and they also suffer deviant sleep cycles that exacerbate their exhaustion (Bigelsen & Schupak, 2011). The fantasy productions are usually directed at self-amusement, pleasure, distraction, or escape (Butler, 2006). When they persist and the individual is no longer able to regulate them, fantasy proneness is regarded as a pathological addictive disorder. Despite increased exhaustion and sleep disorders, compulsive fantasizers report that their fantasies as compelling, exciting, rewarding, and pleasurable.

This classic addiction pattern of seeking rewards despite increased harm is the hallmark of compulsive fantasy behavior; however, in the qualitative studies examining the creative process, many artists described creative experiences as addictive (Nelson & Rawlings, 2009). They need to create; participating in the creative process offers experiential and existential rewards. What separates compulsive fantasizers from the "addicted" artists may simply be that one group never actualizes their mental creations; whereas, the other group is actively manifesting their creative efforts in the real world. For fantasizers, the creative process may be a completely private and internalized experience. On the other hand, performers need to externalize their experiences.

Further investigation needs to take place to determine the differences between compulsive fantasizers and the driven "addictive" behavior of creative individuals, such as performing artists. Although not necessarily ranked as one of the 4% who are regarded as high fantasy-prone individuals, moderate-to-high fantasy-prone individuals are drawn to the dramatic arts, writing, and religion (Dunn et al., 2004). These career paths share similar features; all are concerned with human expression. In a series of studies, dancers, opera singers, and actors all scored higher on fantasy proneness compared to control groups and athletes (Thomson & Jaque, 2013). Unfortunately, fantasy proneness was related to increased shame in performers but not in athletes (Thomson & Jaque, 2013, 2015a), and it predicted increased anxiety (Thomson & Jaque, 2011–2012) and dissociation (Thomson & Jaque, 2013). These findings suggest that fantasy proneness, a multidimensional and complex trait, is greater in performing artists compared to athletes and healthy nonartists.

Employing imagination and fantasy

Educating the imagination is vital. Through this training, performers gain a capacity to intensify the emotional content of images along with enlivening the sensory, visual, and auditory dimensions of them. According to Northrop Frye (1963), an educated imagination enables artists to tolerate perspectives that are remote from their own personal beliefs or values. The imagination facilitates a detachment from these potentially intolerable experiences while enabling active explorations to occur. With an educated imagination, the performer is able to investigate the "heights and depths of what the human mind can conceive" (Frye, 1963, p. 42). There are many books and articles written on visualization work, mental practice, and imagery work, all approaches to improve performance. All of these approaches (and many more) serve to energize the imagination (Carnicke, 2000). What is conjured by the imagination already exists in some form within the lived experience or unconscious mind of the artist; learning to actively interact with the energized imagination is one component in the education of the imagination (Johnson, 1986). The second component is learning to stay emotionally regulated so that the images can be physically and psychologically expressed (Fig. 5.2).

Implementing imagery into training indirectly educates the creative imagination of performers. Typically, *mental practice* is a term that describes a rehearsal

Figure 5.2 "Lean on me." Performers: Jacob "Kujo" Lyons, Anibal Sandoval. Photographer: Christopher Haggard.

strategy that is cognitively based and not physically executed. Performers can improve skill acquisition, rehabilitate injuries, and relearn poorly executed skills (Krasnow & Wilmerding, 2015). This form of mental imagery is fundamentally a motor–cognitive process. The goal is to improve motor skills without the stress of physically repeating the actions. With improved motor skills, aesthetic, and emotional expression can indirectly be improved. *Mental imagery* is viewed as a psychological process that is designed to address the psychological and emotional elements of performance (Krasnow & Wilmerding, 2015). Mental imagery can manifest as a form of external imagery in which performers adopt the view of the audience; they mentally watch themselves perform. Internal imagery occurs when the performer experiences the performance from an "I" perspective and during this process the performer can explore the aesthetic and emotional experiences of performance (Milton, Small, & Solodkin, 2008) (Box 5.1).

High fantasy-prone performers are able to externally and internally imagine the experience through all of their senses; whereas, less fantasy-prone individuals can strengthen one or several of the senses, such as visual, motor, kinesthetic, auditory, gustatory, or tactile. Mental imagery can be experienced as a direct replication of the physical performance but the images can also be more abstract whimsical fantasies that give the imagination a more mercurial quality. These abstract images are often described as metaphoric images such as imagining floating like a feather, moving through water, feeling the wind propel the movement, and taste the sound. Mental imagery and mental practice are powerful imaginative skills and should be incorporated into the training of performing artists, especially because mental imagery and mental practice directly activate neurobiological systems needed to perform (this will be discussed in depth in Section II) (Kleber, Birbaumer, Veit, Trevorrow, & Lotze, 2007).

These mental skills are not only important to improve performance but they are also part of educating the imagination. Actor training has traditionally cultivated these skills, especially mental imagery (Hagen & Frankel, 1973; Hodge, 2000).

Box 5.1 Definitions

Term	Definition
Imagination	Formation of mental image of something that is not present to the senses or never perceived in reality; seeking new ideas
Fantasy	An idea or doing something that is far from normal reality and produced by the imagination; internal escape from reality
Mental practice	Rehearsal strategy that is cognitively based and not physically executed
Mental imagery	Internal imagery to address psychological and emotional aspects of performance; can conjure images from audience perspective (external) or from a self perspective (internal)

Most training practices integrate psychological and physical dimensions that are supported by specific exercises to educate the imagination of the actor (Thomson & Godin, 2011). Actors value their imaginative assertiveness, including encouragement to explore many ideas during the creation of a role (Gluck, Ernst, & Unger, 2002). Traditional Stanislavski training (1924/1963), Stanislavski-based methods (Hodge, 2000), and non-Stanislavski-based training such as those prescribed by Tadashi Suzuki (1986), Ann Bogart (2001), and Eugenio Barba (Barba & Savarese, 2005) all include training exercises that educate the actor's imagination. And all value skills that integrate mind, body, and energy. Actors learn to engage complex internal mental *imagery* during rehearsals and performances and they implement mental *practice* to reinforce the memorization of text and stage blocking.

Musicians and singers also engage mental practice and mental imagery. Many instrumentalists suffer crippling dystonias or muscular injuries that preclude in vivo practice, and singers are frequently forced into prolonged periods of vocal rest (Sandgren, 2002, 2009). Deliberate mental practice, including adding the psychological elements of mental imagery, supports musicians and helps them reduce practice stress while prolonging their performance careers (Halpern, 2012; Zatorre & Halpern, 2005). Mental imagery during performance facilitates planning and executing actions, specifically increased control of timing, intensity, articulation, intonation, and predicting others' actions in the ensemble or orchestra (Keller, 2012). Ultimately, mental imagery promotes artistic perfection and ensemble cohesion. What is difficult to scientifically measure is the complex array of mental images and mental practices operating in professional musicians. Because music involves complex motor, auditory, sensory, and emotional processes, a diverse team of experts is needed to adequately assess the range of imaginal processing activated in preparation for and during a performance (Milton et al., 2008). What is known is that musicians are superior at musical imagery tasks compared to nonmusicians (Aleman, Nieuwenstein, Bocker, & de Haan, 2000). Despite heightened skills in mental practice and mental imagery, these mental skills can never match the physiological and psychological intensity of real-time embodied performance; ultimately musicians and singers must perform (Dietrich, 2008; Meister et al., 2004; Fig. 5.3).

Although mental practice and mental imagery are important skills, few dancers are taught them (Nordin & Cumming, 2006a, 2006b). Most dancers consciously or unconsciously respond to their teachers' imagery, but few receive direct instruction about generating their own images (Overby, Hall, & Haslam, 1997–98). With experience, dancers tend to cultivate more personal complex kinesthetic images and employ them more frequently (Nordin & Cumming, 2006a). Experienced dancers reported that they engaged more metaphorical images to enhance quality of movement during classes and rehearsals. This may be related to increased skill mastery, including the capacity to integrate sensory awareness with motor control (Nordin & Cumming, 2006b). Professional modern dancers also described a multimodal and multidimensional capacity to employ imagery strategies during training and performance. They reported personalized images that served to inspire them, and to evoke atmospheres that enhanced skill, movement quality, and emotional expressivity. They also engaged imagery to enhance feelings of transcendence to a metaphysical

Figure 5.3 "Embodied performance." Performers: Elissa Brock, Hilary Gereaux, Ani Keshishyan, Cameron Shim. Photographer: Chuck Valerio.

realm and the rewarding pleasure of flow experiences (Hanrahan & Vergeer, 2000–01). What is apparent in the studies on dancers is that expertise in dancing promotes more frequent and complex mental imagery; they actively embrace these skills to enrich their performance.

Summary

- Imagination: definition; new images and reinvestigated images; active imagination; imaginative encouragement in childhood and creativity.
- Fantasy and fantasy proneness: definition; fantasy proneness (4% of population); positive effects of fantasy proneness include increased IQ; imaginational overexcitability; talent; openness to experiences such as paranormal experiences. Antecedents for fantasy proneness include personality trait, adverse experiences or both; whereas, compulsive fantasy

and pathology (addiction, sleep disorders, anxiety, dissociation, shame) compromise the positive effects.
- Employing the imagination and fantasy: educating the imagination; mental practice; mental imagery; internal and external perspective-taking and imagery; mental skills for actors, musicians, singers, and dancers.

Concluding remarks

Imagination, fantasy, and educating the imagination are fundamental factors in the performing arts. Acting has directly addressed the inclusion of the imagination into the actor training process. It could be argued that the actor's imagination is the primary source for creating a role (Hodge, 2000). Actors learn to integrate psychological and physical elements to embody a character; they develop a psychophysical imagination that is employed during rehearsals and performances. Musicians, singers, and dancers receive less formal training in educating their imagination. Distinguishing between the imagination and fantasy helps address the essential differences and shared similarities between these two processes, but both are related to creativity. Learning to engage mental imagery, like all skills, can be practiced and deliberately incorporated into instruction (with teachers or coaches) and practice (alone or with ensembles) (Overby et al., 1997–98). Successful skills in mental imagery can also decrease audition and performance anxiety and improve performance overall (Monsma & Overby, 2004). Mental practice can help reduce the physical toll that most performers experience. It can decrease overuse injuries and prolong careers.

Successful engagement of the imagination helps shape the future. The imagination can help us "see" into the future. We can imagine dreams and goals, plan ways and means to achieve them, and anticipate obstacles and challenges (Taylor, Pham, Rivkin, & Armor, 1998). For young performing artists, the imagination helps them envision a future in the performing arts. They can also imagine and create their own theatrical productions, music concerts, and dance performances. The long hours of practice are supported by actively engaging the imagination; it gives depth and dimension to the work or skill being practiced. Mental imaginings can be highly rewarding, but they can also increase shame, anxiety, and dissociation. They can increase flow experiences and enrich the creative process but they can also destroy flow states and taint the creative process, so much so that performing becomes an aversive experience. Sadly, when the imagination is subsumed into the passive activities of compulsive fantasizing then it can provoke sleep disorders and physical and emotional exhaustion; at the extreme, the performer may no longer be able to sustain a career in the performing arts. In the words of Stanislavski (1924/1963), "If you are looking for something, don't go sit on the seashore and expect it to come and find you; you must search, search, search with all the stubbornness in you!" Optimally, the educated imagination supports this search. Performing artists tend to embrace this search; they relish the opportunity to actively imagine and then perform a great work of art.

Section II

Biological Manifestations of Creativity in Performing Artists

Neurobiology, creativity, and performing artists

6

Love art in yourself, not yourself in art.

Konstantin Stanislavski

The brain

The brain operates as a complex open system with multiple regions operating independently and collectively. Like a large symphony orchestra, each section in the brain has its own distinct function. All regions, both cortical and subcortical, can interact to produce novel and powerful works of art but, when things go awry, they can equally manifest in a chaotic cacophony of neural signals. The constant challenge in neuroscience research is to reveal a coherent story that explains how our brains create (Freeman, 2000). How do we discern eminent creative brain activation from everyday creativity? These are important questions that currently remain elusive and tantalizing. Creative thinking is not homogenous; different properties of thinking operate within different regions of the brain drawing upon a complex array of neural networks to coordinate and synthesize creative output (Dietrich, 2007; Park, Kirk, & Waldie, 2015).

In the brain there are approximately 150 billion neurons and each neuron connects with between 1000 and 10,000 other neurons through synaptic (space) connections (Pinel, 2000). A neuron sends a signal from its cell body down through a tentacle-like structure called an axon. Long axons may have myelin cells wrapped around them to increase conduction speed of the neuronal signals. Myelin cells are formed from glial cells and appear as white matter in brain scan images. At the end of the axon there are many axon terminals that transform an electrical signal into a chemical signal (neurotransmitter), which is then released into the synaptic space (Freeman, 2000; Panksepp, 1998). A group of short tentacles called dendrites, located on a neighboring neuron, receives these chemical signals, which are then sent to this neuron's cell body for further processing (Pinel, 2000; Purves et al., 2001). Neurons are constantly firing, with some highly active neurons firing between 50 and 100 times per second; whereas, relatively calm neurons may fire fewer than 10 times per second (Llinas, 2001; Sawyer, 2011; Figs. 6.1 and 6.2).

The brain is organized into discrete structures that have specific functions (Pinel, 2000; Purves et al., 2001). Located at the base of the brain, several structures are responsible for supporting life (pons, medulla oblongata, cranial nerves, respiratory center, tegmentum). Behind this vital life force region, the cerebellum operates as a coordinating center for lower and higher brain structures. Similar tasks are managed

Creativity and the Performing Artist. DOI: http://dx.doi.org/10.1016/B978-0-12-804051-5.00006-8

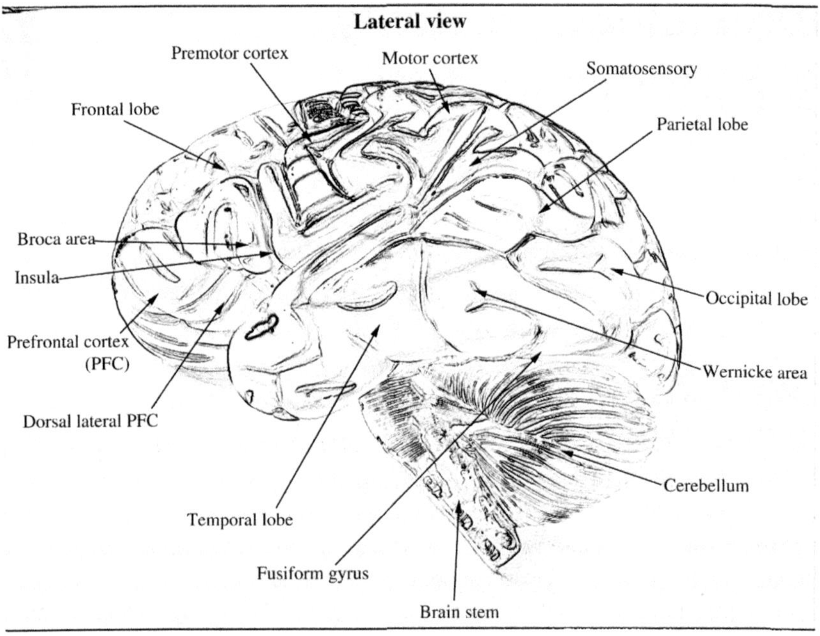

Figure 6.1 Brain image, lateral view. Drawings by Maurice Godin.

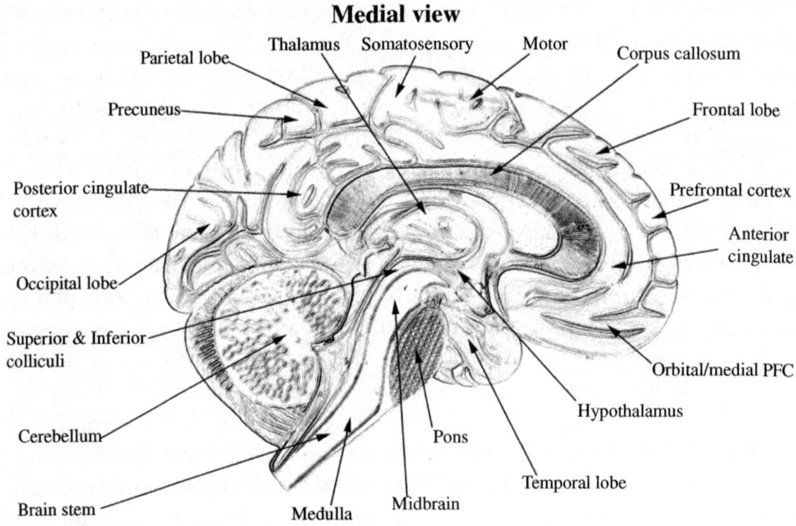

Figure 6.2 Brain image, medial view. Drawings by Maurice Godin.

in the midbrain regions (inferior and superior colliculi, upper tegmentum, substantia nigra). The forebrain of the midbrain region houses the thalamus, hypothalamus, and pituitary gland. These regions are critical for sensory and body processing. They directly interact with both the lower and upper brain regions. Located just above these structures resides the limbic system, a region that is often regarded as a convergence zone between the higher cortical regions and the subcortical regions of the brain. Within this area, the hippocampus, parahippocampus, amygdala, basal ganglia, and olfactory bulb reside. Located at the top of the brain is the neocortex, considered the most recent evolutionary part of the brain (Pinel, 2000; Purves et al., 2001). It is comprised of six layers of cells, is approximately 5 mm thick and is folded into gyri (ridges) and sulci (crevices). This folding is necessary to fit the expansive cortical layers of the brain into the skull. The neocortex is divided into different regions as well. The prefrontal cortex is regarded as the executive center; it is the highest organizational region to manage complex functioning. It is also associated with the most deeply human abilities, including constructing and understanding symbols and deriving meaning from these symbols (Purves et al., 2001; Sawyer, 2011). Located behind the prefrontal cortex is the motor cortex (movement), followed by the parietal cortex (spatial and sensory), and at the back of the skull, the occipital cortex (vision). On both the right and left sides of the brain reside the temporal cortices (sound and memory). Tucked underneath these cortical regions, and still regarded as part of the neocortex, but also regarded as the limbic cortex, are the insula cortex and the cingulate cortex, regions responsible for coordinating internal sensations and external stimuli (Purves et al., 2001; Figs. 6.3 and 6.4).

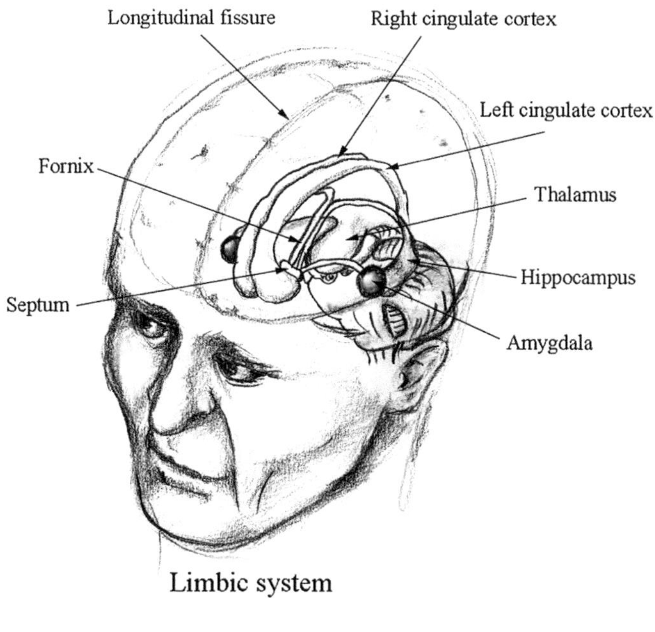

Figure 6.3 Limbic system. Drawings by Maurice Godin.

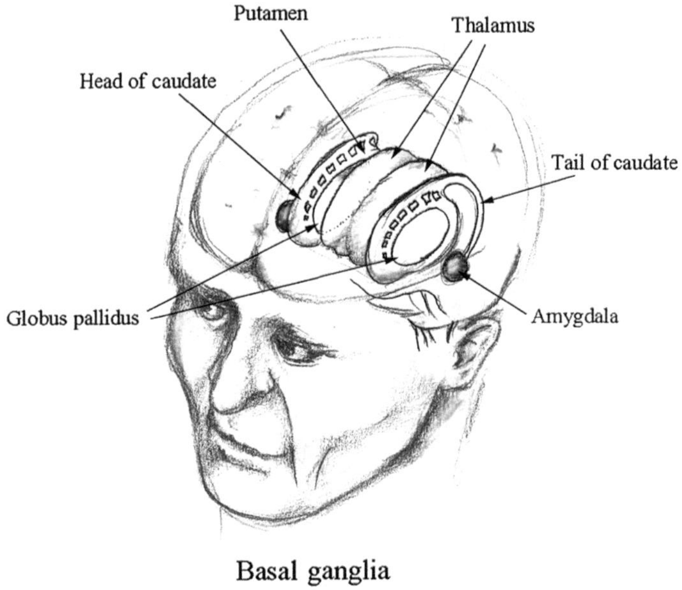

Figure 6.4 Basal ganglia. Drawings by Maurice Godin.

Functional neural networks

Because the brain is so complex, and efforts to isolate regions in the brain to identify specific functions invariably necessitate the inclusion of many other regions, it is more efficient to discuss larger neural networks and how they relate to function (Beaty, 2015; Llinas, 2001). According to neuroscience researchers, the primary function of the brain is to predict and direct actions; these action patterns are organized as complex synergistic networks (Llinas, 2001). Llinas argues that the mind is fundamentally internalized movement within the brain (2001). Freeman (2000) suggests that the brain coordinates vast information as it receives and responds to stimuli (internal and external). Meaning and intentionality are directed by the body/brain; they are deeply interconnected and inform all responses and yet the existential phenomena of experience transcends these narrow parameters, extending outward into the world with the capacity to immediately recognize qualia (experiencing the essence of something such as the redness of red) (Freeman, 2000). Despite many earlier notions that creativity was lateralized to the right hemisphere, the complexity of any cognitive component of creativity is neither right- nor left-hemisphere based; rather it entails cooperation among many different cortical and subcortical areas in both hemispheres (Bekhtereva et al., 2000; Dietrich & Kanso, 2010; Lindt, 2011). These neuroscientists point to the sophisticated and complex models required to describe the performing arts and the performing artists who give life to them. Thus, understanding function via networks, and not just brain regions or hemispheres, is necessary when examining creativity, especially because creativity is one of the most complex neurological activities (Dietrich, 2007; Park et al., 2015).

The first network to be considered is the default mode network (DMN). This network is most commonly active while *internally* focusing, such as during day-dreaming, including thinking about the self (autobiographical information, self-referencing, emotional awareness), thinking about others (theory of mind, emotions of others, moral reasoning, social evaluations, empathy), and remembering the past or thinking about the future (Beaty, 2015; Mayseless, Eran, & Shamay-Tsoory, 2015). It works in tandem with other networks such as the executive network and the salience network and is deactivated during external goal-oriented tasks (Ellamil, Dobson, Beeman, & Christoff, 2012). The DMN functional responsibilities include (1) information about the self and (2) thinking about others (Andrews-Hanna, Smallwoord, & Spreng, 2014).

1. Information about the self is processed in the following regions:
   a. posterior cingulate cortex (PCC) (attention)
   b. precuneus (visual and sensorimotor information)
   c. medial prefrontal cortex (mPFC) (decisions about self, valuing rewards)
   d. angular gyrus (connects perceptions, attention, spatial cognition, and action)
2. Thinking about others primarily takes place in the dorsal medial subsystem, including the medial temporal subsystem. The regions include:
   a. dorsal medial prefrontal cortex (dmPFC) (socially directed thoughts)
   b. temporoparietal junction (TPJ) (theory of mind, also known as beliefs about others)
   c. lateral temporal cortex (retrieval of social knowledge)
   d. anterior temporal pole (abstract conceptual information that is social in nature)
   e. hippocampus (formation of new memories, remembering the past, imagining the future)
   f. parahippocampus (spatial, scene recognition and simulation)
   g. retrosplenial cortex (RSC) (spatial navigation)
   h. posterior inferior parietal lobe (attention and junction of auditory, visual, and somatosensory information).

The *salience network*, also known as the cingulate–frontal operculum network, integrates internal and external stimuli and involves the anterior cingulate cortex (ACC) (attention, pain, reward) and the anterior insula (body sensations, also known as interoception) (Critchley, Wiens, Rotshtein, Ohman, & Dolan, 2004; Menon, 2011). Given these particular regions, the salience network is highly involved in processing sensory stimuli with visceral, autonomic, and hedonic information. Information gathered from the amygdala and the temporopolar cortex binds with input from the cingulate and insula to form visceral and emotional responses. The salience network aids in embodied decision-making, often described as "gut reactions," and increases awareness of subjective feeling states (Critchley et al., 2004; Ellamil et al., 2012; Seeley et al., 2007).

The *central executive network* is responsible for working memory, problem solving, cognitive processing, reasoning, task flexibility, and decision-making (Koechlin & Summerfield, 2007). This network operates during all activities that require cognitive and motor control processing, including filtering out extraneous distractors that can compromise control (Egner & Hirsch, 2005). It comprises the following frontoparietal regions (Alvarez, Emory, & Emory, 2006):

- dorsolateral prefrontal cortex (dlPFC) (planning, problem solving, abstract thinking, verbal and design fluency, maintain and shift cognitive sets)
- anterior cingulate (ACC) (emotional drives and integration, motivated behavior, decision-making, inhibition of inappropriate responses)
- orbitofrontal cortex (OFC) (impulse control, maintain cognitive set, monitoring ongoing behavior, inhibiting socially inappropriate behavior, valuing rewards based on sensory stimuli, and subjective emotional evaluations)
- lateral posterior parietal cortex (sensory integration, perception, and problem solving)

The *action-oriented networks* (both action-execution and action-perception systems) are primarily located in the premotor, supplementary motor, motor, and parietal cortices (Christensen & Calvo-Merino, 2013). The cerebellum, basal ganglia, and lower brain centers (including the autonomic nervous system) also control these actions (Dietrich, 2008). The basal ganglia are truly a complex system on their own, specifically because they support motor, cognitive, motivational, and emotional functioning (Di Martino et al., 2008). Likewise, the cerebellum, containing more neurons than any other region in the brain, receives direct feedback from multiple sensory regions and coordinates all movement responses required to efficiently respond to external and internal stimuli (Dietrich, 2008). For example, studies examining musical expertise demonstrated that the cerebellum influences the processing of temporal and spatial sequencing, critical to comprehending and modulating rhythm structures, tempo shifts, and melodic patterns while coordinating them into movement actions (de Manzano & Ullen, 2012a). Precise musical timing is further enhanced by greater connectivity between the cerebellum, thalamus, premotor cortex and the action-oriented network (Krause, Schnitzler, & Pollok, 2010).

The *reward circuit network* requires the brain to formulate a representation of something desired, a prediction of possibility to acquire the representation, and anticipation about when it will take place, colored by emotional feelings that are either positive or negative (valence) and strongly or weakly anticipated (intensity) (Brattico & Pearce, 2013; Nadal & Skov, 2013; Salimpoor & Zatorre, 2013). The cortical regions include the anterior cingulate (ACC), orbitofrontal cortex (OFC), and the ventromedial prefrontal cortex (vmPFC). The subcortical regions include the tegmentum, caudate nucleus, dorsal and ventral striatum, substantia nigra, and nucleus accumbens; they are integral regions responsible for the rewarding release of the neurotransmitter dopamine (Brattico & Pearce, 2013). Emotional and physiological responses activate the autonomic nervous system as well as the amygdala, insula, thalamus, and hypothalamus (Blood & Zatorre, 2001; Cacioppo, Tassinary, & Bernston, 2007; Salimpoor & Zatorre, 2013). The hippocampus is also activated in reward experiences, a finding that is not surprising given that it is an essential region for memory formation and retrieval (Henke, 2010), emotional regulation (Bechara & Damasio, 2005; Critchley, 2005; Salimpoor & Zatorre, 2013), and anticipation of novelty inherent in all reward responses (Wittmann, Bunzeck, Dolan, & Duzel, 2007). To perceive the aesthetic experience, low and midlevel visual, auditory, and somatosensory cortical processing must occur (Nadal & Skov, 2013). These multiple regions are also integral to the salience network described above (Boxes 6.1–6.3).

Box 6.1 Brain Divisions

Prosencephalon (forebrain becomes telencephalon and diencephalon)	
a. Telencephalon	Cerebral cortex
b. Forebrain of the midbrain/ diencephalon	Thalamus, hypothalamus, pituitary gland, olfactory bulb, basal ganglia, hippocampus
Midbrain	Superior and inferior colliculi, upper tegmentum, substantia nigra
Rhombencephalon (hindbrain divides into metencephalon and myelencephalon)	Cerebellum, pons, medulla, lower tegmentum, cranial nerves, respiratory center
Spinal cord	

Box 6.2 Brain Regions and Functions

Region	**Functions**
Cortical areas	
Medial prefrontal cortex (mPFC)	Decisions about self, valuing rewards
Dorsal medial prefrontal cortex (dmPFC)	Socially directed thoughts
Dorsolateral prefrontal cortex (dlPFC)	Planning, problem solving, abstract thinking, verbal and design fluency, maintain and shift cognitive sets
Orbitofrontal cortex (OFC)	Impulse control, maintain cognitive set, monitoring ongoing behavior, inhibiting socially inappropriate behavior, valuing rewards based on sensory stimuli, and subjective emotional evaluations
Motor, premotor, supplementary cortex	Detection and coordination of movement
Lateral temporal cortex	Retrieval of social knowledge
Anterior and posterior temporal pole	Abstract conceptual information that is social in nature, memory, complex auditory, visual, linguistic, and emotional functioning
Temporoparietal junction (TPJ)	Theory of mind also known as beliefs about others
Angular gyrus	Connects perceptions, attention, spatial cognition, and action
Posterior inferior parietal lobe	Junction of auditory, visual, and somatosensory information and attention
Lateral posterior parietal cortex	Sensory integration, perception, and problem solving
Occipital cortex	Coordinates complex visual and vestibular signals

(Continued)

Box 6.2 (cont'd)

Precuneus	Visual and sensorimotor information
Retrosplenial cortex (RSC)	Spatial navigation
Corpus callosum	Connecting bridge between right and left hemispheres
Subcortical regions	
Anterior cingulate cortex (ACC)	Attention, pain, reward, emotional drives and integration, motivated behavior, decision-making, inhibition of inappropriate responses
Posterior cingulate cortex (PCC)	Attention
Anterior insula cortex	Body sensations, neuroception
Hippocampus	Formation of new memories, remembering the past, imagining the future, emotion regulation, anticipation of novelty, spatial encoding
Parahippocampus	Spatial and scene recognition, simulation
Septal nucleus	Interactional relay center, works in tandem with hippocampus to regulate amygdala
Amygdala	Threat orientation and threat memory encoding
Thalamus	Sensory relay center from body to cortical and subcortical regions of the brain (with the exception of smell)
Hypothalamus	Primary regulator of body and maintains homeostasis (sleep, circadian rhythm, hunger, arousal, direct link to hormonal balance directed by pituitary gland)
Dorsal and ventral striatum (basal ganglia)	Initiating and experiencing rewarding behavior
Caudate nucleus (basal ganglia)	Part of the associative anticipatory loop, detects anything out of order, attention, focus
Putamen & globus palidus (basal ganglia)	Coordinates automatic movement learned by repetition (connected to premotor cortex)
Mesencephalon	
Superior colliculi	Early visual and sensory processing (detecting and analyzing)
Inferior colliculi	Early auditory processing (detecting and analyzing)
Tegmentum	
Nucleus accumbens	Arousal, pleasure, reward, reinforcement, motivation, aversion, addiction, and new motor programming (releases dopamine and projects it to other regions)
Periaqueductal gray	Receives and responds to noxious and painful stimuli, initiates primitive sounds (laughter, crying, groaning, howl)
Substantia nigra	Influences eye movement, motor planning, reward seeking, learning, and addiction, releases dopamine
Locus coeruleus	Responds to physiological stress and panic, releases norepinephrine that influences arousal, sleep–wake cycle, memory, cognitive control, emotions, posture, and balance

Raphe nucleus	Releases serotonin, influences circadian rhythm, memory retention, food intake
Parabrachial nucleus	Influences pleasurable stimuli and hedonic behavior
Nucleus of the solitary tract	Contributes to autonomic nervous system regulation
Nucleus ambiguus	Group of large motor neurons that regulate speech and swallowing; is deeply connected with the vagus nerve
Red nucleus	Motor coordination
Metencephalon	
Reticular formation	Pathway to transmit sensor and motor signals, arousal, attention, and orienting
Pons	Generates neurotransmitters (contains dense nuclei)
Cerebellum	Integration of cognition, emotion, and sensorimotor activation during actual motor responses
Medulla	Lowest part of the brain stem, controls autonomic reflexive functions (vomiting, sneezing, coughing, swallowing) and contains cardiac, respiratory, and vasomotor nuclei that regulate breathing, heart rate, and blood pressure

Box 6.3 Network Systems

1. *Default mode network* = internal focus	A. *Thinking about self*: Posterior cingulate cortex, precuneus, medial prefrontal cortex (PFC), angular gyrus B. *Thinking about others*: Dorsal medial PFC, lateral temporal cortex, anterior temporal cortex, hippocampus, parahippocampus, retrosplenial cortex, posterior inferior parietal cortex
2. *Central executive network* = working memory, problem solving, reason, task flexibility, decision-making	Frontoparietal regions (DLPC, ACC, OFC, lateral posterior parietal cortex)
3. *Salience network* (cingulate-frontal operculum) = integration of internal and external stimulus	Anterior cingulate cortex, anterior insula, somatosensory cortex, autonomic nervous system (hypothalamus), amygdala, temporopolar cortex
4. *Action-oriented networks* = action execution and action perception systems	Premotor, supplementary motor, parietal cortex, cerebellum, basal ganglia, lower brain centers, thalamus
5. *Reward circuit* = anticipation and prediction of pleasure and aversion	ACC, OFC, VMPC, tegmentum, caudate nucleus, dorsal and ventral striatum, substantia nigra, nucleus accumbens, ANS, amygdala, insula, hypothalamus, hippocampus, thalamus, somatosensory and auditory cortices

Measuring brain activity

Physically identifying the active brain regions during the actual creative process is challenging (Fink, Benedek, Grabner, Staudt, & Neubauer, 2007). Movement artifact during brain scanning can destroy signal-to-noise ratios and so participants must remain motionless while engaging in mental thought activities. This restriction eliminates the analysis of real-time embodied creative processing (Dietrich, 2008). What is currently measured is brain activation while listening to an auditory stimulus or viewing images. Individuals being scanned can also internally generate imagery or they can engage small finger or foot gestures, as long as they do not involve head or torso movement. Unlike brain scanning, electroencephalography (EEG) analyses allow more participant movement; however, it is still not possible for performing artists to fully participate in their domain-specific talents while being tested. Given these constrained test conditions, there is very little overlap in study results (Arden, Chavez, Grazioplene, & Jung, 2010). Thus definitive neurobiological statements about creativity are not possible at this time (Dietrich & Kanso, 2010); however, general creative cognitive processes can be assessed and informed extrapolations can be constructed (Abraham & Windmann, 2007) (Boxes 6.4 and 6.5).

Box 6.4 Neuroimaging: Functional (how the brain looks when it is working on a task)[a]

Functional measurements	
Electroencephalography (EEG)	Electrodes are placed on the scalp to gather voltage differences between electrodes over time. The EEG provides clearer functional (temporal) rather than structural/spatial resolution (informational). When an individual responds to a stimulus, the EEG records the event during the response, which is called an event-related potential (ERP). Frequency ranges have been identified and associated with different brain activities. Delta waves (.5–4 Hz) occur during sleep; theta waves (4–8 Hz) are greater during childhood, and are implicated in encoding and retrieving information; alpha waves (8–13 Hz) occur during relaxed wake states and are associated with reduced anxiety and creative inspiration (Horan, 2009); beta waves (13–30 Hz) are associated with increased alertness and focused attention; and gamma waves (>30Hz) are implicated in creating the unity of conscious perception and insight awareness (Horan, 2009; Sawyer, 2011).
Functional magnetic resonance imaging (fMRI)	Measures changes in brain activity while a participant engages in a task inside the scanner. The MRI scanner detects signal changes in blood oxygen level (BOLD).

Near-infrared optical imaging	Measures the relationship of the absorption of water and oxygenated and deoxygenated hemoglobin to "see" inside the brain. Compared to the fMRI, it is a less expensive instrument and can be used to map brain regions engaged in certain cognitive and motor functions.
Positron emission tomography (PET)	Measures metabolic processes in the brain, based on radioactive decay of an isotope (tracer) and assesses amount of regional glucose uptake in the brain. A *single-photon emission computed tomography (SPECT)* operates in a similar fashion as the PET scan but with less power. Compared to the rapid temporal response of EEG studies, PET and SPECT studies require approximately 40 s or more to measure elevated brain activities (Sawyer, 2011).
Regional cerebral blood flow (rCBF)	Can be measured by a PET scan to determine changes in cerebral blood flow.
Magnetic resonance imaging spectroscopy (MRIs)	Measures individual chemicals in the brain and provides an estimate of chemical concentrations operating within the brain at a specific moment in time.

[a] See Arden et al. (2010), Bremner (2005), Dietrich & Kanso (2010), Horan (2009), Sawyer (2011).

Box 6.5 Neuroimaging: Structural (does the task have neuroanatomical associations)[a]

Structural measurements	
Computed tomography (CT)	Multiple X-ray angles are projected through the head and computer algorithms configure a three-dimensional density image of brain regions.
Diffusion tensor imaging (DTI)	DTI uses MRI technology and measures white matter integrity by capturing a three-dimensional diffusion of water in the tissues sampled in a volumetric pixel (voxel). Water diffusion is greater in axons with myelin. This allows researchers to measure connectivity in the brain. High values suggest higher structural coherence and low values indicate lower structural coherence caused by either compromised axonal integrity or less myelination.
Magnetic resonance imaging (MRI)	Measures the physical properties or spectra of brain chemistry, in particular, regions bound by pia mater and gray-white boundaries. Successive images are taken to obtain slices of the brain that are then converted to brain volume. This enables dimensional construction of brain regions.
Brain metabolite activity	Measures N-acetyl-aspartate concentrations (contributes to myelin synthesis), a marker of neuronal health.

[a] See Arden et al. (2010), Bremner (2005), Dietrich & Kanso (2010), Horan (2009), Sawyer (2011).

Generative and evaluative modes of creativity

Based on neuroscience studies, the creative process, including improvisation (Pressing, 1988), is best divided into two modes, the generative and evaluative modes (Ellamil et al., 2012). The generation of ideas offers the potential for novelty; whereas, the evaluation of these ideas helps determine if the ideas are novel and what might be added to make them more novel. This biphasic process is fundamental in all performing arts domains. Every performer generates new associations to any given work they are to perform coupled with an ongoing process of evaluating how to refine these associations. This same process occurs during improvisation, a process that crosses all performing arts domains, although jazz improvisation has received the most scientific attention to date (Beaty, 2015; de Manzano & Ullen, 2012b; McPherson & Limb, 2013).

In general, the generative mode engages stronger activation in the left inferior frontal gyrus, bilateral premotor cortex, inferior and superior parietal cortex, bilateral medial temporal lobe, especially the hippocampus and parahippocampus (Beaty, 2015; Ellamil et al., 2012). This finding is not surprising, given that the role of these regions is for memory retrieval and memory reconstruction, essential ingredients when any new associations form. The DMN has also been identified as participating in the generation of ideas; in particular, divergent thinking tasks (see Box 6.6). Because it is engaged in internally generated thoughts, the medial PFC, posterior cingulate cortex, and the ventral anterior cingulate are more strongly activated (Mayseless et al., 2015). This finding supports low cognitive control theories, also known as decreased latent inhibition, during idea generation processes (Benedek, Franz, Heene, & Neubauer, 2012; Carlsson, Wendt, & Risberg, 2000; Carson, Peterson, & Higgins, 2003). Perhaps the associative network (medial temporal lobe) allows ideas to form while the DMN supports inhibition of irrelevant ideas (Mayseless et al., 2015). Certainly, performing artists are highly skilled at blocking out irrelevant ideas while engaging in the generation of associations relevant to a work that is to be performed.

Curiously, during the evaluative mode the executive, salience, and default networks are all coactivated; evaluating the results of creative output seems to engage a diverse intersection of networks. Both higher cortical regions along with the somatosensory regions influence evaluation—strong evidence for the role of both cognitive and spontaneous “gut” responses that are accessed during the creative evaluation process (Beaty, 2015). The reward networks also influence this evaluative process, including a keen sense of what is regarded as novel and valued.

The concurrent activation of a dual creative process, both the generative (internal) and evaluative (external) response systems, is familiar to all successful performing artists. They know viscerally and cognitively when they succeed or fail and they readily generate new solutions to give life to a performance. Whether during a practice session, rehearsal, improvisational session, or performance, the oscillation between generating new associations and rapidly assessing them is trained into the performing artist (de Manzano & Ullen, 2012b). They develop skills vital

Box 6.6 Thinking Styles

Divergent thinking: problem tackled in different ways.	*Convergent thinking:* single answer to a given problem.

to maintaining a performance, even if unexpected events arise such as getting injured during a performance; costume, prop, or set pieces failing; sound or light cues mistimed; or entrances and exits compromised. The oscillating speed between these two processes supports the intended expression (Freeman, 2000; Llinas, 2001) and gives meaning to both the performer and audience (Box 6.6).

Cognitive processes: divergent, convergent, insight, and working memory

Generally, cognitive psychologists and neuroscientists understand divergent, convergent, and insight thinking within healthy subjects. Two powerful methods to quantify divergent thinking are the Torrance Test of Creative Thinking and the Remote Associates Test. They capture ideational fluency (number of ideas), flexibility (number of categories or types of ideas), and novelty (originality) (Dietrich & Kanso, 2010; Fink et al., 2007). Compared to individuals low in creativity, highly creative individuals such as performing artists (Kogan, 2002; Nemiro, 1997) have unusually high levels of divergent thinking, evidenced by greater cognitive control flexibility (increased frontal cortical activity), persistence, and resistance to distractions (Gibson, Folley, & Park, 2009; Zabelina & Robinson, 2010a). Divergent thinking is a domain-general characteristic of creativity (Beaty, 2015). Interestingly, it has also been associated with the psychodynamic theory of unconscious thought, a process that involves more subcortical bottom-up processing (unconscious) that influences higher cortical conscious thought (Dijksterhuis & Meurs, 2006).

A synthesis of EEG and neuroimaging studies indicates that divergent thinking is not lateralized (Dietrich & Kanso, 2010; Whitman, Holcomb, & Zanes, 2010), although EEG alpha activity seems to be more activated in the right hemisphere during some idea generation tasks (Grabner, Fink, & Neubauer, 2007; Lindell, 2011). Divergent thinking relies on the inhibition of hemispheric dominance; hence decreased corpus callosum activation helps facilitate idea fluency (Moore et al., 2009). In fact, creative individuals tend to have smaller corpus callosum, suggesting a selective developmental pruning to enhance efficient hemispheric functional connectivity necessary for idea generation during divergent thinking (Moore et al., 2009). It has also been demonstrated that decreased dopamine D2 receptor densities in the thalamus lower the thalamic gating threshold; this increases thalamocortical information, a process that is essential during cognitive flexibility activities such as divergent thinking tasks (de Manzano, Cervenka, Karabanov, Farde, & Ullen, 2010).

Although these results have been replicated, it is important to cautiously interpret the results. Most divergent thinking studies are inconsistent and poorly replicated, in part, because of different testing methods. In reality, many different cognitive processes are employed during divergent thinking tasks (Dietrich, 2007; Dietrich & Kanso, 2010).

Creative people also engage in convergent thinking; they employ strategic thinking and select specific solutions to resolve a creative problem (Dietrich, 2007; Gute & Gute, 2015). Convergent thinking shares strong similarities with the evaluation mode of creativity discussed above. Findings suggest that convergent thinking engages multiple network systems in a complex pattern of activation, including the central executive network, salience network, action-oriented network, reward network, and the default mode network.

The third form of creative thinking is insight, often described as "Aha!" or eureka moments. It is commonly regarded as the magical factor in creativity. For performing artists, these moments do not drive great performances; however, they are exhilarating to experience. Studies that investigate insight moments tend to be more consistent, with stronger activation in the superior temporal gyrus, a common convergence region operating during insight problem solving and while making remote associations, plus the anterior cingulate, in part because it detects cognitive conflicts and promotes flexible mental shifts (Dietrich & Kanso, 2010).

Whether employing divergent, convergent, or insight thinking, working memory is always accessed; fundamentally memory networks, including the coordinating efforts of the cerebellum, influence all three cognitive processes (Chavez-Eakle, 2007). Because creativity is complex, measuring these components is ultimately only an indirect indicator of creativity. What occurs is a confluence of action-oriented and process-oriented networks in the brain that operate during creativity (Gute & Gute, 2015).

Neuroaesthetics and performing arts

Neuroaesthetics is considered a separate (although overlapping) field of study that focuses on the neurobiological substrates of aesthetic experience (Nadal & Skov, 2013). This response can be evoked in the creator or the viewer and may differ depending on the demands of a specific domain. Neuroaesthetics investigates performing arts as an expressive art that incorporates motor and emotional responses, as well as the typically studied cognitive processes of creativity (divergent, convergent, and insight) (Brattico & Pearce, 2013). The term neuroaesthetics appeared in the last decade of the 20th century. Once it was recognized as a distinct field of study, research studies proliferated, although the need for more interdisciplinary approaches still remains. Three large tracts of investigation dominate this field of study: (1) the induction or expression of emotions (Brattico & Pearce, 2013; Nadal & Skov, 2013), (2) activating the reward/pleasure response, and (3) aesthetic preferences (Salimpoor & Zatorre, 2013). In general, studies demonstrate that aesthetic preferences are informed by previous exposure and familiarity with the art form, combined with emotional responses (whether positive or negative), expectations and appraisal biases, and the degree of sensory dissonance experienced while listening to or viewing the performance (Brattico & Pearce, 2013).

Aesthetic experiences are a result of a complex interplay of neural processes (Nadal & Skov, 2013). For example, neuroaesthetic findings in music suggest that aesthetic experiences engage both cortical and subcortical structures because aesthetic experiences draw upon cognitive and emotional responses that operate in the default mode network, salience network, and central executive network. Most importantly, the reward circuit network amplifies aesthetic experiences. Whether participating as either a performer or audience member, aesthetic experiences are augmented by forming mental representations, cognitively anticipating and predicting an artistic outcome, and emotionally and physically feeling the positive excitement intensify (Brattico & Pearce, 2013; Nadal & Skov, 2013; Salimpoor & Zatorre, 2013). Actively perceiving and moving all enrich the experience (Christensen & Calvo-Merino, 2013; Nadal & Skov, 2013).

Neuroaesthetics studies have also identified the potential reason aesthetic experiences are positive. Increased release of dopamine, considered part of the mesolimbic dopamine reward system in the brain (Glimcher, 2011) is evidenced. When performing or witnessing a performance, activation of dopamine reinforces the aesthetic experience and tags it as pleasurable and desired (Salimpoor & Zatorre, 2013). The complexity of cognitive processing during these rewarding events involves a performer or audience member who can recognize temporal and spatial patterning; these patterns are then colored by individual emotional, sensory, and memory systems. These are then encoded by the reward neural circuit and identified as pleasurable experiences. The complexity of these interactions makes aesthetic reward processing primarily top-down, cortically directed responses. They are different than addictive behaviors such as addictions to substances, gambling, sex, or eating; they are fundamentally subcortical in nature (Salimpoor & Zatorre, 2013).

Although the studies cited above primarily investigated participants passively observing art works or listening to music, the other performing arts tend to be less studied; however, the findings can be generalized (Cross & Ticini, 2012). Performers simultaneously perceive the work of art as they manifest it. The aesthetic experience is immediately encountered. For example, dancers are constantly listening and responding to music, while concurrently sensing the aesthetic effects they produce when assuming specific body positions and deliberately expressing qualities of movement that may be lyrical or ballistic. Dancers tend to be biased toward an aesthetic expression. In a study that examined genetic differences between dancers and athletes, a strong difference existed between the two. Dancers had serotonergic phenotypes that enhanced their social communication and spiritual expression, factors that propel them toward aesthetic movement experiences (Bachner-Melman et al., 2005). Studies that examine neuroaesthetics reinforce the notion that aesthetic experiences are embodied, a reality that is fundamental to performing artists (Cinzia & Vittorio, 2009). These significant neuroaesthetic advances in the field of art and music have great potential to expand our understanding about creativity in the performing arts (Christensen & Calvo-Merino, 2013). Future studies investigating the specific domains of theater and dance can add to this growing field of study.

Mirror neurons and creativity in the performing arts

Another avenue of neurobiology research that indirectly explains the nature of shared creativity in the performing arts is the mirror neuron system, primarily located in the premotor and parietal cortices (Christensen & Calvo-Merino, 2013; Decety et al., 1997; Jackson & Decety, 2004; Rizzolatti & Craighero, 2004). The insula and anterior cingulate add emotional and visceral nuance (Molnar-Szakacs & Overy, 2006), and the superior temporal sulcus enhances detection of any biological movement (Blakeslee & Blakeslee, 2007). Mirror neurons are activated when observing other individuals perform, whether consciously attending to their actions or peripherally perceiving their movement. For performing artists, the mirror neuron system augments learning and facilitates vital movement entrainment during performance. Interestingly, the mirror neuron response is strongest when performers observe familiar movement styles from their own artistic domain. There is less activation when they watch movement styles in which they received little or no training. For example, ballet dancers' mirror neurons were strongly stimulated while watching ballet video clips compared to when they watched martial arts video clips (Blasing et al., 2012; Calvo-Merino, Glaser, Grezes, Passingham, & Haggard, 2005). These results suggest that matching action execution while observing movement enhances mirror neuron motor representations; this process is practiced in all performing arts domains (Calvo-Merino, Grezes, Glaser, Passingham, & Haggard, 2006).

The mirror neuron system also influences aesthetic appreciation at the proprioceptive level (Keysers & Gazzola, 2009; Montero, 2006). In one study, when appreciation of the movements of a male dancer were observed, the sensorimotor and visual cortices increased activation, in part, because the sensorimotor mirror neuron system in the observer evoked a kinesthetic experience similar to the performer (Calvo-Merino, Jola, Glaser, & Haggard, 2008). Such actions, whether only small traces of internal movements activated in the premotor cortex or complete imitations of the movement that engage the full sensorimotor systems, enhance both perceptual and motor systems in the observer/performer. Watching and imitating movement improves performance and the mirror neuron system is strongly implicated in this age-old practice. This was clearly demonstrated in a study that examined mirror neuron activation with musicians learning to play the guitar. The novice players learned cord progressions via observation coupled with mirror neuron engagement. They were able to successfully play the chords, even without actual deliberate physical practice (Iacoboni, 2009). The mirror neurons provide a feedforward model that predicts motor plans via a perceptual and motor activation.

Figure 6.5 "Visceral performance." Performer: Maurice Godin.

For example, listening to a song activates the mirror neurons, sends signals forward to the premotor cortex; this neural signaling pattern prepares the observer to actually sing the song later (Iacoboni, 2009). When witnessing something that has novelty, the mirror neurons in the observer are even more engaged (Blakeslee & Blakeslee, 2007).

The mirror neurons engage much more than the sensorimotor systems. They can elicit vicarious responses in the observer that are emotionally and viscerally charged (Keysers & Gazzola, 2009). This empathic capacity makes it a powerful neural system for both audience members and performing artists. The contagion effect of shared emotions is experienced on stage among fellow performers and across the footlights with audience members (Jeffers, 2009). It has been proposed that the human mirror system has the capacity to organize language and music information in an effort to communicate meaning and emotion. It constructs much more than a representational model; it constructs a sensorimotor visceral experiential model (Molnar-Szakacs & Overy, 2006; Fig. 6.5).

The body and creativity: real-time creative action

Performing arts rely on the embodied creativity of the performing artists. This reality amplifies the problems in neurobiological studies that examine performing artists. To date, brain studies investigate performers imagining that they are playing, dancing, or singing. An imaginary activity can only activate a subset of brain regions but when the performer is actually fully executing the movements then no brain scans can capture all the regions involved during performance (Dietrich, 2008). Some studies have attempted to approximate real movement and discovered that different, albeit overlapping, neural networks are involved in imagined versus executed movement (Kleber, Birbaumer, Veit, Trevorrow, & Lotze, 2007; Lotze et al., 1999). These studies have also demonstrated that very different muscular activation patterns occurred in real-time performance versus imagined performance (Slade, Landers, & Martin, 2002). For example, critical regions such as the cerebellum are involved in sensorimotor integration during actual motor activity; whereas, the posterior cerebellum inhibits movement execution during imagined movement (Lotze et al., 1999). The feedback loop between the cerebellum and cortical regions prepares and corrects motor responses in real time. Given that the cerebellum houses more neurons than the entire cerebral cortex, including it in creativity studies, especially studies investigating performing artists, is vital (Dietrich, 2008). The role of the cerebellum in real versus imagined movement is just one example highlighting the complexity required when examining the neuroscience of creativity in the performing arts.

To explore this complexity at a larger system level, imagined versus real movement can be organized via the implicit and explicit systems; they have been abundantly identified and described in neuroscience literature (Purves et al., 2001). *Implicit systems* (nonconscious) coordinate real-time movements, including all modifications necessary to successfully complete the targeted movements. For performing artists these implicit responses are powerfully shaped by hours of deliberate practice over the span of thousands of hours, a factor that ensures fluency in performance execution (Ericsson, 2013). Proficient performance is essential for any professional performing artist and for any performers who develop expertise in improvisational skills (Pressing, 1988). Controlling and planning actions, such as opera singing, dancing, playing an instrument, or acting, involves the basal ganglia, primary motor cortex, premotor cortex, supplementary motor cortex, cerebellum, inferior parietal cortex, sensorimotor cortex, dorsolateral prefrontal cortex, thalamus, and lower brain centers (including the autonomic nervous system) (Dietrich, 2008; Siedler, Noll, & Thiers, 2004), plus the auditory cortex when music or language is included (Kleber, Veit, Birbaumer, Gruzelier, & Lotze, 2010). This deliberate practice forms procedural memories (implicit), in particular, in the striatal neurons, including dopamine D2 receptor densities in this region (Karabanov et al., 2010). Dynamic encoding and recoding sequences reinforce these memories. Once they become habits they are difficult to undo, and if they are not practiced for a period of time they can quickly be resurrected with just a few deliberate practice sessions (Barnes, Kubota, Hu, Jin, & Graybiel, 2005). This implicit system also influences nonconscious motor decisions regarding speed or accuracy. Depending on the circumstantial needs of the mover, feedforward processes operate to execute movements directed at a large target (speed); whereas, movements directed at a small target (accuracy) engage feedback control (Siedler et al., 2004), and similar feedforward and feedback processes occur when dancers or musicians negotiate movements that are performed at a fast tempo or when cognitive load is increased due to complex staging or costume demands (Blasing et al., 2012; Corlu, Maes, Muller, Kochman, & Leman, 2015).

Explicit information processing is fundamentally a rule-based higher-order conscious system. It incorporates, organizes and synthesizes knowledge and memory to construct complex responses and involves multiple regions throughout the brain (Dietrich, 2008). It is regarded as a verbally conscious awareness system; it is also engaged in the performing arts. Music and dance are clear symbolic language structures. Performing artists rely heavily on the implicit system for rapid motor responses while simultaneously engaging in higher-order explicit awareness (Blasing et al., 2012; Stevens & McKechnie, 2005). Years of training ensure that the medium of expression (music, dance, acting) is also a fully conscious explicit knowledge-based processing system. Performers organize and synthesize knowledge and memory and construct performances that are detailed and articulated. But they also rely on the implicit system to support their performance when they perform any work of art.

Unfortunately, studying the neurological patterning of performing artists while engaging in the creative process is problematic. Any brain scan study that implements motor imagery is accessing *explicit systems* only because the images are consciously being generated. These studies will never access the speed and efficiency of *implicit motor systems*. Given these limits, it is difficult to claim that current imagery studies that examine jazz improvisation (Beaty, 2015), novice versus professional dancers (Fink, Graif, & Neubauer, 2009), musical performance (Keller, 2012; Meister et al., 2004; Zatorre & Halpern, 2005), or opera singing (Kleber et al., 2010) are portraying accurate information about all the brain functions operating in real time during performance. The implicit system cannot be measured at this time; however, the result of deliberate practice (implicit) versus mental practice (explicit) has been measured. Neurological changes manifested after only one week of deliberate practice, with more changes occurring in the supplementary motor cortex (motor acquisition) for deliberate practice; whereas, more changes occurred in the secondary visual cortex for mental practice. Both forms of practice increased activation in the cerebellum and the motor cortex, with greater enhancement resulting from learning by doing versus learning by imaging (Nyberg, Eriksson, Larsson, & Marklund, 2006). This study result reinforces the notion that the implicit and explicit motor systems are substantially different; however, when a performer is injured, the explicit system engaged during mental practice can partially maintain neural networks essential for performance (Keller, 2012; Meister et al., 2004; Zatorre & Halpern, 2005).

Performing arts also depend on the performers and their audience to recognize and respond to body language, including the emotional meaning conveyed by the performer. Neuroscience studies have demonstrated that the same neurological pathways that assess the emotional meaning expressed in faces are capable of recognizing emotions expressed in body postures and gestures (de Gelder, 2006). Speech and gesture also share the same communication systems; they form a unitary expression of symbolic ideas processed within Broca's area and the correlate motor region located in the right hemisphere (Kleber et al., 2007, 2010), and when dissociated from each other, the union of intended actions and symbols is disrupted (Bernardis & Gentilucci, 2006). Emotional body language, including natural gestures that accompany speech, conveys real-time emotional states of individuals coupled with a set of anticipated behavioral actions (Saarikallio, Luck, Burger, Thompson, & Toivianinen, 2013). Similar to the mirror neuron system, an ability to recognize and respond to body language provides direct implicit knowledge about the person and the situation (Gibbs, 2003). These biologically ingrained pathways are the bedrock of communication in dance and theater, and to a large degree, in music.

The discussion of emotional body language, implicit systems, mirror neurons, and the integration of speech and gesture highlight the reality that motor cognition exists. Because the brain integrates sound and movement, music and dance are obvious partners, with both mediums affecting an audience's perceptions of quality and meaning (Hagen & Bryant, 2003). Cognition is ultimately embodied in action (Freeman, 2000; Llinas, 2001); movements, including the production of sound, are organized by intention and directed toward specific goals. They drive social interactions and basic biological needs (Panksepp, 1998). Understanding the duality of implicit and explicit motor systems as equal players in *motor cognition* captures the experiential reality of performing artists. Action/perception cycles inform an understanding about another person's intentions and shape our sense of self-agency when engaging in motor cognition; these occur primarily in the motor cortex, somatosensory cortex, parietal cortex, and cerebellum (Leube et al., 2003). Emotionally experiencing these motor cognitions, by way of activation in the insula and parietal regions, adds meaning to these motor cognitions (Jackson & Decety, 2004), including experiencing music or performing it (Kleber, Zeitouni, Friberg, & Zatorre, 2013).

Because motor cognition is critical to performing artists, it is also speculated that performers train to enhance an awareness of their inner feelings, a process referred to as interoception in neuroscience literature (Craig, 2004). This capacity engages a complex array of cortical and subcortical regions and is part of the salience network discussed earlier. Interoception incorporates the somatic marker hypothesis—a hypothesis that suggests we have gut responses guiding our decision-making (Damasio, 1999). The creative process relies on interoception, otherwise known as "gut responses" or somatic markers. Through enhanced interoception, performing artists can differentiate congruent and incongruent rhythm structures (Vuust et al., 2005); they have a greater awareness of self-identity via increased multisensory body feedback processes (Tsakiris, Tajadura-Jimenez, & Costantini, 2011). Coupled with implicit motor systems, interoception is enhanced through deliberate practice, a finding that was discovered while training opera singers (Kleber et al., 2010; Patel, 2012), and when identifying and fully imagining musical works that were primed by a brief sound cue (Halpern, 2012). It allows us to perceive emotional resonance within music, dance, and theater, and it informs us that we are subjectively feeling our self while enhancing an awareness of the feelings of others (Craig, 2004) (Box 6.7).

Box 6.7 Historical Development of Neuroscience, the Unconscious, and Creativity

Freud (1908), an early neuroscientist and founder of psychoanalytic theory, speculated that creative individuals engage conscious (secondary) and unconscious (primary) processes in order to express their intentions. "What [the artist] aims at is to awaken in us the same emotional attitude, the same mental constellation as that which in him produced the impetus to create" (p. 254). Freud (1908/1985) claimed that artistic works "make such an impression on us . . . and arouse in us emotions of which, perhaps, we had not even thought ourselves capable" (p. 131). Freud was aware that conscious thought was necessary during the creative process, but the depth of artistic expression originated within the primary unconscious, where primal instincts and repressed memories resided (1942/1985).

Later, *Holt* (2002) divided primary unconscious into two processes, one associated with creativity and the other with neurotic/psychotic thoughts. He analyzed creative individuals' responses to the Rorschach inkblot test and compared them with the responses of individuals suffering from mental illness. He determined that high primary process responses incorporated a degree of control and were associated with creativity, versus low primary process responses, which lacked cognitive control, hence manifesting in chaotic neurotic/psychotic thoughts. Holt's high primary unconscious thought resembles the "active" unconscious activities described by Dijksterhuis and Meurs (2006). According to these researchers, passive unconscious thought did not augment creative incubation; however, active unconscious thought, believed to be associated with subcortical activation that influences cortical activity, is associated with the incubation process of creativity and spontaneous insight.

Supporting *Holt's* work of high versus low primary unconscious thought, Chavez-Eakle, Lara, and Cruz (2006) demonstrated a strong negative correlation between creativity and psychopathology, with imagination, humor, fantasy, cognitive flexibility, abstraction, emotional expressiveness, and resistance to premature closure most affected by the presence of psychopathology. Creative achievement, compared with psychopathology, was associated with the serotonin transporter gene 5′SLC6A4 and the dopamine receptor DRD4 gene (Chevaze-Eakle, 2007). These researchers also demonstrated simultaneous activation in both brain hemispheres, including the cerebellum (Chavez-Eakle, 2007). Later, cognitive psychologists suggested that unconscious thought was basically divergent thinking (broadening associations); in contrast, conscious thought equaled convergent thought (narrowing associations) (Dijksterhuis & Meurs, 2006). Creative insight, especially the instantaneous "Aha!" moment, fascinated many cognitive neuroscientists. This was also described as a form of unconscious cognition, a process that was deeply

(Continued)

Box 6.7 (cont'd)

related to the incubation of new and diverse ideas and insights (Dijksterhuis & Meurs, 2006).

Despite the ongoing debate regarding the exact nature of unconscious processes and creativity, it has guided substantial research inquiry (Sawyer, 2011). Unfortunately, the majority of cognitive neuropsychologists do not include high and low primary unconscious thought as variables when studying divergent, convergent, and insight thinking, and so the work of the early psychodynamic theorists remains relatively unexplored in neuroscience.

Neurofeedback training and the performing artist

Using neuroscience findings to augment training has drawn the attention of many educators and clinicians. Neurofeedback is rapidly becoming an invaluable tool to enhance neurological performance. For example, in an empirically supported evidence-based training protocol devised by Gruzelier and colleagues (2009), the alpha/theta neurofeedback protocol increases the ratio between theta and alpha wave lengths. The results of this protocol included improved performance along with reduction in anxiety, depression, and posttraumatic stress disorder (PTSD) symptoms (Gruzelier, 2009). Theta states are common during deep relaxation and meditation. These states modulate both cognitive and emotional/motivational arousal, in part because theta waves are widely distributed and integrate neural circuits throughout the cortical and subcortical regions of the brain. In this neurofeedback training protocol, the goal is to increase theta activity. When alpha/theta neurofeedback is consistently trained (Gruzelier, Thompson, Redding, Brandt, & Steffert, 2014), musicians improve instrumental competence, musicality, rhythmic and stylistic accuracy, interpretive imagination, and expressive range. A similar gain was found in a group of dancers and actors (Gruzelier, 2014). Musicians, dancers, and actors who received the alpha/theta neurofeedback training also increased their commitment, confidence, emotional expression and enjoyment while performing.

In a different set of studies, enhanced alpha activity was associated with creative thinking, in particular, creative idea generation (Grabner et al., 2007). In a study that examined a group of ballroom dancers compared to novice dancers (Fink et al., 2009), more right-hemisphere alpha synchronization was found in the posterior parietal brain region in the professional dancers. This pattern was more predominant when the dancers visualized improvising a dance compared to visualizing a learned dance such as a waltz, suggesting that more alpha synchronization is engaged in novel activities that require creative idea generation (Grabner et al., 2007; Fink et al., 2009).

Although more investigation is needed, the alpha/theta neurofeedback protocol offers intriguing benefits for performing artists. Shaping wave patterns in the brain certainly facilitates complex connections between the key neural networks that are activated in performing artists. This approach is also easier to measure because EEG signaling assessments are less constricting than current brain scan imaging devices.

Summary

- The brain: complex open system, neurons, overview of brain structures.
- Functional neural networks: (1) default mode network that focuses internally (self and other), (2) salience network that integrates internal and external stimuli, (3) central executive network, action-oriented networks (action-execution and action perceptions), (4) reward circuit network.
- Measuring brain activity: box of structural and functional imagery.
- Generative and Evaluative Modes of Creativity: coactivation, low/high cortical activation, gut responses and reward circuits, visceral/cognitive engagement.
- Cognitive processes: Divergent, Convergent, Insight, and Working Memory: hemispheric inhibition, decreased corpus callosum activation, decreased thalamic gating, complex coordination and integration of diverse action-oriented and process-oriented neural networks.
- Neuroaesthetics and performing arts: neurobiological regions and aesthetic appreciation, induction or expression of emotions, activating reward/pleasure responses, aesthetic preferences.
- Mirror neurons and creativity in performing arts: imitation, aesthetics, emotional contagion, vicarious responses.
- The body and creativity: real-time creative actions: motor imagery and motor actions in the performing arts, implicit versus explicit processing, measuring implicit and explicit systems, body language, emotion and interoception.
- Neurofeedback training and the performing artist: increased ratio between alpha and theta neurofeedback protocol, integrate diffuse neural circuits.

Concluding remarks

To summarize, some consistent neurobiological patterns have been identified in the performing arts. Consistent across EEG studies is a relationship between alpha synchronization and creative cognition (Arden et al., 2010); however, like all studies on creativity, whole-brain intra- and interhemispheric synchronization between theta, beta, and gamma waves is also essential during all aspects of the creative processing (Horan, 2009). Although MRI study results are determined by the specific task being tested, consistent findings report the engagement of heteromodal cognitive regions of the frontal, temporal, and limbic lobes (Arden et al., 2010). In general, brain structure and metabolism findings suggest that total brain volume and gray matter moderately predicts intelligence and creative cognition. Neurobiology findings reinforce the hypofrontality theory suggesting a disinhibitory network of brain regions but only in the generative phase of creativity (Arden et al., 2010; Benedek et al., 2012; Ellamil et al., 2012).

Understanding creativity through neurobiological studies is exciting, but what needs to be embraced by researchers is that creativity is complex. It is not a unitary construct; consequently, it involves all regions of the brain (Dietrich, 2007; Park et al., 2015). We can not point to any single brain pattern and claim that the individual who possesses this activation is creative. Nor can we state that a specific set of brain networks can define a musician, dancer, or actor. What has been consistently demonstrated is the training effect in the brain of performing artists. Whether the performer trains to perform an existing repertory work or participate in an improvisational session, the brain requires long hours of deliberate practice to acquire expert-related skills. This practice effect promotes increased activation, connectivity, and information exchange efficiency between bilateral dorsolateral prefrontal cortices, dorsal premotor cortices, and presupplementary motor areas, regions that are involved in both the generative and evaluative creative process (de Manzano & Ullen, 2012a; Pinho, de Manzano, Fransson, Eriksson, & Ullen, 2014). The brain gives us life and it gives us opportunities to engage creatively. The study of creativity and the study of the brain are not simplistic enterprises. Searching for the underlying neuromechanisms is important, especially when an artist suffers from a neurological or psychiatric disease. Training and treating performing artists more effectively can only come with more advances in neurobiological research.

Physiology and creativity

7

There is a vitality, a life force, an energy, a quickening that is translated through you into action, and because there is only one of you in all time, this expression is unique. And if you block it, it will never exist through any other medium and will be lost.

Martha Graham

Introduction

Physiology is the study of mechanical, physical, and biomechanical functions of the body. It provides a foundational base to understand how the human body adapts to stress, physical activity (performance), and disease. It is an integrative field of study examining the human body at the micro (cells, biochemistry) and macro (organ systems) levels (Sherwood, 2010). In general, studying the functioning of the brain (neuroscience) is part of the larger field of physiology, especially because the brain and body are mutually dependent on each other. Because the previous chapter addressed neurobiology and creativity, less attention is given to neuroscience in this chapter; however, when necessary, neurobiology will be included in the discussion. The intention of this chapter is to focus on the human body, including some of the interconnections between the body and the brain. An in-depth discussion on all the systems of the body (i.e., digestive, reproductive, circulatory, skeletal, muscular) is beyond the scope of this book. Attention will be directed toward physiological functioning that promotes creativity in performing artists. Because the study of physiology evaluates the body's response to stress, for the purpose of clarity, "normal" daily performance stress is discussed in this chapter. Later, an in-depth examination is offered regarding stress associated with childhood adversity (see chapter: Attachment, Parenting, and Childhood Adversity) and trauma (see chapter: Unresolved Mourning and Posttraumatic Stress Disorder in Performing Artists; Box 7.1).

Physiology and creative stress: a dynamic process

The study of physiology is rooted in understanding the effects of stress on body systems. Stress is generally not regarded as a disease or an assault on the body. It is an adaptive process in our body that allows mobilization in order to successfully respond to daily life, including a performance career (Baumann & Turpin, 2010). Homeostasis is the term that describes the physiological regulation of normal daily

Creativity and the Performing Artist. DOI: http://dx.doi.org/10.1016/B978-0-12-804051-5.00007-X

Box 7.1 Human Physiological Systems (Sherwood, 2010)

System (Alphabetical Order)	Clinical	Physiological
Circulatory system (cardiovascular system): heart and blood vessels to transfer oxygen and nutrients and remove waste	Cardiology (heart) Hematology (blood)	Cardiovascular
Digestive system/excretory system: includes mouth, tongue, gastrointestinal tract (small and large), liver, pancreas, gall bladder, salivary glands	Gastroenterology	Gastrointestinal
Endocrine system: pituitary, thyroid, adrenals, pancreas, parathyroid, gonads	Endocrinology	Endocrinology
Immune system: white blood cells, thymus, lymph nodes, lymphatic system	Immunology	Immunology
Integumentary system: skin, sweat glands, sebaceous glands	Dermatology	Cell physiology Skin physiology
Lymphatic system: extract and metabolize lymph fluids	Oncology Immunology	Oncology Immunology
Musculoskeletal system: skeleton, muscles, ligaments, tendons, cartilage, bone marrow	Orthopedics	Cell physiology Musculoskeletal physiology
Nervous system: central, somatic, and autonomic nervous systems (brain and spinal cord)	Neuroscience Neurology Psychiatry Ophthalmology (vision) Otolaryngology (hearing, taste, smell)	Neurophysiology
Renal system/urinary system: kidneys, ureters, bladder, urethra	Nephrology Urology	Renal physiology
Reproductive system: sex organs, fetal development, gonads	Gynecology Andrology Sexology Embryology Obstetrics	Reproductive physiology
Respiratory system: nose, nasopharynx, trachea, lungs, gas exchange	Pulmonology	Respiratory physiology

living (stress). It defines the maintenance of a relatively constant condition within the body. Recently, scientists have employed the term *homeodynamic*; this term reflects the incredible variance operating within the body and mind. The concept of homeodynamic regulation is extrapolated from complexity theory. In complex systems, multiple physiological processes simultaneously interact (feedback and feedforward) and dynamically respond and adjust to moment-to-moment demands of living (Berntson & Cacioppo, 2007). Homeostasis and homeodynamic regulation concepts imply that a physiological set point (or range) is maintained for optimal performance, albeit more variance is accounted for in the homeodynamic term. Allostasis, on the other hand, suggests that regulatory levels are not fixed, especially under conditions that demand increased physiological exertion, such as during traumatic events (Berntson & Cacioppo, 2007; McEwen, Gray, & Nasca, 2015). Allostasis accounts for regulatory strategies that operate outside the range of normal daily living.

Basic physiological responses to stress (homeostasis, homeodynamic, allostasis) are directed by the brain, autonomic nervous system (ANS), neuroendocrine system, and immune system. Each influences the timing and effectiveness of the other physiological systems. All are powerfully shaped by genetic background, early childhood experiences, and current exposure (Baumann & Turpin, 2010). Each individual responds to stress in unique ways. For example, some performers struggle with the stress of success while others struggle with the stress of failure (Conroy, Poczwardowski, & Henschen, 2001). How these situations are psychologically appraised influences physiological responses; however, sometimes the body experiences a situation as excessively stressful even though the psychological/cognitive assessment is underestimated (Costa, Ferreira, Orsini, Silva, & Felicio, 2016; Foxman & Brugel, 2006). This physiological response to stress, despite the performer's lack of psychological or cognitive awareness, often sets the stage for performance-related musculoskeletal overuse injuries or burnout, topics that will be discussed later in the book (see chapter: Injuries and Rehabilitation, and chapter: Health Concerns and Burnout).

In general, performing artists' physiologic responses are dictated by the body's ability to withstand emotional and physical challenges required to regulate performance demands. Physiological levels of activation are acutely affected by factors such as psychological state, perception of challenge, level of skill, domain-specific training, type of emotional expression required, and level of physical conditioning. The intensity of the aesthetic experience is a uniquely subjective experience for each performer. For example, individuals who are listening to or performing Samuel Barber's *Adagio for Strings* may respond to it in a calm, bored, or highly excited state. Individuals have their own unique emotional and physiological responses to the work they are performing (Salimpoor, Benvoy, Longo, Cooperstock, & Zatorre, 2009). These individualized responses add complexity when discussing physiological responses within performing artists. One way to address performer differences is to continually recognize that physiological engagement has great variance within the individual and between groups of performers.

Cardiac and respiratory systems and performance

The interplay between heartbeats and breathing patterns occurs naturally every moment of our life; however, trained performers also learn to modulate these interacting systems to give aesthetic expression to music, singing, dance, and acting. The regulation of the heart is directed by neural connections (i.e., nucleus of the solitary tract and nucleus ambiguus) that innervate pacemaker cells located within the heart. The primary pacemaker is the sinoatrial node, located in the right atrium (top portion of the heart). Tiny oscillators in the node are phase locked to enable regulatory contraction commands (Vickhoff et al., 2013). Respiration also influences heartbeat and heart rate. Respiratory function balances arterial blood gases, either oxygenating blood cells or removing carbon dioxide and waste products from the blood. The breathing rate and tidal volume of the breath adjust depending on the metabolic intensity levels and their influences on arterial blood gases (Hahnengress & Boning, 2010).

Efficient cardiorespiratory systems are vital for all performing artists. For example, musicians who play wind and reed instruments demonstrate intense cardiopulmonary strain. While performing, heart rate reaches moderate-to-heavy levels, such as those achieved during moderate-to-intense physical exercise (Hahnengress & Boning, 2010). A similar work intensity is found in conductors (Jaque et al., 2015), and musicians in general (Inesta, Terrados, Garcia, & Perez, 2008). This heart rate response is further intensified in performance settings versus rehearsal settings. The cardiopulmonary response is greater, although more tightly regulated in highly trained professional musicians in comparison to less-trained students. Advanced players modulate ventilation better, whereas students tend to hyperventilate or breathe more shallowly, especially during difficult musical passages (Hahnengress & Boning, 2010). Despite these intense cardiopulmonary demands the ventilatory and circulatory strain of most musicians never resembles pathophysiological levels (Hahnengress & Boning, 2010). However, many wind- and reed-playing musicians experience enhanced ventilatory drive, a condition caused by carbon dioxide buildup. Inefficient breathing patterns and dyspnea, a subjective appraisal of unpleasant or uncomfortable breathing, exacerbates ventilatory drive. Hypercapnia, a pulmonary condition marked by abnormally elevated carbon dioxide, further compromises cardiopulmonary performance. This condition may require specialized attention, especially if it diminishes performance levels (Iltis, 2003; Fig. 7.1).

Autonomic nervous system

Newberg and d'Aquili (2000) suggest that the body is the direct source that communicates with the brain while the ANS, housed in the brain, communicates with the body. These instantaneous recursive loops between the body and the ANS regulate all physiological responses. The ANS is an involuntary system that regulates a number of complex tasks in the body. The functions of its two

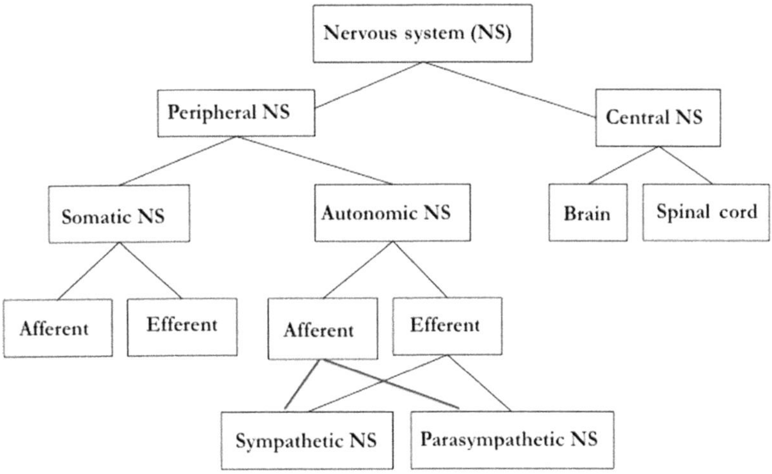

Fig. 7.1 Major divisions of nervous system.

branches, the sympathetic nervous system (SNS) and parasympathetic nervous system (PNS), are complicated and diverse. The SNS is associated with physiologic activation, such as increasing cardiovascular functions, which is apparent in both preparation for and participation in a variety of physical and psychological tasks. It traverses the spinal cord, the ganglia that run outside the spine, and then innervates both vessels and visceral organs within multiple systems. The primary neurotransmitter of the SNS is norepinephrine, although the SNS also releases acetylcholine during the thermoregulatory process of sweating (Goldstein, Robertson, Esler, Straus, & Eisenhofer, 2002). One of the target organs activated by the SNS is the central adrenal gland, which releases epinephrine (adrenaline) and cortisol. The SNS action is rapid, regardless of whether the response is positive joy or negative pain and terror.

Conversely, the PNS is associated with the maintenance of the body's resting functions, such as digestion, cardiovascular, and respiratory systems; its primary neurotransmitter is acetylcholine (Baumann & Turpin, 2010). The vagus (vagal) nerve is often equated with the parasympathetic branch, although other cranial nerves are involved in parasympathetic responses. For example, the vagal efferent tract (away from the brain) extends to the viscera (throat, heart, lungs, gastrointestinal tract, reproductive system). One key role of the vagus nerve is to control and regulate heart rate. Because 80% of the vagal complex also contains afferent fibers (toward the brain), it operates as a major feedback loop to the brain (i.e., brain stem, reticular formation, locus coeruleus) (Ghacibeh, Shenker, Shenal, Uthman, & Heilman, 2006). These brain centers relay information to higher subcortical and cortical regions for further regulatory processing.

The two branches of the ANS directly influence all regions of the body (see Box 7.2). Both branches of the ANS strongly influence the release of other

Box 7.2 Autonomic Nervous System

Organ	Sympathetic Nervous System	Parasympathetic Nervous System
Lungs	• Faster respiration, inhibits mucous secretion	• Slower, deeper respiration, stimulates secretion
Heart	• Quicker heart rate, increased blood pressure	• Slower heart rate, decreased blood pressure
Eyes	• Pupils dilate, tear ducts dry	• Pupils constrict, tears secreted
Throat	• Vasoconstriction (dry mouth)	• Secretion of saliva
Skin	• Pale skin color, sweating and cold skin, piloerection	• Blushed/flushed skin color, skin warm, and dry
Muscles	• Muscles tonic	• Muscles relaxed
Gastrointestinal	• Digestion (peristalsis) decreases	• Digestion (peristalsis) increases
Adrenal gland	• Catecholamine secretion	• No effect
Bladder and colon	• Bladder relaxes	• Colon relaxes
Penis	• Ejaculation	• Erection

neurotransmitters and neuropeptides (corticotropin-releasing hormone or CRH, neuropeptide Y, somatostatin, encephalin, neurotensin, substance P); this complex cascade of neurochemicals, including adenosine triphosphate (ATP) and nitric oxide, refine responses in peripheral organ systems and promote optimal adaptive flexibility (Tsigos & Chrousos, 2002). The continual interaction of these two branches allows for autonomic regulation vital to meeting the demands of both physiological and psychological stressors (Yu, Zhang, & Zhang, 2014). Although these branches operate largely outside of awareness, they can be somewhat influenced by conscious volition. For example, while listening to or playing music, the ANS mediates the entrainment of the heartbeat with the musical beat. Studies demonstrate that musicians' heart rate matches musical tempi, regardless if they are conducting, performing, or listening to it (Cervellin & Lippi, 2011; Jaque et al., 2015). It is a physical scientific reality; in fact, this relationship with music, brain, and heart has been described since the beginning of recorded history. Ancient cultures employed music to soothe the soul, to lull babies to sleep, and to induce deep spiritual states of reverence, practices that continue to the present time (Cervellin & Lippi, 2011). The effects of music on the body and brain are immediate, whether soothing or stimulating. This cardiac entrainment is a complex interaction between cortical and subcortical regions of the brain, the ANS, and the heart.

Individual efforts (direct or indirect) can increase or decrease activity of both branches of the ANS. For instance, when engaging in creative cognitive tasks, particularly during the evaluative phase of creativity, or when learning new repertory, greater coupling occurs between the heart (ANS) and higher cortical regions in the brain (i.e., parietal, sensorimotor, insula, prefrontal, hypothalamus). As the cerebral cortex activation increases so does heart rate (sympathetic response); whereas, when calming mental practices such as yoga, tai chi, or meditation take place, heart rate decreases (parasympathetic) (Yu et al., 2014). Physical training not only enhances cardiovascular and pulmonary functioning, it also alters ANS patterns of response. With increased physical fitness, tolerance to withstanding brief states of hypoxia increases. Performing during hypoxic states is common in singers, dancers, and many musicians and actors. The by-product of performing during hypoxic states is enhanced creative cognitive flexibility. Distant associations can be formed to generate new ideas, while well-rehearsed physical activities remain stable. The only negative side effect of performing during hypoxic states is compromised memory encoding; however, performers are experts at pacing aerobic and anaerobic states while training and performing and so compromised memory is not problematic (Razumnikova, Kovtun, & Krivoshchekov, 2013). Given these findings, the positive effects of ANS/cardiopulmonary fitness on creativity and performance are clearly demonstrated.

Assessment of the ANS during creative activities is possible; by understanding the interaction of the two branches of the ANS, indirect information is gleaned regarding higher subcortical and cortical activation patterns. Basically, the downstream influences of neural networks can be deduced during creative performance via measuring ambulatory cardiovascular, respiratory, and ANS functioning (Thomson & Jaque, 2011b). For example, experimental stimulation of the tenth cranial vagus nerve directly influences the parasympathetic branch of the ANS. Study findings demonstrate that vagal stimulation impairs cognitive flexibility and creativity. It is speculated that increased vagal stimulation is associated with increased activation in the locus coeruleus, a region in the brain that narrowly focuses on sensory input while inhibiting global network connections (Ghacibeh et al., 2006). Creativity is dependent on global network connections, a response pattern that is more closely associated with sympathetic arousal.

Administering a peripheral beta-adrenergic antagonist medication (nadolol) also decreases creative processing; this medication behaves in a similar manner as stimulation of the vagal nerve. Both inhibit sympathetic arousal while enabling more parasympathetic regulation. The administration of propranolol, a central and peripheral beta-adrenergic antagonist, actually enhances cognitive flexibility. It is believed that decreased central adrenergic arousal increases global neural network connections, a process that replicates neural latent inhibition (hypofrontality) (see chapter: Neurobiology, Creativity, and Performing Artists for more details) (Ghacibeh et al., 2006). Administration of beta-blocker medication (propranolol) to highly anxious performing artists is a common practice. Given the massive feedback and feedforward loops during the creative process, providing a medication that prevents binding to norepinephrine neurotransmitters not only reduces heart rate, but also reduces narrow-focused cognitive processes that are activated during high-arousal anxious situations (Ghacibeh et al., 2006). Another positive by-product of beta-blocker administration is altered bone turnover. When individuals are in chronic states of sympathetic arousal, bone health is compromised. Adequate levels of beta 2 adrenergic activity influences both osteoblastic and osteoclastic cells required for neural regulation of local bone metabolism. Combined central and peripheral beta-blocker medication ultimately operates as a protective factor for bone health (i.e., decreases osteoporosis), as well as creative flexibility (Togari & Arai, 2008). The use of beta-blocker medication by musicians and athletes has been researched extensively, with positive results consistently reported; however, this treatment practice for performance anxiety has not been examined as thoroughly in dancers, singers, and actors (Harris, 2001).

Heart rate variability and creativity

To understand the influences of the ANS at rest and during creative performance, heart rate variability (HRV) is assessed. HRV refers to the variability of time between heartbeats (defined as the interval between R waves). HRV is an indirect measurement of the overall activity of the ANS, as well as the specific contributions of the PNS and SNS (Thayer, Ahs, Fredrikson, Sollers, & Wagner, 2012). The concept of homeostasis suggests that HRV is stable over time; however, all adaptive systems are essentially highly variable. This is also true for heart rate. Generally, the greater the variance in heart rate the more adaptive the organism. This variability is typically shaped by a reciprocal relationship between the sympathetic and parasympathetic branches of the ANS. As one branch increases activity the other branch counterbalances with a diminished response. High-SNS and low-PNS activation causes cardiac acceleration; whereas, low-SNS and high-PNS activity effectuates cardiodeceleration (Acharya, Joseph, Kannathal, Lim, & Suri, 2006). Variance within this reciprocal pattern is commonly examined; however, the ANS has many other patterns to meet the demands of stress. For example, coactivation or coinhibition of both branches of the ANS reduces cardiac balance, while an uncoupled response of the two branches enables maximum adaptation (Berntson, Norman, Hawkley, & Cacioppo, 2008a). HRV, in particular, parasympathetic activation, is also directly influenced by both respiratory rate and volume (Berntson et al., 2008a).

During adulthood, circadian rhythm directly influences HRV, with greater HRV evidenced during nighttime hours and plateauing during daylight. This shift in day–night patterning is relevant to performing artists, especially because they typically perform in the evening. Also, intense workload during the daylight hours reduces parasympathetic regulation during slow wave sleep at night (Dupuy, Bherer, Audiffren, & Bosquet, 2013). With normal aging, HRV decreases, and there is less cardiac vagal modulation and nocturnal parasympathetic activation (Yeragani, Sobolewski, Kay, Jampala, & Igel, 1997). Younger males have greater day–night differences; however, this gender difference decreases with age (Bonnemeier et al., 2003). When examining ethnicity, African American compared to Caucasian individuals show a more rapid decline in HRV variance with aging, suggesting signs of premature aging in their ANS (Choi et al., 2006). Future studies on performing artists are recommended to determine if performers follow normal vagal activation of PNS during nocturnal hours or if their circadian PNS pattern is altered (Bonnemeier et al., 2003). This information may offer an explanation regarding increased sleep disorders found in performing artists (Crnivec, 2004; Fjellman-Wiklund, & Chesky, 2006; Guptill, 2012; Wilhelm, Kovess, Rios-Seidel, & Finch, 2004).

Future studies on performing artists, in particular, examining the effects of aging, gender, and ethnicity on HRV, are needed. For example, cardiac responses to stressful events (both physical and mental) differ in men and women. Females generally mobilize greater sympathetic activation. This response is not due to an overactive SNS; rather it is due to decreased parasympathetic regulation that buffers the effects of the sympathetic branch of the ANS. This is an example of an uncoupled response in the two ANS branches. Consequently, more catecholamine is released by the SNS leading to enhanced cardiac contractility. This stress response is adaptive; however, if it persists then cardiotoxic myocytes are released (Lee et al., 2013), a condition that increases female risk for heart disease.

Reduced HRV and compromised ANS functioning are associated with many health conditions. Compromised cardiac regulation is strongly associated with higher incidence of morbidity and mortality. Some of the health conditions associated with reduced HRV and altered ANS regulation include cardiovascular disorders (Floras, 2009; Triposkiadis et al., 2009), depression plus coronary heart disease (Carney, Freedland, & Veith, 2005), posttraumatic stress disorder (Gillie & Thayer, 2014; Lee et al., 2013), and panic disorders (McCraty, Atkinson, Tomasino, & Stuppy, 2001). Other disorders include bipolar mania (Henry, Minassian, Paulus, Geyer, & Perry, 2010), irritable bowel syndrome (Heitkemper et al., 2001), and metabolic syndrome (Licht et al., 2010; Lindmark et al., 2005; Tentolouris, Argrakopoulou, & Katsilambros, 2008).

Although the ANS can be negatively changed, it can also be positively altered. The healing quality of music has a direct effect on the ANS. For example, professional singers demonstrate greater HRV after completing a singing lesson (Ardeleanu & Ardeleanu, 2014). When singing in an ensemble, HRV significantly increases, along with greater respiration and heart rate entrainment to the music and to the other singers (Vickhoff et al., 2013). With increased exposure to music, performers cultivate more HRV, which is a buffer to cardiorespiratory diseases (Ellis & Thayer, 2010). It is postulated that the healing qualities of musical interventions provide increased neurovisceral (neurophysiological) integration. For example, repetitive exposure to music (for both trained musicians and nonmusicians) influences the ANS; in particular, excitatory music inhibits parasympathetic activation (Iwanaga, Kobayashi, & Kawasaki, 2005). Although the effects are subtle, engaging with music clearly elicits more physiological resilience (Ellis & Thayer, 2010).

During a public performance, a different ANS picture emerges in performers. Greater sympathetic activation is required to successfully perform (and to access creative flexibility); consequently reduced HRV occurs while performing. Alterations in HRV allow higher energy output needed for performance. Even when performers take beta-blocker medication to suppress the sympathetic branch of the ANS, during performance they still have more sympathetic compared to parasympathetic activation (Jaque et al., 2015). They mobilize energy to meet performance demands, a marker of decreased ANS balance (coactivation of the two ANS branches). This pattern does not persist during the recovery phase postperformance. Reduced ANS balance during performance is also highly correlated to heightened flow states in performers (Jaque et al., 2015; Thomson & Jaque, 2011b).

The ongoing stress of a performing career can take its toll on performers. When homeodynamic balance converts to prolonged allostasis, the stress systems of the body are overly recruited (see chapter: Unresolved Mourning and Posttraumatic Stress Disorder in Performing Artists for more details). Alterations in the ANS, and, more specifically, the parasympathetic branch, are associated with glucose dysregulation, increased inflammation, and elevated cortisol levels (Thayer & Sternberg, 2006). Higher resting heart rate is a major marker of allostasis in performers (unless they are deconditioned) (Thayer & Sternberg, 2006). During normal stress, major physiological adaptive strategies are mobilized, such as accelerated cardiac output and respiration, increased catabolism (energy production), and blood flow, all to provide the highest perfusion and fuel to support the performance of the brain, heart, and muscles (Tsigos & Chrousos, 2002). When these responses persist, the rate of energy supply is not able to meet the metabolic demands; the body is no longer able to adapt to stress. Assessing allostasis is complicated, especially when using resting heart rate as an indicator. Increased conditioning in performers can mask psychological stress; physiological stress is not displayed despite high levels of "cognitive" anxiety (Studer, Danuser, Wild, Hildebrandt, & Gomez, 2014). For example, professional ballet dancers typically have altered cardiovascular physiology. Dancers have thickened interventricular septum and ventricular walls, male dancers have increased incidence of third heart sounds, and female dancers have more sinus bradycardia and sinus arrhythmia (Cohen, Gupta, Lichstein, & Chadda, 1980). These cardiovascular alterations may be related to greater incidence of smoking in dancers, elevated stress levels, nutritional deficits, or adaptive fitness demands (Fig. 7.2).

Fig. 7.2 "Shared emotion." Performers: Susan McManus, Bethana Rosenblatt, Evan Rosenblatt. Photographer: Lee Choo.

Emotional expressivity and autonomic nervous system physiology

The association with the body and emotional expressivity remains consistent throughout time. In Ancient Greece and Rome, the god Bacchus (Dionysus) ruled embodied freedom; whereas, Apollo reined over reason. The early years of psychology reflected these ancient mind–body divisions. Some theorists claimed that emotion and the body were inseparable, while other theorists denied this concept. For example, William James described visceral signals as the source of distinct emotions; Walter Canon claimed that somatic responses were too diffuse and slow to initiate emotions (Rainville, Bechara, Naqvi, & Damasio, 2006). Psychophysiologists and neuroscientists are now refining these binary concepts (Friedman, 2010). Rather than an either-body-or-mind concept, neurophysiology describes an integration of both via rapid feedback and feedforward signals between the peripheral and the central nervous systems. It is now understood that this neural interaction is responsible for specific somatic states and emotions, in particular, the primary emotions of anger, sadness, happiness, and fear.

This neurophysiological integration is evident in the deep interconnections within the cardiorespiratory systems. Inhalation increases sympathetic activation and heart rate while exhalation increases parasympathetic responses, which decreases heart rate. Specifically, the intraabdominal pressure during inhalation activates the baroreceptor reflex, which provokes a rapid increase in heart rate. With each respiratory cycle there is stimulation of the vagus nerve during exhalation, resulting in a slowing of the heart (Rainville et al., 2006). Emotional states are reflected in the cardiorespiratory systems; anger is associated with increased heart rate, high sympathetic activation, and low vagal inhibition. Fear, sadness, and happiness are associated with high sympathetic and high vagal tone. A reduction in HRV is also found in sadness and happiness; fear is marked by high sympathetic activation combined with an ineffectual increase in parasympathetic response (coactivation of both branches). Like fear, coactivation of both branches occurs during states of sadness and happiness, but at lower threshold levels (Rainville et al., 2006). Other studies have demonstrated similar ANS patterning during emotional responses (Kreibig, 2010).

Performing artists, in particular, actors, singers, and dancers, are well aware that when they activate motor facial muscles (including the eyes) they can elicit specific emotional states. Other performers and the audience recognize these emotional states conveyed by the performer's face and they also experience similar emotional states. These responses influence ANS patterning in the performers and those watching (Balconi & Bortolotti, 2012). In early studies, trained actors were able to evoke significantly higher physiological responses when expressing emotional states compared to untrained performers (Stern & Lewis, 1968). As well, specific patterns in heart rate and galvanic skin conductance (measure of sympathetic activation) were associated with anger, disgust, fear, and happiness (Levenson, Ekman, & Friesen, 1990). These early studies have been supported over the years (Kreibig, 2010).

For performers, goal relevance and goal appraisal are significant factors in physiological emotional expressivity. Cardiac autonomic reactivity, a measure of sympathetic and parasympathetic coactivation, is associated with desired goal relevance and goal support (Kreibig, Gendolla, & Scherer, 2012). Performer cardiorespiratory responses are congruent with changing emotional states, such as when goals are obstructed or supported, and when goals are deemed too challenging or readily accessible. For example, when performing in situations that support achievable goals, performing artists manifest coactivation patterns of sympathetic, and to a lesser extent, parasympathetic arousal (Thomson & Jaque, 2011b). A similar pattern is found in viewers who emotionally and physiologically attune with performers, regardless of whether they view scenes that are emotionally cooperative or conflictual. This physiological and emotional entrainment is intensified when the audience is more attentive and involved while viewing the performance (Balconi & Bortolotti, 2012).

Physiological activation via musical mood induction has been studied frequently, with mixed results. Often, the time domain measures of HRV do not differentiate specific moods evoked by music listening; however, respiratory patterns, and to some degree, heart rate, are influenced by musical mood inductions. For instance, shorter respiratory cycles and increased heart rate are associated with happiness, intermediate patterns for fear, and slower patterns for sadness (Etzel, Johnsen, Dickerson, Tranel, & Adolps, 2006). One of the challenges when investigating physiological emotional responses to music is the normal entrainment between respiration and musical tempo (Etzel et al., 2006). Although this is a problem for scientific investigation, most musical composers and performers capitalize on this entrainment. Tempo is frequently varied in an effort to engage the audience at a visceral and emotional level. Tonal modulation (pitch and minor/major keys) for happy versus sad music further influences physiological responses, including diastolic blood pressure (higher for happy music), electrodermal activity (greater sympathetic activation for happy music), and zygomatic activity (greater response for happy music) (Khalfa, Roy, Rainville, Bella, & Peretz, 2008).

Physiological reactivity is an integral experience in "aesthetic chill" responses; "goose bumps," shivers, or piloerection are directly related to sympathetic arousal in the ANS (Grewe, Nagel, Kopiez, & Altenmuller, 2007). While listening to music, individuals with aesthetic chills hear music with more discrimination compared to those who do not experience chills. Distinct musical structures, the entrance of a voice, and personal musical preference are general factors that evoke chills, whereas, loudness, pitch, and rhythm changes are less important (Grewe et al., 2007). Although not necessarily experiencing chills, when musicians perform with more emotional expressivity, greater sympathetic physiological reactivity is evident. Compared to conditions in which they perform in a technically accurate manner, the emotionally expressive performances are marked by increased heart rate, sweating, minute ventilation, and tidal volume. They have lower parasympathetic activation, and yet they also have reduced respiratory rates, indices of uncoupling in the autonomic control of the cardiorespiratory systems (Nakahara, Furuya, Francis, & Kinoshita, 2010). This physiological profile reflects extreme autonomic variance, one in which both branches are no longer operating in a reciprocal pattern (Berntson et al., 2008a). Similar responses are found when musicians perform compared to when they listen to a recording of themselves playing the same piece of music. These study findings reinforce the obvious reality that performing arts is an embodied expressive art form. Complex neural network integration occurs while performing music in a deeply expressive manner (Nakahara, Furuya, Masuko, Francis, & Kinoshita, 2011).

Despite findings that demonstrate emotional profiles within cardiorespiratory systems, the complex ANS and HRV patterns suggest that higher cortical, subcortical, and peripheral neural networks influence emotional expressivity. Certainly emotional states influence the cardiovascular system; emotions evoke high variability within stroke volume, cardiac output, and total peripheral vascular resistance. Emotional states directly affect hemodynamic processes (Yu, Nelesen, & Ziegler, 2001). For example, the aesthetic effects of music versus nonmusical sounds directly engage the cardiovascular systems, either initiating calming or activating responsiveness in heart and ventilatory rate, cardiac and tidal volume, ANS patterning, and HRV (Gomez & Danuser, 2004). Despite rich cardiorespiratory–emotional study findings, the assumption that a single measure, such as HRV, can capture complex emotional responses is foolhardy. A convergence of multiple psychophysiological measures derived from interdisciplinary research designs will further understanding about embodied aesthetic experiences (Collet, Vernet-Maury, Delhomme, & Dittmar, 1997; Friedman, 2010; Kreibig, 2010; Rainville et al., 2006).

Physiological performance demands

Performing artists exert great physical effort while training, rehearsing, performing, and on tour. The cardiorespiratory systems are directly affected by fitness level and workload demand. These systems require substantial energy reserves to function optimally. Determining how to enhance energy production to perform at elite levels has been one of the major topics examined in traditional sport and exercise science studies. A substantial base of knowledge exists regarding athletic fitness. Dance science is just starting to investigate aerobic, anaerobic, strength, and endurance fitness in dancers; minimal research is conducted on fitness needs of musicians, singers, and actors and how to improve their energy reserves (Clark, Holmes, & Redding, 2011; Inesta et al., 2008).

To meet the rigors of a career in the performing arts, adequate nutritional energy must be ingested. Dietary needs vary depending on rehearsal and performance demands. Unfortunately, dancers often do not acquire sufficient nutritional fuel to meet their energy expenditure. Dancers consume below 70–80% of the recommended daily allowance of energy intake, a significant problem given the daily rigor of their career (Koutedakis & Jamurtas, 2004). Adding to this nutritional deficit, dancers have lower body fat percentages. They also have higher incidence of eating disorders and secondary amenorrhea. Despite nutritional deficits, compared to nondance individuals, dancers' bone mineral density is higher. The mechanical loading and increased strength required in dance may be key factors in promoting bone health, even when nutritional deficits exist (Friesen et al., 2011).

The ability to access energy stores is enhanced through cardiorespiratory training. Increased physical activity promotes the production of ATP, the primary energy source located in every cell of the body (Sherwood, 2010). Performers are deemed fit when they can sustain dynamic muscular activity at moderate/high-intensity for prolonged periods of time (Martyn-Stevens, Brown, Beam, & Wiersma, 2012). To determine aerobic fitness, maximal oxygen uptake (VO_2 max) is assessed on a progressive incremental treadmill or cycle ergometer. This test yields information regarding individual ventilatory threshold, heart rate, and gas exchange (Guidetti, Gallotta, Emerenziani, & Baldari, 2007). Individuals who are aerobically fit can maintain appropriate blood lactate levels during exercise. When measuring strength, protocols typically involve standard lower and/or upper body strength assessment that requires participants to exert one repetition of maximal force (1RM). Trials progressively increase in weight load until task failure. Endurance is derived from repeated resistance exercises that are set at approximately 60–65% of maximal strength for the upper and lower body. Participants engage in the repetition task until muscle fatigue leads to task failure (Guidetti et al., 2007). Anaerobic exercise, sometimes referred to as nonrobic, indicates a condition in which the metabolic demands exceed the capacity of aerobic metabolism to supply energy required. In this situation, increased metabolite production and elimination result in accumulation of lactate in muscle fibers and blood (Guidetti et al., 2007). During this process, lactic acid, potassium, arachidonic acid, and adenosine levels increase. This metabolic alteration initiates muscle metaboreceptors (small chemoreceptors located in muscle tissue) to stimulate responses in arterial blood pressure and cardiac output required to restore homeodynamic balance (Crisafulli et al., 2003). Anaerobic activities draw upon phosphocreatine and glycolytic systems; through anaerobic training these systems promote muscle adaptation and anaerobic endurance (Martyn-Stevens et al., 2012).

Short intense bursts of anaerobic activity typify the performing arts training and performance process. Training anaerobic capacity, along with muscle adaptation to withstand these metabolic changes, is essential for performers. Dancers also engage in bursts of isometric muscular activity (slow adagio movements) and dynamic explosive power movements (combination of speed and strength required for jumping, and rapid extreme level changes) (Martyn-Stevens et al., 2012). For all performing artists, cardiorespiratory demands may reach levels associated with heavy workloads. For example, elevation in heart rate during rehearsal, and even more so during performance, can move into high workload ranges (Inesta et al., 2008; Liiv et al., 2014), but performers generally do not sustain high heart rates long enough to reach an aerobic training effect (Rodrigues-Krause, Krause, & Reischak-Oliveira, 2015). Given this pattern of cardiorespiratory activation, performers resemble fitness profiles of nonendurance athletes. They engage in high-intensity intermittent activities suggesting greater anaerobic capacity rather than aerobic fitness (Wyon et al., 2007).

Although not necessarily evaluated as aerobic or anaerobic fitness, singers have greater respiratory efficiency. Effective vocal training develops consistent lung volume, ribcage movement, and respiratory patterns. Consistent exhalation patterns in singers not only support vocal production but also phonation (Thomasson & Sundberg, 1999). Inhalation patterns have more individual variance, particularly in abdominal wall movement, in part, related to immediate vocal demands, such as preparing for a long sustained vocal passage versus a short staccato passage (Sundberg & Thomasson, 2001). These respiratory patterns appear in all highly trained singers; however, small variances exist in singers with different body types. Singers with more body fat tend to breathe lower (umbilicus area) than lean (lateral chest) or muscular (ribcage/costal) singers. All beginning and poorly trained singers employ vertical chest movement, a pattern that is inefficient and not recommended (Cowgill, 2009). Actor vocal training also enhances respiratory and vocal fitness. Compared to untrained actors, trained actors have greater sound pressure levels, phonatory airflow, subglottal pressure, volume, and prolonged inspiratory airflow. Pitch is also lower in trained actors (Master, Guzman, Azocar, Munoz, & Bortnem, 2015). Even though professional voice users demonstrate stronger respiratory muscle use, further respiratory training can enhance maximum expiratory pressure, maximum inspiratory pressure, and maximum phonation time. Ultimately, increased vocal and respiratory training improves pulmonary function as well as vocal production (Nam, Lim, Ahn, & Choi, 2004).

Musculoskeletal systems and kinesthetic creativity

Motor activation is contingent on multiple physiological systems, such as cardiovascular, pulmonary, endocrine, and neurological (central, peripheral, and autonomic). Evaluating muscular activity involves determining muscular force, velocity, power, goal intention, reward expectation, perceived effort, sex, health, and previous training (Batson, 2013). Because of the complex interplay of factors involved in executing highly skilled motor actions inherent in the performing arts, it is difficult to discuss isolated motor responses while excluding other key components that influence movement. Examining isolated motor groups is relevant when diagnosing an injury, but is not sufficient when creating rehabilitation and training programs.

Muscle force is produced when sufficient signals are transmitted from the central nervous system motor network to the descending (efferent) tracts (ventral horn) in the spinal cord. This initial signaling is called the "central drive" (Williams & Ratel, 2009). The signals are then transferred to peripheral motor neurons that mobilize or inhibit muscle groups. Ascending sensory signals (afferent) are sent back to the central nervous system to refine motor processing (Batson, 2013). These feedforward (efferent/descending) and feedback (afferent/ascending) sensorimotor loops continually direct all voluntary and involuntary movement. Strengthening these sensorimotor networks requires ongoing muscular engagement. Sensorimotor synchronization of whole-body movements is strengthened by repetition, especially when it is trained at different tempos (Miura, Kudo, & Nakazawa, 2013; Miura, Kudo, Ohtsuki, & Kanehisa, 2011). For example, the ability to learn auditory rhythm structures (discriminating and memorizing) requires movement integration; singing, dancing, or playing an instrument are all motor activities that reinforce rhythm learning (Phillips-Silver & Trainor, 2007). Elite performance motor training also promotes greater proprioceptive integration and multijoint coordination (Jola, Davis, & Haggard, 2011; Thullier & Moufti, 2004).

Assessing musculoskeletal functioning in performing artists requires complex neuromuscular understanding. Domain-specific training deeply refines sensorimotor integration, hence musicians discriminate and respond to tasks differently than dancers (Karpati, Giacosa, Foster, Penhune & Hyde, 2016), and dancers employ different balancing strategies compared to athletes (Schmit, Regis, & Riley, 2005). It is critical to account for these disparities when examining skill performance. Examining musculoskeletal interactions within the entire kinetic chain ensures targeting training for elite performers. For example, measuring biomechanical pressure shifts while a performer is working reveals individual balancing strategies (Pedersen, Erleben, & Sporring, 2006). Simple perturbation (disturbing equilibrium) can expose compensatory kinetic chain patterning. For example, vestibular stimulation tasks such as turning can perturb balancing systems (Hopper, Grisbrook, Newnham, & Edwards, 2014). Likewise, performing balancing tasks with eyes closed can enhance neuromuscular balancing systems (visual, vestibular, proprioceptive) (Hutt & Redding, 2014; Perez, Solana, Murillo, & Hernandez, 2014). These integrative kinetic chain assessments indirectly reflect peripheral and central nervous system interactions.

Muscular fatigue results from a combination of perceived exertion and repetitive strain (Batson, 2013). Peripheral muscular fatigue is the inability to voluntarily contract muscles. Some factors that lead to peripheral muscular fatigue include metabolic substrate availability, as well as accumulation of metabolites. Central fatigue is associated with the central nervous system; during muscle fatigue the cortical and subcortical brain regions that coordinate motor action also become inefficient. The combination of peripheral and central fatigue accounts for overall fatigue that manifests in musculoskeletal responses (Batson, 2013). Typically, physical muscular fatigue results when a muscle is unable to contract to meet the force required to execute a motor skill. Many performers may perceive they have reached maximum fatigue; however, muscular contraction is still possible. Likewise, performers may perceive they can push harder, and to their surprise, voluntary, and involuntary muscular contraction diminishes. This state is termed "task failure." The central drive from the central nervous system no longer sends signals to the peripheral musculoskeletal system (Batson, 2013). During these states, excess exercise-induced production of reactive oxygen species, proteolysis, and changes in calcium balance all reduce muscle function and may lead to muscle damage (Rodrigues-Krause, Krause, et al., 2014).

Muscular fatigue may be associated with excess cardiorespiratory demands. In this case, insufficient oxygen and fuel are delivered to muscles to initiate or maintain muscular contractility. For all performers, insufficient rest between performances leads to muscular fatigue. Lack of rest prohibits a return to resting blood lactate concentrations, and increased lactate levels are associated with greater muscle fatigue and risk of injury (Baillie, Wyon, & Head, 2007). Insufficient recovery

Fig. 7.3 "Physiological regulation." Performers: Sara Adler, John Ghera. Photographer: Lee Choo.

time can cause a failure to resynthesize ATP (Wyon & Koutedakis, 2013). Inadequate ATP supply to calcium and potassium pumps is directly related to muscle fatigue. Fundamentally, ATP deficits are associated with the type, duration, and intensity of an activity. For example, dance requires increased muscular cocontraction (compared to athletes). Cocontraction requires more metabolic energy, which makes dancers more vulnerable to muscular fatigue (McEldowney, Hopper, Etlin-Stein, & Redding, 2013). Dancers may also experience more oxidative stress and muscle cell membrane damage due to excessive eccentric and isometric muscular activities (Rodrigues-Krause, Krause, et al., 2014). Increased peripheral and central fatigue can elicit sensorimotor incongruence in all performers. When sensory feedback and motor output are no longer synchronized, performers are no longer able to coordinate highly refined skills (Nijs, Wouters, Cras, & Daenen, 2015). Reducing muscular fatigue is essential in the performing arts. Sufficient physiological stamina must exist in order to perform at elite levels (Fig. 7.3).

Fitness and performers

Currently, more studies are conducted on cardiorespiratory and musculoskeletal physiological fitness levels in dancers, in part, because they are often considered both artists and athletes (Koutedakis & Jamurtas, 2004). This general assumption precludes studying other performing artists who could easily be considered "athletes." For example, rock musicians exert extreme physical effort during performance (Clark et al., 2011). Although not all performance repertory demands high cardiorespiratory and musculoskeletal physical exertion, most performers experience these states at some point during a performance season. Across all the performing arts, physical task demand varies depending on the situation (class, rehearsal, performance), and the rank of the performer. For instance, soloists and ensemble performers frequently differ in their physiological performance requirements; however, this may vary depending on the work performed (Wyon et al., 2007). A career in the performing arts also requires different daily physiological output. Classes focus on high skill development (less physical exertion), while rehearsals require greater physical exertion, especially near the end of a rehearsal process. Performances demand the highest physical output (Martyn-Stevens et al., 2012). In general, workload during classes and rehearsals are insufficient to reach fitness training levels, thus it is recommended that performers augment their practice with aerobic and strength training to better prepare them for the physical demands required during performance settings. Unlike athletic training programs, the ability to preserve aesthetic expression while enhancing fitness levels is critical for performers (Guidetti et al., 2007; Martyn-Stevens et al., 2012; Wyon & Redding, 2005).

Studies have demonstrated that participating in the performing arts does meet basic fitness levels; although fitness levels are often insufficient for the physiological rigor required in elite performances. For example, while playing, musicians increase their heart rate and sustain it at levels typical of brisk walking, thus they are deriving some fitness benefits (Clark et al., 2011; Hahnengress & Boning, 2010; Inesta et al., 2008). Participating in recreational dance improves cardiorespiratory fitness and reduces disease risk (Janyacharoen, Phusririt, Angkapattamakul, Hurst, & Sawanyawisuth, 2015). Engaging in recreational ballroom dance (Lankford et al., 2014) or Latin dance meets recommended weekly fitness standards (Domene & Easton, 2014). However, the physical demands of professional dance requires even higher levels of fitness, including aerobic, anaerobic, strength, endurance, flexibility, lean body composition, and lower body mass. Different dance styles and ranking within a dance company (principle, solo, corps) also require different levels of energy expenditure and fitness (Twitchett, Angioi, Koutedakis, & Wyon, 2010). Despite these fitness gains, unlike highly trained athletes, dancers (and other performers) are frequently deficient in meeting standards in one or multiple fitness areas (Koutedakis & Jamurtas, 2004). Studies recommend that dancers, musicians, singers, and actors augment traditional training with more aerobic and anaerobic exercise components. These fitness recommendations are based on uneven, and often extremely elevated, physiological workload demands experienced in classes, rehearsals, and performances. For example, during performance (but not during class training) professional Flamenco dancers exert 25% greater levels of blood lactate than anaerobic athletes while also working at higher aerobic work levels (Pedersen, Wilmerding, Kuhn, & Encinias-Sandoval, 2001). Ballet dancers share a similar combination of aerobic and anaerobic fitness demands (Twitchett, Angioi,

Koutedakis, & Wyon, 2009). Conductors must sustain moderate cardiovascular levels of exertion, with workload demand increasing as musical tempo increases (Jaque et al., 2015). Adding supplemental fitness programs into performers' regular training regimes may offset fitness deficits. Essentially, performance-specific domain training effectively builds high skill execution, including isometric strength and integrated coordination patterns, but not a well-balanced fitness profile (Rodgrigues-Krause et al., 2014).

Enhancing physiological conditioning

Optimal physical conditioning enhances health, performance, and creativity. It also reduces muscular fatigue. Currently, the technical demands of performers continue to increase; they are expected to meet these new requirements without sustaining injury (Bronner & Rakov, 2014). To help performers understand the importance of fitness, screening assessments (aerobic, anaerobic, strength, endurance) at the beginning of training programs or preseason (Bronner & Rakov, 2014) can offer baseline fitness levels. Specific training recommendations can be offered along with individualized fitness programs to address any deficits (Bronner, Ojofeitimi, Lora, Southwick, Kulak, & Gamboa, 2014). Developing fitness assessments that are appropriate for each performing arts domain is equally critical (Twitchett, Nevill, Angioi, Koutedakis, & Wyon, 2011). Repertory demands of each company, and working conditions in performing venues, all influence fitness needs and should be factored into exercise programming. Assessing performers' baseline fitness levels will offer invaluable information about their cardiorespiratory and musculoskeletal capacities; however, adding assessment protocols to measure the effects of fatigue on performers will identify problem areas that may increase injury risk. This form of assessment should include measuring cognitive–attentional fatigue as well (Coutts et al., 2006). An incremental fatigue protocol is a common assessment to determine athlete fitness; performing artists share similar demands as athletes and should be evaluated in a similar process, especially because they must coordinate, integrate, and rapidly change sensorimotor and cardiorespiratory responses to address performance intensity, dynamic, and tempo (Coutts et al., 2006).

Employing more complex measurements such as three-dimensional motion analysis, electromyography, or force-plate analysis may reveal structural anomalies or biomechanical discrepancies; misuse patterns can lead to muscular inefficiency, fatigue, and/or injury. Deriving a refined analysis, whether from sophisticated musculoskeletal instrument assessments or from highly trained kinesiologists or medical practitioners, can promote even more individualized fitness programs and targeted training (Fjellman-Wiklund, Grip, Karlsson, & Sndelin, 2004; Pata, Welsh, Bailey, & Range, 2014; Sherman, Mayall, & Tasker, 2014; Turner-Stokes & Reid, 1999). For example, direct, objective, and scientifically based feedback can be given to musicians to augment body awareness, timing, coordination, and physical skill execution (Fjellman-Wiklund, Grip, Andersson, Karlsson, & Sundelin, 2004; Rodger, O'Modhrain, & Craig, 2013).

Training recommendations for nonendurance athletes usually includes high-intensity interval training protocols. This training approach positively affects autonomic regulation, especially for individuals who have more vagal tone during base rest periods (Kiviniemi et al., 2015). Multiple repetitions of high workload intensity, with short recovery periods, enhance anaerobic lactate capacity and promote performance when fatigue increases (Rodrigues-Krause et al., 2015). This form of training also increases tolerance for sustained sympathetic activation. High-intensity interval training is also ideally suited for performer needs. One major problem with this form of training for performing artists is that recovery time is longer between interval training sessions (Stuckey et al., 2012). Extended recovery time is not always possible for performers. Beyond anaerobic fitness, performers require cardiorespiratory endurance developed via low-intensity exercise protocols. Cardiorespiratory endurance is essential to sustain slow movements, and for dancers, slow movement sequences also involve a wide range of motion (Rodrigues-Krause et al., 2015).

Along with interval and endurance training, increased flexibility, range of motion, postural stability, and proprioceptive awareness must be included, especially during warm up (Morrin & Redding, 2013; Rubini et al., 2011). One area of focus in training is addressing the influence of the Golgi tendon organ, a sensory receptor located within the muscle. This organ detects active contractions throughout a wide range of muscular motion. It influences nonconscious proprioceptive processing via the cerebellum. An integration of Golgi tendon organ feedback with cerebral proprioceptive centers enhances the learning of new movements and improves previously learned actions. Adequate muscle flexibility further augments Golgi tendon organ receptivity, which ultimately promotes more refined proprioceptive awareness (Moore, 2007).

Performer fitness needs can be addressed by implementing supplementary training approaches such as strength and power training (Martyn-Stevens et al., 2012). Core stability and strength training are vital for all performing artists, both to enhance performance quality and diminish risk of injury (Kline, Krauss, Maher, & Qu, 2013). There are many training methods that address these needs, such as Pilates, yoga, conditioning with imagery (CI-training), gyrotonics, traditional strength training, and Alexander technique (Rafferty, 2010). Training programs specifically designed for the needs of performers yield extremely positive results. For example, introducing a fitness program that incorporates yogic breathing, muscle strengthening, and flexibility exercises enhances musician awareness of postural alignment, as well as improving performance expressivity (Lee, Carey, Dubey, & Matz, 2012). Incorporating a variety of fitness components during classes or coaching sessions can further improve muscle adaptation required for eccentric and isometric muscular activity.

The advantages of well-trained performers are that they experience less oxidative stress and muscle damage (Rodrigues-Krause, Krause, et al., 2014). A popular training approach implemented in sport is periodization. The overarching philosophy in periodization is to train the "whole" athlete/performer. Training involves psychophysiological approaches designed to increase, and then decrease, workload intensity in preparation for a performing season (Wyon, 2010). Another approach is plyometrics, a form of weight training that builds strength (ability to exert maximal force) and power (ability to exert force quickly). Plyometrics requires participants to jump off and onto raised platforms. It involves bouts of repetitive sequences, similar to interval training. This form of exercise evokes similar strength and power results as traditional weight training; both enhance performer fitness (Brown, Wells, Schade, Smith, & Fehling, 2007).

Addressing daily dietary needs will also ensure sufficient glycogen stores necessary for adequate neuromuscular signaling; deficits in glycogen ultimately increase muscular fatigue and task failure (Rodrigues-Krause et al., 2015). During training seasons, most dancers, especially female dancers, manifest decreased weight along with decreased percentage of body weight as fat. These losses may be due to insufficient nutritional intake or increased workload demands, especially as demands increase as the performance season progresses (Micheli, Cassella, Faigenbaum, Southwick, & Ho, 2005). Attending to a well-balanced diet is essential for all performers; it supports vocal health and musculoskeletal wellbeing.

Assessing performers' baseline physiological cardiac vagal tone and nutritional practices can guide individualized fitness programs. Intensity and frequency need to be factored into the creation of a fitness program. Each performer, regardless of performance demands, will respond differently to training protocols, including amount of time for recovery. Gathering a comprehensive neurophysiological profile of the performer will facilitate the creation of an appropriate fitness protocol; it will also serve as a screening assessment for potential musculoskeletal, cardiac, and respiratory disorders (Hautala, Kiviniemi, & Tulppo, 2009). For example, physiological resilience is clearly identified by enhanced ANS responsiveness and regulation (Radtke, Kriemler, Eser, Saner, & Wilhelm, 2013). If a performer has poor autonomic regulation (vagal tone), then aerobic training may be recommended. After several weeks of aerobic training resting cardiac vagal tone improves (Radtke et al., 2013). For obese performers, this is vital due to alterations in their ANS (greater sympathetic and decreased parasympathetic activation). Carefully designed exercise programs, coupled with weight loss, can help improve autonomic functioning in obese performers (Amano, Kanda, Ue, & Moritani, 2001; de Jonge, Moreira, Martin, & Ravussin, 2010). Consistent exercise in older performers also positively affects autonomic functioning; in fact, aging performers must maintain cardiorespiratory health to safely meet the demands of a professional career (Wichi, de Angelis, Jones, & Irigoyen, 2009).

Evidence-based fitness studies demonstrate substantial positive gains in performers who are fitter. Musicians with better cardiorespiratory health and higher levels of fitness have less performance anxiety and perform with less cardiorespiratory strain (Wasley, Taylor, Backx, Williamon, 2012). Dancers who augment their training with aerobic and strength training programs are fitter, without losing their artistic expressivity and aesthetic appeal (Angioi, Metsois, Twitchett, Koutedakis, & Wyon, 2009; Koutedakis et al., 2007; Wyon, 2005). Similar results are found when ballet dancers add low-intensity aerobic training; they gain aerobic capacity without loss of psychomotor speed and accuracy (Smol & Fredyk, 2012). Regardless of fitness level, performers are sensitive to tempo and rhythm patterns in music. After bouts of exercise, classes, rehearsals, or performances, listening to music with a decreasing tempo helps restore autonomic balance (Wallert & Madison, 2014). This reflects the cool-down practices provided in most dance classes. The cardiorespiratory system rapidly entrains to the musical stimuli, a fact that can promote physiological health and resilience.

Summary

- Introduction: definitions of physiology, systems of the body.
- Physiology and creative stress: a dynamic process: homeostasis, homeodynamic, allostasis, responses to stress, brain–body interconnections, and individual differences.
- Cardiac and respiratory systems and performance: terms (heartbeat, heart rate, tidal volume, pacemaker), work effort and metabolic demands, ventilatory drive.
- Autonomic nervous system: SNS, PNS; cortical, subcortical, and physiologic system integrations; beta-blockers and stress.
- HRV and creativity: HRV synergistic, coactivation, coinhibition, influences on HRV (circadian rhythm, age, weight, conditioning, performance).
- Emotional expressivity and ANS physiology: influences of mood, music, movement and the complex interactions between the brain and body; influence of music tempo and dynamics on respiration and heart rate; aesthetic chills and physiologic responses.
- Physiological performance demands: metabolic demands, energy requirements, aerobic and anaerobic activity, muscle adaptation.
- Musculoskeletal systems and kinesthetic creativity: motor activation dependent on multiple systems (cardiovascular, pulmonary, endocrine, neurological), sensorimotor network integration and synchronization, muscle fatigue and recovery.
- Fitness and performers: physiologic demands in training, rehearsals, and performance, basic fitness needs and general fitness levels of performers.
- Enhancing physiological conditions: screening fitness levels in performers, training requirements for nonendurance athletes/performers, high-intensity interval training, flexibility, supplementary training and somatic practices, daily dietary needs, evidence-based training protocols.

Concluding remarks

Further research is needed to gather a deeper understanding regarding physiology and creativity (Zhang, Zhang, Yu, & Zhao, 2015). Physiological responses directly and indirectly influence individual creative engagement and capacity. Carefully designed studies that address the extreme physiological workload variability during class, coaching, rehearsal, and performance will increase understanding about performers, as well as aid in the creation of optimal training protocols. Performing artists are typically trained with bursts of activity followed by instructional feedback. Rehearsals are usually very uneven in terms of physical demand, and this unevenness persists throughout the full rehearsal period. The most consistent physical demands occur during performances; they typically require greater physiological exertion compared to rehearsals, classes, and coaching sessions. Repertory choice also has unique physiological requirements. Even when performers seemingly engage in low physical work intensity, their cognitive demands are often very high. Given this extreme variance, studies should account for a multitude of physiological factors that influence performers (Beck, Redding, & Wyon, 2015).

There is a strong relationship between physiological health and creativity (Oppezzo & Schwartz, 2014). The creative life of performers necessitates physical engagement and fitness. For example, moderate levels of aerobic exercise, a typical level for most performers, not only improves creativity immediately following the aerobic bout, but also several hours after the activity is terminated (Blanchette, Ramocki, O'del & Casey, 2005). Participating in exercise, including the physical experience of performance, increases a sense of revitalization (Bartholomew, Moore, Todd, Todd, & Elrod, 2001), a common experience cited by performers postperformance. Ultimately, physiological and motor activation powerfully influence embodied cognition and underpin social cognition and social awareness (Sevdalis & Keller, 2011). For example, the major communication of musical conductors is their ability to gesture expressively. High gestural intensity by a conductor evokes high-quality musicianship from ensemble players (Bender & Hancock, 2010). Performing artists rely on their embodied cognition to give life to their art form. Performers' physiological wellbeing promotes their creative wellbeing as well as optimizes their prodigious artistic skills.

Flow and neurophysiological responses

8

The key to the mystery of a great artist is that for reasons unknown, he will give away his energies and his life just to make sure that one note follows another . . . and leaves us with the feeling that something is right in the world.

Leonard Bernstein

Introduction

The nature of flow has been directly studied in the fields of neuroscience and physiology (de Manzano et al., 2010; Peifer, Schulz, Schachinger, Bauman, & Antoni, 2014). Flow is a state of enjoyment marked by full concentration on a task with decreased attention directed at self-conscious awareness. In all conditions that elicit flow, skill and talent must match the challenge presented in the activity; a sense of control and an ability to respond effectively during a performance facilitate flow experiences (Csikszentmihalyi, 1990). Because flow is a subjective experience, flow research studies are dependent on self-reports offered by participants (Klasen, Weber, Kirchner, Mathiak, & Mathiak, 2012). Flow has been investigated with many different performing arts groups, and the results demonstrate that heightened flow, also known as optimal performance, is embraced and endorsed by performers (Robb & Davies, 2015; Thomson & Jaque, 2011b; 2011–12; 2012c; 2016a). They recognize these states and are intrinsically motivated to reexperience them (Martin & Cutler, 2002).

As discussed in Chapter 4, Creative Experiences, flow is comprised of nine discrete factors, although these variables often reinforce each other (Jackson & Eklund, 2004; Kawabata & Mallett, 2011). Factors that directly relate to physical performance include (1) adequate skill level to match the demands and challenges of the physical activity, (2) full concentration on the task, (3) clear goals directed toward accomplishing the activity, (4) feedback during the performance of the activity that is unambiguous and immediate, (5) a sense of control while performing the activity, and (6) an ability to merge an awareness of the activity with the actions being performed. Personal experiences that manifest during a flow experience include (1) a loss of self-conscious awareness, which enhances concentration and the ability to attend to feedback during the execution of the activity; (2) a sense that time is distorted, either slowing down or accelerating; and (3) an autotelic

Creativity and the Performing Artist. DOI: http://dx.doi.org/10.1016/B978-0-12-804051-5.00008-1

experience, one that is personally meaningful, positive, and strongly indicative of a flow-prone personality trait.

Achieving flow in elite performers necessitates that they receive training that addresses these nine factors; they must optimize performance via conscious and calculated physical, dietary, and psychological skill acquisition (Smith, 2003). Learning to perform under pressure is key to a successful career; it requires skills that ensure stress management and self-preservation despite perceptions of negative judgment and high performance expectations (Hallett & Hoffman, 2014). Repeatedly achieving peak performance is not contingent on random magical moments. Elite performers, like elite athletes, nurture powerful abilities such as keen concentration and kinesthetic awareness, along with a willingness to embrace challenge and flourish in high-pressure situations (Hallett & Hoffman, 2014).

Embodied flow, genetic heritability, and flow proneness

Although not all performing artists achieve high flow states, some consistently report these experiences. When performers claim a heightened flow experience, unlike other professions, their flow states always encompass an integration of physical and emotional expressivity. Singers, actors, and dancers are acutely aware of their own emotional sensitivity as they physically embody and give voice to their creative imagination. They also sense the emotional receptivity of their audience members, while concurrently integrating their own technical mastery and artistic sensibility (Spahn, Echternach, Zander, Voltmer, & Richter, 2010). They respond with full physiological arousal during performance; all performers engage their body and voice, even if their voice is channeled through a musical instrument. Performing demands great physiological exertion; this physiological state is subjectively appraised as experiences of either flow or anxiety. Only psychological and cognitive thoughts determine whether a performer will suffer from anxiety or channel physiological arousal into flow states (Spahn et al., 2010). This holds true for performing artists in all domains.

Studies suggest that some individuals are genetically predisposed to flow (Mosing, Magnusson et al., 2012; Mosing, Pedersen, et al., 2012); these are the individuals with an autotelic personality (Csikszentmihalyi, 1990). They demonstrate greater ability to control their own future (internal locus of control), they are less anxious (Mosing, Pedersen, et al., 2012), more outgoing, and conscientiously attend to tasks and responsibilities (Ross & Keiser, 2014). The genetic heritability profile also suggests that proneness to high flow experiences is substantially related to personality traits, with minimal additive effects from environmental factors (Mosing, Magnusson, et al., 2012). Further, findings indicate that genetic and personality traits for flow proneness are not related to intelligence or intellectual aptitude (Ullen et al., 2012). Individuals, regardless of intelligence, can experience the pleasure of flow states, a finding that offers hope to many disabled individuals, whether from traumatic brain injury or developmental disorders (Fig. 8.1).

Figure 8.1 "Dancer and cellist." Performers: David Aks, Cheyenne Spencer. Photographer: Lee Choo.

Neuroscience and flow experiences

Both flow and the creative process in the performing arts are embodied experiences. Sensorimotor systems, including the action-oriented network and the salience network (see chapter: Neurobiology, Creativity, and Performing Artists for more details), all provide essential feedback information; these neural activation patterns occur similarly during the creative process and flow states (Chirico, Serino, Cipresso, Gaggioli, & Riva, 2015). The implicit sensorimotor system is the primary system operational during flow experiences (Dietrich, 2004). During flow experiences, the prefrontal cortex, especially the explicit motor system, which is regulated by the prefrontal cortex, is in a state of transient hypofrontality; literally this region is temporarily quiet. As a result, it does not inhibit the implicit motor and emotional systems. This disinhibited state allows deeper brain structures to interact without interference; hence a flow state is potentiated. For well-trained professional performing artists, the ability to induce a transient hypofrontality state during performance heightens flow states (Dietrich, 2004). Unfortunately, performers can equally experience performances that are not flow-like. They describe situations when the conscious explicit system inhibits and constrains their performance, as evidenced by statements such as "not being in the moment," "being stuck in their head," "overthinking," or "not being in the zone." During these moments, they kinesthetically know they are unable to access their implicit sensorimotor system. They are fully aware that their performance is off; despite their best efforts, they are unable to mobilize the freedom found in flow states (Hefferon & Ollis, 2006; Martin & Cutler, 2002; Robb & Davies, 2015).

Complex interconnections are usually engaged between the frontoparietal (sensorimotor) cortices and between implicit and explicit systems (de Manzano, 2010). The error-detecting regions of the basal ganglia, in conjunction with the fundamental sensorimotor regions, are directly involved in flow states (Klasen et al., 2012). These brain regions operate in phasic activation patterns. They fluctuate between intense activation and quiescence, suggesting a complex neurological interplay during performance. These activation patterns help performers rapidly evaluate immediate goals, adjust to direct moment-to-moment feedback, and shift concentration and attentional foci while sustaining efforts to mobilize physical and psychological skills (Klasen et al., 2012). All performers, regardless if they are actors, musicians, or dancers, must organize rhythmic and melodic patterns to give expressive depth to their performance; these demands enhance the challenges of flow activities. To succeed in recruiting these complex skills, performers require integrative neurological responses in the prerostral dorsal premotor regions (rhythm), presupplementary motor regions (melody) (de Manzano, 2010), and within different regions of the cerebellum (de Manzano & Ullen, 2012a). These regions are key areas in the implicit system, adding further support that the explicit system is quiet during flow experiences. For example, studies that investigate musicians indicate flow states are associated with deactivation in the prefrontal cortex, hence deactivation in the explicit system. Most studies suggest that attentional processes that reside in the prefrontal cortex are minimally engaged during musical performance; consequently, activation within this region would be redundant (Dietrich & Kanso, 2010). This flow activation pattern may also be associated with task familiarity. Performing familiar tasks during performance can elicit high intrinsic rewards (and high flow); these tasks engage the mesolimbic (subcortical) structures where the reward circuit is located. Without excessive prefrontal cortical activation, the deeper brain structures can freely evoke aesthetic pleasure (Menon & Levitin, 2005).

The unique characteristics of flow ultimately reflect a complex and simultaneous firing pattern in multiple regions of the brain. In one study, activation in the basal ganglia, in particular, the putamen, increased the ability to anticipate outcome results, a factor related to unambiguous feedback and clear goal setting (Ulrich, Keller, Hoening, Waller, & Gron, 2014). A sense of control was supported by increased arousal in the inferior frontal gyrus, while decreased self-awareness was associated with decreased activation in the medial prefrontal cortex (hypofrontality hypothesis). Lastly, positive feelings elicited during flow experiences were related to decreased activation in the amygdala, a subcortical limbic structure that immediately responds to real or imagined threat (Ulrich et al., 2014).

Not surprisingly, findings demonstrate that flow is a rewarding experience, and as such, activation of the reward circuit is evidenced. Along with the reward circuit, somatic and autonomic responses regulate flow factors (i.e., challenge of the task while merging action and awareness). Unlike other professions, performers' reward experiences must also include a heightened awareness of beauty, including highly trained abilities to produce beauty (Hargreaves, 2012). When they describe the experience of flow, this fundamental aesthetic component is embedded within it. Specifically, the rostral anterior cingulate, insula, hypothalamus, ventral tegmentum, and nucleus accumbens all engage and interconnect during flow experiences (Menon & Levitin, 2005; Ohira et al., 2010). This neurological response pattern has been clearly demonstrated when participants listen to or perform music; they describe the experience as aesthetically rewarding and meaningful (Menon & Levitin, 2005). These neurological networks influence complex associational processes such as personal, cultural–contextual, and imaginative. Collectively, the intricate recursive looping of sensorimotor, cognitive, emotional, and imaginative information produces aesthetic rewards for performers and their audiences; they positively promote flow states for all (Hargreaves, 2012).

Neurochemistry and flow experiences

When examining flow from a neurochemical lens, studies have demonstrated that higher dopamine (D2) in the dorsal striatum enhances flow (de Manzano et al., 2013). This finding reinforces the notion that individuals who achieve high flow states are also less impulsive and more emotionally stable. As well, they ascribe meaning to these experiences and identify them as positive and pleasurable (Csikszentmihalyi, 1990). Individuals with a genetic predisposition for flow also manifest a more robust dopaminergic system (Mosing, Pedersen, et al., 2012), one that increases both reward experiences and flow activities (de Manzano et al., 2013).

Flow states are also enhanced by activation in the locus coeruleus and its production of norepinephrine. The release of norepinephrine optimizes task performance, including exploratory behaviors that are typical in flow activities (Aston-Jones & Cohen, 2005). Because participating in performing arts activities inevitably involves performing with other artists, it is not surprising to discover that oxytocin, the bonding neuropeptide, has been implicated in social flow experiences among singers (Keeler et al., 2015). In this same study, along with increased oxytocin, a concurrent decrease in adrenocorticotropic hormone was demonstrated. This suggests that the bonding experience during singing helped alleviate the stress response, but only when singers endorsed that they experienced the positive effects of flow. Singers also have a genetic bias toward more affiliate neurochemistry as evidenced by enhanced vasopressin and serotonin—factors that may promote greater collective group flow (Morley et al., 2012). These findings lend support to previous studies illustrating that flow experiences are enhanced by cognitive, motivational, and emotional regions in the brain via the release of neurotransmitters and neuropeptides from these regions (Klasen et al., 2012).

Physiology and flow experiences

Physiological responses during flow are deeply related to neurological activation patterns. Unlike measuring brain activation, the autonomic nervous system (ANS) can be indirectly measured while artists perform and experience flow states (Thomson & Jaque, 2011b). What is evident is that more activation is manifested in the sympathetic branch of the ANS. But determining whether the aesthetic experience was positive or negative necessitates that performers subjectively report emotional states. Physiological arousal occurs during both positive and negative emotional states, and even more importantly, aesthetic flow responses are highly individualized. For example, individuals who are listening to or performing Samuel Barber's *Adagio for Strings* may respond to it in a calm, bored, or highly excited state. Individuals have their own unique emotional and physiological responses to the work they are performing (Salimpoor et al., 2009).

Newberg and d'Aquili (2000) suggest that the body is the direct source that communicates with the brain, while the ANS housed in the brain communicates with the body. Like the ANS, there are reciprocal arousal and "quiescent" systems operating synergistically or coactivated/coinhibited within the brain, and these shifting activating/quiescent responses inform creativity, aesthetic appeal, heightened spiritual bliss, and flow experiences. For example, cardiovascular responses are either active or quiet during aesthetic reward experiences. Increased vagal activity, directed by the rostral anterior cingulate, modulates pleasurable states induced during flow (Ohira et al., 2010). Greater neurological connectivity (ventral tegmental area, nucleus accumbens, hypothalamus, insula, cingulate, and prefrontal cortex) mediates autonomic responses during the aesthetic pleasure of listening to music (Menon & Levitin, 2005). Further, greater autonomic heart rate variability supports self-regulatory functions. This variability is present during the flexible responsiveness associated with all flow states; literally parasympathetic coactivation modulates sympathetic activation to ensure optimal arousal to meet the challenges of flow activities (Segerstrom & Nes, 2007).

An accumulation of research indicates that flow, despite its exhilarating and pleasurable sensation, is actually physiologically stressful. Adaptive biological responses are mobilized to meet performance demands, evidenced by increased muscular cocontraction and elevated heart rate (Yoshie, Kudo, Murakoshi, & Ohtsuki, 2009). These musculoskeletal and cardiovascular responses indicate readiness for unpredictable environments. Achieving these high physiological states requires great metabolic output; adequate cardiovascular and muscular fitness levels ensure sufficient energy to meet performance demands (Hallett & Hoffman, 2014). Performers learn to manage these physiological responses; this promotes increased flow opportunities, as opposed to performance anxiety. This activation pattern is heightened even more when performers move from the rehearsal room to public performances (Yoshie et al., 2009).

Flow activities increase activation in the fast-acting sympathetic branch of the ANS along with greater excitation in the hypothalamus–pituitary–adrenal (HPA) axis. These physiological patterns reflect the natural excitement generated during challenging situations, especially when sufficient skills can be mobilized to achieve success (Peifer et al., 2014). Further, activation of the HPA axis increases the release of cortisol, a stress hormone that also facilitates greater auditory detection and increased blood-glucose levels necessary for elevated energy demands. Enhanced auditory attention and energy mobilization are essential during performance, including performances that are flow-like (Peifer et al., 2014). During experimental studies, increased sympathetic and HPA axis responses heightened flow states, but in a curvilinear pattern (inverted u-shape). Low or high physiological arousal hindered flow experiences (Hallett & Hoffman, 2014). This response pattern may indicate subjective appraisal differences between challenge and threat. Flow manifests during challenge, a condition that mobilizes moderate levels of arousal from the physiological stress systems; whereas, threat increases stress responses and constrains flow (Peifer et al., 2014).

Moderate sympathetic arousal is typical in any situation that requires cognitive attentional processes; hence decreased parasympathetic activation is appropriate. When studies demonstrate increased parasympathetic activation during flow, the findings may not be fully measuring flow; rather, they are assessing decreased cognitive and physical demands, high absorption states (Bruya, 2010), or in some cases, increased conditions of stress (Peifer et al., 2014). Flow states may actually be associated with both sympathetic and parasympathetic activation; literally the parasympathetic branch of the ANS is modulating the sympathetic branch, a response pattern indicative of coactivation of both ANS branches rather than the usual reciprocal relationship between the two branches. Coactivation of the ANS increases adaptation during high demand situations and is indicative of an active coping response pattern. These coactivation patterns are also associated with better health, because parasympathetic activation modulates sympathetic responses (Berntson, Norman, Hawkley, & Cacioppo, 2008b). This coactivation pattern was observed in dancers participating in improvisational authentic movement. When dancers claimed that they felt as if they were "being moved," rather than being the initiator of the movement, their sympathetic and parasympathetic nervous systems were coactivated (Tantia, 2012). The experience of "being moved" is literally a flow experience for all performers.

During piano playing, expert musicians manifest an entrainment of respiration with musical tempo, a factor that influences parasympathetic activation. Respiratory rate is associated with musical tempo, and respiratory depth is related to metabolic workload demands (de Manzano, Theorell, et al., 2010). For example, "hard work" demand was evident in conductors. During performance they reached high workload levels (higher heart rate and greater respiratory depth) and their heart and respiratory rate patterns matched the tempo of the music being performed (Jaque et al., 2015). Greater sympathetic and parasympathetic activation, evident during flow states, reflects the complex interplay between the body and the subjective appraisal of flow (de Manzano, 2010; Thomson & Jaque, 2011b). Heightened flow states engage a complex interplay between physiological and emotional systems, in particular, coactivation of the sympathetic and parasympathetic branches of the ANS facilitates greater concentration during performance and heightened autotelic (meaningful) experiences (de Manzano, Theorell, et al., 2010).

The ability to receive unambiguous feedback while performing is one of the nine factors in heightened flow states. This is achieved by a complex integration of neurophysiological activation patterns. One of the major players in this feedback loop is the proprioceptive system (Smitt & Bird, 2013). This neuromuscular system detects joint angles, muscle force, velocity of movement, and spatial relationship between the environment and the coordinated movements that are initiated. Through training and deliberate practice, this system is strengthened; thus elite performers are able to maximize muscular efficiency with great accuracy and speed. Proprioceptive awareness contributes to postural control and balance and refines the integration of sensory

Figure 8.2 "Drumming flow." Performers: Ric Aviso, Malik Sow. Photographer: Lee Choo.

stimuli and motor expression (Smitt & Bird, 2013). When proprioceptive systems operate efficiently and accurately, flow states are optimized (Fig. 8.2).

Neurophysiological approaches to enhance flow experiences

Despite robust research demonstrating that a predisposition to experience high flow states is rooted more in genetic heritability and personality traits, current training protocols to enhance performance flow are underway. Some promising training approaches include neurofeedback, a protocol designed to enhance sensorimotor rhythms (Gruzelier, 2009; Gruzelier, Inoue, Smart, Steed, & Steffert, 2010). Slow wave alpha/theta neurofeedback training, a form of deep relaxation training, enhances performance under stress in elite conservatory-trained musicians, actors, singers, and dancers. Studies demonstrate that this supplementary training protocol (increased theta/alpha ratio during neurofeedback) is more efficacious than psychological performance skill training, Alexander Technique (alignment-based training), and physical exercise (Egner & Gruzelier, 2003). Theta wave states in the frontal regions of the brain are associated with increased focused attention, meditative concentration, decreased sympathetic activation, and reduced anxiety. Unlike the faster wave training protocols (beta and sensorimotor rhythm), the combination of alpha/theta, especially theta wave protocols, increases creative flow states, including greater indirect integration in the sensorimotor system (Gruzelier et al., 2014). Performers who receive this form of training improve technically, musically, and expressively. They are also more adept at communicating. The success of this training approach may be related to the fact that theta waves influence neural connections in widely distributed regions of the brain, connections that enhance creativity in actors, dancers, singers, and musicians (Gruzelier, 2009).

Other approaches to increase flow and creativity reveal positive results as well. For example, enhanced heart rate variability (autonomic resilience) was observed when performers participated in authentic movement explorations, a process designed to freely explore creative impulses. During the movement exploration and while speaking about the experience, dancers manifested increased parasympathetic activity while maintaining sympathetic activation, a hallmark pattern of autonomic regulation during the creative process (Tantia, 2012). A similar result was found in musicians who participated in a slow breathing protocol along with biofeedback (Wells, Outhred, Heathers, Qunitana, & Kemp, 2012). The musicians who participated in this training also experienced reduced performance anxiety when they returned to performance situations. The alpha/theta neurofeedback training also enhanced heart rate variability, decreased anxiety, and increased creativity in a group of dancers (Gruzelier et al., 2014). These changes directly promote flow states in performing artists.

In a study that examined the effects of an intensive opera singer training program, flow states increased while psychological distress (anxiety, shame, and overly constricting perfectionism) decreased. These results were achieved via a program that provided an integrative and psychologically supportive environment. Singers actively explored and trained musical, dramatic, and physical skills; the educational mission was to train singers without fear of criticism (Thomson, Jaque, & Baltz, 2017). They reclaimed their passion to sing and they rekindled their self-esteem, confidence, and motivation, essential factors to evoke flow states and to effectively sustain a career in the performing arts. Conditions conducive to positive flow experiences reside within supportive environments. In these environments threat is decreased, challenges are embraced, and social bonding is valued; confidence, pleasure, and optimal flow states are nurtured (Keeler et al., 2015).

Summary

- Introduction: general remarks on flow and the factors that evoke flow experiences.
- Embodied flow, genetic heritability, and flow proneness: increased physiological demands balanced by increased pleasure, autotelic personality.
- Neuroscience and flow experiences: brain regions and networks; hypofrontality hypothesis reinforced; neurochemical influences on flow; genetic heritability and flow-prone personality traits.
- Neurochemistry and flow experiences: dopamine, oxytocin, serotonin.
- Physiological and flow experiences: reciprocal and coactivation of the ANS.
- Neurophysiological approaches to enhance flow experiences: training approaches to augment flow proneness; neurofeedback; intensive training environments that promote safe exploration.

Concluding remarks

A deeper understanding about the neurobiological and physiological responses during flow states can inform optimal approaches to train and coach performers.

Flow is essentially a mind–body experience; sensorimotor systems are coupled together in a complex feedback loop (Chirico et al., 2015). It also involves strong activation of the reward network that influences the sensorimotor systems (Klasen et al., 2012). The complexity of real-time flow processes is reflected in the shifts between attentional, emotional, and physiological activation patterns (Klasen et al., 2012).

Performing arts are challenging activities; they are also stressful for many performers. Learning to modulate the stress systems (sympathetic nervous system and HPA axis) will heighten flow experiences (Peifer et al., 2014). Despite the subjective experience that flow is a state of ease, the complex neurophysiological demands require high metabolic output. Performing is a high-energy activity; to be vulnerable and expressive is neurologically and physiologically costly (de Haas, 2013). Ultimately, flow states are effectuated by heightened neuromuscular and cardiovascular activation, focused and flexible cognitive appraisal, positive emotional valence, and sustained motivation. The capacity to perform optimally can be trained, both physically and psychologically (Hallett & Hoffman, 2014). Performing artists generally work in groups, and these groups have the possibility of amplifying flow experiences. Future studies in the performing arts should consider assessing group flow states (networked flow), rather than just focusing at the individual level (Chirico et al., 2015). Performers frequently experience heightened flow states; they derive meaning from these experiences, especially when their skills match performance demands (Robb & Davies, 2015).

Section III

Development of the Performing Artist

Early development

9

I always had a repulsive need to be something more than human. I felt very puny as a human. I thought, Fuck that. I want to be a superhuman.
David Bowie

Introduction

Human development is a complex interplay between biology, psychology, and society (Newman & Newman, 2015). Fundamentally, development happens throughout the lifespan, and is shaped by epigenetics (gene and environment interaction) and by active individual contributions that influence development. When evaluating human development, the focus can narrowly assess such things as motor skill acquisition, language development, cognitive and intellectual maturation, height and weight gain, competency to regulate emotional responses, and cultural assimilation (Newman & Newman, 2015). Each subsystem offers a glimpse into the maturation of an individual, including whether that individual is progressing at an accelerated rate or is developmentally delayed (Bornstein & Lamb, 2005). Further, subsystems do not always mature in tandem, as is evidenced when a child may walk at an early age but have language skill delays, or an adolescent may show signs of uneven physical growth such as a nose that grows faster than the rest of the face. Despite these particular anomalies in development, there is a relative window of time in which specific sensorimotor and cognitive abilities emerge, and all are dependent on ongoing interactions between the genetic transcription of information that guides maturation coupled with optimal experiential interactions with the social and physical environment (Newman & Newman, 2015; Tsekhmistrenko & Vasil'eva, 2001). For human development, the interaction with the environment includes the essential relationship with primary attachment figures. The nature of attachment experiences and how they shape development will be discussed in detail in Chapter 11, Attachment, Parenting, and Childhood Adversity; the focus of this chapter is to investigate the early developmental stages of performing artists.

Any discussion that focuses on talented/gifted individual development inevitably addresses the challenges experienced by creatively gifted people and those who support their development. Talented children tend to develop outside the parameters of the typical stages of development (Mika, 2005; Roazen, 1976; Tannenbaum, 1992). Their unique talents set up a cascade of uneven developmental landmarks; consequently it is more difficult to predict developmental sequencing and more challenging to offer supportive developmental environments (Miller & Cohen, 2012; Morelock, 1996; Subotnik et al., 2011). For example, they are often

Creativity and the Performing Artist. DOI: http://dx.doi.org/10.1016/B978-0-12-804051-5.00009-3

misdiagnosed with some form of psychopathology such as attention deficit hyperactivity disorder or anxiety (Wellisch & Brown, 2013). This is more likely to happen if the child is endowed with the trait of psychomotor overexcitability or emotional/sensual overexcitabilities, common traits exhibited in talented children who are drawn to the performing arts (Ackerman, 2009; Alias et al., 2013; Mika, 2005). During early development, their unique talents need to be recognized and educated so that their developmental trajectory is not thwarted by lack of support or encouragement (Alias et al., 2013; Piirto, 1995). For example, dance students frequently respond with heightened psychomotor and emotional intensity; whereas, musicians typically engage increased sensual responses (Chang & Kuo, 2009). These children manifest unique creative-productive giftedness that has its own developmental trajectory (Renzulli, 2005). All these variants need to be factored into an understanding of the typical stages of development and into the family support system and educational curriculum that nurtures their development (Fornia & Frame, 2001).

Studies have demonstrated that optimal early support of talented performing artists resides within family structures. Complex family systems provide a balance between establishing ethical and material limits with ready support of the individual's domain-specific interests (Gute, Gute, Nakamura, & Csikszentmihalyi, 2008). In these families, adequate time was spent together, along with alone time where free exploration or deliberate practice took place. Complex families accepted failures and yet encouraged the performing artists to continue exploring and pursing their talents. These families provide opportunities to build skills to meet challenges, while encouraging a growing sense of autonomy and independence against a background of ongoing support (Gute et al., 2008).

Arts educational programs, although ideally tailored to meet individual talent development (Morelock, 1996), incorporate curriculum that influences the maturation of creativity, especially factors such as knowledge, problem identification, problem solving, evaluation, and contextual awareness of the talent domain (Raston, 1997). Cognitive and affective processes mature throughout development, consequently enhancing all creative skills including divergent thinking abilities (Claxton, Pannells, & Rhoads, 2005). Awareness about the unique above-average development of specific talents helps educators assess and identify talented children. These children are typically strongly committed to their domain-specific interests, often stating that it is easy for them or they learn quickly in this domain. Behaviors such as independently generating new ideas, whether it is a new musical phrase, a song lyric, or a new dance combination are frequently displayed by these burgeoning performing artists. And these talented children also evaluate their novel ideas; they are actively engaged in the biphasic creative process without formal instruction on how this operates (Rasmussen & Rasmussen, 2015; Renzulli, 2002). Further, at a very early stage of development talented children begin to establish a sense of identity that is tied to their domain-specific interests (Kogan, 2002). Musicians, dancers, and actors claim that their early interests were frequently directed toward career aspirations (MacDonald & Wilson, 2005); with early support these interests can be reinforced (Rasmussen & Rasmussen, 2015).

Erik Erikson's psychosocial model of development

To organize this massive topic, Erik Erikson's psychosocial developmental model will provide a theoretical framework to illuminate unique features in each stage of early development. Embedded within these developmental stages is the cognitive development outlined by Piaget (Newman & Newman, 2015). Piaget's theory is closely aligned with Erikson's stages of development (see Box 9.2). The motor, language/communication, and cognitive milestones are also integrated into Erikson's staged model (see Boxes 9.1, 9.3, 9.4). For the purpose of this chapter, only the

Box 9.1 Cognitive Milestones (Piaget, 1951)

Infancy: 1–2 years	Sensorimotor learning modality, exploring world through tastes and touch, sensorimotor representational schemes formed, object permanency
Toddlerhood: 2–4 years	Piaget's preoperational thought (internalize sensorimotor schemes into semiotic thinking/symbols), able to talk about feelings and express thoughts and ideas to others (including some ability to speak about intense angry feelings), awareness of distress in others, aware they have self-agency and desire to direct others' behaviors, able to use language in goal-directed desires
Early childhood: 4–6 years	Understand gender, gender constancy, and gender schemes; early moral development emerges (judgments based on reward and punishment, consequences affecting being loved, approval or disapproval from authority); empathy awareness increasing, friendship groups formed
Middle childhood: 6–12 years	Perspective taking and cognitive flexibility growing; aware of social norms and peer pressure; awareness of friendships, loneliness, and rejection; Piaget's stage of concrete operational thought; aware of physical matter as stable and no longer magically changing; aware actions can be reversed; increased awareness of categorization and computational skills; metacognition emerges (thinking about thinking); domain-specific skill strengths and weaknesses clearly manifested; reading literacy and fluency rapidly developing; social and cultural contexts assimilated through reading; aware of social and parental pressures
Early adolescence: 13–18 years	Piaget's formal operational thought; approach problem solving with probabilistic thinking; able to mentally manipulate more than two categories simultaneously; draw upon many variables to explain behavior; think about things changing in the future and aware that things will not remain the same; predict logical outcomes based on behavioral actions and choices; detect logical inconsistencies and distressed by statements that contradict reality; can think in relativistic terms about self, other individuals, and the world; fully aware of cultural norms and differences (including group norms, boundaries, and identities)

Box 9.2 Erikson's Psychosocial Developmental Stages (Erikson, 1959/1980)

Infancy: 0–24 months	Trust versus mistrust
Toddlerhood: 2–4 years	Autonomy versus shame and doubt
Early school age: 4–6 years	Initiative versus guilt
Middle childhood: 6–12 years	Industry versus inferiority
Early adolescence: 12–18 years	Group identity versus alienation
Late adolescence: 18–24 years	Individual identity versus identity confusion
Early adulthood: 24–34 years	Intimacy versus isolation
Middle adulthood: 34–60 years	Generativity versus stagnation
Late adulthood: 60+ years	Integrity versus despair

Box 9.3 Motor Milestones (Newman & Newman, 2015)

Infancy: 0–24 months	Raises head, brings hands to mouth, grasps and shakes toys, rolls both ways, sits with and without support, supports weight on legs, reaches with one hand, transfers objects from hand to hand, gets to sitting position without help, crawls forward on belly, assumes hand-and-knee position, gets from sitting to crawling position, pulls up to stand, walks holding furniture, walks alone, pulls toys behind while walking forward, carries large toys or several toys, begins to run, stands on tiptoe, kicks a ball, climbs onto furniture, walks up and down stairs
Toddlerhood: 2–4 years	(2–3 years) Walks rhythmically with opposite arm and leg swing, jumps down from a step, jumps into the air, hops 1–3 times on same foot, throws ball with forearm extension only (feet remain stationary), catches with rigid arms, pushes a riding toy with feet, walks up stairs (3–4 years) Walks upstairs alternating feet, walks downstairs leading with one foot, jumps with coordinated arm action, broad jumps about 1 foot, hops 4–6 times on same foot while coordinating upper body and nonhopping leg, throws ball with slight body rotation but little or no torso rotation or transfer of weight, flexes elbows in anticipation to catch a ball, but catches by trapping ball against chest, steers and pedals tricycle
Early school years: 4–6 years	(4–5 years) Walks downstairs alternating feet, runs more smoothly, gallops and skips with one foot, improved upward and forward jumps and covering greater distance, throws ball with increased body rotation and some transfer of weight forward, catches ball with hand (sometimes trapping ball against chest), rides tricycle rapidly, steers smoothly (5–6 years) Increased speed of run; gallops easily and skips; increasing jumping and broad jumping; hops rhythmically, changing feet and rhythm patterns; mature throwing and catching pattern; adjusts body to accommodate pathway and size of ball, rides bicycle with training wheels
Middle childhood: 6–12 years	Domain-specific skills clearly identified and valued by child (prior to this stage, talented children present precocious abilities)

Box 9.4 Language and Communication Milestones (Newman & Newman, 2015)

Infancy: 0–24 months	Smiles as greeting; startles when hearing loud sound; makes cooing sounds; quiet attentive listening; recognize voices and faces; cries differently to signal needs; makes gurgling sounds; babbles repetitive syllables; uses voice to express pleasure and pain; eye tracking to follow sounds and objects; responds to tone of voice; pays attention to music and toy sounds; tries to imitate words; says a few words; understands words and increasingly able to speak them (50–180 words); understands "no"; turns and looks in the direction of sounds; points to objects and pictures and tries to name them; recognizes people, objects, and body parts; follows simple directions and gestures
Toddlerhood: 2–4 years	(2–3 years) Speaks between 50 and 300 words, enjoys listening to stories, uses some adjectives to describe environment and able to name everything in the environment, uses 2–3 word sentences, comprehension by others unfamiliar with child still problematic (3–4 years) Vocabulary of 500–1000+ words; speaks in 3–4 word sentences, and by 4 years able to speak in full sentences; answers simple questions and others can understand the child; bilingual acquisition limits number of words within each language; comprehension for multiple languages slower but this evens out in early childhood
Early childhood: 4–6 years	Language acquisition continues to expand, early stages of reading and mathematical comprehension
Middle childhood: 6–12 years	Rapid growth in reading and mathematical skills
Early adolescence: 13–18 years	Complex abstract concepts expressed in language and mathematics

early developmental stages will be outlined; the later adult stages will be addressed in the discussion on careers.

Erik Erikson proposed a psychosocial theory to explain the trajectory of human growth. This theory was based on an interaction between the psychological maturation of the individual and the societal context they are raised within (Newman & Newman, 2015). According to Erikson (1959/1980), each developmental stage is marked by psychosocial crises that necessitate some form of resolution; the individual is then propelled into the next phase of development. Erikson outlined eight major stages of development throughout the lifespan; however, he also claimed that if a particular stage was poorly navigated, the individual could address these deficits and complete the developmental tasks of that stage later in life. For example, if an infant is raised in an environment that is inconsistent and unreliable, an abiding belief forms that the world and the caregivers in the world are untrustworthy. This belief also influences a sense of self. All become untrustworthy. This first developmental stage, according to Erikson's theory, can be repaired in later stages of development. In this example, the infant who completed the first stage of development with a solidified mistrust in self, other, and world, can, as a young child, adolescent, or adult, experience relationships, education, or career stability that cultivates growing trust in the world and the self. Erikson's theory offers developmental landmarks and yet there are always possibilities to change; he viewed humanity as fundamentally resilient and able to adapt to the world and circumstances surrounding them (Erikson, 1959/1980; Newman & Newman, 2015).

According to Erikson, the first stage of development, trust versus mistrust, describes the infant and young toddlers' task as one of forming a sense of trust about the burgeoning self and the world surrounding the self. In the second stage, autonomy versus shame and doubt, the toddler acquires a sense of independence, while in the third stage, initiative versus inferiority, the child between 4 and 6 years of age learns to interact in school and group environments, promoting a greater sense of purpose and self-identification. In middle childhood, the fourth developmental stage, industry versus inferiority, education and competencies are highlighted. The fifth stage of development, group identity versus alienation, defines the psychosocial tasks of early adolescence, one that addresses the need to solidify and stabilize peer interactions and form a sense of group identity. The sixth stage of development, individual identity versus identity confusion, takes place during late adolescence and early adulthood, a time when a sense of self is clearly defined and guides all future plans and decisions. If these stages are unsuccessfully traversed then the negative developmental task is confirmed. For example, toddlers who are unable to establish a sense of autonomy are left feeling doubt; a growing self-conscious awareness that others may reject the very essence of their being fills them with intense internalized shame (Budden, 2009; Newman & Newman, 2015; Schoenleber & Berenbaum, 2012).

Prenatal development and epigenetics

Although Erikson did not include prenatal development in his psychosocial development theory, it is now regarded as one of the most critical periods in human development. During this earliest stage of development, the prenate (a term that encompasses the developing embryo and fetus) coexists with the mother, sharing her world, whether it is full of pleasure or agony (Thomson, 2004, 2007, 2012a). Her lived experiences influence the physiological maturation of her prenate. If she is physically (interoceptive) or psychologically (exteroceptive) stressed it is communicated to her prenate, consequently a biological stress response is mobilized in the prenate (Porges, 2001, 2004). Literally, a predictive adaptive response, or survival phenotype, prepares the prenate's physiology to survive in the world into which it will be born, for example into a world of chronic poverty, war and political upheaval, domestic violence and abuse, or unremitting maternal depression, anxiety, or psychosis (Rickard &

Lummaa, 2007). The stress axis (hypothalamic–pituitary–adrenal axis), growth axis (hypothalamic–pituitary–thyroid axis), and other cortical and subcortical brain regions are modified to prepare the prenate to enter a world filled with stress or one that offers security and ease (McDougall, Widdop, & Lawrence, 2005). Likewise, the developing organs, bones, and muscles grow in response to this predicted environment (Rickard & Lummaa, 2007). This early prenatal pattern of development sets in motion a physiological, and ultimately psychological, capacity to regulate stress. During pregnancy, the prenate builds a physiological foundation of neuroendocrine, immune/inflammatory, and cardiovascular processes essential for stable self-regulation (Card, Levitt, Gluhovsky, & Rinaman, 2005; de Weerth, van Hees, & Buitelaar, 2003). Postnatally this regulatory predictive set-point can be somewhat modified both positively and negatively by future experiences encountered throughout the lifespan (Luo et al., 2006). Although some stress may increase resilience during prenatal development, chronic and intense stress is deleterious, and this reality remains true throughout the lifespan (Fujioka et al., 2001).

As a result of the significant research findings on prenatal development, disease vulnerability and the capacity to regulate physiological and psychological responses are better understood. Discerning heritability is now more difficult to identify. Although soundly debated (Schlesinger, 2009), studies demonstrate higher incidences of psychological distress in artists (Kaufman, 2005); however, what is not known is whether these individuals were conceived in a chronic stressful environment, including whether their mother suffered unremitting depression, anxiety, or psychosis during pregnancy. Was her mental illness genetically inherited by her offspring or was it transmitted to her offspring? These are important background themes to consider when examining the early development of performing artists, especially given knowledge that the environment can be embedded into genetic expression via DNA methylation (Essex et al., 2013; Foilb & Christianson, 2016; Szyf & Bick, 2013). Despite these sobering findings, it has also been suggested that creativity is also genetically inherited (Piffer & Hur, 2014). Perhaps maternal stress is communicated to her offspring but her creative predilection may also be transmitted.

Trust versus mistrust

Returning back to Erikson's psychosocial model, his first stage of development, trust versus mistrust, includes the baby's expanding sensory/perceptual and motor skill acquisition. This is a major stage for brain development; the brain is undergoing massive neurological synaptic connectivity as myelination growth proliferates (Schore, 1994, 2003). It is fundamentally an extension of prenatal development. For example, reflexes and "fixed action patterns" (Llinas, 2001) emerged during prenatal maturation. These reflexes and fixed action patterns support postnatal survival and include the grasping reflex, which reinforces attachment behaviors; sucking and swallowing reflexes, which operate independently of respiratory mechanisms; stepping reflex in anticipation of walking; and the Moro or startle reflex activated in response to loud sounds or predatory behaviors. Complex emotional behaviors are present at birth and continue to mature through experience. Babies can signal distress and communicate pleasure, important skills that persist throughout life. During this time period, via sensorimotor exploration, babies rapidly understand the nature of objects, that they are permanent and do not simply disappear, and that they can be categorized into coherent models. This is evident in the rapid speed that babies comprehend different animal species, recognize different faces, and grasp intention and causality (Newman & Newman, 2015). Language and symbol processing correspond with enhanced memory, attention, processing speed, and representational awareness.

Although discussed later, attachment to primary caregivers form and consolidate during this critical period (Sroufe, Egeland, Carlson, & Collins, 2005). Temperament, often a strong hallmark for personality traits, is present at birth (see chapter: Personality and Motivation for details) and is clearly expressed as the baby gains more sensory/perceptual and motor skills. Emotions rapidly differentiate along with the ability to socially reference the world through observing and responding to the reactions of others (Schore, 2003). Cultural and subcultural behaviors and beliefs are internalized and assimilated by babies; by the end of this stage of development, babies can discern familiar versus "foreign" foods, sounds, and family practices (Newman & Newman, 2015). This rapid first stage is shaped by an ongoing coordination between caregivers and babies. Failed attempts at soothing or engaging the baby are optimally augmented by rapid efforts to repair the mismatch. What emerges is a sense that the world is reliable and inviting; however, if the repairs are not adequate or they are too slowly administered, the baby loses a sense of hope and withdraws in anticipation of future disappointments.

During this early phase, giftedness and domain-specific talents emerge and preliminary identifications can be postulated (Tannenbaum, 1992). Sensitive caregivers observe their infants attending to specific objects or activities with greater intensity and interest. They can facilitate an expansion of these interests, including providing opportunities to allow the infant to explore through free-movement play sessions (Hammond, 2009). Further, they can enhance language acquisition around these interests by naming them. This early awareness and support is especially evident within families who have multigenerational family members who display talents and/or giftedness (Perrone et al., 2010). Many infants respond with active interest to music and movement, and for those with inherent talents in these domains, their interests frequently persist throughout development (Tannenbaum, 1992).

Babies are fully dependent on a social support system to ensure survival. Parents, along with family and community members, provide babies with either a safe and trusting environment or one that is unsafe and frightening (Sroufe et al., 2005). The temperament of the baby can ameliorate or exacerbate these lived experiences. Emerging giftedness and talents can also influence these early experiences. It is a complex interaction between babies and their environment that influences the successful or failed trajectory of this early first stage of development (Kagan & Snidman, 2004; Tannenbaum, 1992; Thomas & Chess, 1977).

Autonomy versus shame and doubt

In Erikson's second stage of development, autonomy versus shame and doubt, the developmental tasks for the toddler (2–4 years of age) are to expand motor coordination, including more stable and efficient locomotor patterns as well as muscular regulation of urinary and bowel movements essential for removal of diapers. With greater motor control, the child's expressive language skills markedly improve from a few words that only close caregivers understand to full sentences that all can understand. In conjunction with these motor skills, cognitive abilities also shift to more symbolic and representational thoughts (Newman & Newman, 2015). Increased fantasy and pretend play emerge, essential components that help children practice and improve emotional communication and regulation.

Greater socializing behaviors are introduced, a process that can introduce intense social shame if discipline is harsh and alienating (Erikson, 1959/1980). Optimal parenting helps children understand the need for discipline as a way to keep them safe while still encouraging exploration. All these skills facilitate a growing sense of autonomy, a sense of separateness while still maintaining the essential attachment relationships that ensure survival. If children are harshly disciplined, shame escalates and doubt increases. They will only feel safe in highly structured environments and if clear parameters are not established they feel insecure and doubtful (Newman & Newman, 2015). This negative psychological state increases internal stress that can manifest as internalizing disorders (anxiety and fear) or externalizing disorders (behavioral and emotional labiality and volatility) (Shirtcliff, Granger, Booth, & Johnson, 2005; Sroufe et al., 2005).

Initiative versus guilt

The third stage, initiative versus guilt, transpires during early childhood (4–6 years). This stage is marked by a growing awareness of the interpersonal world that surrounds children (Erikson, 1959/1980). Although they still engage in magical thinking their external world begins to become more structured and they acquire a greater sense regarding how they fit into this world. They recognize that they will remain a girl or boy for life and they understand what it means to be good or bad. A growing sense of morality stabilizes, including awareness of caring for others, empathic understanding, and perspective taking (Newman & Newman, 2015).

During this stage of development, children who are drawn to the performing arts will realize that others are watching them dance or play music. This scrutiny by others, although intimidating, does not thwart these children; whereas, other children with less intense domain-specific desires often withdraw from further exploration (Chaplin & Norton, 2014). During this phase of development, children acquire greater competence in motor, language, cognitive, and social skills and they are able to use these skills to motivate and initiate plans. They can collaborate within group activities and create games and pretend play fantasies that enhance their sense of efficacy. The negative correlate to these new skills is the awareness of responsibility for actions and the feeling of guilt when they have failed. Children in this stage of development feel remorse and have a great desire to repair mistakes or inappropriate behaviors. During the initiative versus guilt stage, children gain a sense of purpose and learn how to inhibit impulses that will interfere with their goals (Erikson, 1959/1980).

Industry versus inferiority

The fourth stage of development, industry versus inferiority, between the ages of 6 and 12 years, is a period when most children who express an interest in the performing arts are solidly identified. Prior to this stage, their interests could be observed by their self-selected play within their chosen domain, including possible participation in classes to support their interests. However, with the developmental changes that occur during this fourth stage, children in the performing arts rapidly mature and their domain-specific talents are abundantly clear (Rasmussen & Rasmussen, 2015). The overarching developmental tasks of cultivating friendships and enhancing formal educational skills are facilitated by the cognitive shift from preoperational thinking to concrete operational thought, a capacity to comprehend categorization, computational skills, reversal of concepts, and metacognition (Newman & Newman, 2015). During this stage, children interact in diverse social groupings and appreciate the cultural significance of these groups. They understand social norms and expectations including feeling the support from close friends and the emotional sting when bullied, rejected, or isolated.

According to Erikson (1959/1980), at this stage of development, the acquired skills enhance intrinsic motivation and draw children closer to adult behaviors. They experience a sense of industry via growing cognitive, behavioral, and emotional competencies. They can self-direct behavior as they move toward intrinsically motivated goals, skills that are strongly implemented by young performing artists (Wellisch & Brown, 2013). However, if they falter, as a result of lack of parental, peer, or educational support, then a growing sense of inferiority develops (Newman & Newman, 2015). They have the capacity to understand that they are falling behind, and with a sense of inferiority, a pattern of "learned helplessness" emerges. Children who do not successfully meet the challenges during this developmental stage believe that they have no ability to change their destiny; they perceive control residing outside of them (Seligman, 1975/1992; Seligman, Steen, Park, & Peterson, 2005). These children generally have great difficulty regulating their emotional reactions, often evident in disruptive behaviors. Not surprisingly, failure to traverse this developmental stage compromises creative maturation as assessed in divergent thinking skills. The "learned helplessness" response and their inability to inhibit disruptive behaviors significantly limited their creative responses (Butcher & Niec, 2005).

During this developmental stage, parenting styles can promote creativity or augment loneliness. When parents accept children with a passion for the performing arts, they enhance creative expression and ameliorate any feelings of loneliness that might be experienced among the children's peer groups (Lim & Smith, 2008). These parents can successfully support the child's initiative to pursue talent domain activities. School programs are no different than parental support. During this stage of development, educational institutions can reliably assess and predict creative and artistic potential in children (Oreck et al., 2003). Ideally these identified children can be placed in rigorous and accelerated arts education programs, whether that occurs during the daily school curriculum or as an after-school program.

Both parental support and school programming can enhance talent development. For example, 9-year-old students who expressed interest in acting had already self-identified their interests via actively engaging in role play and pretend play, making up performances and playing dress up, and these behaviors were evident as early as 5 years of age (Goldstein, 2015). Parents enrolled them into acting programs at the age of nine based on these observed behaviors and the child's request. For children with a keen interest in dance, music, or acting, encouraging them to participate in these domain-specific activities during this developmental stage enhances their creative abilities, and this persists throughout life (Kogan, 2002: Pagona & Costas, 2008). In fact, engaging in motor creativity also augments the sturdy formation of self-concept, a factor that is integral to a sense of industry (Bournelli, Makri, & Mylonas, 2009). When nurturing talent during this developmental stage, it is not only important to provide unstructured opportunities to creatively explore movement self-expression, it is equally advantageous to provide a balance with structured organized activities that facilitate skill mastery. Both enhance creativity and this will persist into adulthood (Bowers, Green, Hemme, & Charlip, 2014). For young performing artists, during this developmental stage, it is advised that they be given equal opportunities to move creatively while also receiving rigorous training in their specific art form.

Group identity versus alienation

In the fifth stage of development, group identity versus alienation, adolescents between the ages of 12 and 18 years experience massive biological and psychological changes. This stage of development is second only to the period of prenatal and early infancy for neurological changes and resculpting of the brain (Spear, 2000; Sturman & Moghaddam, 2011). Puberty is a term that only defines changes of the physical secondary sexual characteristics. Hormonal changes affect the brain and the body, and all influence psychological responses. Adolescents must manage changes in circadian rhythms that influence wake–sleep cycles, eating behaviors, plus rapid and uneven growth spurts that alter balance, strength, flexibility, and coordination (Krasnow & Wilmerding, 2015; Spear, 2000). Early-onset puberty differentially affects gender self-esteem, with males experiencing greater self-esteem whereas females suffer a loss of self-esteem (Newman & Newman, 2015; Spear, 2000).

Because of the significant neurological changes during this time period, adolescents experience a decrease in dopamine, a finding that may promote increased risk taking, novelty seeking, and exploration away from the natal family unit (Spear, 2000; Sturman & Moghaddam, 2011). This significant change can propel adolescents toward greater exposure to competition, whether that is in sport or the performing arts, but it can also nudge them toward criminal behavior, including increased drug abuse (Casey & Jones, 2010). Along with changes in dopamine production, the autonomic nervous system is in a transitional state, hence adolescents physiologically respond paradoxically to many medications and illicit drugs (Spear, 2000). They are also more vulnerable to increased sympathetic activation and less vagal regulation from the parasympathetic branch, a fact that renders adolescents vulnerable to sudden death as a result of tachycardia (induced by either drugs or environmental stressors) (Spear, 2000).

Emotionally, adolescents are also biased toward interpreting events as more negative; their limbic system, especially the amygdala, and their interoceptive system directed by the insula cortex and anterior cortex, all respond with more activation when appraising social threat (Monk et al., 2003). These neurological changes may counter the urge for risk taking; however, for adolescents who are prone to increased anxiety, this compounds their difficulties navigating group identity. Often these adolescents feel alienated from their peers, a reality that adds emotional pain given that this is a time period in which they are driven to spend more time with their cohort (Spear, 2000). As they explore these relationships, sexuality, sexual orientation, and dating take center stage (Newman & Newman, 2015). Cognitively, adolescents acquire an ability to formally operationalize thought. They have the cognitive capacity to comprehend abstract reasoning and probabilistic thought. They can manage complex thought with relative ease, and under nurturing environments they can incorporate complex self-reflective thought (Casey, Jones, & Hare, 2008; Newman & Newman, 2015). Unfortunately, along with the maturation of cognitive thinking, they must also manage emotional responses that are experienced more intensely than in adulthood or when they were in the earlier stages of development (Casey, Jones, & Somerville, 2011; Spear, 2000). This imbalance makes adolescents vulnerable for increased psychopathology such as eating disorders, anxiety, depression, and psychosis (schizophrenia), and they are at increased risk for suicidality (Newman & Newman, 2015; Romeo, 2010; Fig. 9.1).

Individual identity versus identity confusion

The final stage of early development discussed in this chapter, individual identity versus identity confusion, takes place during late adolescence (18–24 years). Based on substantial brain research, it is believed that early adulthood does not begin until after the brain is fully matured, a time period occurring after the age of

Figure 9.1 "Identity." Performers: CSUN Dancers. Photographer: Lee Choo.

24–25 years (Casey et al., 2008; Spear, 2000). This late adolescent time period is marked by greater individual determination to realize goals augmented by increased legal rights and privileges. It is also a time period when career paths are explored, although often not fully acknowledged or solidified. Greater autonomy and self-sufficiency is attained and a deeper understanding regarding cultural norms and values are examined and either accepted or rejected in accordance to a burgeoning sense of an adult identity (Newman & Newman, 2015). The earlier need to identify with peer groups shifts to identifying with individual values and desires. These are weighted more than group norms. This process can prove challenging if individual identity status runs counter to family or cultural beliefs. For example, this is a period when sexual orientation is fully acknowledged; many struggle with concerns about acceptance or rejection from family, church, and/or community. Others may struggle with career choices that contradict family beliefs. Many performing artists struggle with a desire to embrace a career in the performing arts, despite financial instability and disapproval from family members. During this phase of development, increased ability to manage stress is evident, especially when balancing educational, financial, and personal needs. If these challenges are not met, the late adolescent enters adulthood without a clearly defined sense of individual identity. Consequently, early adulthood becomes substantially more confusing, in particular, there is greater difficulty in defining social roles, establishing intimate relationships, and engaging in meaningful careers (Newman & Newman, 2015). If successfully navigated, Erikson's sixth psychosocial stage of development clearly prepares the individual to enter adulthood.

Figure 9.2 "Constraint or freedom." Performers: Sahara Ahal, Ashley Campbell, Tiffany Davis, Rene Garcia, Kelsi McRee. Photographer: Lee Choo.

Along with individual identity consolidation, the impulse to seek a mate potentially enhances creative productivity, although gender differences exist. Men tend to work more creatively regardless if the romantic relationship is potentially long- or short-term; whereas, women tend to display greater creative productivity when they are in safe long-term relationships (Griskevicius, Cialdini, & Kenrick, 2006). This gender relationship pattern is developmentally normal during late adolescence, although it often persists into adulthood. Clearly, establishing intimate relationships influences career stability but also career stress for many performing artists. This topic will be addressed in later chapters that discuss attachment formation and career development (Fig. 9.2).

Creativity development through the lens of other theorists

Vygotsky's theoretical developmental model significantly influenced educational practices, offering greater appreciation for sociocultural influences on maturation. According to Vygotsky, creativity matures in tandem with the cognitive development of the child, although the origins of creativity can be seen within early childhood play activities (Ayman-Nolley, 1992). During adolescence, the emergence of creativity fully manifests, in part because the adolescent has greater control over abstract thinking and access to more sensorimotor memories that coalesce to shape complex fantasy constructs. During adolescence, richer and deeper creative expression is influenced by the capacity to fully implement the biphasic generative and evaluative creative processes, or in Vygotsky's language, imagination and reason (Ayman-Nolley, 1992).

Based on Vygotsky's belief that talent is supported by an interaction between talented individuals and the sociocultural environment that they reside within, the talent development model was proposed (Morelock, 1996). It highlights the need to identify talent at a young age and promote this talent during the formal educational years. Efforts to nurture talent require accommodating each individual's pace for skill acquisition, including skills in emotional regulation and expressivity, social interactions, motor abilities and demands, and cognitive maturation of abstract thought and problem solving. Following Vygotsky's recommendation, educating talented artists requires educating and assimilating sociocultural symbol constructs, along with physical and psychological skills. All these inform the inner construction of creative products. As the individual moves through the developmental stages, creative products are expressed in an ever-growing complexity. As stated earlier, talented individuals manifest a precocious and rapid growth pattern in their domain-specific talent(s), especially if they are given educational opportunities to master sociocultural symbols at their own pace (Tannenbaum, 1992). The interface between a supportive environment and the child's internal desire to gain mastery becomes the timekeeper for their development (Morelock, 1996).

Research studies that use divergent thinking tasks to determine developmental stages of creativity demonstrate that creativity generally matures concurrently with cognitive and affective maturation (Claxton et al., 2005; Urban, 1991). This pattern is strongly evident when assessing children in fourth, sixth, and ninth grades (Claxton et al., 2005). Despite this consensus, some studies suggest that creative maturation doesn't progress in a linear direction, nor is it equal with gender (Lau & Cheung, 2010). This may be related to the fact that children become more constrained in divergent thinking tasks during fourth grade, perhaps because during preoperational cognitive development children are more concrete in their thinking (Charles & Runco, 2000–01). The magical thinking that existed earlier and the later ability to engage more abstract thought are not evident in fourth grade. But by ninth grade both boys and girls demonstrate similar creative abilities and creative development continues to progress in a linear pattern (Lau & Cheung, 2010).

Because the sociocultural lens is critical for understanding the early development of creativity, it remains disturbing to realize that this same lens reveals societal inequalities. To date, white upper-middle-class populations are disproportionately represented when identifying creativity, including early performing arts talents (Gaztambide-Fernandez et al., 2013; Rostan, Pariser, & Gruber, 2002). Contributing to this disparity is the reality that different cultures advocate parental behaviors or cultural values that negatively affect the development of creativity, such as promoting authoritarian controlling parenting styles or encouraging submissive, quiet behaviors in children (Fearon, Copeland, & Saxon, 2013; Lim & Smith, 2008). Further, gender differences within the arts persist (Miller & Cohen, 2012), with more females dominating dance whereas more males dominate music (Kogan, 2002). Unfortunately, talent development is still contingent on economic means, especially because nurturing talent requires capital investment (Rasmussen & Rasmussen, 2015).

Promoting development in the performing arts requires a keen awareness of the domain-specific skills necessary for a successful career and the ability to create programs that promote successful attainment of these skills (Subotnik et al., 2011). Also, talented children do not all look the same, nor do children in domain-specific talents develop at the same rate (Subotnik et al., 2011). For example, dancers and musicians tend to participate earlier than actors in their specific art forms, although that is not a categorical given. Many successful performers can begin their training much later. Once again, depending on family, financial, and educational opportunities, performing artists develop at their own unique pace. It is also important to know that performing artists do not all present the same way during development. Some are immediately identified as distinguished and highly gifted, some are quiet and compliant, some are precocious and manifest multitalented versatility, and some are industrious and focused on high-level achievement (Rasmussen & Rasmussen, 2015). Informing parents and educators about the myriad factors that determine talent identification can help young performing artists realize their potential.

Summary

- Introduction: comments about developmental theory and the development of talented individuals, epigenetics, optimal early development (parental and educational support).
- Erik Erikson's psychosocial model of development: general outline and integration of Piaget's cognitive stages of development, recursive process.
- Prenatal development and epigenetics: fetal development, effects of fetal stress on later development, physiologic set-points for environmental stress, DNA methylation.
- Trust versus mistrust: physiologic and psychological development of infant, influences of early attachment, early identification of talent and giftedness.
- Autonomy versus shame and doubt: 2–4 years of age, physiologic development (motor, sensory, language), cognitive representational thoughts, fantasy play, socialization.
- Initiative versus guilt: 4–6 years of age, increased awareness of interpersonal world, increased skill acquisition (language, motor, cognitive), sense of purpose.
- Industry versus inferiority: 6–12 years of age, talent clearly evident, intrinsic motivation enhanced, learned helplessness versus self-efficacy, parental and school support.
- Group identity versus alienation: 12–18 years of age, puberty versus adolescence, alterations in sleep-wake cycles, eating cycles, and neurological pleasure responses.
- Individual identity versus identity confusion: late adolescence and early adulthood, defining social roles and sexual orientation.
- Creativity development thorough the lens of other theorists: Vygotsky, talent development model, gender, age, complex factors influence development.

Concluding remarks

Engendering talent is critical for the burgeoning performing artist. The family, educational systems, and society at large need to support and understand talent development (Miller & Cohen, 2012). Offering support also includes facilitating interactions with other developing performing artists (Chua, 2015). Group inclusion positively influences self-efficacy and identity formation (Chua, 2015). It is essential that support networks recognize the myriad developmental tasks that young artists must achieve while being trained in their specific art domain. The challenge when working with talented children is addressing their uneven maturational journey. For example, a young pianist may be highly advanced in skills such as reading complex musical scores, motor dexterity to play advanced musical repertory, and depth of understanding to give emotional expression to the music, and yet this same child may be socially immature when interacting with his peer group, and demonstrate average developmental skill acquisition in other cognitive and motor activities. It is a combination of young artists, their families, schools, and cultural communities that facilitate talent development (Rasmussen & Rasmussen, 2015).

Play and theory of mind

10

Art is the only way to run away without leaving home.

Twyla Tharp

Introduction

Ideally, play begins early in development. The playful face-to-face interaction between caregivers and infants establishes a mutual experience of pleasure; together they build a bond of trust and joy (Bornstein & Lamb, 2005). During the first 2 years of life the baby explores the world through sensorimotor play, a process that enhances the formation of mental maps or schemes that will mature into symbolic representations. According to Piaget (1951), play allows children to explore, imitate, and rehearse; play is a form of cognitive assimilation (elaborating on a given scheme) and accommodation (altering the scheme to allow new schemes to form). Ultimately, play promotes the ability to flexibly adapt to environmental demands (Bornstein & Lamb, 2005).

During toddlerhood, children learn to engage in solitary play, parallel play (solitary play that coexists with another child engaged in solitary play), associative play (shares with others), and cooperative play (social play with defined roles and interactive sequencing) (Bornstein & Lamb, 2005). At this stage of development, children begin to engage in active role-playing, including the rules that emerge to support the role-playing activity and the imaginary situation that is acted out (Bodrova, Germeroth, & Leong, 2013). These situations may or may not include objects, but if they include objects then these objects can rapidly transform supporting symbolic functions required for the imaginary situation that is construed (Moller, 2015). To accomplish play tasks, children demonstrate an ability to follow the rules and restrain from breaking or ignoring them. Paradoxically, this form of free play actually contains many restraints that children readily impose and value. Not only are they playing, they are also building essential self-regulatory skills that will be implemented throughout life, particularly because the quality of their play matures in tandem with role-playing skill acquisition and cognitive, emotional, and sociocultural development (Ayman-Nolley, 1992; Bodrova et al., 2013).

Social development and educational support

Through shared play interactions young children begin to identify and show preference for specific friends and this pattern, although gaining more nuance and

Creativity and the Performing Artist. DOI: http://dx.doi.org/10.1016/B978-0-12-804051-5.00010-X

complexity, persists through life. In childhood, friend selection can also include imaginary companions, a phenomenon that is commonly recognized in children with high imaginative skills (Firth, Alderson-Day, Woods, & Fernyhough, 2015). These children are well aware that their imaginary companions are not real, but through their pretend play, they are able to navigate social and emotional problem solving with an imaginary other (Taylor & Mottweiler, 2008). As normal development progresses, shifts in playmate preferences occur, including a relinquishing of their imaginary companions. By early school age development, children typically show preference for same-gendered friends. Later in development, children gradually interact with different gendered playmates (Bornstein & Lamb, 2005). Although this is a general pattern, young migrant children do not demonstrate a gender bias in friendship selection during play activities. It is believed that gender role preferences are altered as a consequence of displacement stress during migration (Mathur & Parameswaran, 2015).

Unlike gender preference, children do not demonstrate patterns of seeking same-age playmates. Today's educational culture unfortunately continues to segregate children into age-specific cohort groups, and sometimes into gender segregation. This reality compromises child play development (Bodrova & Leong, 2015). According to Vygotsky, cognitive and creative maturation is promoted via a supportive environment coupled with opportunities that exceed the immediate developmental skills of the child. By interacting with older children during play, more developmentally complex cognitive, sensorimotor, social, and emotional skills are indirectly absorbed (Ayman-Nolley, 1992; Bodrova & Leong, 2015). Through scaffolding (support) and a zone of proximal development (modeling and exposure to developmentally more advanced skills), children assimilate skills that promote emotional, cognitive, and creative development (Newman & Newman, 2015). Sadly, many educational programs ignore the model of zone of proximal development proposed by Vygotsky. Young children are no longer able to learn from the modeling play behavior offered by older children. Consequently, play skill acquisition is more impoverished in today's children (Bodrova & Leong, 2015).

Although this is a general trend in educational systems, within the performing arts, children learn their domain-specific skills in a model that resembles Vygotsky's scaffolding and zone of proximal development. The long-practiced apprentice–master model persists within the performing artists (Bennett, 2008). Aspiring performers are routinely exposed to peers, older students, preprofessionals, and professional performers. Many programs actively integrate all age and skill levels into recitals, concerts, and performances. For example, the children's youth orchestra might be included in a mixed program that also features adult professional musicians. Young dancers annually participate in productions of the ballet, *The Nutcracker*. Theater, television, and film productions regularly cast performers in roles that may cross the full developmental lifespan. Major opera companies regularly include children as performers within larger chorus scenes. In the streets, young hip-hop and break-dancers are routinely mentored by older street dancers. This practice offers powerful modeling to young artists, including informal opportunities to be mentored and coached in their chosen art form (Fig. 10.1).

Figure 10.1 "Street training." Performers: Joseph Abesamis, Ryan Consebido, Timothy Hooker, Kenneth Miles, Clint Nilo. Photographer: Lee Choo.

Performing arts: worldplay and other talents

A child's preferred play activities frequently predict future careers in the creative arts, including the performing arts (Crozier, 2003). Fantasy proneness, a natural trait in performing artists (Thomson et al., 2009), facilitates vivid and intense imaginings that enrich make-believe play (Sanchez-Bernardos et al., 2015). These abilities are strongly related to personality traits marked by openness to experience and neuroticism, traits that are typical of many highly creative individuals (see chapter: Personality and Motivation for more details). One of the early signs indicating acting talent is a desire to live in alternate realities and inner worlds (Goldstein, 2015; Goldstein & Winner, 2009). This process of creating imaginative worlds helps shape causal reasoning skills, a process that enables a child to determine "what if" consequences in any situation (Gopnik & Walker, 2013). While exploring elaborate fantasy play worlds, young performing artists develop important executive functioning skills (Berk & Meyers, 2013). These young performers often spontaneously prepare productions that are performed to an audience of their siblings, peers, and/or parents (Goldstein & Winner, 2009; Piirto, 1998). Although their imaginative worlds may take on the potency of "worldplay," also termed "paracosms" (Taylor et al., 2015), young performers actively move these worldplays into performance reality. Most worldplays reside within the minds of the individual, an imaginative ability that may persist into adulthood. These elaborate constructs may be populated by fantasy cultures with their own languages and rituals, an imaginative process that can continue for days, months, or years.

Individuals who report childhood histories of worldplay fantasies are often recognized in adulthood as highly creative, as evidenced by a high proportion being honored with the MacArthur Fellows award (Root-Bernstein & Root-Bernstein, 2006). Worldplay constructs are always in the volitional control of the individual, unlike maladaptive daydreaming, an imaginative behavior that lacks the focused elaboration of a worldplay construct (Somer, Lehrfeld, Bigelsen, & Jopp, 2016). What is unknown to date is whether performing artists engage in worldplay fantasies more frequently than other populations; however, their early pattern of performing their imaginative scenarios suggests that they are driven to concretely realize these worldplays rather than allow them to remain quietly residing within their imagination.

Other identifying performing artist traits that emerge during early childhood include a keen ability to memorize language, movement, or sounds (Sawyer, 2006a). Young actors' natural love of mimicry and relative ease in memorizing text are all evidenced, even before formal training begins. Young dancers quickly assimilate movement sequences and can mimic older performers' dance styles and skills, and young musicians demonstrate a strong ability to replicate sounds based on hearing a musical piece performed. This can happen long before they learn to read music. Young performing artists are literally playing within the natural creative process of play-making (theater, dance, music) and play-performing. And while they interactively engage in performance play they acquire a growing awareness of the impulses and intentions of others who are confronted with obstacles within the imaginary scenes they create (Goldstein & Winner, 2009). Pretend play and role-playing is not only a childhood activity, it persists into adulthood and becomes a professional career for many. Performing artists naturally gravitate to pretend role-playing, constructing elaborate worldplays, which are then realized as performed play-productions (Piirto, 1998). Assuming an imaginary role, one that is elaborated and acted out, is a major indicator of childhood creativity, and small children drawn to the performing arts frequently participate in these spontaneous activities (Mottweiler & Taylor, 2014). Incorporating dramatic play into classroom settings promotes creative thinking, and more importantly, it provides young talented performing artists an opportunity to explore within the creative domain that they naturally embrace (Gupta, 2009). The integration of objects (imagined or real), adoption of role-playing, and interacting in a situation provides rich opportunities to exercise cognitive, affective, social, and cultural skill development (Bateman, 2014), a process that is intrinsic in the play of young performing artists.

For performing artists, pretend play not only foreshadows a career in the performing arts, it also enhances divergent thinking as well as increasing academic achievement, in particular math (Wallace & Russ, 2015). During childhood, opportunities to engage in imaginative pretend play promote better emotional processing. Children who easily engage in pretend play display greater comfort expressing a wider range of emotions, they are able to manage greater emotional expressivity, they easily construct stories that depict more interpersonal complexity, and demonstrate more problem solving solutions, especially when more distressed (Hoffman & Russ, 2012). With expert teaching, play skills can be enriched that further increase divergent thinking skills, academic achievement, creative interests, and emotional regulation (Moore & Russ, 2008; Russ, 2003; Russ & Wallace, 2013). Subjective wellbeing and happiness are strongly correlated to pretend play experiences in childhood and they tend to persist over time (Fiorelli & Russ, 2012). All these gains are directly rooted in environments that promote play exploration, such as parental support or educational programming that incorporates play experiences. Research demonstrates that these optimal supportive environments promote psychological wellbeing (Pearson, Russ, & Spannagel, 2008); a lack of play opportunities are associated with increased psychopathology (McKinney & Power, 2012; Singer & Singer, 2015).

Play benefits in the performing arts

Exploring play that is rooted in performing arts skills directly facilitates the emergence of self-regulatory behaviors (Zachariou & Whitebread, 2015). These behaviors include self-monitoring and controlling emotional, social, and motivational impulses, a process that demonstrates an ability to think about thinking (metacognition). Metacognition enables the child to reflect on behaviors and modify behavioral responses. For example, self-regulation is directly enhanced when children and adolescents participate in musical play (Lill, 2014). Making up songs, responding to rhythmic patterns, and perceiving harmonies and melodic phrases are all components of musicality, and musicality is a fundamental aspect of human functioning. Research studies have demonstrated the effectiveness of musical play and its relationship to increased self-regulation. It is present when children perform handclapping games, circle games, movement play, instrumental play, or a combination of musical play activities. During these activities, metacognitive knowledge (knowledge of persons, tasks, strategies), metacognitive regulation (planning, monitoring, control, evaluation), and emotional/motivational regulation (monitoring, control) are all enhanced (Zachariou & Whitebread, 2015). Perhaps this form of music play, one that is inherent in the performing arts, is what promotes self-regulation, including motivation and persistence, in these young artists.

The play of performing artists is both an orientation to the other players, whether that is young musicians playing together, actors creating scenes to perform, singers harmonizing and creating choral music, or dancers exploring ways to coordinate rhythmic and spatial elements essential for any dance performance. During play experiences these performers are oriented to each other and only secondarily are they concerned about an audience. Early play in the performing arts is a convergence between the imaginative and real worlds that exist both outside and within the performative realm (Marjanovic-Shane & White, 2014).

Theory of mind

The capacity to infer what someone else is thinking is fundamental for human survival (Bornstein & Lamb, 2005). By the end of the first year of life, infants can sense the mind of another and by 3 years of age they are able to predict another's behavior and can even engage in deception (Leudar, Costall, & Francis, 2004). For example, small children quickly learn that it is best to disavow any responsibility for an event if they detect that the other is upset. The ability to engage "theory of mind" is a rapid, and largely a nonconscious response. The reason it is called a "theory" is simply because inferences are made about the state of mind of another, including emotional states of being, and behavioral intentions. This capacity is sometimes referred to as "mind-reading" or "mentalizing"; it is essentially self-reflection while actively socially referencing and interacting with others (Leudar et al., 2004).

Theory of mind rapidly develops in early childhood, with elaboration and complexity merely enriching an inherently established capacity to understand the mind of another. This developmental trajectory is altered in children with developmental disorders such as autism or deafness (Wellman, Fang, & Peterson, 2011). When examining children raised in collectivist cultures versus individualist cultures, subtle cross-cultural sequential developmental patterns manifest; however, by 6 years of age no differences in theory of mind abilities are evident (Shahaeian, Peterson, Slaughter, & Wellman, 2011). Beyond cultural influences and special needs of children, one of the most influential factors that enhance young children's acquisition of theory of mind is an elaborated discursive interaction between young children and their mothers (Ontai & Thompson, 2008). Open communication, whether verbal or nonverbal, enriches the experience of the child and offers a context to understand behaviors and intentions. Gradually children assimilate shared experiences and they learn to elaborate on their own understanding about others' mental states. Regardless of cultural values, the early relationship between mothers and babies/young children influences the development of theory of mind (Ontai & Thompson, 2008).

One major concern related to the term "theory of mind" is that it implies a separation between mind and behavior, and yet perceiving intentionality in another is an interactive mixture of behavior and mind (Leudar & Costall, 2004). Through discourse, whether verbal or nonverbal, we are able to approximate an understanding about the intentions of another. This process of mentalizing is readily perceived between performing artists engaged in the creation or performance of a work of art, and between audience members and the performers. They rapidly sense the mind of another, with all the emotional, cognitive, and behavioral nuances and intentions that are communicated. It is a discursive interaction that operates in real time; the minds of others are sensed via a deeply social survival mechanism (Leudar & Costall, 2004).

Narrowing the parameters of theory of mind is difficult because it incorporates many mental processes vital for social interactions; however, theory of mind is generally regarded as something separate from empathy and perspective taking. Empathy is regarded as a capacity to sense (or match) another's emotional state (Goldstein & Winner, 2012) via a sensorimotor activation in an individual that "matches" the psychophysiological state of another. Perspective taking is a distancing process that enhances understanding about another's response within a specific

Figure 10.2 "Flamenco." Performers: El Nino Miguelito, Monica Ramos. Photographer: Lee Choo.

situation. It is believed that theory of mind may be a precursor to both empathy and perspective taking (Chaplin & Norton, 2014). Self-perspective, on the other hand, is the ability to represent a sense of self, a process that engages metacognition and metarepresentation. This important aspect of perspective taking about the self, and not of the other, allows us to engage a form of theory of mind about our own mind (Vogeley et al., 2001; Fig. 10.2).

Neurobiology and theory of mind

The universal ability to comprehend another mind is a fundamental factor operating in the performing arts. It is assumed that performers and their audiences will have the capacity to comprehend storytelling at the level of behavioral prediction of others (Fletcher et al., 1995). Neural circuits in the brain accommodate this level of social comprehension; for example, the ventral frontoparietal network rapidly orients to potential advantageous and threatening stimuli, whereas, the dorsal frontoparietal network selects and integrates stimuli and responses in a focused concentrated manner. Inhibiting the ventral network allows focused attention but when an external stimulus demands attention the ventral network will be rapidly recruited to reorient focus and coordinate responsive actions. These two networks, ventral and dorsal, operate reciprocally, a phenomenon that enables resilient adaptive attentional processes. It is speculated that these networks are part of theory of mind (Corbetta, Patel, & Shulman, 2008). Another key region that influences theory of mind is the temporoparietal junction, a region that has preferential responses to human faces, hence the ability to construct a representation of another person while also constructing a representation of the other person's mind (Saxe & Kanwisher, 2003). With the capacity to reorient and to focus we can assimilate information that shapes awareness of another's mind; a process that is essential for all social interactions, and one that is particularly vital in the performing arts. Further, while engaging theory of mind we can inhibit our own motor and emotional responses in order to integrate information about the other. These self-inhibitory processes share similar neural mechanisms as stated above (van der Meer, Growenewold, Nolen, Pijnenborg, & Aleman, 2011).

Performing artists and theory of mind

Performing artists tend to demonstrate strong abilities to engage theory of mind, whether in the process of creating characters to be performed or in the working relationship with other performers during rehearsals and performances. There is clear evidence that acting training augments theory of mind skills (Goldstein & Winner, 2010–11), although not necessarily empathy (Goldstein, Wu, & Winner, 2009–10). This study finding reinforces the fact that empathy operates independently from theory of mind (Chaplin & Norton, 2014). Despite the fact that theory of mind and empathy are separate variables, in an earlier study investigating personality profile and empathy in actors it was found that actors demonstrated strong empathic abilities (Nettle, 2006), a finding that contradicted Goldstein et al.'s 2009 study. Dancers were also found to have higher empathic abilities compared to a control group (Kalliopuska, 1989). These results may highlight the difficulties inherent in differentiating theory of mind from empathy (Chaplin & Norton, 2014). Another difficulty in the performing arts is differentiating theory of mind from self-perspective, especially when performers are claiming the role of a character. Boundary blurring between the performer and the character can increase emotional distress, especially if empathy for the character is viscerally experienced during role-playing (Burgoyne, Poulin, & Rearden, 1999).

Acquiring theory of mind gradually inhibits children from engaging in once-joyful play activities such as singing, dancing, and acting (Chaplin & Norton, 2014); however, this inhibitory pattern is not as pronounced in young performers. Although they are aware that the mind of another may be judgmental and rejecting, young performers persist in their love of the performing arts. Perhaps their early play behaviors override the inhibitory process that is connected with theory of mind, and more importantly, they manage to implement an understanding of another's mind as they develop characters and portray them in front of an audience (Goldstein et al., 2009–10). Performing artists have strong theory of mind abilities; however, they are also vulnerable to intense internalized shame (Thomson & Jaque, 2013, 2015a), a process that operates via theory of mind. Further, mood disorders such as depression impair theory of mind. For performers who suffer clinical depression, they are still capable of seeing someone else's feelings and mind but they are not able to respond; they have difficulty integrating this information

Figure 10.3 "Another's mind." Performer: Maurice Godin. Photographer: Robert Ragsdale.

(a topic to be discussed in more detail later in the book) (Wolkenstein, Schonenberg, Schirm, & Hautzinger, 2011; Fig. 10.3).

Theory of mind is critical for social survival. It is a capacity that is strengthened during role-play activities, especially when experienced in the performing arts. Engaging in the creation and performance of theatrical dramas further enhances a depth of understanding about others' minds (Black & Barnes, 2015), a process that preoccupies actors, opera singers, and many dancers. Performing artists have a keen sense of theory of mind, and with this heightened awareness they learn to override the inhibitory impulses that constrain the desire to perform in front of others. Along with learning to override these inhibitions, they also must learn to tolerate potential shaming experiences, and if they suffer any form of psychopathology, they must implement even stronger internal impulses in order to continue a career in the performing arts. Theory of mind is a powerful ingredient in the performing arts. These artists must develop complex strategies to manage the information gleaned from theory of mind.

Summary

- Play: early development and play, social development and educational support, worldplay and other talents in the performing arts, play benefits in the performing arts.
- Theory of mind: definitions, early developmental trajectory, neurobiology, theory of mind in performing artists.

Concluding remarks

Theory of mind and the importance of early play behavior are powerful ingredients in the performing arts. Through play, young performing artists explore their natural performing desires. Play experience, especially when it includes pretend play, role-play, and the inclusion of real or imagined objects that are incorporated into imagined situations or scenarios, helps children develop self-regulatory skills. Children are able to inhibit emotional distress; they can adhere to rules that they created. They learn valuable information such as impulse control, assertion, self-expression, conflict exploration, and how to seek solutions to resolve conflicts. They gain a growing understanding about the theory of mind of another, including related abilities such as empathy toward others, perspective taking, and self-reflection. These skills are essential for any performing artist who wishes to communicate effectively to other performers and the audience. Through play engagement, performing artists hone their skills and enhance their imagination. With practice they increase the complexity of their pretend play, role-playing, and scenario creations. Detailed elaboration and symbolic representation steadily mature. Abstract thought, including theory of mind, emerge as children play in their ever-developing fantasy play activities. This is the early terrain that nourishes young performing artists, and it remains the source of nourishment throughout their career in the arts.

Attachment, parenting, and childhood adversity

11

A lot of parents tell their children that if they want to be an actor, that's fine, but they should do something else first, so they've got something to fall back on. It doesn't work like that, as far as I'm concerned.

Ewan McGregor

Introduction

Attachment relationships are biologically essential for infant safety; infants must attach to primary caregivers to survive (Bowlby, 1988). Attachment theory describes the unique patterns of infant behavioral responses with primary caregivers, usually mothers and fathers, and how these behavioral patterns develop into internal working models. Internal working models define all future relational responses to important individuals, and they influence the formation of a self and how that self experiences the world at large (Sroufe et al., 2005). Establishing internal working models results from repeated predictable interactions with caregivers during the first year of life. These interactions shape the child's future expectations and operate as anticipatory defensive strategies to decrease surprise and disappointment. New relationships that develop throughout the lifespan are all powerfully informed by these infant attachment experiences (Main, Hesse, & Kaplan, 2005; Waters, Merrick, Treboux, Crowell, & Albersheim, 2000; Weinfield, Whalley, & Egeland, 2004). Although adolescents and adults do not need primary relationships for biological survival, establishing significant relationships with others does ensure emotional connections that can enhance life satisfaction (Bowlby, 1988; Sroufe et al., 2005).

Primary attachment relationships during infancy are determined by the infant and influenced by the parent. Infants set up a hierarchy of attachment figures based on how their need for security is met (Bowlby, 1988). Usually the primary attachment figure is the mother, however, this is not essential or always accurate. A primary attachment figure can be a father, grandmother, older sibling, adopted parent, or nanny. One of the key ingredients for primary attachment figures is their sensitivity to the needs of the infant. This quality of responsiveness shapes infant ego development and the trajectory of social–emotional growth (Bowlby, 1988). Ultimately, the nature of these interactions between the primary attachment figure and infant predict later social, emotional, and cognitive development (Stams, Juffer, & van IJzendoorn, 2002).

Creativity and the Performing Artist. DOI: http://dx.doi.org/10.1016/B978-0-12-804051-5.00011-1

Figure 11.1 "Exploration." Performer: Lilia Kibarska, John Ghera. Photographer: Lee Choo.

In order to survive, infants need attachment figures but they also need to explore their environment (Bowlby, 1988). Exploration is the hallmark of Piaget's earliest cognitive stage of development (Piaget, 1951). Through sensorimotor exploration, infants and toddlers learn about their environment. This exploratory behavior also promotes motor skill development (Newman & Newman, 2015). The drive to explore is a powerful motivating and pleasurable force, one that augments competency and self-efficacy (Elliot & Reis, 2003). Infants and toddlers approach the world with curiosity and joy; however, if they feel frightened or threatened then they cease to explore and actively seek their primary attachment figure(s) for safety. Only when they feel safe are they able to explore and this safety is provided by responsive parental support (Bowlby, 1988). When internal working models are constructed, infants and toddlers not only construct representations of their attachment figures but also of whether the world is safe to explore or threatening. They rapidly learn whether they can freely investigate their surroundings or if they need to keep close proximity to their attachment figure to ensure safety. These two biological systems, attachment and exploration, operate throughout the lifespan; as children mature into adults these early attachment experiences are internalized and influence subsequent exploratory behaviors (Elliot & Reis, 2003; Fig. 11.1).

Infant and adult organized attachment classifications

According to attachment theory, infants form attachment strategies in one of three organized patterns of behavior. The first pattern is described as secure. Securely attached infants are raised in environments that consist of parental sensitivity, availability, and responsiveness (Main, 2000). These caregivers offer protection and safety and promote infant exploratory behavior. They accept the infant's negative emotions and provide assistance so that negative emotions can be reduced. They also welcome positive emotions and respond reciprocally with their infants during play interactions. Generally, securely attached infants receive warmth and affection from their caregivers; these caregivers, often mothers, perceive a full range of emotional cues and respond appropriately and within a time span that the infant can tolerate (Milligan, Atkinson, Trehub, Benoit, & Poulton, 2003). Fundamentally, the emotional attunement between the securely attached infant and caregiver is attuned, appropriate, timely, and flexible (Elliot & Reis, 2003; Main, 2000). For example, an infant or toddler that is distressed receives adequate comfort from the parent and is quickly soothed. The parent recognizes the child's distress, comforts the child, and is sensitively empathic to the child's needs. When this happens consistently, children grow to trust that their parents can help them recover from distress (Corriveau et al., 2009). With repeated interactions these experiences are internalized and generalized into self-perceived effective regulatory responses. Self-regulatory capabilities are then mobilized during child exploratory play, especially when a task may present as difficult or frustrating. The young child learns to tolerate frustration and flexibly seek solutions to resolve problems that arise during play. Even sleep behaviors are more regulated when infants are securely attached, a pattern that persists into childhood (Belanger, Bernier, Simard, Bordeleau, & Carrier, 2015). Over time, these early attachment experiences between infants and primary caregivers become internalized attentional and emotional regulatory strategies that operate during all exploratory activities. As securely attached children develop into young adults these same effective self-regulatory strategies persist, even during threatening or stressful situations (Main, 2000).

Secure adults are no longer dependent on attachment relationships for safety; they now have a state of mind that is secure and autonomous. They value attachment figures; are honest, open, and compassionate; can readily acknowledge regrets and recognize early difficulties; and can contextualize childhood experiences based on parental needs, circumstances (including financial, social, and political), and their own personality dispositions and temperaments (Main, Goldwyn, & Hesse, 2003). Studies have demonstrated that attachment security enhances better social understanding (Fonagy & Target, 1997). Securely attached children do better on "theory of mind" tasks, a finding that continues into adulthood (Symons & Clark, 2000). Secure adults actively explore and are curious (Johnston, 1999). They seek sensory stimulation and are better able to process and integrate these experiences (Jerome & Liss, 2005). When engaged in discourse such as during the Adult Attachment Interview (Main et al., 2003), they can easily elaborate when asked for information (McCabe & Peterson, 2011–12), while maintaining awareness that they are in a discursive relationship with another (Main et al., 2003). They don't dominate conversational interactions nor do they minimize and withdraw from discourse (Main et al., 2003). Secure/autonomous adults are also able to reflect on the minds of others and elaborate on the emotional states that others may experience. They do this while maintaining self-awareness about their own feelings and thoughts (Bouchard et al., 2008). To be considered secure/autonomous on the Adult Attachment Interview does not mean all secure adults experienced idyllic childhoods. They may have suffered rejection, role-reversing relationships with parental figures, family dysfunction, and/or abuse. Despite these adversities, as adults they are able to reflect and speak with coherence and balance. Secure/autonomous adults tend to parent children who are also securely attached (Main, 2000). Even if these adults suffer a diagnosis of posttraumatic stress disorder as a result of experiencing great losses and/or traumatic experiences, they will not transmit frightened or frightening behaviors to their infants (Hesse & Main, 2006; Schuengel, Bakermans-Kranenburg, & van Ijzendoorn, 1999). They can provide better care to their infants, children, and adolescents (Zajac & Kobak, 2009), and these attachment experiences will be transmitted across generations; mothering does beget mothering (Fleming et al., 2002).

The second and third organized attachment classifications are not inherently secure attachment experiences; rather the infant remains anxiously attached to the caregiver. Such infants do not form an abiding faith that they will be soothed when distressed, nor do they feel safe to freely explore (Sroufe, 2005). Often insecure children are parentified; they assume the role of parenting their parents simply because they sense the inadequacies of their parents (Byng-Hall, 2002). Because these experiences persist over time, they influence self-regulatory strategies; however, these strategies are inadequate to manage negative emotions of the child and adult. The organized insecure attachment experiences and behaviors shape internalized working models that are more rigid and less flexible. They powerfully influence perceptions of the external world and the self; insecurely attached infants and adults lack a sense of self-awareness and attentional flexibility (Main, 2000), and they are less able to construct a theory of mind about another (Bouchard et al., 2008). Insecure attachment increases the risk for memory consolidation and retrieval errors, especially when the child is placed in stressful situations (Melinder, Baugerud, Ovenstad, & Goodman, 2013). They also have increased risk for developing some form of behavioral problem in childhood and psychological disorder in adulthood (Sroufe, 2005), especially when the insecurity is with both parents (Kochanska & Kim, 2013). Although these organized insecure behaviors are adaptive attempts to manage the strained relationship between the infant and caregiver, both infants and adults demonstrate difficulties expressing the value of attachment experiences and struggle to explore the surrounding environment (Main, 2000). Like secure attachment, which endures throughout the lifespan and is transmitted across generations, organized insecure attachment patterns tend to endure throughout the lifespan and are also transmitted across generations (Carlson, Sroufe, & Egeland, 2004; Fleming et al., 2002; Weinfield et al., 2004).

The avoidant insecure attachment classification is given when infants have formulated internal working models that promote an active turning away from a caregiver when they are distressed. Through repeated experiences, infants have learned that their caregivers cannot tolerate their negative emotions, in particular, anxiety and anger. Although these feelings are present in infants, avoidant-attached infants have learned to hide them. Rather than stress an already stressed relationship, these infants adopt a defensive strategy to protect themselves from further disappointment. They no longer seek comfort when distressed and they display less trust and reliance on their parents (Corriveau et al., 2009). Their caregivers' failure to respond sensitively and in a timely manner to their emotional needs ultimately leaves them more distressed and isolated. They have great difficulty socializing with their peers and struggle with situations that introduce novelty (Sroufe, 2005). Consequently, avoidant-attached infants do not develop adequate self-regulatory strategies to tolerate frustration, nor are they able to explore their environment with confidence that they can adequately and flexibly seek solutions to problems (Elliot & Reis, 2003; Main, 2000).

The parents of avoidant-attached infants tend to demonstrate dismissing states of mind. These parents exhibit a restricted range of emotions; they frequently withdraw from negative emotional interactions and display less sensitivity to the needs of their children (Whipple, Bernier, & Mageau, 2011). When discussing early attachment experiences, dismissing individuals consistently minimize any negative experiences, and endorse rejecting or neglecting parental behaviors as appropriate responses in their childhood. During discourse, they frequently respond with brief and poorly elaborated discourse and remain unaware that the listener requires more detail to understand a situation. Individuals with dismissing states of mind commonly turn their thoughts away from discussions on early attachment experiences, as is evidenced by comments such as, "I don't remember." When dismissing adults discuss early childhood experiences they will describe them as ideal, loving, and happy but they either fail to provide any examples of such experiences ("I don't remember") or will give contradictory examples that demonstrate outright rejection or cruelty by their parents (Main et al., 2003). These contradictory examples remain uncorrected and reveal a lack of self-awareness of any negative feelings; painful memories have been denied or repressed. Both a dismissing parent and an avoidant-attached infant actively turn away from each other in distressing situations. Like the enduring pattern of security that persists throughout development and is transmitted generationally, the avoidant child frequently becomes the dismissing adult who parents an avoidant-attached infant (Main, 2000; Sroufe et al., 2005). The infant and the adult are not able to tolerate frustration or cultivate curiosity (Milligan et al., 2003); ultimately they use a deactivating emotional regulation strategy to reduce memories of attachment-related experiences and stimuli (Dykas, Woodhouse, Jones, & Cassidy, 2014). Dismissing states of mind that deactivate emotional distress are adaptive during daily situations, and even during exposure to trauma (Ortigo, Westen, DeFife, & Bradley, 2013); however, they severely constrict creative exploration (Elliot & Reis, 2003).

The third organized insecurely attached infant classification is the ambivalent/resistant pattern of attachment. These infants are unable to freely explore simply because they are unable to withdraw attention from their caregivers. They are repeatedly frustrated by the lack of emotional attunement and poorly timed responses from their parents (Milligan et al., 2003). For example, when infants feel threat while exploring and seek comfort from their parents, they are inconsistently soothed. The caregivers are often preoccupied by their own emotional needs, they fail to recognize positive emotional responses in their infants, or they overly identify with their infants' negative emotions. They are ultimately unable to soothe them, as is evidenced by behaviors such as picking up a distressed baby and then putting the baby down before being comforted. They often misjudge the infant's needs and respond inappropriately. With repeated failure to recognize infant needs, the infants respond with tantrums, increased negative emotions, greater distress behaviors, and constant bids for parental attention. The emotional inconsistency of the parents provokes profound states of anxiety and uncertainty in the infants. For example, this is evident in infant resistance to accept ineffectual parental efforts to soothe them or they ambivalently desire and simultaneously reject parental comfort. Insecure ambivalent/resistant babies have great difficulty with emotional regulation and they are unable to maintain cognitive attention vital for exploratory behaviors (Main, 2000). They do not trust their parents and yet they constantly seek their parents' attention (Corriveau et al., 2009). As they develop, ambivalent/resistant children lack self-reliance, are overly dependent on parents and teachers for problem solving, lack tolerance for frustrations, and have great difficulty socializing (Sroufe, 2005).

The counterpart to the ambivalent/resistant infant is the preoccupied adult. These adults reveal a state of mind that is entangled with past insecure attachment experiences. They are unaware of appropriate relational discourse as is evidenced by long and confused passages of dialog. They forget that someone else is listening and begin to address attachment figures from the past as if they were present now. Their discourse is often filled with preoccupied anger, muddled confused speech, thoughts that persistently oscillate between two viewpoints, incomplete thoughts that are left unclarified, or continually and repeatedly recount frightening experiences with parental figures (Main et al., 2003). Such preoccupied states of mind preclude sensing another point of view, recognizing that someone else may have a different mental state or emotional need, or being able to clearly and succinctly provide information relevant to a dialogical interaction (Bouchard et al., 2008). Like the dismissing adults, preoccupied adults also have great difficulty with emotion regulation and social interactions (Sroufe, 2005). Because they are entangled with negative emotions, they are less able to flexibly shift awareness to positive emotions; consequently their creative problem solving is severely constrained (Kerns, Abraham, Schlegelmilch, & Morgan, 2007; Mikulincer & Sheffi, 2000). Preoccupied attachment is associated with increased and unmonitored memories of attachment-relevant experiences that interfere with open exploration; these hyperactivating strategies ultimately heighten emotional responses despite overt efforts to diffuse them (Dykas et al., 2014). Preoccupied individuals remain hypervigilant to attachment-related memories, thus there is decreased attention to open exploration (Elliot & Reis, 2003).

All three organized attachment patterns of behavior and states of mind tend to persist throughout the lifespan and are transmitted generationally. These attachment strategies powerfully influence exploratory behavior, cognitive flexibility, and creativity. A sense of self is formed through early attachment experiences and these influence how the world is perceived, including how important relationships are organized and maintained. Early attachment is one of the key factors influencing emotion regulation, and secure/autonomous individuals are the most successful at self-regulation, regardless if it occurs during stressful situations or pleasurable exploratory events (Box 11.1).

Box 11.1 Attachment Organized Classifications: Infant/Adult

Infant Patterns of Attachment	Adult State of Mind Regarding Attachment
Secure: Actively seeks parents when distressed and is readily soothed, easily engages with parent in joyful play, and able to explore freely	*Secure/autonomous*: Values attachment experiences, easy and balanced discourse when discussing early attachment experiences, demonstrates compassion and an ability to place events in context
Avoidant: Actively turns away from parent when distressed, subdued play, hides negative emotions	*Dismissing*: Minimizes negative childhood experiences, idealizes parental behavior (gives positive adjectives to describe childhood relationships with parents without the ability to provide supportive examples), turns away from thinking about attachment-related experiences by an inability to remember events or by very succinct responses that are devoid of elaboration
Ambivalent/resistant: Unable to shift attention away from parent and yet not able to be soothed by the parent when distressed, appears angry, and resistant or passively distressed, unable to focus on play or explore	*Preoccupied*: Entangled preoccupation with early attachment and current experiences, long entangled discourse that is filled with vague phrases, angry blaming, or frightened perseveration about early experiences

Parenting styles, attachment, and creativity

Parenting styles reflect how a parent disciplines and socializes a child, whereas attachment is an early biological survival drive between an infant and the primary caregiver (mother) (Doinita & Maria, 2015). Attachment experiences directly influence the formation of a sense of self and the behavioral responses that operate within significant relationships. Parenting styles, on the other hand, describe the emotional climate created as parents attempt to socialize their children (Doinita & Maria, 2015; Neal & Frick-Horbury, 2001). The three major parenting styles, as described by Baumrind (1991), indicate three different pathways for child development; however, they are not related to the early attachment internal working models established within the first year of life (Neal & Frick-Horbury, 2001; Whipple et al., 2011). Ultimately, it is the interaction of early attachment and parenting styles that influences perceptions about relationships, self-esteem, and self-discipline (Doinita & Maria, 2015; Neal & Frick-Horbury, 2001).

According to Baumrind (1991), the three parenting styles are authoritative, authoritarian, and permissive. Authoritative parenting is identified by high responsiveness to the child as well as high demandingness regarding discipline and socialization. Authoritarian parenting occurs when the parent is highly demanding but lacks responsiveness to the child. Contrary to the authoritarian parent, the permissive parent is highly responsive to the child but is also undemanding. Responsiveness is evaluated based on the parents' ability to recognize the needs of the child and provide support, warmth, and affection. Responsive parenting does not include harsh discipline or excessive criticism. Demandingness describes parental requirements for their children so that they can develop into mature and responsible adults. The establishment of rules and limits and how parents will discipline children if the rules are not met are examples of demandingness.

Children raised by authoritative parents develop self-confidence; they trust that their needs will be met (Baumrind, 1991). Because authoritative parenting is fundamentally responsive, warm, and supportive, children raised by authoritative parents are frequently the same children who formed secure attachments with their primary caregiver (Doinita & Maria, 2015). On the other hand, authoritarian parents tend to display low self-esteem, anger, aggression, and anxiety, traits that are transmitted to their children. This parenting style severely constrains creativity (Fearon et al., 2013). Permissive parents are considered lax, inconsistent, and tend to withdraw their love as a way of punishment. Children raised by permissive parents have low self-control and self-reliance (Baumrind, 1991). Both authoritarian and permissive parenting styles are associated with insecurely attached children (avoidant and ambivalent/resistant), a finding that suggests these parenting styles reflect the parents' inability to self-regulate emotional responses, consequently promoting the construction of more negative self-views in their children (Doinita & Maria, 2015; Karavasilis, Doyle, & Markiewicz, 2003). Further, permissive parenting can be viewed as either indulgent or negligent (Milevsky, Schlechter, Netter, & Keehn, 2007). Permissive-negligent parenting style is associated with more avoidant and fearfully attached children (Karavasilis et al., 2003), whereas, indulgent parenting frequently results in children who are more vulnerable to drug addiction and misconduct (Milevsky et al., 2007). More insecurity and decreased self-esteem emerge in the adolescent years when permissive parents withdraw their love as a form of punishment combined with their general negligence. Authoritarian parents who are harsh, controlling, and punitive also provoke more adolescent insecurity and poorer self-esteem (Karavasilis et al., 2003; Milevsky et al., 2007; Box 11.2).

Parenting style and behavior are powerful influences on the successful trajectory of talented children. In a longitudinal study examining the careers of creative women over a 43-year span, a consistent pattern of parental behavior emerged (Helson, 1999). Compared to creative women who did not maintain a career in the arts, the women who did succeed were parented by mothers who were supportive and sensitive to their needs, although they were often described as frail, dependent, and had low self-esteem. Despite their mothers' insecurities, the essential variables to foster a secure attachment, support and sensitivity, were present in these mothers. The fathers, on the other hand, were directive, intelligent, and dynamic. These

Box 11.2 Baumrind's Parenting Model (1991)

Parenting Style	Description
Authoritarian	Values obedience, forceful assertion of parents' will (high demandingness and low responsiveness)
Authoritative	Encourages independence, sensitive to needs of children and maximizes resources to support them, provides interpersonal interactions with children, models effective engagement with others, balances reason and punishment, instills social and cultural values while fostering independence (high responsiveness and high demandingness)
Permissive	Values nonintrusive behavior but offers support and resources (high responsiveness and low demandingness)
Nonconformist	Opposed to authority, less passive than permissive parents and exerts more control in their parenting

successful women had the opportunity to receive love and security from their mothers while also deriving strength and confidence from the more authoritative parenting provided by their fathers. In a different longitudinal study of artists (Feist & Barron, 2003), the long-term creative success over a 44-year span was associated with the strength and solidity of self-identity established early in life. This sturdy artistic identity was shaped in part by the early recognition of talent in childhood and the ensuing achievements later in life. The successful artists' careers were influenced by an increased ability to tolerate ambiguity and the employment of psychological mindedness (Feist & Barron, 2003), factors that form within secure/autonomous attachment experiences (Main, 2000).

Ultimately, parenting styles that foster creativity, and the joys inherent in creative experiences, are marked by open acceptance of children's interests while establishing moderate parental control (Mesurado & de Minzi, 2013). With parental support and acceptance, creative children feel less isolated and lonely (Lim & Smith, 2008). Similar to authoritative parenting styles, when parents are able to balance seemingly contradictory tasks such as establishing rules and limitations while also supporting children's talents and interests, spending time with the child and yet allowing quiet time for exploration, and holding high expectations but still accepting failure, children are able to acquire the skills needed to succeed in their artistic domain. Such parenting styles are optimal for promoting creativity and nurturing talent (Gute et al., 2008; Tannenbaum, 1992). Generally, parents state that they spend time with their talented children. They explore the interests of the child, offer new information to cultivate their understanding, and share in problem solving tasks. They tend to engage in discourse that is more grammatically correct, infused with more metaphors and imagery, and more abstract complexity and variety (Perrone et al., 2010). These parenting approaches help children manage emotional frustration, tolerate uncertainty and ambiguity, and expand skills and knowledge

Figure 11.2 "Support and recovery." Performers: Hilary Gereaux, Ilana Torres, Arianna Douglas, Darrell Hardaway, Cheyenne Spencer. Photographer: Lee Choo.

beyond the limits of formal education. Parents who are most successful at raising talented children provide exposure to talent-promoting activities while ensuring that the child is safely embedded within a family structure (Snowden & Christian, 1999; Tannenbaum, 1992). This parenting approach reinforces the early attachment security of the child while facilitating optimal talent exploration and development (Fig. 11.2).

Performing artists, parenting, and attachment patterns

As stated above, the burgeoning performing artist is best served when they have internalized secure attachment working models and their parents adopt an authoritative parenting style (McPherson, 2009). Parental support, encouragement, and sensitive attunement foster self-confidence and promote creative exploration in young performers. Parents who successfully nurture the talents of their gifted children tend to be well-educated, respectful of their children's desires, value their role as parents, and take seriously the responsibility of raising talented children (Rasmussen & Rasmussen, 2015; Snowden & Christian, 1999). Parents also need to be open-minded, adaptable, and able to appreciate abstract thought and complexity in order to understand and encourage their talented children (Bournelli et al., 2009). These traits, however, are not equally valued or endorsed across cultures (Lim & Smith, 2008). For example, in Jamaica, the predominant parenting style is authoritarian and consequently creativity is decreased in Jamaican children compared to an equally matched sample of children from other cultures (Fearon et al., 2013).

Unlike the effects of parenting style and cultural biases, attachment security and insecurity are not culture-bound (Bakermans-Kranenburg & van IJzendoorn, 2009), with security contingent on maternal sensitive attunement and insecurity influenced by maternal intrusiveness or remoteness (Tomlinson, Cooper, & Murray, 2005). The probability of talented children experiencing secure/autonomous attachment is great if parents are supportive, warm, and encouraging (Wellisch & Brown, 2013). These secure/autonomous unconscious internal working models powerfully influence performing artists throughout their careers.

Early attachment internal working models directly influence the appraisal of social communication such as facial emotional expression. Securely attached children are generally able to make better emotional appraisals and mobilize appropriate emotional responses when viewing emotionally laden scenarios (Steele, Steele, & Croft, 2008). Dismissing-avoidant individuals have greater difficulty matching their own negative facial expressions with others who are suffering negative emotional experiences (Sonnby-Borgstrom & Jonsson, 2004). They tend to react inappropriately in social contexts such as smiling when someone is angry. Their deactivating strategy to nullify negative emotions operates without their conscious awareness. For performing artists, this noncontingent response may be misconstrued as the performer lacking sensitivity or being arrogant and uncaring. Dismissing-avoidant individuals are unconsciously recapitulating their early infant–caregiver patterns of behavior; however, in a professional adult setting it is no longer adequate or appropriate. Despite these noncontingent emotional expressions, dismissing-avoidant individuals monitor attachment-relevant social stimuli with intense vigilance (Maier et al., 2005). Their speed at perceiving distressing emotions is essential if they are to successfully avoid negative information or situations. There is a preconscious alertness to detect potential painful rejection of attachment-related needs, a response based on an overactive fear-signaling system. Compared to the dismissing-avoidant individuals, secure/autonomous individuals were also able to detect threat (however, at a much lower intensity). In contrast, the preoccupied grouping, although they tend to operate with hyperactive arousal strategies, did not respond quickly to positive or negative emotions. Perhaps this response was related to their constant emotional entanglement and bias toward negative emotions. They may have detected the negative stimuli but did not marshal a defensive strategy in a timely fashion simply because their attention is primarily internally focused, or narrowly directed toward a significant attachment figure or important relationship. This emotional preoccupation excludes all other external stimuli such as appraising social signals in others who are not deemed important in their lives (Maier et al., 2005). Performing artists who are deemed preoccupied have great difficulty navigating the professional demands inherent in a career in the performing arts, primarily because they are not as adept at participating in group settings and their creativity is compromised (Elliot & Reis, 2003).

Musicians

Young musicians are commonly raised in an environment that values the aesthetic and emotional qualities of music; music is part of their life and is infused into their early memories (Standley, 2008). Their parents accept and support their musical goals. Parents who promote musical talent are able to integrate the needs of their child and family while maintaining awareness and meeting the needs of a larger sociocultural context. They help their children achieve musical competence, including the integration of emotional expression through music (McPherson, 2008; Standley, 2008). Most importantly, children establish a musical identity based on the accomplishments they experience and the emotional support and physical resources provided by their parents, family, and the teachers or mentors that are introduced to augment their musical expertise (McPherson, 2009). Positive expectations grow from parental support during early musical performance experiences. For example, more securely attached young musicians habituate to the anxiety of performing much easier than those young musicians who were anxious and insecurely attached (Boucher & Ryan, 2011). In fact, music performance anxiety often stems from early attachment failures (Kenny, Arthey, & Abbass, 2014)—failures that can lead to deactivating or hyperactivating strategies to manage anxiety (Dykas et al., 2014). Establishing an identity as a musician is powerfully determined by parental expectations and perceptions of the musical talent in their children (Davidson & Borthwick, 2002). This musical identity is strengthened through sensitive caregiving and authoritative parenting; the talented musicians develop autonomy and security as well as self-perceived competence, ego boundaries, self-esteem, purpose, and musical community (Davidson & Borthwick, 2002; McPherson, 2009; Swart, 2015). Secure attachment-related responses and experiences along with authoritative parenting styles cultivate a capacity for emotional self-regulation and self-reliance, essential ingredients for career success in music (McPherson, 2009; Raeburn, 2007). These factors prove to be even more important for successful female musicians, in part, because they work in a male-dominated profession. These female musicians typically have highly musical parents who support and encourage their musical aspirations and they are often raised as only children or the oldest sibling with younger male siblings (Stremikis, 2002).

Actors

Unlike musicians, actors frequently recount memories of parental discouragement regarding their desires to become professional actors (Goldstein & Winner, 2009). They describe minimal family support, including few family members who were involved in acting (Kogan & Kangas, 2006). Despite this seeming lack of support, actors recognize parental support when they engaged in childhood role-play and the creation of alternate worlds (Goldstein, 2009, 2015). Indirect evidence suggests that a career as an actor may be enhanced by ongoing parent–child discourse, including more elaboration when recalling attachment-related memories and constructing personal and family narratives (Ontai & Thompson, 2008). For example, secure/autonomous experiences between parents and children cultivate theory of mind capabilities (Ontai & Thompson, 2008), and increased theory of mind is a feature of acting talent and acting skills (Goldstein & Winner, 2010–11, 2012; Goldstein et al., 2009). Engaging in the rehearsal and performance of theatrical and television dramas can actually enhance greater understanding of others; in essence theory of mind is promoted during these activities (Black & Barnes, 2015). In a study by Thomson and Jaque (2012b) there was a greater distribution of secure/autonomous states of mind in actors compared to nonactors. The actors in this study demonstrated a marked ability to recount early attachment memories that were balanced, valuing, compassionate, and open. Whether they derived this security from early attachment experiences, from optimal authoritative parenting styles, or from the actual training and working within a community of other actors is unknown; however, it is suspected that all these factors may contribute to their security (Gute et al., 2008). Secure/autonomous individuals, in general, are more self-regulated, better at using problem-focused coping, and maintain close relationships with more ease compared to dismissing or preoccupied individuals (Scharf, Mayseless, & Kivenson-Baron, 2004), a pattern that was also evident in the actors in the Thomson and Jaque (2012b) study.

Unlike musicians and dancers, many children work as professional actors. Given the importance of early attachment experiences and effective parenting, these children may be placed in situations that compromise the ongoing parental support essential for optimal development. Often they spend long hours on film or television sets or in rehearsal halls and theaters, hence they are removed from the complexity of their nurturing family environments (Anderson, 2011). They are also exposed to public scrutiny, including the general misattribution that the child actor is the character they are playing (Tal-Or & Papirman, 2007). This misappraisal promotes greater boundary blurring and identity confusion (Hannah et al., 1994). Consequently, they are at great risk of developing weaker self-concepts and poorer emotional regulation (Anderson, 2011).

Dancers

Dancers excel in motor creativity, and like musicians and actors, they thrive when their desire to express creatively through movement is accepted and supported (Bournelli et al., 2009). Studies have demonstrated that a positive mother–child interaction is the key factor in promoting a healthy self-concept in the young dancer and consequently enhancing creative exploration (Bournelli et al., 2009; Helson, 1999). Creativity is positively influenced by early secure attachment and the attachment-related experiences of acceptance, attunement, and love. Because a career in dance is often short-lived and financially challenging, emotional, and physical support enhances career longevity. Most dancers who enter college have substantial early dance training and they draw upon this training as they prepare for a career postgraduation (Montgomery & Robinson, 2003). Without early parental support they would not be as well prepared; in fact, dancers are largely dependent on adult support during their early years of training. The essential role of parental support transcends all cultures and all dance forms (Chua, 2015); young dancers rely on their parents, with many parents making difficult decisions such as relocating to other cities to provide training for their talented children and adjusting work schedules to ensure effective training opportunities (Oreck, Baum, & McCartney, 2000). Once again, acceptance of dancers' interests and promotion of their choices help facilitate their career success (Kalliopuska, 1989). Effectively parenting talented dancers includes providing resources so that they can excel and emotional "cheerleading" support that endures regardless of the many setbacks or victories that transpire during their development (Chua, 2015). Parents guide their young dancers and offer ongoing reassurance; this helps young dancers develop confidence and resilience, and ultimately make a successful transition from conservatory dancer to professional dancer.

Adverse childhood experiences, disorganized attachment, and performing artists

Childhood adversity is fundamentally a result of maladaptive family functioning, such as abuse (physical, sexual, emotional), neglect (physical, emotional), and family dysfunction (divorce, mental illness, addiction, imprisonment, domestic violence). Childhood maltreatment is associated with long-term negative consequences (Dong et al., 2004), including poorer health and increased frequency of psychiatric disorders (Felitti & Anda, 2010; Springer, Sheridan, Kuo, & Carnes, 2003). In a World Mental Health Survey, childhood family dysfunction and maltreatment accounted for 29.8% of all disorders in the world (Kessler et al., 2010). This staggering worldwide pattern indicates the deleterious consequences of early relational maltreatment. Although all forms of maltreatment are potentially damaging, specific types of adversity influence later behaviors and disorders in particular ways. For example, emotional abuse has been associated with greater illicit drug use and deliberate self-harm behavior such as cutting (Goldstein, Flett, Wekerle, & Wall, 2009). All forms of adverse childhood experiences are related to nonsuicidal self-injurious behavior, but maternal emotional neglect is the strongest influencing factor (Kaess et al., 2013). The combination of childhood emotional abuse and emotional neglect is associated with later symptoms of anxiety, depression, and dissociation. Individuals who suffer early emotional abuse and neglect also tend to believe that they are more vulnerable to harm, they have greater feelings of shame, and they exhibit more self-sacrificing behaviors (Wright, Crawford, & Del Castillo, 2009).

Generally, early childhood adversity experiences have a dosing effect on psychiatric disorders, with more adversity leading to more serious and enduring psychiatric diagnoses (Pietrek, Elbert, Weierstall, Muller, & Rockstroh, 2013). For example, it has been posited that depression may have slightly different symptom profiles depending on whether the roots are purely genetic or if childhood adversity influenced genetic expression (Heim, Plotsky, & Nemeroff, 2004). It is well documented that early life stressors profoundly influence the development of the major stress response system in the body, the hypothalamic–pituitary–adrenal (HPA) axis (Heim et al., 2004). Alterations in the HPA axis are directly related to the influence of early life stress on gene methylation in children (Romens, McDonald, Svaren, & Pollak, 2015). Along with a cascade of neurotransmitter, neuropeptide, and hormone activation, more cortisol, and norepinephrine are specifically secreted to meet the demands of the stressor. These neurochemicals alter major regions in the brain, such as the amygdala and hippocampus, regions that interact with the HPA axis during stress (Heim et al., 2004). With repeat exposure to childhood maltreatment the stress system becomes sensitized to preemptively initiate a stress reaction in anticipation of other adverse experiences (Heim et al., 2004). This childhood neurobiological sensitization reaction to stress directly influences increased frequency and severity of depression and anxiety in later adulthood (Heim et al., 2004). Adults with a history of childhood adversity will respond faster and with greater HPA activation, even when a stressor is seemingly mild. For performing artists with such a history, this heightened HPA response may amplify their stage fright and performance anxiety (Kenny, Driscoll, & Ackermann, 2014; Fig. 11.3).

Figure 11.3 "Reaching." Performers: Dana Fukagawa, Logan Schyvynck. Photographer: Chuck Valerio.

With aging, this hyperresponsive stress system further compromises multiple biological systems (Danese & McEwen, 2012). Enduring changes take place in the nervous, endocrine, and immune systems. Changes in brain volume (i.e., amygdala, hippocampus, prefrontal cortex) impair emotion regulation, memory retention and recall, judgment, and problem solving. Changes in the endocrine system are directly reflected in alterations in the HPA axis, and the immune system is severely limited as evidenced by increased inflammatory levels leading to more autoimmune disorders (Danese & McEwen, 2012). With aging, the long-term effect of early childhood adversity leads to poorer health. According to McEwen, when the stress system is mobilized at an early age, there is a metabolic demand that takes its toll later in life; the debt that was incurred in childhood is paid in adulthood. McEwen describes this process as allostasis (Danese & McEwen, 2012). The biological mandate to survive until after procreation years is directly associated with the delayed presentation of physical and psychiatric symptoms. For performing artists, this delayed symptom presentation can be career ending, especially for musicians and actors who potentially have much longer professional careers compared to dancers.

Beyond the neurobiological changes related to childhood maltreatment, four psychological pathways are also altered. Changes in emotional, behavioral, social, and cognitive functioning can all lead to compromised physical and psychological health outcomes in adulthood (Kendall-Tackett, 2002). For example, as discussed above, substance abuse is strongly associated with early emotional abuse and neglect. The development of drug addiction alters the behavioral pathway as evidenced in patterns of high-risk actions required to acquire illegal drugs. Alterations in the emotional pathway occur when the individual turns to drugs to manage negative distressing feelings, which then harms social interactions and impairs cognitive judgment. Over time, drug addiction inevitably compromises the physical health of the individual. Generally, drug addiction is cited as the root cause for alterations in these four psychological pathways; however, the hidden precipitating factor was early childhood emotional abuse and neglect. Identifying the precipitating factor(s) associated with this cascading pattern of addiction is vital when treating performing artists, primarily because they have a higher incidence of addiction compared to other populations (Kenny, Driscoll, & Ackerman, 2014). For example, actors reported a higher incidence of mental illness in their families compared to nonartists (Ayers et al., 1999), and a group of performing artists experienced more childhood adversity compared to athletes (Thomson & Jaque, 2016c). Early childhood maltreatment may be the underlying cause for addiction in performing artists.

Adverse childhood experiences are deeply embedded into attachment relationships, especially when the parent is the perpetrator of the abuse. If parents are the perpetrators of abuse, then they are frightening figures to the child. Fear activates the attachment system, as well as the stress system, in children; they need to run to the parent for safety and yet the parent is also the source of the fear (Hesse & Main, 2006). These intense contradictory circumstances cause a collapse in the organized attachment strategies to seek security (Hesse & Main, 2000). The attachment system is momentarily disorganized and the child is disoriented and frightened. In attachment theory, this response is recognized as the fourth infant classification, disorganized/disoriented. All infants have one organized strategy (secure, dismissing, preoccupied) but when raised in families that maltreat them, they also manifest profound disorganized attachment behaviors. Consequently they respond with a disorganized collapse in attachment strategies when frightened by the maltreating parent; whereas, in other attachment-related situations, they employ their primary organized attachment strategy. For example, when bullied by a schoolmate, the securely attached child will negotiate solutions to repair the situation, the dismissing child will actively avoid or minimize the situation, and the ambivalent-resistant child will become enmeshed in emotional conflict that is difficult to disentangle. But when abused by the parent these coping strategies are inadequate; they will freeze, cower, or run toward the parent while falling down as they try to simultaneously run away (Hesse & Main, 2000).

Even when parents are not directly abusive to their children, if they experienced childhood adversity in their own childhood, their failure to integrate these memories and experiences into a coherent sense of self can be transmitted to their children (Hesse & Main, 1999). The child will ultimately be frightened by the parents' momentary lapses in managing their own childhood adversities. Parents who remain unresolved are both frightening to the young child and they may be momentarily frightened by their offspring. For example, parents who were sexually abused as children may still manifest unresolved trauma symptoms such as exaggerated startle response, dissociative amnesia, or dissociative compartmentalization. These behaviors can be frightening to a small child. Sometimes, when parents suffer momentary trauma flashbacks triggered by hearing a child's loud cry or seeing a child's distressed gesture, the child may believe that they caused the parent's terror response. The child rapidly sees the fear in the parent's eyes and unconsciously assumes that he or she is the source of fear. In all these situations, the child is confronted with an irreconcilable problem. The child longs to be comforted by the parent and yet the parent is unable to feel secure enough to comfort the child. The parent's behavior ultimately evokes disorganized and disorienting responses in the child. When parents remain unresolved for past adverse childhood experiences, they may directly or indirectly transmit these experiences onto their children (Hesse & Main, 1999; Box 11.3).

Assessing unresolved trauma and loss experiences can only be determined by a careful coding of the Adult Attachment Interview. When there are sufficient indices for lack of resolution, the individual is given an unresolved classification. Not surprisingly, an unresolved classification is also strongly associated with greater adverse childhood experiences (Murphy et al., 2014). In an attachment study that examined actors and nonactors, a significantly larger proportion of actors had unresolved past trauma and loss (Thomson & Jaque, 2012b). Because actors must draw from their own past experiences to portray characters this may provoke greater difficulty in their successful ability to resolve past adverse experiences. Consequently actors may be more vulnerable to the deleterious effects of their childhood maltreatment simply because they are accessing them more frequently. In a second study that examined performing artists as a group, they also had a greater distribution of unresolved mourning compared to the general population (Thomson & Jaque, 2012a).

Box 11.3 Disorganized Attachment and Unresolved Mourning

Infant Disorganized/Disoriented Attachment	Adult Unresolved Trauma/Loss (Mourning)
In the presence of the caregiver, the infant manifests one or several marked behaviors indicating a collapse of an organized attachment strategy. The behaviors include: sequential contradictory actions (moving toward and then away), simultaneous contradictory actions, slowing, stilling or freezing, cowering, disoriented wandering.	During discourse there is a marked lack of awareness and monitoring that reveals an inability to resolve past loss or traumatic events. Lapses include denying an experience and then making a statement about it, feeling possessed by the abuser, feeling that the dead person is still alive, extreme behavioral reactions (suicidal attempt, severe avoidance of areas that remind the person of the event), eulogistic speech, collapsing speech patterns that become filled with detail and absorption, long silences in the midst of a sentence and then a failure to complete the thought and sentence.

Like the actor group, performing artists across domains may have greater difficulty resolving childhood adversity experiences because they draw from past memories to give emotional expression to the work that is to be performed. Adverse childhood experiences inevitably add to the psychological pressures beyond the usual demands of performing (Sataloff, Rosen, & Levy, 1999), especially for those performers who remain unresolved.

Despite these negative effects of childhood maltreatment, studies have also suggested that some early adversity can actually enhance resilience; however, greater frequency and intensity of adversity remains detrimental (Seery, Holman, & Silver, 2010). According to Seery et al. (2010), a small exposure to adversity may enhance stress management, including augmenting emotional regulation. Significant dose-related early stress does alter genetic methylation leading to different expressions of genetic potential (Heim et al., 2004), but some children who suffered childhood maltreatment remain unharmed as adolescents and adults. The interaction of a more resilient genetic profile combined with adequate parenting from one parent serves to protect these children (van Ryzin et al., 2015). The quality of relationships in childhood, adolescents, and adulthood is another consistent resilience factor (Collishaw et al., 2007). Good quality relationships can buffer the negative effects of abuse, a factor that protects performing artists when they are trained and work in supportive environments (Grape, Sandgren, Hansson, Ericson, & Theorell, 2003; Jarvin & Subotnik, 2010). Another factor that predicted greater resilience was gender, with females demonstrating more resilience compared to their male counterparts (McGloin & Widom, 2001). Some childhood adversity, particularly unpredictability rather than harsh childhood adversity, promotes cognitive flexibility. Children may augment cognitive flexibility and attentional shifting as a coping strategy essential to survive in unpredictable environments (Mittal & Griskevicius,

Figure 11.4 "Resilience." Performer: Maurice Godin. Photographer: Blake Heathcote.

2014; Mittal, Griskevicius, Simpson, Sung, & Young, 2015). Lastly, despite greater incidence and type of childhood adversity in performing artists, resilience may be associated with the positive experiences derived while performing, especially because performing can serve to promote wellbeing and enhance quality of life (Thomson & Jaque, 2016c; Fig. 11.4).

Summary

- Introduction: description of attachment theory, internal working models, exploratory system.
- Infant and adult organized attachment classifications: secure/autonomous, avoidant/dismissing, ambivalent-resistant/preoccupied.
- Parenting styles, attachment, and creativity: primary parenting styles (authoritarian, authoritative, permissive-negligent, indulgent), trajectory of talent development and parental style outcome.
- Performing artists, parenting, and attachment patterns: secure/autonomous culture-bound responses, influence of internal working models on performing artist careers and talent development (musician, actors, dancers).
- Adverse childhood experiences, disorganized attachment, and performing artists: definitions, unresolved mourning, deterrents, and resilience.

Concluding remarks

Talented performing artists do not develop in a vacuum. Early attachment formations powerfully influence subsequent development, in particular, secure/autonomous states of mind optimize the development of talent; whereas, avoidant/dismissing and ambivalent-resistant/preoccupied attachment compromise talent development. Performers must work within a community of other artists and their early internal working models inform the quality of relationships that they cultivate. As a group, they tend to be more secure/autonomous than the general population, even though many also struggle with more unresolved trauma and loss experiences (Thomson & Jaque, 2012a, 2012b). Parental support, in terms of both emotional acceptance and providing physical resources, promotes skill mastery and confidence. These are essential for young children to move into successful careers as performers.

Many talented children come from a lineage of family talent, with both parents and grandparents identified as talented or gifted (Perrone et al., 2010). Perhaps this multigenerational talent distribution enhanced understanding and support and fostered more secure/autonomous attachment. It is also noteworthy that many talented children did not come from a family lineage of talent; however, the general approach to parenting these children included providing opportunities to explore their talent and supportive acceptance of the child's needs (Perrone et al., 2003). Consistent research findings indicate that optimal parenting styles, secure/autonomous attachment, and minimal or no childhood adversity, form a platform on which artists can flourish as they successfully move into careers as performers.

Personality and motivation

12

Great dancers are not great because of their technique, they are great because of their passion.

Martha Graham

Personality

Introduction: temperament and personality

Temperament describes our genetic inheritance. It is innate and enduring, although it can be modified by early attachment, as well as psychosocial and biological environmental influences (Rothbart, Ahadi, & Evans, 2000). Temperament and disposition are often overlapping terms; whereas, personality and traits share similar meaning. Both temperament and personality tend to remain stable over time, they differ across individuals, and they influence behaviors. Identifying personality traits and attributes, especially individual similarities and differences, is the fundamental focus in the field of personality psychology, and of this chapter (Ivcevic & Mayer, 2006–07).

Efforts to identify specific temperaments and personality patterns that typify performing artists can enhance understanding across all the performing arts; however, it is a challenging endeavor. As discussed in previous chapters, personality traits such as fantasy proneness (Sanchez-Bernardos et al., 2015) and dispositional factors such as overexcitability (Alias et al., 2013) influence talent development, as do early developmental experiences and attachment relationships. This chapter will explore personality traits and motivational behaviors, and, in particular, how they relate to performing artists.

It is important to clarify that temperament is typically regarded as a biological heritable trait (see chapter: Self-Regulation, Emotion, and Resilience for detailed discussion), whereas, personality character (traits) is/are acquired by environmental, social, and cultural interactions (Chavez-Eakle et al., 2012; McAdams & Pals, 2006; Rothbart et al., 2000). Although this distinction is easily named, it is often difficult to disambiguate the effects of temperament from personality traits. Further, personality studies examine multiple areas that are also influenced by biological heritability such as cognitive processes, emotion and motivational responses, social interactions, and self-regulatory processes (Ivcevic & Mayers, 2006–07). The complicating factor in this field of study is the overlapping influence of temperament on all these subareas of personality study, in particular, the capacity for self-regulation, a topic that will be discussed in detail in Section IV, Chapter 14, Self-Regulation, Emotion, and Resilience. Personality, although generally stable, does

Creativity and the Performing Artist. DOI: http://dx.doi.org/10.1016/B978-0-12-804051-5.00012-3

develop throughout the lifespan. As described in Chapter 9, Early Development, early development influences and consolidates identity formation. Essentially, personality traits stabilize during these important developmental milestones (McAdams & Olson, 2010).

In an effort to identify individuality (McAdams & Pals, 2006), adult personality models describe psychological attributes such as how we feel, think, interact, and regulate behaviors (Ivcevic & Mayer, 2006–07). There are multiple approaches to identify traits within personality studies; however, two overarching models have clustered a myriad of traits into larger factors, and these factors remain stable over time (see Box 12.1). Eysenck's three-factor model (neuroticism, extraversion, and psychoticism) describes personality as a series of traits, whether strongly present or minimally evident, and all can be clustered successfully into three factors (Eysenck & Eysenck, 1991). Like Eysenck's approach, the Big Five personality model identifies five large factors that encompass related traits (neuroticism, extraversion, agreeableness, open to experience, conscientiousness) (McCrae & Costa, 1987). Each of these factors contains multiple attributes that influence the larger identified factor. For example, extraversion (opposite of introversion) indicates that an individual is outgoing, sociable, and gregarious. If individuals do not claim that these traits are similar to them then they are excluded from the extraversion grouping. Open to experience, a major factor related to creativity, describes someone who has wide interests, and is intelligent, curious, easily absorbed in a task, novelty seeking, and accepting of spiritual transcendence. Those who display high open to experience traits tend to have better emotional wellbeing and greater intrinsic motivation, a factor that mobilizes them into exploratory behaviors (McCrae & Costa, 1987). Low scores on open to experience suggest extrinsic motivation, emotional distress, avoidance of novelty, and narrow interests.

The creative personality

The traits or attributes that describe a creative personality are large and often contradictory (Selby, Shaw, & Houtz, 2005). Further, each creative individual will manifest a unique array of traits. One consistent feature is that creative individuals tend to be more androgynous, a trait identified by the MMPI-2 masculinity/femininity scale (Harrington & Anderson, 1981; Selby et al., 2005). Polarities are evident in all creative personalities. For example, they fluidly engage in convergent and divergent thinking, spend pronounced periods of great physical exertion contrasted with long spells of quiet, freely explore imaginative realms while maintaining a stable awareness of reality, display rebellious or defiant behavior and yet are conscientiously aware of rules and conventions, are sensitive and empathic and yet ruthless and aloof when required, and are emotionally insecure and also independent and assertive (Selby et al., 2005). In contrast to scientists, artists generally demonstrate more openness to experience, coupled with a self-assessment that they are creative (Kaufman, Pumaccahua, & Holt, 2013). They are more neurotic and less agreeable and extraverted (Feist, 1998), and they tend to be less conscientious, although this trait changes as artists get older (Feist, 1998).

Box 12.1 Personality Factor Clusters: Terms and Descriptions (Eysenck, 1997; Feist, 1998)

Factor	Description
Agreeableness (high) *From five-factor model (FFM)*	Agreeable, relational, communal, easygoing, cooperative, empathic, friendly, nurturing, warm, supportive, generous
Agreeableness (low)	Aggressive, argumentative, cynical, hostile, exploitative, headstrong, suspicious, psychoticism
Conscientiousness (high) *From FFM*	Careful, conscientious, controlled, orderly, persevering, reliable, responsible, self-controlled
Conscientiousness (low)	Disorderly, impulsive, needy, psychopathic deviance
Extraversion (high) *From FFM and three-factor model (3FM)*	Active, adventurous, ambitious, assertive, autonomous, confident, dominant, energetic, expressive, exhibitionistic, hypomanic, impulsive, independent, positive emotions, self-sufficient, sensation-seeking, sociable
Extraversion (low)	Dependent, depressed, introverted, reflective, reserved, submissive, unambitious, unsociable, unadventurous, cautious, sensation-avoiding
Introversion (high) *From 3FM*	Low tolerance threshold for arousal, desire to withdraw from social settings (especially if overstimulating), withdrawal behavior
Introversion (low)	Higher tolerance for arousal, sensation-seeking, approach behavior
Neuroticism (high) *From FFM and 3FM*	Anxious, depressed, emotional, excitable, hypochondria, insecure, nervous, worried
Neuroticism (low)	Calm, happy, stable, well-adjusted
Open to experience (high) *From FFM*	Aesthetic appreciation, achievement, independence, creative, curious, exploratory, flexible, humorous, imaginative, intelligent, open-minded, original, preference for complexity, sensitive, tolerance for ambiguity, supernatural beliefs, wide interests
Open to experience (low)	Conventional, inflexible, rigid, socialized, narrow interests
Psychoticism (high) *From 3FM*	Susceptibility to develop psychotic symptoms, creative, overinclusive thinking
Psychoticism (low)	Narrow-focused thinking, low creativity

When only including the Big Five factor model, creativity is identified strongly by greater extraversion, openness to experience, and conscientiousness (Furnham & Bachtiar, 2008; Mesurado & de Minzi, 2013). These personality traits were independent of intelligence, a finding that contradicts many studies suggesting a relationship between intelligence and creativity (Furnham & Bachtiar, 2008). When examining talented and gifted individuals, the Big Five factors have been associated with overexcitability (OE); in particular, sensual OE was related to openness to experience and psychomotor OE was associated with extraversion. Interestingly, all the OE responses in the gifted/talented group were negatively related to neuroticism (Limont, Dreszer-Drogorob, Bedynksa, Sliwinska, & Jastrzebska, 2014; see Section I, chapter: The Person: Talent and Performing Artists for details).

Following a different model, Chavez-Eakle et al. (2012) found that a creative profile included high exploratory behavior in novel situations, optimism about finding a solution, feeling unafraid in uncertain situations, and being energetic and persistent despite adversity. Creative individuals assumed responsibility for their goals and directed their behavior to accomplish these goals. They readily cooperated with others, drew upon a large resource base, and were empathic and tolerant toward others and themselves. In summary, they demonstrated a complex ability to integrate behaviors and responses while engaging creatively (Chavez-Eakle et al., 2012). Despite research findings that associate highly creative individuals with greater psychopathology (Kaufman, 2005), Chavez-Eakle et al. (2012) did not find creative individuals suffering more pathology, although factors that may increase pathology such as fantasy and emotional expressiveness were elevated in creative individuals. Their results resemble the findings by Limont et al. (2014), in particular the lack of significance between neuroticism and overexcitabilities in talented/gifted individuals.

When Martinsen (2011) organized personality traits within and beyond the Big Five factors, creative personality features included seven affective-motivational traits (associative orientation, ambition, emotional instability, motivation, desire to seek originality, agreeableness, and flexibility): (1) Associative orientation was augmented by fantasy proneness, playfulness, absorption and idea production traits. All these associative traits were strongly correlated with the Big Five factor, openness to experience. (2) Ambition in creative people suggested a need for attention, recognition, assertiveness, seeking a position to influence others, and engaging in socially stimulating environments. These traits strongly correlate with extraversion and negatively with agreeableness. (3) Emotional instability, strongly correlated with neuroticism, was related to poor self-confidence, anxiousness, mood swings, and feeling ineffectual. (4) Motivation, associated with conscientiousness in the Big Five factors, included autonomy, persistence, goal-directedness, and being achievement focused. Creative people like to set their own goals and make independent decisions on how to achieve these goals. (5) A desire to seek originality, a factor that did not correlate with any Big Five factors, indicated that creative people distrusted rules and regulations, held a preference for complexity, and were liberal-minded. (6) Agreeableness was obviously strongly correlated with the Big Five factor, agreeableness. Creative people described themselves as friendly; valuing good intentions and moral integrity regarding right and wrong; and less critical, including unwillingness to find fault in others. (7) Flexibility was weakly associated with agreeableness and openness in the Big Five factors. It indicated that creative people tend to restructure events, understand the complexity of situations, and display tolerance when engaged in uncertain or ambiguous situations.

Figure 12.1 "Nurtured." Performers: Hilary Gereaux, Tharini Shanmugarajah. Photographer: Lee Choo.

In summary, consistently identified general traits of creativity that endure across the lifespan include openness to experience, intellectual curiosity, persistence, self-confidence, and tolerance (Feist & Barron, 2003; Ivcevic & Mayer, 2006–07). Trait hypomania was also identified as a stable and significant predictor of a creative personality (Ivcevic & Mayer, 2006–07; von Stumm, Chung, & Furnham, 2011). Creative people described themselves as imaginative, complex, innovative, artistic, high energy, inquisitive, with strong beliefs in their abilities, wide interests, and a preferential bias for complexity (Ivcevic & Mayers, 2006–07). Although these traits are common to creative individuals it is important to remember that all personality traits can be enhanced or constrained by environmental influences, including early attachment experiences (Feist, 1998). It is paramount that opportunities are provided so that creative individuals acquire domain-specific skills within a nurturing environment, such as positive mentoring. This environment can promote the maturation of the creative personality, one that can endure into adulthood (Feist & Barron, 2003; Ivcevic & Mayers, 2006–07; Fig. 12.1).

Autotelic personality, flow, and the creative personality

The autotelic personality trait, as described by Csikszentmihalyi (1990), is characterized by the increased ability to experience flow (pleasurable state of optimal performance). Research findings demonstrate that this predilection is partially genetically inherited, although environmental influences can enhance or curtail this genetic potential (Mosing et al., 2012). This autotelic trait has also been aligned with the Big Five factors, most notably in a positive direction for extraversion and conscientiousness and in a negative direction for neuroticism and agreeableness. What is striking is that an increased propensity to experience flow was not associated with openness to experience (Ross & Keiser, 2014). It was posited that the lack of significance between autotelic flow and openness to experience might indicate that openness is a more cognitive attribute; whereas, autotelic flow depends on more sensorimotor feedback.

Because a propensity to experience flow is related to creativity (Csikszentmihalyi, 1996), it is somewhat surprising to see a different Big Five profile in Ross and Keiser's study. Less agreeableness and conscientiousness coupled with high openness to experience and extraversion are consistent findings in creative personalities (Feist, 1998; Furnham & Bachtiar, 2008; Mesurado & de Minzi, 2013). The autotelic personality shared greater extraversion with the creative personality profile but, unlike the creative personality, there was no significant association with openness to experience and a positive relationship to conscientiousness. Although these two profiles, the autotelic personality profile and the creative profile, seem to be different, future studies should investigate performing artists to see if an autotelic personality resembles a performing artist personality profile. Current findings demonstrate that individuals who possess autotelic personalities have greater capacity to meet challenges (Asakawa, 2004), achieve goals (Ishimura & Kodama, 2009), derive a sense of fulfillment, have more active strategies to cope with challenges, and claim satisfaction with life (Asakawa, 2010; Hirao & Kobayashi, 2013). These are all important features for performing artists, especially because these features will help them meet the challenge of managing careers that are inherently stressful (Marchant-Haycox & Wilson, 1992).

Personality in performing artists

Rather than lumping performing artists into a general group of artists, a slightly different picture is revealed when examining personality traits. The identifiable personality traits for performing artists tend to remain consistent across studies, although more studies are needed (Feist, 1998; Feist, 1999). Marchant-Haycox and Wilson (1992) determined that performers, as a group, were more emotional, a finding that predicted greater depression, panic, and migraines. Further, performers who were more introverted tended to suffer from more shoulder aches; whereas, performers who were more adventurous had greater frequency of liver disease and alcoholism (Marchant-Haycox & Wilson, 1992). Martinsen's (2011) seven-factor model suggests that performing artists displayed greater flexibility, ambition, strong need for originality, and agreeableness. These findings differed with artists in general (visual artists, writers) who displayed less agreeableness, intermittent motivation, and less adaptive flexibility. Differences between performing artists and other artists may be related to their need to please others. In order to remain in a performing arts career, they must cultivate an ability to flexibly adapt to changing environments. Intermittent motivation may also plague performing artists, as it does with artists in general, but in order to maintain sufficient skills they must continue to engage in deliberate practice on a regular basis. Due to the physical demands, including constant drills and practices, it is speculated that performers may be more like athletes, hence sharing more personality traits with athletes (Kogan, 2002). But, unlike athletes, they must express emotionality and give aesthetic form to their movement; in fact, they manifest significantly different psychological profiles than athletes, including greater imagination, emotionality, and psychological vulnerability (Thomson & Jaque, 2013, 2015a, 2016a, 2016b, 2016d).

Rather than assuming that all performing artists share similar personality traits, research focus has shifted to examine personality differences between musicians, actors, dancers and singers, and also between different subdomains within these genres such as differences between rock musicians and classical musicians, string versus brass players (Benedek, Borovnjak, Neubauer, & Kruse-Weber, 2014; Marchant-Haycox & Wilson, 1992). Studies also examine differences between a specific performing artist genre and a control sample, or between other artists such as writers or visual artists (Ayers et al., 1999; Haller & Courvoisier, 2010). Personality research has attempted to identify performing artist talent to ensure they are given opportunities to train in optimal learning environments (Woody, 1999). For example, although not consistent across all musical genres, openness to experience proved an excellent personality predictor for individuals who desired musical training and who had innate musical ability (Corrigall, Schellenberg, & Misura, 2013). This personality information is invaluable for curriculum development and improving training programs.

Musicians

In the seminal study by Marchant-Haycox and Wilson (1992), musicians as a group were more introverted, cautious, empathetic, cynical, resigned, world-weary, inactive, unsociable, submissive, and unambitious. Contrary to Marchant-Haycox and Wilson's gloomy findings (1992), when compared to the general population, musicians were more extraverted (Shuter-Dyson, 2000), emotionally stable, sensitive and conservative (Buttsworth & Smith, 1995). When musicians were compared to visual artists they were also more extraverted, agreeable (they preferred harmony in their relationships), and conscientious, although they were not as open to experience as their visual artist counterparts (Haller & Courvoisier, 2010).

Domain-specific musicians such as singers, jazz, popular, and rock musicians were more open to experience compared to folk and classical musicians (Benedek et al., 2014; Cameron, Duffy, & Glenwright, 2015; Dyce & O'Conner, 1994; Gillespie & Myors, 2000). Singers and popular musicians were also more extraverted compared to rock musicians and orchestral players (Cameron et al., 2015; MacLean, 2011). The general stereotypes of rock and popular musicians were substantiated by empirical research. They possess personality traits such as greater neuroticism and arrogance (Dyce & O'Conner, 1994), less agreeableness and conscientiousness, and average extraversion (Gillespie & Myors, 2000). Orchestral musicians, especially at professional levels, are more homogenous within their talent specialties (strings, brass, woodwinds) (Langendorfer, 2008; Langendorfer, Hodapp, Kreutz, & Bongard, 2006). When studying student musicians, they vary somewhat, with string players being more conscientious (Langendorfer et al., 2006), keyboard players more warm-hearted, emotionally stable, and shrewd (Buttsworth & Smith, 1995), bass players tending to be more extraverted and agreeable and less anxious compared to string players (Buttsworth & Smith, 1995; Cameron et al., 2015). Personality traits in rhythmic musicians included greater sensation seeking and less tolerance for boredom (Vuust et al., 2010). Collectively, these variances in personality traits reinforce the fact that musicians are not easily lumped into one group based on personality traits.

This same variance can be found between male and female musicians, with females displaying more self-directedness and motivation, and less conformist behaviors to stereotypes (Stremikis, 2002). They were more tender-minded and neurotic compared to males; however, the males were more religiously oriented (Shuter-Dyson, 2000). Males were also found to be more suspicious, apprehensive, and radical, when compared with female singers (Buttsworth & Smith, 1995). Regardless of gender and musical domain differences, musicians with more negative emotionality were more likely to suffer from performance anxiety (Sadler & Miller, 2010). Classical musicians also tended to experience more performance anxiety (Vuust et al., 2010), perhaps because they tend to be less extraverted and open to experience (Benedek et al., 2014; Cameron et al., 2015; Dyce & O'Conner, 1994; Gillespie & Myors, 2000; MacLean, 2011).

One major finding that may unify all personality traits in musicians is the reality that they tend to be more ruminative and reflective (Jones et al., 2014). When combined, these traits allow musicians to maintain their career goals, including practicing with deliberate intention and attention (practice strategies vital for performance enhancement) (Jones et al., 2014). Reflective rumination may be associated with talent ability, better active analytic problem solving, and self-regulation. Ruminating nonmusicians tended to engage in more maladaptive brooding, a form of rumination that undermines self-regulation and reflection. It is possible that reflective rumination enhances musicians' emotional stability (Buttsworth & Smith, 1995), agreeableness (Haller & Courvoisier, 2010), and introversion (Marchant-Haycox & Wilson, 1992; Fig. 12.2).

Actors

Actors' personality traits, when compared to singers, dancers, and musicians, included greater extraversion, adventurousness, expressiveness, dogmatic behavior, aggressiveness, impulsiveness, sensation seeking, ambition, reflectiveness, risk-taking, irresponsibility, and guilt proneness (Marchant-Haycox & Wilson, 1992). Compared to the general population, actors also had greater capacity to experience mystical experiences, a trait that may be related to openness to experience (Ayers et al., 1999). Like singers and dancers, they demonstrate strong abilities to memorize, especially when it is related to emotional content (Kogan, 2002). They have the ability to converge visual and linguistic feedback, a process that promotes their memorization skills (Palmiero, Nakatani, Raver, Belardinelli, & van Leeuwen, 2010). The Big Five factors within an actor group included

Figure 12.2 "Expressive exploration." Performers: Susan McManus, Evan Rosenblatt, Courtney Ryan. Photographer: Lee Choo.

greater extraversion, openness to experience, agreeableness, and a tendency toward neuroticism (Nettle, 2006). They also had greater empathy toward their characters (Kogan, 2002; Nettle, 2006), a fact that also renders them vulnerable to blurring their personality with that of their character during rehearsals and performances (Hannah et al., 1994). Given these factors, actors may be ideally suited for the performance demands of pleasing an audience, a willingness to explore characters while being emotionally and physically vulnerable, and adapting to multiple performance settings. One of the unique hazards for actors is that their audiences often make personality attribution errors; they tend to believe that the actor's personality is identical to the character's personality that they are portraying (Tal-Or & Pairman, 2007). This increases potential emotional distress and can contribute to actors' feeling that their personality identity is less stable (Hannah et al., 1994; Thomson & Jaque, 2011a).

When looking at stand-up comedians, compared to actors, in general, and comedy writers, in particular, a slightly different personality profile emerges (Greengross et al., 2012). They are more open to experience and extraverted and less conscientious, although comedy writers were even more open to experience (Greengross & Miller, 2009). Contrary to stereotypes, stand-up comedians were more introverted than comedy writers and actors, a finding that may suggest they are not seeking fame to the same degree as actors, and that they use comedy and performance to disguise who they really are in daily life (Greengross & Miller, 2009). When looking at the Big Five factor of neuroticism, stand-up comedians, actors, and comedy writers were similar (Greengross & Miller, 2009).

Dancers

Unlike musicians and actors, dancers present as more emotional, cautious, unhappy, anxious, dependent, hypochondriac, obsessive, depressed, and inferior (Marchant-Haycox & Wilson, 1992). Compared to athletes and sedentary persons, dancers were considered less hardy. They tended to deny negative emotions, avoided personal conflicts, and tightly controlled their emotional responses and behaviors (Biasi, Bonaiuto, Giannini, & Chiappero, 1999). In ballet dancers, greater anxiety was associated with more hypochondriasis and obsessive-compulsive personality traits (Taylor, 1997). Based on the MMPI-2 scoring, ballet dancers had elevated scores on schizophrenia, hypochondriasis, and obsessive-compulsiveness (Taylor, 1997). Collectively, these negative findings may be driven by the rigors of dance training and the aesthetic demands of a professional career. Dancers must interact in a social environment regardless of their personal feelings; negative feelings are generally not welcomed in class, rehearsal, or performance (Biasi et al., 1999). Contrary to study findings that demonstrate dancers' vulnerability, there are studies that show positive effects of being a dancer. For example, dancers who trained four to five times per week had good self-esteem and enjoyed many creative hobbies outside of dancing (Kalliopuska, 1989). Although dancers were found to be more neurotic, they displayed strong achievement motivation (Bakker, 1991), and they were highly empathic (Bakker, 1991; Kalliopuska, 1989; Marchant-Haycox & Wilson, 1992). Professional modern dancers, compared to professional ballet and jazz dancers, exhibited more openness to experience, psychoticism, and lower conscientiousness. They were also deemed more creative compared to the other two groups of dancers (Fink & Woschnjak, 2011).

Research conducted on dancers usually includes investigations into body satisfaction, anorexia, and preoccupation with dieting (Bachner-Melman, Zohar, Epstein, Elizer, & Constantini, 2006). Dancers who were more neurotic and less agreeable demonstrated greater negative eating attitudes and poor self-regulation (Scoffler-Meriaux, Falzon, Lewton-Brain, Filaire, & d'Arripe-Longueville, 2015). High neuroticism has been associated with increased depression and anxiety, psychological disorders closely related to eating disorders (Scoffler-Meriaux et al., 2015). Interestingly, internalizing a belief in thinness norms mediated the effects of neuroticism and negative eating attitudes, thus high neuroticism is a risk factor only when thin-ideal beliefs are adopted. These can lead to an eating disorder (Scoffler-Meriaux et al., 2015). Because of the aesthetic nature of dance, embracing thinness norms is a strong risk factor, especially because dancers tend to also possess more neuroticism in their personality structure.

When differentiating gender, it was found that nondancer women were more attracted to male dancers who possessed increased conscientiousness and social agreeableness, and to a lesser degree, extraversion (Fink et al., 2012). Male dancers with greater neuroticism and openness to experience were not deemed to be as appealing to the same group of women. Although greater neuroticism is understandable as a deterrent to attractiveness, the nondancer women also rejected openness to experience, suggesting a distrust for male dancers who were potentially interested in art and were more emotional, adventurous, and actively exploring unusual ideas (Fink et al., 2012).

It is suggested that dance attracts persons who possess the particular personality traits outlined above; traits that persist long after the dancer may have retired (Bakker, 1991). Dancers express through kinesthetic and visual means (Alter, 1984). They are intelligent, flexible, and demonstrate strong achievement motivation (Alter, 1984; Bakker, 1991), although compared to other performing artists, they are the most vulnerable to negative emotionality (Marchant-Haycox & Wilson, 1992). Compared to athletes they have increased imagination, but some dancers are more vulnerable to emotional distress such as anxiety (Thomson & Jaque, 2011a), dissociation (Thomson & Jaque, 2012c), and shame (Eusanio, Thomson, & Jaque, 2014; Thomson & Jaque, 2013, 2015a). Sadly, personality research-based profiles of dancers suggest they are one of the most fragile groups of performing artists.

Singers

Singers' personality profiles tended to fall between those of musicians and actors, a finding that is not surprising given the overlapping demands of this profession (Marchant-Haycox & Wilson, 1992). However, they were more cautious than the other artist groups and less adventurous. Singers were also more hypochondriacal, but only related to vocal health (Sandgren, 2002, 2009). Compared to dancers, singers tend to be more imaginative and emotional (Thomson & Jaque, 2016a). Singers' personality traits indicate that they may generally share more positive traits and fewer negative traits compared to their fellow musicians and actors.

Motivation

I don't want people who want to dance, I want people who have to dance.

George Balanchine

Introduction

Although the study of personality held a primary focus in early research, it was always understood that motivation was an inherent part of personality and creativity (Ivcevic & Mayers, 2006–07; Prabhu, Sutton, & Sauser, 2008). In order to be successful in the performing arts, strong motivation, along with a love or passion for the art form, propels artists (Collins & Amabile, 1999). Amabile proposed a model to describe the production of creative products. She stated that a creative individual must possess "intrinsic task motivation, domain-relevant skills (expertise and talent in the task domain), and creativity-relevant processes (cognitive skills and work styles conducive to the production of novelty)" (Collins & Amabile, 1999, p. 299). The concept of intrinsic and extrinsic motivation still dominates the

Box 12.2 Motivation Types (Collins & Amabile, 1999; St-Louis & Vallerand, 2015)

Intrinsic motivation: Engage in an activity for its own sake; activity is interesting, engaging, challenging, pleasurable, and satisfying	*Extrinsic motivation*: Participate in an activity based on meeting an external goal, garnering praise and approval, winning a competition, or receiving an award or payment
Approach/avoidance model: Built on emotional conflict model, individuals seek (approach) success and fear (avoid) failure	*Achievement goal theory*: Three distinct types of achievement goals: (1) mastery goals (gaining skills to accomplish tasks and the resultant sense of competence that follows skill mastery), (2) performance-approach goals (reinforces competence), and (3) performance-avoidance goals (motivational strategy that addresses a desire to avoid feelings of incompetence relative to others)
Harmonious passion: Commit to an activity freely, internalize and identify a sense of self through the activity (i.e., I am an actor), easily engages in activity without conflict with other activities in life	*Obsessive passion*: Uncontrollable urge to commit to the activity, internalize and identify a sense of self through the activity (i.e., I am an actor), conflicts with other activities in life

research field; however, other models, such as flow motivation (Martin & Cutler, 2002), approach and avoidance motivation (Elliot & Covington, 2001; Elliot & Thrash; 2002), and harmonious versus obsessional passions are now included (St-Louis & Vallerand, 2015; Box 12.2).

Intrinsic, extrinsic, and synergistic models of motivation

Intrinsic motivation is defined as a motivation to engage in an activity for its own sake. It holds interest, is engaging and challenging, and elicits feelings of pleasure and satisfaction. Extrinsic motivation is defined as a motivation to participate in an activity based on meeting an external goal, garnering praise and approval, winning a competition, or receiving an award or payment. Extrinsic motivation is not driven by the desire to engage in an activity for its own sake. Amabile originally proposed the intrinsic motivation hypothesis, a hypothesis that described how intrinsic motivation promotes creativity; whereas, extrinsic motivation compromises creativity (Collins & Amabile, 1999). This hypothesis is no longer regarded as an absolute truth. Certainly intrinsic motivation must be present if a performing artist is to excel, however, appropriate external motivators can also enhance creativity

(Collins & Amabile, 1999). All performing artists prefer audience approval, which is by definition an extrinsic motivator. Many enter competitions to further their careers, and all desire financial remuneration if they are to succeed as a professional artist. Intrinsic and extrinsic motivations are seemingly dichotomous factors, but maintaining a binary concept marginalizes the complexity of intrinsic and extrinsic motivation and how they operate within creative individuals (Collins & Amabile, 1999).

When integrating Atkinson's need achievement theory into intrinsic and extrinsic motivation, it further nullifies a binary concept, especially because this theory includes trait-like behaviors such as approaching success and avoiding failure (Covington & Mueller, 2001). The *approach/avoidance theory* is built on the emotional conflict view that individuals seek (approach) success along with the accompanying feelings of pride and efficacy and fear (avoidance) failure, including feelings of shame and humiliation (Covington & Mueller, 2001). With repeated experiences, individuals who are biased toward high approach motivation may be success-oriented or overachievers; whereas, individuals who adopt low approach motivational strategies either are failure avoiders or assume and accept that they will fail (failure acceptors). On the other hand, high avoiders either may be driven by the need to avoid failure or fear that their overachieving efforts may result in failure. Low avoiding individuals will be either success-oriented, or paradoxically, fully accepting of failure. The failure acceptors will display low motivation for both approach and avoidance; they lack motivation, including an inability to respond to goals that are either intrinsic or extrinsic (Covington & Mueller, 2001).

The *approach/avoidance model* inherently assumes that a desire to achieve goals drives these strategies (Elliot & Thrash, 2001). To achieve goals, mastery of skills is essential, coupled with an awareness of how one performs relative to others (Elliot & Thrash, 2001). *Achievement goal theory* identifies three distinct types of achievement goals. The first, mastery goals, includes gaining skills to accomplish tasks and the resultant sense of competence that follows skill mastery. The second component in achievement goal theory is performance-approach goals. This reinforces competence, whereas, performance-avoidance goals indicate a motivational strategy that addresses a desire to avoid feelings of incompetence relative to others (Elliot & Thrash, 2002). Including approach/avoidance motivation and achievement goal drives adds complexity to the primary motivational strategies. For example, a successful performing artist must achieve skill mastery and also be aware of performance achievement. Ultimately, they must engage an approach motivational strategy in most circumstances; however, avoidance strategies, when used judiciously, may protect them from unnecessary experiences of failure and feelings of incompetence (Elliot & Covington, 2001). These strategies are deeply ingrained biologically. Appetitive goals are pleasurable; we approach them often without conscious awareness while aversive goals are instinctively avoided. Performing artists intuitively and instinctively recognize appetitive versus aversive goals. For example, they willingly approach aesthetic pleasures and they withdraw from works that lack an aesthetic appeal (Kharkhurin, 2014; Martindale, 2007; Osipovich, 2006).

Returning to the prevailing theories of intrinsic versus extrinsic motivation, robust research demonstrates the power of intrinsic motivation in creativity. It is a powerful mediating factor in predicting the personality trait openness to experience, as well as self-esteem and creativity (Prabhu et al., 2008). Goal achievement was directly related to intrinsic motivation in performing artists (Lacaille et al., 2007). In the study by Lacaille et al. (2007), intrinsic motivation predicted positive performances and heightened enjoyment; whereas, mastery did not significantly influence goal achievement. Certainly the ability to meet a challenge influences goal achievement; however, intrinsic motivation to acquire skills to meet a challenge influences goal-directed activities and ultimately goal achievement (Abuhamdeh & Csikszentmihalyi, 2012). Individuals who experience more positive, rather than negative, emotions are able to maintain intrinsic motivation as they seek enjoyable activities. They are also able to meet external demands, even if these demands temporarily delay their engagement with their desired activities. This finding demonstrates how intrinsic motivation can be sustained across a variety of positive and negative demands (Isen & Reeve, 2005). Intrinsic motivation is certainly improved by social–environmental factors, such as support from family, teachers and coaches; cultural values that are aligned with the desired activities; and opportunities to participate in performing arts activities (Hallam, 2002; Hassandra, Goudas, & Chroni, 2003). When performers are allowed to pursue their ideal standards and set goals, rather than responding out of a sense of obligation, greater creativity is possible (Zabelina, Felps, & Blanton, 2013). They are intrinsically motivated to excel.

Extrinsic motivation, on the other hand, has generally been regarded as a deterrent to creativity. External judgment and expectation can constrain performance; they ultimately divide attention and operate as a distraction. When these external forces are amplified, performance can be severely compromised; however, if they are constructive and informative, goal achievement is enhanced (Collins & Amabile, 1999). Consequently, appropriate extrinsic motivators can improve competency and facilitate both intrinsic motivation and goal achievement. Optimally, extrinsic motivators operate synergistically with intrinsic motivation; at moderate levels of intensity they promote sustained efforts to master skills and solve problems (Prabhu et al., 2008). They can reinforce an intrinsic motivation to engage in a desired activity (Collins & Amabile, 1999).

Performing artists bring a unique variable to the *synergistic intrinsic–extrinsic model*. Because they usually perform within group settings, their obligation and responsibility to the group is great. In these conditions, although obligation might be regarded as an extrinsic motivator, it operates as an intrinsic motivator and enhances efforts to meet external performance goals. Giving a strong performance is highly rewarding for performing artists, thus a synergistic balance between extrinsic, intrinsic, and obligation motivation facilitates this outcome (Cooper & Jayatilaka, 2006). For example, this synergistic melding of motivating factors was strongly present in the success of groups such as the Beatles. They were strongly intrinsically motivated, driven by external rewards such as success and fame, and they were deeply obligated to one another, even if they were competitively rivaling one another (Clyesdale, 2006). The group interaction of the Beatles is one example that typifies most successful performing arts groups, whether in music, theater, or dance (Fig. 12.3).

Figure 12.3 "Group flow." Performers: Dayna Charlot, Kenneth Mayfield, Diontay Odom, Janelle Whittaker. Photographer: Lee Choo.

Flow and motivation

Flow experiences, by definition, are intrinsically motivated, although external factors influence whether flow is attained or not (Csikszentmihalyi, 1990). Empirical evidence demonstrated an underlying source for intrinsic motivation, the possession of sufficient skills to meet the demands of a challenging task. When sufficient skills are present to succeed in an activity, the potential to experience flow is great, and the induction and maintenance of intrinsic motivation reinforces the quality of this experience (Keller & Bless, 2008). Intrinsic motivation also operates as the source for achieving flow, beyond the need to balance skills with challenges (Busch et al., 2013). These interacting factors are certainly inherent in individuals with autotelic personalities. They are able to draw upon intrinsic motivation and optimistically meet external demands, including when these external demands become extrinsic motivators (Busch et al., 2013). This was evident in male ballet dancers who were strongly intrinsically motivated; they also experienced high flow states (Levin, 2006). A similar result was found in theater actors (Martin & Cutler, 2002) and musicians (Bernard, 2009). These results are amplified when intrinsic motivation is matched with self-determined extrinsic motivation. Coupled with the intrinsic and extrinsic conditions, greater flow is achieved when the performers perceive that they possess greater autonomy and competence (Kowal & Fortier, 1999; Mills & Fullagar, 2008).

Harmonious and obsessional passion

Performing artists have a deep passion and dedication to excel in the aesthetic expression of their particular performing art. They have a deep love for it. This is clearly expressed by Merce Cunningham, one of the leading figures in early modern dance. He stated, "You have to love dancing to stick to it. It gives you nothing back, no manuscripts to store away, no paintings to show on walls and maybe hang in museums, no poems to be printed and sold, nothing but that single fleeting moment when you feel alive" (retrieved from www.goodreads.com/quotes/tag/dancing). Passion captures a deeper emotional state than what is traditionally defined in intrinsic motivation (St-Louis & Vallerand, 2015). Indeed, passion drives the emotional experience of the creative process. Being passionate is a specific relationship that is developed through innate talent, desire to explore the talent, and opportunities to excel in the talent (Mageau et al., 2009).

The *dualistic model of passion* delineates two opposing forms of passion, harmonious versus obsessional passion (St-Louis & Vallerand, 2015). Harmonious passion implies that performing artists shape part of their identity around the activity they desire. For example, most performing artists will self-define as musicians, singers, dancers, or actors (Kogan, 2002). They internalize the activity they love into their identity. The activity is a significant part of their life but they are also able to maintain and find meaning in other interests and relationships with others. Engaging in their passionate activity does not provoke conflict in their life; hence it is defined as harmonious passion. Obsessive passions indicate that the performer experiences an uncontrollable need to participate in the activity; however, their internalization of the activity and their self-identity is also contingent on acceptance. Their self-esteem is diminished by negative emotions, rumination, physical and psychological tension, and difficulties maintaining relationships (Mageau et al., 2009).

Evidence demonstrates the positive effects of harmonious passion and the deleterious effects of obsessive passion (St-Louis & Vallerand, 2015), but just how did one form emerge as the dominant factor? Earlier studies identified two different pathways for talented individuals, such as a young child who actively seeks playing the violin. With early exposure, dedicated time to excel in playing the violin, and family and teachers who value the child's interest (autonomy-supportive social environments), the desired activity grows into a passion and is readily assimilated into self-identity. This child becomes a violinist. These early repeated experiences are in concert with the performing artists' interests and talents, and through development, they become integrated into an harmonious sense of self. A passionate love for the art and a life filled with this artistic medium emerges (Mageau et al., 2009). Obsessive passion, unfortunately, develops from a less positive developmental trajectory. Controlling behaviors exerted on the child, such as parental and teacher demands and expectations, pressure to achieve that is contingent on parental and teacher acceptance, and coercive actions to select a particular activity and sacrifice other genuine interests all provoke an identity formation shaped by the activity. Obsessional passion is integrally related to early insecure attachment formations (discussed in chapter: Attachment, Parenting, and Childhood Adversity). The child begins to self-define as a musician but intense pressure is also internalized. It becomes a passion but is driven by an obsessional need to be accepted and valued without an internalized sense of an autonomous self (Mageau et al., 2009).

Positive emotions reinforce and are promoted by harmonious passions. Likewise, negative emotions reinforce and are evoked by obsessive passions. Both the emotional states and the passion orientations influence performance, either positively or negatively, respectively (Vallerand et al., 2008). These dualistic passions delineate two pathways: one that moves directly toward goal mastery, deliberate practice, and successful performance, and the other that leads to goal avoidance, inefficient practice modalities, and poor performance (Vallerand et al., 2008). For musicians, harmonious passion positively predicted subjective wellbeing and adaptive elite performance (Bonneville-Roussy, Lavigne, & Vallerand, 2010). The same pattern was found in dancers; whereas, obsessive passion was strongly associated with perfectionism, eating disorders, and poor self-esteem (Padham & Aujla, 2014). Obsessional passion was also found in performing artists who displayed greater internalized perfectionism, dysfunctional approaches to seek achievement, and higher rates of burnout (Hall & Hill, 2012).

When injured, dancers who are harmoniously passionate tend to display problem-focused and health-promoting strategies. They are more flexible in their rehabilitation process and actively seek future injury prevention skills. Obsessive passionate dancers, on the other hand, suffered longer with their injuries, frequently developed chronic injuries, in part because they ignored early signs of pain and discomfort, and rigidly insisted on participating in dance training and performance despite injuries. They had difficulty accepting injury and suffered greater loss of psychological self-esteem when injured (Rip, Fortin, & Vallerand, 2006).

Based on substantial research findings, harmonious passion is clearly the preferred pathway to achieve success in the performing arts. Harmonious passion positively enhances innovative cognitive styles (Luh & Lu, 2012). It is also related to the Big Five factor openness to experience, a finding that reinforces its association with creativity. Obsessional passion, on the other hand, was associated with the personality factor neuroticism, and an excessive need for control (Luh & Lu, 2012). Although harmonious and obsessional passions have been identified as intrinsic motivational factors, like motivation, they are also powerful variables in personality (Luh & Lu, 2012). These dual passions are powerfully formed by early attachment experiences and early childhood developmental influences. Family, teachers, and cultural values directly influence the trajectory toward harmonious passion and an autonomous open sense of self versus an obsessional passion and a controlling neurotic sense of self. For performing artists with a strong passion for a particular art form, they can thrive within their talent domain or they can struggle, depending on the manifestation of their passion.

Summary

- Introduction: temperament and personality: definitions (genetic and environmental influences), challenges of identifying types of personality and temperament, Eysenck's three-factor model, Big Five personality model.
- The creative personality: integration of personality models (extraversion, openness to experience, conscientiousness), overexcitability traits, Martinsen's seven-factor model of affective-motivational traits.

- Autotelic personality, flow, and the creative personality: alignment of autotelic personality with Big Five personality traits, optimal responses during stress.
- Personality in performing artists: general traits, musicians, actors, dancers, singers.
- Motivation: introduction (definitions).
- Intrinsic, extrinsic and synergistic models of motivation: complex interaction of internal and external rewards and drives, approach/avoidance model, goal achievement, synergistic intrinsic–extrinsic model.
- Flow and motivation: intrinsic motivation, influence of flow dimensions (e.g., skill–challenge balance, clear goals, unambiguous feedback).
- Harmonious and obsessional passion: definitions and influences of both forms on talent, influence of positive and negative emotions on passion, injury and recover.

Concluding remarks

When discussing personality factors it is critical not to lump performing artists into a larger category such as "artists" (Haller & Courvoisier, 2010). Although some personality traits may be shared, performing artists display great individual differences (Marchant-Haycox & Wilson, 1992). Performing artist personality profiles are complex, often manifesting inherently contradictory traits such as engaging high levels of fantasy and yet maintaining a strong sense of reality (Haller & Courvoisier, 2010). Research studies have identified unique differences and similarities between musicians, actors, dancers, and singers. These variances offer some insight about career stressors that face these artists and how they may navigate these stressors given their unique talents, personality traits, and the motivation that propels them forward. Motivation in performing artists is a complex synergy of intrinsic, extrinsic, approach-avoidance, master-performance achievement, and obligatory factors, and is augmented or compromised by harmonious or obsessional passion. Both personality and motivation are key factors that influence performing artist identities. Ultimately, early desires to explore music, acting, dance, or singing are shaped by these variables.

Training the performing artist

13

I do not try to dance better than anyone else. I only try to dance better than myself.

Mikhail Baryshnikov

You don't enter a dance studio and say "I can't do that." If you do, then why are you in the studio in the first place?

Judith Jamison, Dancing Spirit: An Autobiography

Introduction

The training of performing artists is a long process that frequently involves several teachers, different training environments, and a variety of performance experiences necessary to cultivate skill development, including the aesthetic expressiveness vital to any performing art. Even before training begins, children must be exposed to the arts. The experience of seeing live performance enables them to creatively explore their own situations and shape these experiences in structures similar to the plays, operas, or dances that they witnessed (Schiller, 2005). Children rapidly assimilate cultural diversity as they incorporate these experiences into their own creative performances. They readily acknowledge increased meaning-making during their participation in the arts and this is directly enhanced by witnessing professional performances (Schiller, 2005). These positive experiences foster a desire for further engagement in the training process. However, without exposure to the performing arts, none of these positive results emerge (Schiller, 2005).

Regardless of age, all talented performing artists progress from beginners who demonstrate ability, to students gaining competence, and eventually to experts who work professionally (Olszewski-Kubilius, Subotnik, & Worrell, 2015). In order to acquire expertise, all performers must cultivate skills to sustain the rigors of training. Motivation and persistence are key factors; teachers, coaches, mentors, and family members all help the maturing performer sustain motivation during times when training inevitably becomes tedious and frustrating or when the young performer falters during a performance (Olszewski-Kubilius et al., 2015). Musicians, singers, actors, and dancers learn technical skills that must be practiced in three very different environments: class, rehearsal halls, and theaters (or film/television sets or sound stages). These three settings are uniquely different and require skills specific to each setting. Other components that support performing artists' training include the performers' ability to assume responsibility for learning, accurately evaluate and address strengths and weaknesses, regulate moods and emotions

Creativity and the Performing Artist. DOI: http://dx.doi.org/10.1016/B978-0-12-804051-5.00013-5

so that they will not interfere with training, and cultivate support from others while building a network with other performing artists (Olszewski-Kubilius et al., 2015).

Training creativity

Performers require long-term focused training to achieve domain-specific skills; however, acquiring domain-general creativity skills such as ideational fluency and originality of ideas is also needed (Hennessey & Amabile, 2010). Gaining more domain-specific knowledge enhances an ability to discern originality. Educating performers typically includes studying a canon of work within their field. This formal education includes how to analyze and interpret these works. For example, a musician learns about compositions written specifically for their musical instrument or voice type, actors learn about past and present playwrights and their specific theatrical genres, and dancers learn the repertoire of major choreographic figures and their associated dance styles. Direct exercises to enhance ideational fluency, in particular, divergent thinking exercises, can be introduced during these compositional analyses. Encouraging performing artists to generate remote associations related to a composition (score, dance, script) enhances greater divergent thinking; however, these exercises will not specifically increase originality (Hennessey & Amabile, 2010). Performing artists also need to train the capacity to assess creative ideas for originality and value. Through repeated experiences, the ability to generate images and, most importantly, appraise them, becomes the key to enhancing creativity (Perez-Fabello & Campos, 2007). Performers must also be able to use these skills to enhance performance expressivity (LeBoutillier & Marks, 2003).

Creativity is augmented when performers are trained in environments that balance organized and unstructured exercises (Bowers et al., 2014). Performers cultivate greater creativity when they are given structured training in their domain along with opportunities to explore freely. Improvisational exercises are fundamentally unstructured exercises and they enhance cognitive fluency and originality (Beaty, 2015; Matare, 2009; Sawyer, 2000), including performance-specific kinesthetic and cognitive fluency (Bournelli et al., 2009; Jackson & Decety, 2004; Scibinetti, Tocci, & Pesce, 2011). Music, theater and dance improvisation exercises increase self-confidence, promote creative abilities, and provide opportunities to engage in divergent thinking. These positive results do not require long improvisational explorations; rather, brief periods of exploration, interspersed within structured skill training, optimize creative abilities (Sowden, Clements, Redlich, & Lewis, 2015). For example, musicians and singers can play with musical ideas while learning a new score, actors can engage in group improvisation or generate self-scripted monologues during a rehearsal process, and dancers can freely explore movement ideas in solo or group settings when they begin learning a new dance. Performing artists need to generate cognitive and motor ideas, as well as integrate emotional expression into these ideas. This integration of cognitive/motor idea generation and emotional expressivity enhances self-reflection, self-awareness, self-confidence, and promotes a ritualized experience; they are all associated with performance creativity (Nagy, 2015; Pagona & Costas, 2008).

During structured and unstructured training exercises, teachers and artistic directors (i.e., coaches) frequently offer criticism to performers. The nature of their criticism is to enhance the quality of performance given by the performer. Mutual respect and trust between the performer and the expert develops through interactions that promote creative thinking and problem solving. When an expert offers criticism it is founded on years of experience as well as a creative response that occurs during the interaction (Gibson & Mumford, 2013). For example, a stage director may identify a problem and then challenge an actor to explore a different set of ideas during the delivery of a monologue. In order for the stage director to conceive of this new idea, the director had to examine the actor's skills, the content of the monologue, and the desired effect required to engage an audience. The critique produced by the director demonstrates a depth of understanding, specificity to the immediate situation, and usefulness for the actor. This form of complex criticism enhances creative responses in the actor and the director; however, if this same director offers a large number of criticisms then the actor will not be able to assimilate the suggestions. Criticism then becomes a constraint to creativity and to performance (Gibson & Mumford, 2013). Ultimately, great teachers, coaches, and directors have cultivated skills to generate criticism that contains a depth of knowledge, a creative capacity to enhance the performer's creative expression, and an ability to perceive the utility of the criticism (Gibson & Mumford, 2013).

Self-evaluation, or self-judgment, is integral to the appraisal process of idea generation and creativity. Training performers to be creative artists should include tools to regulate and optimize self-judgment. Just as too much expert criticism can constrain creativity, harsh self-judgment can also constrain creative originality. Self-compassion can counter-balance harsh or punitive self-judgment. It is associated with intrinsic motivation as well as an orientation toward skill mastery (Neff, Hsieh, & Dejitterat, 2005). Self-compassion, as a skill to augment the creation of original ideas, is essential to counter perfectionistic and shame-based traits that often plague performing artists (Eusanio et al., 2014; Thomson & Jaque, 2013, 2015a, Thomson, Jaque, & Baltz, 2017). Just as complex, depth-oriented criticism enhances creativity (Gibson & Mumford), self-judgment can also be trained so that it offers depth-oriented appraisals, and not simplistic self-censure (Zabelina & Robinson, 2010b; Fig. 13.1).

Deliberate practice and other factors leading to expertise

Becoming an expert in any field typically requires 10 years of training; however, improvement continues beyond the initial 10 years (Hass & Weisberg, 2015). Certainly talent is important to become an expert in the performing arts, but expertise that persists is dependent on the quality of practice (Ericsson, 2007, 2014). Deliberate practice is identified as efficient and optimal practice. For example,

Figure 13.1 "Stage challenges." Performer: Lilia Kibarska. Photographer: Lee Choo.

performing artists learn how to parse out basic skills that support complex skill performance. They divide skills into smaller "chunks" to practice on a daily basis (Krasnow & Wilmerding, 2015). Focused attention and concentration make the essential repetition of skills productive. Performers acquire cognitive and motor learning strategies, while also learning how to self-regulate emotional and physical demands (Ericsson, 2007). Deliberate practice, and all the attributes associated with it, support skill expertise and career longevity. This pattern is evident in older musicians who retain performance speed and memory retrieval; they match the speed and memory that are present in younger performers. This is not the same in older amateur musicians who engage in shorter periods of deliberate practice (Krampe & Ericsson, 1996). Expert musicians practiced longer than less accomplished performers (Ericsson, 2014). Unfortunately, more obsessional-oriented performing artists may practice for the same length of time, or even longer, but fail to develop self-regulatory qualities inherent in deliberate practice; consequently they are typically more injured and often struggle with goal avoidance (Vallerand et al., 2008).

Other factors that contribute to professional expertise include the starting age of a performer. Usually training at a young age enables more time for deliberate practice; however, many performers start in late childhood or adolescence and they can still excel (Hambrick et al., 2014). Key ingredients for success include general intelligence. It is required in order to comprehend the complexity in the performing arts. A strong working memory capacity is vital during deliberate practice, as is cognitive flexibility and problem solving. Genetic heritability also influences aptitude (Hambrick et al., 2014). Domain-specific innate talents are vital, such as greater musical audiation for musicians, singers, actors, and physical aptitude in dancers (Ruthsatz et al., 2008). Along with deliberate practice, combinations of these factors converge to develop expertise. Unfortunately, a professional career in the performing arts is not solely due to hard work.

According to the achievement goal theory, motivating performers to excel includes instilling feelings of competence (Hancox, Quested, & Duda, 2015). The ability to perceive competence is contingent on degree of task involvement, goals, and self-comparison with other performers. While training, performers must view themselves as successful. They need to recognize that they have tried their best and have improved as a result of these efforts. They are also simultaneously aware that they must achieve superior performance abilities if they are to become a professional performer (Hancox et al., 2015). These factors, along with a strong sense of self-determination, powerfully influence goal achievement (Kowal & Fortier, 1999). Because performing artists must succeed in public domains, intrinsic motivation and appropriate goal setting enhances achievement. As Baryshnikov stated, "I do not try to dance better than anyone else. I only try to dance better than myself." Intrinsic goals promote better performances, longer career commitment, and greater life satisfaction; whereas, performance-approach goals tend to increase anxiety while performance-avoidance goals decrease life satisfaction (Lacaille et al., 2007). This pattern of goal success is unique to performing artists. Athletic and general student educational goal achievement is typically contingent on mastery-performance goal setting. Performers must set mastery-goals to become accomplished but, surprisingly, these goals do not transfer to situations when they perform publicly. Intrinsic goals enhance public performance and reinforce the aesthetic experiences essential to their art form (Lacaille et al., 2007).

Training performing artists requires educating them about their domain-specific field as well as augmenting their creative capacity. They must successfully achieve their goals, including a powerful ability to perform with aesthetic and emotional expression. Gaining a complex ability to evaluate their performance with a depth of understanding that is matched by nurturing self-compassion promotes success as performing artists. Incorporating all the components of deliberate practice, coupled with structured and unstructured exercises, will further increase performance success. Ultimately, the intricate relationship between the well-informed educator and the aptitude of the performer facilitates the complex layers required to promote creativity in the performing arts.

Training practices within domain-specific fields

Musicians

Professional training programs for musicians tend to resemble each other, a reality that exists in all major countries (Bartle, 1990). In all programs, curriculum includes training in at least one instrument, as well as solo and ensemble courses and performance opportunities. All offer training in aural perception, musical theory, and history; however, few offer formal courses focusing on "survival skills" needed to secure and maintain a musical career (Bartle, 1990). Very few programs address the psychological skills necessary to manage performance stress and anxiety, and even fewer provide or encourage physical conditioning. Some programs

Figure 13.2 "Mentoring chamber orchestra." Performers: David Aks, James Dorsa, Megan Gilbey, Amara Gyulnazarova, Ramon Osa. Photographer: Lee Choo.

attempt to address injury rate, primarily due to physical misalignment and poor physical usage. The most common college course that addresses misalignment is the Alexander Technique, a technique that successfully changes inefficient motor habits. Because musicians are vulnerable to potential career-ending injuries and burnout, they need courses in psychological skill training and physical conditioning (Bartle, 1990; Fig. 13.2).

In a series of interviews, music conservatory faculty identified very specific criteria to ensure professional success (Jarvin & Subotnik, 2010). They claimed that students must have the ability to learn quickly, have substantial musicality, and have strong abilities to recognize and analyze patterns and structures. The ability to learn must be matched by a desire to learn. According to conservatory faculty, the musician must be teachable, and this can be optimized when the student–teacher relationship is strong and positive. Students who enter conservatory programs are typically screened based on technical proficiency and this proficiency will be further developed during their years of study in the conservatory. Parental or mentor support helps offset economic and emotional stress, essential factors to help musicians engage fully in their training. Beyond external support, student musicians must persist and commit to the learning environment offered in conservatory programs. Most importantly, they must demonstrate persistence when things seem to falter. Intrinsic motivation must be present from the beginning, a trait that helps during times of difficulty. Self-awareness, self-confidence, and social skills all help music students identify their strengths and weaknesses, mobilize efforts to address weaknesses, and cultivate promotional opportunities necessary to prepare for a professional career. Beyond musical ability and training, to succeed in a musical career, conservatory students require training in social and personal skills vital to overcome frustrations and failures, as well as promote and cultivate career opportunities (Bartle, 1990).

The effects of musical training are profound, especially when the training occurs in childhood. Increased interhemispheric connections form as a result of playing and processing music (Patston, Kirk, Rolfe, Corballis, & Tippett, 2007; Schlaug et al., 2009; Schlaug, Norton, Overy, & Winner, 2005). Not surprisingly, the most consistent brain regions to benefit from musical training include the auditory, visual, and sensorimotor cortices (Watanabe, Savion-Lemieux, & Penhune, 2007). These integrative regions respond directly to experience-dependent learning, especially when musical skills are developed in the sensitive period of motor development during early childhood (Ohnishi et al., 2001; Watanabe et al., 2007). Musical practice and training enhances integration in neural systems in the developing brain of the child; this enrichment persists into adulthood (Hyde et al., 2009). For example, early musical training, in particular before the age of seven years, promotes pitch recognition and timing sensitivity (Kraus & Chandrasekaran, 2010; Levitin & Rogers, 2005), which enhances speech recognition and verbal memory (Jakobson, Cuddy, & Kilgour, 2003; Franklin et al., 2008; Patston & Tippett, 2011; Strait, Parberry-Clark, Hittner, & Kraus, 2012). Even when musical training begins later in the lifespan, similar neural networks are formed suggesting the powerful effects musical training has on neural plasticity and neural integration; however, this training must go beyond simply listening to music. Visual, auditory, and motor regions must be engaged during musical training to achieve neural integration (Montinaro, 2010). For example, two major convergence regions, the supramarginal gyrus and dorsolateral cerebellum, are directly involved in pitch memory; consequently musical training (pitch recognition) enhances both verbal and musical memory (Gaab, Gaser, Zaehle, Jancke, & Schlaug, 2003). Musical training enhances the processing of emotional resonance in subcortical regions; including increased perception of vocally expressed emotion, and faster auditory processing of emotional cues (Musacchia, Sams, Skoe, & Kraus, 2007; Strait, Kraus, Skoe, & Ashley, 2009). With repeated training, the cerebellar hemodynamic responses not only offset the need for increased motor cortical activation, but also establish a unique pattern of dominant/nondominant handedness activation. These cerebellar adaptations enable trained musicians to direct cortical activation toward more refined aesthetic expression while maintaining motor control during complex motor activities (Abdul-Kareem et al., 2011; Koeneke, Lutz, Wustenberg, & Jancke, 2004). Fundamentally, musical training promotes auditory fitness (Kraus & Chandrasekaran, 2010), a resource that enables faster processing of verbal and emotional cues, and enhances memory retention and retrieval (Jakobson et al., 2003). Coactivation of the auditory and motor cortices only occurs while playing music, and this coactivation facilitates greater cortical integration in general (Lotze, Scheler, Tan, Braun, & Birbaumer, 2003). Coactivation is also elicited when reading and playing music; greater integration in visual and motor systems are evidenced in trained musicians (Stewart et al., 2003). Furthermore, these training effects are not related to preexisting differences in children's brains; rather, musical training promotes these functional and structural brain changes, especially in the frontal corpus callosum, a region that connects the premotor and supplementary motor cortices in the two hemispheres (Schlaug et al., 2009).

Long-term musical training facilitates brain plasticity, a phenomenon that is directly related to increased cognitive adaptation and neural protection in aging (Hagen & Bryant, 2003; Herholz & Zatorre, 2012). Years of training directly shape functions and structures within the auditory and motor cortices. The complex multimodal systems in the brain are also directly enhanced, as seen in better auditory, motor, and visual integration (Conde, Altenmuller, Villringer & Ragert, 2012; Harris & de Jong, 2014). During a musical performance, stimuli from several modalities must be processed rapidly and precisely, including cognitive abilities such as processing speed and executive functioning (Patston & Tippett, 2011). For example, a violinist must simultaneously integrate visual stimuli from the musical score, interpret and respond to the conductor's gestures, attune to the music produced by other members of the orchestra, and organize motor activation and emotional expressivity while playing. Integrating all these responses is achieved by rapid feedforward and feedback signals produced in unisensory regions that are then integrated in the multisensory regions of the brain (Herholz & Zatorre, 2012).

A unique training protocol has been developed that capitalizes on the neural feedforward and feedback integrative processes. The real-time sound visualization computer program provides immediate visual feedback related to a student's performance, in particular, accuracy of pitch, harmony, and loudness. This visual feedback loop system is a powerful training tool and reinforces the natural neurological changes that occur during musical training (Ferguson, Moere, & Cabrera, 2005). Like all musical training, the act of performing a musical work requires a convergence of visual–motor, auditory–motor, emotional–motor, and cognitive–motor networks. Any training that enhances feedforward and feedback integrative processes enriches musical training (Parsons, Sergent, Hodges, & Fox, 2005).

Although long hours of deliberate practice can be tedious, and sometimes frustrating, hearing music is also a reward-inducing activity. These activities increase the brain's reward circuit, including increased profusion of dopamine, a neurotransmitter that not only promotes pleasure but is also critical in facilitating motor coordination (Herholz & Zatorre, 2012). Seeking music experiences also strengthens creativity, a result that adds to the neurological changes in the reward circuit (Goncy & Waehler, 2006). This complex neurobiological response to playing music augments social bonding and empathy, phenomena that occur between fellow musicians, musicians and their audience, and in student–teacher interactions. Powerful bonds are built during training and are reinforced by the reward circuit activation while playing music (Blood & Zatorre, 2001; Wollner, 2012).

Listening to and playing music enhances emotional regulation, offsets boredom, and augments intellectual activities (Chamorro-Premuzic, Furnham, Goma-i-Freixanet, & Muro, 2009). During years of training and performing, musicians commonly cultivate a positive relationship with the instrument they play. This positive relationship enhances emotional wellbeing and optimal (flow) experiences and decreases performance anxiety and perceived performance pressure (Simeons & Tervaniemi, 2013). Teachers and coaches must help musicians nurture this positive relationship with their instrument; if their instrument is regarded as an obstacle, then musicians suffer greater anxiety and decreased flow. High responsiveness to beauty and excellence emerge as a consequence to years of musical practice and performance, and soloists are even more sensitive to artistic beauty and excellence compared to music teachers and orchestra players (Gusewell & Ruch, 2014). In a more global sense, musical training reinforces enculturation. Young musicians assimilate deep understanding about cultural tonalities and compositional structures; this knowledge enriches intrinsic cultural aesthetic values, all processed in the attentional and cognitive regions in the brain (Hannon & Trainor, 2007).

Effectively training musicians requires emotional and psychological support from teachers and coaches, especially because musicians spend long hours alone practicing (Nagel, 2009). With adequate emotional support and psychological skill training, young musicians learn to manage performance stress, a key factor in training musicians who will become professionals (Boucher & Ryan, 2011; Osborne, Greene, & Immel, 2014). Music schools that provide psychological services can also help student musicians who suffer severe performance anxiety (Kenny et al., 2014).

Singers

Opera singer training remains similar at all major institutions around the world. The curriculum includes courses in vocal production, music theory, language studies (primarily French, German, Italian, and sometimes, Russian and Spanish), acting, stagecraft, and some form of movement education (Bartle, 1990). Like musician training, curriculum for opera singers often excludes psychological skill training and they have limited education in career preparation, including recording experience (Bartle, 1990; Bennett, 2009). During the training years, singers gain greater precision in auditory processing; they derive similar neurological auditory and motor integration capacities as musicians. They also have greater ability to discern auditory modulation and emotional expression in speech, plus greater memory retention for sound patterning (Herholz & Zatorre, 2012; Patel, 2012; Strait et al., 2012).

Singers train to modulate primary musical elements such as pitch, intonation contour, loudness, timing, vocal quality, articulation, vibrato, and timbre. These vocal manipulations enable emotional expressivity and musicality, all factors in vocal aesthetic appeal. Acoustic patterns and sequences exist within both spoken and musical communication and both convey accurate emotional states (Juslin & Laukka, 2003). Singers are in a unique position because they express emotions vocally and musically; the integration of voice and music makes singing one of the most emotionally expressive performing arts (Juslin & Laukka, 2003; Fig. 13.3).

Training singers alters their brain structures and functioning. The complexity of human song (i.e., harmonization, monophonic, polyphonic, and antiphonal imitation) indicates that it requires neural specialization that has evolved over human evolution (Brown, Martinez, Hodges, Fox, & Parsons, 2004). Singers have enhanced bilateral feedback integration in the vocal–motor and sensory–motor regions of their brain (Kleber et al., 2013). Years of practice also alter bilateral prefrontal processes and subcortical regions, with increased activation in the basal ganglia, thalamus, and cerebellum. Singing expertise (i.e., sound production, harmony and melody discrimination) increases enhanced kinesthetic motor control and sensorimotor integration (Brown & Martinez, 2007). Fundamentally, singing skill

Figure 13.3 "Alone." Performer: Annie Sherman. Photographer: Emma Rosenblatt.

acquisition promotes greater interoceptive awareness and facilitates greater implicit motor engagement (Kleber et al., 2010). Singers cultivate a keen ability to simultaneously perceive sound production and initiate motor activation.

Unlike musicians, opera singers develop skills to perform dramatic roles within the opera canon. Similar to actors, they must explore the nature of the character they are to portray, a process that can increase psychological distress (Burgoyne et al., 1999). Each time singers prepare a role they must participate in a renarration process, one that requires musical fidelity to the score and inclusion of current values and aesthetics (Beech et al., 2012). Beyond character work, singers must acquire complex attentional skills to manage intense cognitive demands placed on them (Corlu et al., 2015). While on stage they must give authentic life to the character, interact dramatically with fellow singers, adhere to the stage directions and blocking established by the director, and constantly maintain eye contact with the conductor so that they remain musically integrated with the orchestra. Years of training are essential to prepare them for these performance challenges.

Singers form a professional identity via their ability to produce vocally resonant emotional sounds, and through their unique bonds established between their teachers and between other singers (Beech et al., 2012). Through years of practice, singers remain focused on singing technique, their vocal apparatus, and their physical posture. Unlike amateur singers who experience singing as a means to reduce tension, and promote wellbeing, professional singers and conservatory students do not derive these same positive results. They actually experience less wellbeing and more physiological and psychological arousal and tension. They are achievement-oriented and they have an acute awareness of the rigors of their profession (Grape et al., 2003). If training programs do not address these negative outcomes, singers will struggle with performance tension; however, when training includes skills to regulate achievement-oriented goals, including debilitating perfectionism and shame, singers are able to optimize performance and increase self-esteem (Thomson et al., 2017). Because singing places great demands on the vocal apparatus, singers also need to be given very specific directions on how to manage vocal care, both self-directed and with the guidance of singing teachers, coaches, and medical professionals (Timmermans, Vanderwegen, & De Bodt, 2005).

Actors

Like singers, actors are dramatically and vocally trained to express emotion. They are equally able to convey a myriad of emotional states through their vocal intonation, tempo, pitch, vocal quality, loudness, and articulation (Juslin & Laukka, 2003). Actor training also includes character and script analysis, movement, improvisation, stage combat (i.e., martial arts, fencing), alignment work (i.e., Alexander Technique), and acting techniques (Lese, 2014; Wilkie, 2015). Actor training requires student actors to explore and enact monologues and scenes that are fraught with situational conflict and intense emotions; consequently acting students are extremely vulnerable during their training. Their acting, voice, and movement teachers profoundly influence them, positively or negatively (Seton, 2010); thus ethical practices must be implemented in acting classrooms, departments, and schools in order to create an environment that supports talent development (Seton, 2010). Student actors flourish when trained in environments that foster trust, respect, freedom, challenge, collaboration, and clarity; however, they can also fail when trained in environments that are not supportive (Nemiro, 1997).

Emotional expression is one of the primary requirements in actor training. Although actors assume a dramatic role and embody a character, the reality is that they physically and psychologically experience the emotions (Hetzler, 2008). Boundary blurring between the actor's self and character result, especially when the actor must perform intensely emotional scenes. This boundary blurring occurs regardless of professional expertise and prior training (Burgoyne et al., 1999). Optimal training programs provide tools to successfully prepare roles, including managing the emotional intensity required to portray the role (Noice & Noice, 2002). When acting training begins in early childhood, children directly assume roles and engage in dramatic play. These educational opportunities foster creativity and emotional regulatory skills to manage stress and anxiety (Gupta, 2009; Russ, 2007). Later, actors learn how to analyze scripts before they begin to actively experience the role. This analytic phase promotes cognitive–emotive–motor processing and integration (Noice & Noice, 2002). While analyzing the script and actively experiencing the role, actors cultivate an ability to empathize with the character to be performed (Verducci, 2000); during actor training they gain greater awareness of another person's/character's mind (theory of mind) (Goldstein et al., 2009; Goldstein & Winner, 2012; Fig. 13.4).

Actors frequently struggle with a fidelity toward the text they are to perform and their dramatic instincts. Ultimately, the tension between the playwright's words and the actors' living portrayal coexists, either harmoniously or discordantly (Pavis, 2008). One significant element that influences the text/performer relationship is the actor's vocal capacity (Master, De Biase, Chiari, & Laukkanen, 2008). Through actor training, actors learn to project their voices, and modulate volume and pitch

Figure 13.4 "The other." Performers: Maurice Godin, Rose Williams. Photographer: Cylla von Tiedemann.

to enrich emotional expressivity. This ability requires intense training for several years, especially if a student actor begins with limited vocal production, or worse, vocal disorders (Walzak, McCabe, Madill, & Sheard, 2008). Another factor that influences fidelity to script is the actor's ability to memorize. Actors learn how to analyze scripts; they section the text into units of action and these units of action are then memorized verbatim. Actors also incorporate the memorized units of action into real-time sensorimotor actions, which reinforces their cognitive memorization (Noice & Noice, 2002, 2006). For actors, language and action are merged, a process that is also evidenced in singers, dancers, and musicians. Increased neurological feedback responses between the cortical and subcortical regions of the brain facilitate greater integration of complex activities (Lotze et al., 2003), including memorizing and performing large amounts of dialogue. This ability to memorize large amounts of dialogue far exceeds the general population (Noice & Noice, 2002, 2006). To manage complex concurrent demands they cultivate attentional flexibility and socially sensitive awareness (Thomson & Godin, 2011; Thomson & Jaque, 2012b).

Like singers and musicians, vocal–motor and sensorimotor neural network integration is promoted during actor training. Actors learn to discriminate and decode pitch in verbal discourse (Gaab et al., 2003; Levitin & Rogers, 2005). Auditory–motor integration is experience-dependent; consequently actor training reinforces implicit motor neural responses to auditory feedback. This integration enables actors to differentiate task-relevant and task-irrelevant auditory and visual signals (Conde et al., 2012). Actor training reinforces auditory–motor expertise (Dick, Lee, Nusbaum, & Price, 2011), and it also fosters ready access to emotional neural circuits. Actors are able to self-generate specific emotional states that draw upon an integration of complex array of cortical and subcortical regions (Pelletier et al., 2003). Despite common beliefs that actors' are "faking" emotional responses, neural research studies demonstrate that they are activating primary emotional neural circuits. This enables them to perform with intense emotional expressivity; ideally their training cultivates skills in emotional regulation so that they can sustain the vocal–motor and emotional demands of performance.

Acting programs are generally successful at training actors; however, there are currently too many actors for too few jobs. A major shortfall in actor training curriculum includes no (or limited) psychological skills training, career preparation, and wider generic entrepreneurial skills that are transferable beyond an acting career (Wilkie, 2015). The reality that talented actors seek training regardless of parental encouragement and career stability does not alter the ethical dilemma facing actor training programs (Goldstein & Winner, 2009). Also, actors tend to enter formal actor training at a later age than musicians and dancers, in part, because of the complexity and multidimensional demands required to be a successful actor. To fully portray multidimensional characters, actors must comprehend this layered complexity and embody it. Consequently, many actors consciously elect to study this art form at a later stage in development, and they do so despite awareness that there are limited career opportunities (Kogan & Kangas, 2006).

Dancers

Dancers share many similar training experiences as musicians. They tend to start their studies at a young age; as children they are guided by a teacher and spend hours practicing on their own under the supervision of their parents. Later, they increase classes during the week, thus increasing contact time with teachers (Hutchinson, Sachs-Ericsson, & Ericsson, 2013). Females tend to begin dancing at an earlier age than male dancers; however, with accumulated deliberate practice dancers who begin at a later stage of development can reach expert status (Hutchinson et al., 2013). Providing adequate resources to support the training of dancers is essential, whether funds are provided by families or by state-run programs such as those in Russia, Cuba, or China. Other constraints to becoming a successful dancer, especially a ballet dancer, include body type, kinesthetic awareness, coordination, and musicality. Without these features, a dancer will not achieve expertise (Hutchinson et al., 2013).

All dance programs must train dancers to work effectively within class, rehearsal, and performance settings. Although overlapping, each setting has its own set of policies and requirements; dancers respond differently, both physically and psychologically, in all three settings (Quested, Duda, Ntoumanis, & Maxwell, 2013). Training environments, whether high-pressured conservatory programs or local dance schools, profoundly influence dancer anxiety and self-esteem; this can be exacerbated during performance seasons. Like acting teachers (Nemiro, 1997),

Figure 13.5 "Kinesthetic feedback." Performers: Cheyenne Spencer, Christian Vidaure. Photographer: Lee Choo.

dance educators can ameliorate anxiety and increase dancer self-esteem if they are respectful of the dancer, particularly when they challenge dancers to perform at their optimal level (Nordin-Bates, Quested, Walker, & Redding, 2012). Feedback from teachers and coaches is critical for improvement and does not necessarily mean that dancers' self-esteem will be compromised. Effective feedback must be accurate, objective, and informative. Dancers perceive that they can improve based on this quality of feedback, and as a result, they not only improve in performance skills but also they gain a greater sense of self-efficacy (Garcia-Dantas & Quested, 2015; Fig. 13.5).

Because dance aesthetic continually shifts over decades and centuries, teachers must be aware of these changes. They must train dancers to perform older repertory works while simultaneously preparing them for the physical rigors of today. These changing aesthetic values, such as greater verticality in today's dancers compared to dancers even 50 years ago places different physical demands on the dancer and the training programs that prepare the dancers. These diverse requirements may increase injury rates, in part because the audience expects to see dance performed at standards driven by a changing aesthetic environment (Daprati, Iosa, & Haggard, 2009).

Body image is a major concern in most dance training programs (Bettle, Bettle, Neumarker & Neumarker, 2001). Dancers spend long hours in highly revealing clothes and their technique is perfected in part from the watchful eye of the teacher but also from the reflection of their image in the mirror (Price & Pettijohn, 2006). These variables can promote pathological levels of perfectionism and shame and compromise a sense of self (Eusanio et al., 2014). Dancers are also acutely aware of cultural and media demands for slenderness (Heiland et al., 2008). Most dancers self-define based on dance stereotypes, such as the belief (and reality) that ballet dancers, ballroom dancers, folk dancers, and modern dancers require different physical proportions. Dancers within their own genre also tend to exaggerate the stereotyped requirements, a fact that increases anxiety and decreases self-esteem (Neiminen, 2000).

Promoting creativity within dance training requires a procedure of integrating structured classes along with unstructured exploratory movement investigation (Watson, Nordin-Bates, & Chappell, 2012). Dance teachers who engender creative responses from students often approach this task in a unique manner. Unlike cognitive measures to determine increased creativity, dance educators and students must cultivate an embodied practice of self-reflection. They think in action that operates at a procedural and declarative level (Stevens & McKechnie, 2005). Both teacher and student respond from a kinesthetic knowledge base that draws upon layered sensorimotor perceptual experiences (Chappell, 2007; Lussier-Ley & Durand-Bush, 2009). Dance teachers develop an ability to appraise turning velocity, balance duration, range of motion, and jump height. Whether implicit or explicit, expert trainers can evaluate each element independently, as well as how they are integrated during complex movements (Torrents, Castaner, Reverter, Morey, & Joffre, 2015). They also determine how these variables integrate to enhance aesthetic appeal, a factor critical for a successful career in dance. Dance educators and dance students ultimately cultivate a developed capacity for motor creativity, a process that powerfully enhances a sense of self during these creative experiences (Bournelli et al., 2009).

Dance training enhances organized perceptual–cognitive structures, which facilitates memory retention. Competent dancers have formed a lexicon of movement vocabulary that operates similar to spoken language (Blasing et al., 2009; Stevens & McKechnie, 2005). Further, body positioning helps encode autobiographical memories; all dancers recognize the ease of memory retrieval when they perform dance positions and these experiences reinforce their dancer identity (Dijkstra et al., 2007; Sheets-Johnstone, 1999). Dancers also learn to coordinate dance movement vocabulary to precise rhythmic structures and tempos. Their motor expertise is evident in neurobiological differences (compared to nondancers) that are formed during years of training. For example, greater activation is found in regions such as the premotor cortex, temporal–parietal lobes, and the mirror system (Calvo-Merino et al., 2005). Expert dancers are neurologically more efficient at integrating dance sequences to music; over time, their integrative processing of feedforward and feedback information is rapid and direct (Bar & DeSouza, 2016). These integrative neural networks directly influence balance, equilibrium, and coordination, factors that profoundly influence the efficiency of dance performance (Golomer, Mbongo, Toussaint, Cadiou, & Israel, 2010). Like the other performing arts domains, dancers rapidly engage complex regions in the brain that coordinate cognition, emotion, movement, and sensory stimuli. Sensorimotor convergence regions that discriminate, encode, and extract information, such as proprioception, integrate information from cortical and subcortical brain regions; they then mobilize responses that emerge as intricate dance movement (Brown, Martinez, & Parsons, 2006). Dancers also exhibit a greater genetic proclivity to release and transmit more serotonin, a neurotransmitter that enhances spiritual and communal experiences; experiences that dancers also subjectively report (Bachner-Melman et al., 2005).

Although dancers generally stop performing by their mid-forties, like older musicians, continued dance training postretirement enables the maintenance of balance, mobility, cognitive flexibility, and positive mood (Alpert et al., 2009; Coubard et al., 2011; Verghese, 2006). Little or no degradation in sensorimotor regions is evident in brain imaging studies of aging dancers (Kattenstroth et al., 2010). These positive gains are not exclusive to former professional dancers. Similar neurological and physiological training effects can be seen in older nondancers as well (Kattenstroth, Kalisch, Holt, Tegenthoff, & Dinse, 2013). The communal experience of dance may also increase serotonin release in nondancers. In short, dance improves quality of life, a finding that is currently influencing treatment protocols recommended for adults diagnosed with Parkinson's disease or dementia (Hackney, Kantorovich, & Earhart, 2007).

Within the dance-training environment, whether that is in private dance schools, conservatories, or college programs, high expectations are placed on dancers. Students are expected to attend multiple classes each day, rehearse long hours, and perform in a myriad of settings such as theaters, shopping malls, gymnasiums, and street fairs. This high-intensity workload increases physical and psychological stress and provides insufficient recovery time (Grove, Main & Sharp, 2013). After university or conservatory training most dance majors rely on nondance jobs to supplement wages as a dancer (Montgomery & Robinson, 2003). As with training of musicians, singers, and actors, a similar curricular problem exists in student dance training; courses in career preparation, skill training to cope with stress, and transferable skills to manage fiscal responsibilities while working professionally are all missing (Bennett, 2009). All dancers require education in injury prevention and injury rehabilitation (Grove et al., 2013). Private sector dance schools operate with even fewer curricular demands (Posey, 2002); consequently most professional dancers are unevenly prepared for a career in dance.

Summary

- Introduction: process, obstacles, and challenges.
- Training creativity: structured and unstructured exploration, domain-general and domain-specific creative training, criticism (external and self-evaluation/judgment) and influence on creativity.
- Deliberate practice and other factors leading to expertise: focus and attention, memorization skills, skills required to achieve expert status, achievement goal theory, self-determination.
- Training musicians: formal training programs and curriculum, social and personal skills, early versus late beginning in training, neurological changes due to long-term musical training, rewards and challenges.
- Training singers: formal training programs and curriculum, brain structures and singing, socialization and character work, identity as singer.
- Training actors: formal training programs and curriculum, emotional expression and brain development, memorization and fidelity to text, character development.

- Training dancers: formal training programs and curriculum, injury rates and kinesthetic aesthetics, perceptual–cognitive structures and memory retention, shorter careers and transitions.

Concluding remarks

Training performing artists is a substantial commitment for the performers, teachers, and training schools. The majority of programs successfully prepare talented artists to become professionals. Intensive and extensive training programs prepare them for the rigors of a performance career, and during this training, the performer's brain and body are altered. Years of training nurture cognitive and motor resources that ensure deep familiarity with the art form. At the neurological level, deliberate practice facilitates neural networks that are deeply integrated. In order to execute complex motor actions such as playing a musical instrument, singing, or dancing, neural networks in multiple sensorimotor regions of the body overlap and fuse during the training years. This is advantageous for the performance of complex movements; however, the fused networks can increase the potential for future overuse injury or dystonia (Shin, Kang, Hallett, & Sohn, 2012).

Effective physical, emotional, and psychological training prepares performers to optimally regulate physiological states of hyperarousal evoked during performance (Morris, Cleary, & Still, 2008). Through deliberate practice and expert teaching, performers adapt to the demands of their performance domain. Unfortunately, many of these highly successful training programs fail to include courses in career preparation, physical conditioning, and psychological skills (Bartle, 1990; Bennett, 2009). Performing artists need to be physically fit in order to decrease risk of injury and they need coping skills to manage the stress and tension that are part of their professional career (Olszewski-Kubilius et al., 2015).

Section IV

Emotion: Regulation, Resilience, and Pathology

Self-regulation, emotion, and resilience

14

Once in a while I experience an emotion onstage that is so gut-wrenching, so heart-stopping, that I could weep with gratitude and joy. The feeling catches and magnifies so rapidly that it threatens to engulf me.

***Julie Andrews Edwards from* Home: A Memoir of My Early Years**

Self-regulation

Self-regulation is a term that suggests an ability to control responses within the self. It is both the ability to regulate internal physiological responses outside of conscious awareness and psychological responses initiated "by" the self within conscious awareness (Vohs & Baumeister, 2004). Fundamentally, self-regulation is a process in which an individual's reactivity to internal and external stimuli is managed, including exerting control over impulses, emotions, and thoughts in an effort to accomplish tasks and interact with others (Vohs & Baumeister, 2004). The higher cortical regions of the brain, coupled with the anterior cingulate, operate as the self's executive function; these regions actively, consciously, and intentionally direct planning, initiating, monitoring, and revising ongoing responses (Hall et al., 2008). They enhance awareness of thoughts and emotions, providing an experience that the "self" is controlling and modulating behavior (Schmeichel & Baumeister, 2004). Self-regulation increases during the hierarchical maturation of the brain; a sense of self emerges during this developmental process (Schore, 1994).

Executive function relies on the integration of higher cortical regions that inhibit and govern subcortical structures, including the autonomic nervous system (ANS) (Banfield, Wyland, MacRae, Munte, & Heatherton, 2004; Segerstrom & Nes, 2007). Optimal executive functioning requires high metabolic resources and when these resources are depleted, self-regulation is also compromised (Schmeichel & Baumeister, 2004). Greater heart rate variability (HRV) also promotes self-regulation and this physiological regulating system is fully dependent on metabolic resources. HRV resources deplete during high performance demands and are minimally available when an individual is fatigued. Unfortunately, we have minimal awareness that our self-regulatory stores are depleting, despite great awareness that we are physically fatigued (Segerstrom & Nes, 2007). Greater metabolic resources ensure greater self-regulation. For performing artists, enhanced metabolic resources are nurtured during the long years of training (Baert, Casier, & De Raedt, 2012). Through deliberate practice over years, performers strengthen feedforward and

Creativity and the Performing Artist. DOI: http://dx.doi.org/10.1016/B978-0-12-804051-5.00014-7

feedback integrative self-regulatory systems within the ANS, the cortical associative regions, and prefrontal cortex (Bar & DeSouza, 2016; Herholz & Zatorre, 2012). When training decreases, whether as a result of injury or career change, rapid loss of metabolic resources results. Poorer self-regulation is manifested in increased pain, fatigue, and negative mood (Glass et al., 2004). In contrast, ongoing training supports these physiological and neurological adaptations, optimizing metabolic processing and strengthening self-regulation in performing artists.

Self-regulation is both behavioral (personality) and biological (temperament/disposition) (Rothbart et al., 2000; Rothbart, Ellis, & Posner, 2004). Reactivity is an individual response that is measured by the speed and intensity of any response to a stimulus, in particular, novel stimuli. For example, when performing a familiar work, performers can readily self-regulate; the metabolic demands are low. However, if they encounter an unexpected change such as a missed entrance by an actor, dancer, or singer, self-regulating demands increase. Greater physiological reactivity is manifested by acceleration in heart rate, profuse sweating, and racing thoughts. Ideally, through training effects, environmental support, and/or genetic predisposition, greater physiological regulation, via robust vagal tone (parasympathetic regulation), leads to decreased anxiety and increased self-regulation (Movius & Allen, 2005). Performers with higher vagal tone are less psychologically defensive and are more engaged in rewarding activities. Optimal self-regulation is directly related to how well performers manage novel events, a capacity that is influenced by temperament, early developmental experiences, and personality traits (Chavez-Eakle et al., 2012; Rothbart et al., 2000; Fig. 14.1).

Figure 14.1 "Revealing." Performer: Liezel DeGuzman. Photographer: Lee Choo.

Organize–disorganize–reorganize: the role of self-regulation in the creative process

The organize–disorganize–reorganize pattern described in Chapter 1, Understanding Creativity in the performing arts, outlines how this process accommodates complexity and the emerging properties inherent in seemingly chaotic patterns of thought or behavior (Lewis & Granic, 2000; Montuori, 2003). This recursive process of organizing ideas, deconstructing them, and reorganizing them prevails during the creative process. Throughout training and a performance career, performers must recognize organized structures or ideas. They must be willing to dismantle these structures, such as intricate motor patterns, to create new ways of executing skills. They also follow this sequence when reinvestigating familiar repertory prior to performance. Direct experience can make this process growth-enhancing, but it can also be growth-destroying. For those that prefer the stability of familiar organized states, the entrance into disorganization may be very unsettling and the urge to prematurely reorganize behaviors or responses may curtail the exploration of novelty. Other performers may relish the disorganization phase, linger in this state comfortably, and not exert efforts to reorganize and adapt to new patterns of behavior or thought. How these stages are managed represent the degree of self-regulation during creativity (Kirschenbaum, 1998).

When self-regulation works optimally, the ability to easily and fluidly shift states is evident. The performer encounters positive and negative experiences with a sense of integrated cohesion. A sense of self-efficacy prevails. The individual has an abiding confidence that a solution will be found (Montuori, 2003). Even when it is painful and difficult, the performer senses the task will be successfully completed. However, when self-regulation is not operating optimally, the ability to shift between organize–disorganize–reorganize states is compromised. Either a rigid adherence to a single state or a chaotic fluctuation between all three states is manifested (Siegel, 2012). Performers with a rigid stance are not able to self-regulate; managing change becomes a process of psychological and physiological strain. Not only do they struggle with regulating their creativity, in extreme cases, they also have great difficulty attending to the simple events of daily life. Likewise, the chaotic fluctuations of a dysregulated self stresses the entire self-system. These chaotic performers manifest diffuse emotional and physiological states that change abruptly and cannot be maintained for protracted periods of time. This pervasive chaotic state leads to extreme emotional and professional instability.

The organize–disorganize–reorganize concept describes open complex systems such as the universe, the human brain, and creativity (Montuori, 2003). When preparing for a career that involves engaging the creative process, the question of self-regulation must be considered. By achieving adequate self-regulation performers can persist and cope with the challenges that inevitably emerge in their careers. Although during states of disorganization, performers may appear dysregulated, they are able to tolerate these states as they move toward greater reorganization (Feist, 1998). Most performers have great determination to pursue a creative life and most tenaciously endure the inevitable ups and downs during years of training and performing.

Temperament

Talent is an accident of genes, and a responsibility.

Alan Rickman

Temperament is a term that generally describes stable biological individual differences in self-regulation, differences that appear in early infancy and remain throughout life. Two factors typically receive consensus among temperament researchers. First, temperament influences personality and motivation, but all three are regarded as different constructs, and two, temperament is primarily identified as an affective system that influences regulation (Clark, 2005; Goldsmith et al., 1987). Studies by Goldsmith suggest that infant emotional valence and intensity (i.e., anger, fear, pleasure, disgust) are early indicators of temperament; these emotional biases influence behavioral reactions (Goldsmith et al., 1987). Buss and Plomin identified three traits that indicate temperament: (1) emotional arousal (high- versus low-threshold reactivity to stimuli, particularly distress responses), (2) activity (tempo and vigor of movement such as lethargy or hypomanic energetic bursts of activity), and (3) sociability (preference to be alone or with others) (Goldsmith et al., 1987). Thomas and Chess described three clusters of infant temperamental behaviors that persist throughout adulthood, although sometimes in a modified version: (1) easy, (2) slow to warm up, and (3) difficult (see Box 14.1; Goldsmith et al., 1987; Thomas & Chess, 1977). Rothbart and colleagues outlined four temperamental dimensions that are clearly manifested in early and late infancy: (1) negative affective reactivity (behavioral and attentional aversion expressed as distress); (2) positive extraverted reactivity (behavioral and attentional approach as evidenced by positive affect and active movement toward a stimulus); (3) emerging near the end of the first year of infancy, behavioral inhibition and effortful control in order to respond to novel and/or intense stimuli; and (4) also emerging in late infancy, capacity to focus and shift attention (Goldsmith et al., 1987; Rothbart, Ahadi, Hershey, & Fisher, 2001). Based on these differing models of temperament, it is evident that researchers still have much to investigate.

Early infant temperament shapes how the developing baby will interact with the environment. For example, infants who are classified as "easy" or "behaviorally uninhibited" readily habituate to stimuli; they actively explore novelty, and are quickly soothed when distressed. Infants who are "slow to warm up" have more difficulty responding to new stimuli; whereas, infants who are classified as "difficult" or "behaviorally inhibited" demonstrate great difficulty adapting to new stimuli; they are generally anxious and fretful (Kagan & Snidman, 2004; Thomas & Chess, 1977). Performing artists who have "difficult" or "behaviorally inhibited" temperaments present as more introverted, anxious, or shy in childhood, a pattern of

Box 14.1 Temperament: Two Major Theories and Self-Regulation

Thomas and Chess (1977)		Kagan and Snidman (2004)	
Easy temperament	Readily approaches new situations; general mood is positive; good physiological functioning; adaptable; low–moderate stress reaction responses	***Behavioral inhibition temperament***	Low threshold of arousal; rapidly reactive to novel situations despite passive behavior; fearful, shy, and anxious predisposition
Slow-to-warm up temperament	Tendency to withdraw from new situations; somewhat negative mood; low physiological intensity reactions and low activity level responses; slow to adapt	***Behaviorally uninhibited temperament***	High threshold of arousal; responds positively to novel stimuli; active exploration of environment without fear or anxiety
Difficult temperament	Generally withdraw from new situations; negative emotions pervade; unusually intense emotional reactions; irregular physiological responses; slow to adapt and difficulty managing adaptation		

behavior that remains stable throughout adulthood. Without adequate support, the instability of the performing arts may be too much for the temperamental disposition of some talented artists classified as behaviorally inhibited or "difficult."

Because temperament is regarded as a biological trait, research has focused on identifying neurological patterns related to the different temperamental traits. Gray proposed a theory that encompasses two systems, one that regulates aversive stimuli and the other that regulates appetitive motivations (Carver & White, 1994). The aversive system has been labeled the *behavioral inhibition system* (BIS). It activates cortical and subcortical inhibitory regions in the brain, especially when anticipating negative or painful outcomes. Greater BIS sensitivity indicates a temperament that is inhibited, fearful, negative, and introverted (Carver & White, 1994; Derryberry & Reed, 1994). Individuals with this temperament tend to avoid harm as a coping strategy and are often regarded as shy (Cloninger, 1994; Miskovic & Schmidt, 2012).

The *behavioral approach system* (BAS) engages more dopaminergic pathways in the brain, especially because this system is biased toward reward anticipation and novelty seeking (Cloninger, 1994). Greater BAS sensitivity is evident in temperamental behaviors of extroverts who actively move toward goals, pleasure, and novelty (Carver & White, 1994; Miskovic & Schmidt, 2012).

According to Rothbart et al. (2001), the BAS system can be described as a temperament profile that includes extraversion, low shyness, high levels of motor activity, more impulsivity, more social affiliated behaviors, and intense positive emotional (pleasure, laughter, smiling) and anticipatory activation and responsiveness (Rothbart et al., 2001). A more global perspective of the BAS system suggests that it is a positive affective–emotional approach system (Clark, 2005). The BIS system on the other hand is a negative affective–emotional withdrawal system and is marked by negative affectivity (fear, frustration, and discomfort) and shyness (Clark 2005; Miskovic & Schmidt, 2012). Rothbart (2007) includes the development of a third temperamental trait, effortful control, as a way to self-regulate the BAS and BIS systems. Effortful control involves attention, focus, perceptual sensitivity, and the cultivation of low-intensity pleasure as a way to self-regulate. The positive–negative affective–emotional and approach–withdrawal temperaments are considered strong predictors for later psychopathology; however, environmental interactions have direct influences on protecting or propelling individuals toward psychopathology (Carter, 2005). During development, effortful control capacities will also influence the degree of acquiring externalizing or internalizing disorders (Carter, 2005).

The BAS (approach) temperament is a significant factor in establishing achievement goals, in particular, mastery and performance-approach goals. This temperament profile is ideal for performing artists; whereas, the BIS temperament significantly influences performance-approach and performance avoidance goals (Elliot & Thrash, 2002; see chapter: Personality and Motivation for more details). Achieving mastery requires moving toward challenges and persevering during the challenges in order to acquire mastery. The inhibited (avoidance) system is operational when negatively appraising performance mastery against the competencies of others. For example, inhibited performing artists may approach auditions or performances when they are certain they have achieved an appropriate level of competence that will ensure success; however, if they deem that they may not succeed then the inhibitory avoidance strategy will be mobilized to avoid failure. Performers with BAS temperaments will view situations as an opportunity to acquire or reinforce mastery, regardless of success or failure. They positively approach the situation and can emotionally frame and manage success or failure (Covington & Mueller, 2001; Elliot & Thrash, 2001).

When examining temperament related to creativity, a similar pattern emerged. Approach, including actively seeking mastery, was directly associated with creativity in a group of artists; whereas, the emotional reactivity manifested in the BIS was negatively related to creativity (Necka & Hlawacz, 2013). Although not necessarily advantageous for creativity, harm avoidance, as manifested in the BIS temperament, may operate as a protective factor, especially later in development if individuals are diagnosed with a psychiatric disorder (Akiskal & Akiskal, 2005). This temperament pattern in performing artists may support success or increase failure. Adopting harm avoidance strategies may actually protect them from injury, career-ending performance anxiety, or burnout, especially if they have a BIS temperament. Harm avoidance, a typical strategy that many shy (fearful) individuals adopt, can be minimized during performing arts training (Boucher & Ryan, 2011; Osborne et al., 2014). Supportive exposure to performance settings promotes effective self-regulation skills; shy performers can learn to manage their fear and harm avoidance responses. They may be reserved and introverted but they can learn how to engage socially and perform in public settings (Cheek & Buss, 1981).

Fundamentally, all temperament models describe approach and avoidance. The approach–avoidance systems are evolutionary survival systems; they are evident in all life forms and are embedded in all emotion, motivation, and personality models. They are predispositions that motivate behavioral responses based on implicit or unconscious appraisals of the environment. The earliest manifestation of infant temperament reveals biases toward more approach or avoidant responses, and environmental and early attachment experiences further shape these temperamental traits (Elliot & Covington, 2001). This two-factor model incorporates key variables inherent in motivation and personality studies as well as emotion regulation investigations. Approach temperament includes the BAS, extraversion, behavioral activation, and positive emotionality and the avoidant temperament contains the BIS, introversion, behavioral inhibition, and negative emotionality (Elliot & Thrash, 2002). These powerful biological responses influence all human behavior. Despite temperamental biases, the primal nature of approach and avoidance responses informs all interactions throughout the lifespan (Elliot & Thrash, 2002).

Temperament, attachment, and self-regulation

Temperament and attachment are different constructs (Sroufe, 1985). Although temperament is a stable trait that manifests in early infancy, attachment experiences are directly established by the quality of the relationship with primary parental or caregiving figures. One of the most formative influences shaping early self-regulation occurs in secure attachment experiences (Calkins, 2004); attachment security enhances optimal attentional flexibility and emotional regulation (Main, 2000). Ultimately, attachment experiences can ameliorate or exacerbate temperament traits and can facilitate or compromise self-regulatory capacities. For example, securely attached babies with temperaments that are considered shy, "difficult," or "behaviorally inhibited" can develop effective self-regulatory strategies to manage their low-threshold responsiveness to aversive or novel stimuli. These children have an abiding faith that they are safe to explore and when distressed they can be effectively soothed and protected by their attachment figures (parents, caregivers). Anxiety is reduced when encountering novel situations and they learn to habituate to internal and external stimuli more quickly (Sroufe, 1985). Through attachment security, temperamental dispositions, although indicating greater difficulty regulating internal and external stimuli, are no longer dictating behavioral responses. Securely attached children learn to draw upon self-regulatory strategies shaped by repeated interactions with their caregivers; they learn how to manage their distress through the quality of their interactions with parents (Stright, Gallagher, & Kelley, 2008). Sadly, insecurely attached children, even if they have easy or uninhibited temperaments, will present with more anxiety and greater difficulty self-regulating. Despite temperamental dispositions that should promote ease in self-regulation, their attachment experiences leave them unable to effectively explore the world (Sroufe, 1985; Stright et al., 2008).

Secure attachment is established during infancy; it is fundamentally formed when the caregiver is sensitively attuned to infant needs. The temperament of an infant can make caregiving more or less challenging (Stams et al., 2002; Stright et al., 2008). For example, parents experience pleasure when their infants have temperaments that are positive, approach-biased, and uninhibited. These infants move toward novelty and they are generally less distressed; their attachment security reinforces their innate temperament to approach situations. When parents are not sensitively attuned to their infants, even infants who have a predispositional approach temperament, they are distressed and anxious. These infants present behaviorally as avoidant, negative, and inhibited. On the other hand, parents who are shy and harm avoidant will not encourage their children to explore freely. The parents' temperamental disposition, driven by greater fear, apprehension, and introversion, may be transmitted behaviorally to their children. This can exacerbate the genetically inherited inhibited behavioral/temperament systems in children even further (Beekman et al., 2015). Such caregiver–infant interactions are powerful predictors for later social–emotional and cognitive development (Stams et al., 2002).

Despite the powerful influence of temperament on creative exploration and personality development, when children form secure attachment relationships with their caregivers, they are able to manage conflicts with less distress compared to insecurely attached children (Laible, Panfile, & Makariev, 2008). This ability to tolerate conflict augments creative exploration and enhances talent development. Regardless of infant temperament, secure/autonomous parents can promote attachment security in their children. Even children with the most difficult temperaments (inhibited, shy, withdrawing, difficult) can be given the confidence and reassurance to explore their worlds if they are parented by sensitively attuned caregivers (Cantero, Alfonso-Benlliure, & Melero, 2016). These secure/autonomous parents provide a safe haven for their children, and their children internalize this haven into a working model that supports creativity and exploration within novel situations (Cantero et al., 2016; Laible et al., 2008). Certainly a more inhibited child will approach novelty with more trepidation. For example, shy children tend toward conformity and decreased creative behavior; however, attachment security (or insecurity) intensifies the quality of creativity and problem solving exploration (Cantero et al., 2016). The securely attached, yet shy or inhibited, child will slowly approach and draw courage from parental support; whereas, the insecurely attached shy child

Figure 14.2 "Commitment." Performer: Shelly Livingston. Photographer: Nester Garcia.

will be more hesitant or may fail entirely to approach a novel situation (Belsky & Rovine, 1987; Sroufe, 1985; Fig. 14.2).

Stimulus hunger: temperament, overexcitability, and flow

The inborn survival need for stimulation is expressed differently within individuals based on their temperament. Temperament propels them to either react or act in an effort to increase or decrease stimulation (Necka & Hlawacz, 2013). For example, highly reactive individuals are easily overstimulated and overwhelmed; they are more emotionally reactive and employ deactivating strategies to self-regulate. This temperamental bias can decrease motivation and task-tenacity. On the other hand, individuals who are highly active tend to have lower reactive thresholds; they are often persistent and less emotionally reactive. Generally, creative individuals have lower reactive thresholds and are highly active; they are stimulus hungry. They tend to display temperament traits that are low in sensory and emotional reactivity and high in activity. They readily explore their environment as they seek and engage in numerous creative activities (Necka & Hlawacz, 2013). The intense desire to explore is strongly associated with creativity, behaviors that are often described as hypomanic or cyclothymic (Akiskal, Savino, & Akiskal, 2005; Strong et al., 2007; Vellante et al., 2011). Although these traits are often associated with mood disorders, they are also markers of increased involvement in pleasurable creative experiences, experiences that necessitate emotional and cognitive approach activation (Strong et al., 2007; Vellante et al., 2011).

Similar to the stimulus hunger temperament, the biologically rooted overexcitability disposition is another way to describe temperament in creative individuals (Chavez-Eakle et al., 2012; Mika, 2005). The five overexcitability dimensional traits (psychomotor, sensual, imaginational, intellectual, emotional) are present in early childhood and remain throughout development (Ackerman, 2009; Mendaglio & Tillier, 2006; Piirto & Fraas, 2012). Like the infants classified as "difficult," many talented children with heightened overexcitability responses are easily distressed in situations but only when these situations do not allow expression within their overexcitability dimension (Tannenbaum, 1992). For example, many singers and musicians demonstrate elevated sensual responsiveness (Thomson & Jaque, 2016a). When they hear music they may not be able to attend to other activities because they are fully absorbed in the music and fail to notice other social interactions (Mika, 2005). When allowed to sing or play music, they do not present as behaviorally difficult; rather their stimulus hunger for the sensory experience of music is satiated. Within their overexcitability talent (sensory) they have a high approach temperament, despite the fact that they may also have high avoidance biases in other settings.

Likewise, individuals whose temperament propels them to engage in activities that elicit flow states are driven to seek novelty and to persist in activities for protracted periods of time. Similar to the two-factor BIS and BAS model, Cloninger's temperament model (Cloninger, Svrakic, & Przybeck, 1993) describes four basic temperament traits that can be partially subsumed into the two-factor model. Harm avoidance is part of the BIS system. Individuals with this temperament are unlikely to experience flow states until they reach expert status. Once mastery of a skill is achieved then the harm avoidance (BIS) trait is no longer directing behavioral responses (Teng, 2011). Many performing artists may have harm avoidant (BIS) temperaments, but with adequate support, including secure attachment experiences, they can readily experience flow states. Like the BAS system, novelty seeking and reward dependence encourage flow experiences. Cloninger's fourth temperament trait, persistence, is critical to maintain approach activation. It reinforces engagement in activities despite frustration or fatigue, and is one of the key ingredients in achieving mastery, as well as experiencing flow states (Teng, 2011).

Temperament is an important factor in talent development. Understanding this variable can increase awareness of individual needs and how to tailor appropriate training approaches. Psychological skills can be implemented to manage positive and negative stressors and promote self-regulation in performing artists.

Coping strategies under stress

Core coping strategies to manage stressful situations include problem-focused, appraisal-focused, emotion-focused, avoidant-focused, plus failing to cope (Poczwardowski & Conroy, 2002). Problem-focused and appraisal-focused coping can be subsumed into one coping strategy, task-oriented coping. This form of coping directly addresses the task or the situation and provides optimal results during stressful situations such as in performance settings (Endler & Parker, 1990). Task-oriented coping is also related to decreased psychopathology (Myers, Fleming, Lancman, Perrine, & Lancman, 2013). Emotion-focused coping tends to increase emotional distress, in part, because attention is directed toward the emotional experience rather than appraising and problem-solving during the stressful situation (Sears, Urizar, & Evans, 2000). Individuals who use emotion-oriented coping as their primary coping strategy tend to have higher levels of depression, depersonalization, and emotional exhaustion, and lower levels of personal accomplishment (McWilliams, Cox, & Enns, 2003; Sears et al., 2000). Despite these biases toward a particular coping strategy, individuals are active agents; they are capable of implementing a complex array of coping strategies to address the needs of complex situations (Moos & Holahan, 2003).

Performing artists learn to cope with the stressors that accompany both success and failure. Successful outcomes generally elicit positive emotional responses; however, they can also increase stress. Fear of success is common as evidenced by subjective feelings of greater internal and external pressure, such as fearing social isolation, guilt regarding excelling beyond levels of friends and colleagues, fear of realizing their full talent potential, and the responsibilities associated with their talent (Conroy et al., 2001). Failure, on the other hand, necessitates that performers absorb emotional disappointment while continuing to train, appraise, and problem-solve (Poczwardowski & Conroy, 2002). Performers must incorporate both emotional and task-oriented coping and manage both effectively (Thomson & Jaque, 2016d). They rely on their emotional impulses to gauge their performance, and appraise the outcome of their performance, regardless of success or failure (Conroy et al., 2001).

Performers rely on their creative abilities; employing complex coping strategies supports their talent. They learn to increase creative output in positive circumstances and can continue to produce original ideas and insights under negative circumstances by activating self-regulatory strategies (Baas, De Dreu, & Nijstad, 2011). They engage in a wide repertory of coping skills such as fluidly appraising their performance, both during and after the performance, and emotionally "letting go" while performing. Before and after a performance, they actively plan and train. They identify the demands of a role or score, acknowledge the emotional and physical demands of an upcoming performance, and accept and embrace anxiety inherent in any performance. Sometimes they rely on negative coping strategies such as avoiding or blocking out conflict or distress, escaping into fantasy worlds, denying feelings of distress, or detaching from the support offered by others. Many performers also fall prey to excessive self-blame, panic, and fear of collapse or loss of control. Coping is a complex endeavor in the performing arts with mixed successes and failures (Poczwardowski & Conroy, 2002; Box 14.2).

Box 14.2 Emotion Regulation Strategies for Performing Artists

Category	Activity
Self-regulation	1. Awareness of cognitive and physical fatigue 2. Adequate sleep and good sleep hygiene 3. Nutritious daily diet (including portions and timing of foods) 4. Develop skill mastery and talent competency 5. Cultivate persistence and increase frustration tolerance 6. Recognize and manage stimulus thresholds to avoid boredom or overwhelming states 7. Recognize and accept temperamental dispositions and engage supportive strategies to optimize career success 8. Actively cultivate self-esteem 9. Identify and seek activities that promote meaning, fulfillment, and flow experiences
Emotion regulation	***Physical*** 1. Compassion meditation 2. Relaxation techniques 3. Deep breathing exercises 4. Maintain fitness levels (regular exercise) 5. Adequate preperformance preparation allows emotionally "letting go" during the performance ***Cognitive*** 6. Cognitive reframing and perspective-taking 7. Nurture relationships and build a supportive community 8. Recognize and accept successes and failures as a normal part of a performing arts career 9. Accept performance anxiety as normal 10. Challenge negative self-talk and implement positive thoughts and emotions 11. Challenge shame- and anxiety-based thinking (i.e., "it is all my fault"—personalizing, "it is devastating"—catastrophizing, "it will never get better"—overgeneralizing, "I am helpless"—learned helplessness, "I am rotten"—black and white thinking) 12. Visualize successful performances/rehearsal outcomes
Coping	1. Task-oriented (assess the situation and seek solutions to accomplish the task) 2. Emotion-oriented (assess when to express, suppress, or modulate emotions and learn to identify and regulate negative emotions) 3. Avoidance-oriented (distraction or diversion) 4. Appraising and problem solving 5. Recognize when to inhibit or activate approach—withdrawal behaviors

Emotion regulation

Coping with emotional distress and effectively regulating it is part of self-regulation. Emotion regulation implies that either individuals are regulated by their emotions or they are the regulator of their emotions (Gross & Thompson, 2007). Emotions engage physiological, behavioral, and cognitive processes, and these same processes can be directed to heighten or reduce emotional intensity (Bridges, Denham, & Ganiban, 2004). Fundamentally, emotion regulation requires an awareness, understanding, and acceptance of emotions; managing impulses in order to achieve goals is the hallmark of emotion regulation (Gratz & Roemer, 2004). For example, enhancing emotional expressivity can alleviate distress; however, it can also exacerbate distress. On the other hand, emotional suppression may reduce distress initially but it requires greater psychophysiological resources to suppress intense distress feelings. This is the essence of self-regulation, an approach that accesses expression and suppression and modulates emotions according to internal and external demands (Bonanno, Papa, Lalande, Westphal, & Coifman, 2004). Regardless of gender or temperament, emotional expression is contingent on the social context (Zakowski et al., 2003) and culture norms of the individual (Mesquita & Albert, 2007). For example, during a performance, full emotional expressivity is welcomed, as long as it is relevant to the role of the character or to the nature of a musical or dance work. Ultimately, adopting a flexible approach that incorporates complex variables such as cultural norms, societal expectations, and family biases promotes greater understanding about emotion regulation.

Although studies indicate that emotion-oriented coping strategies are poorer than task-oriented coping strategies (Endler & Parker, 1990), assuming a binary either-or approach is just as problematic (Bonanno & Burton, 2013). Flexible emotional coping strategies include evaluating and responding appropriately to the context, selecting regulatory strategies that can meet the demands of the situation, and monitoring and modifying throughout the stressful situation (Bonanno & Burton, 2013; Bridges et al., 2004). Differentiating between emotions that are evoked during a specific situation and moods that tend to persist regardless of situational demands is a significant variable when addressing emotion regulation. Moods directly influence coping strategies; for example, coping with bereavement alters emotional regulation and coping (Gross & Thompson, 2007). Achieving this flexible approach is the highest application of self-regulation while under duress.

Emotional appraisal of others is achieved by highly complex neurological feedback/feedforward signals. Fine-grained rapid assessment of facial muscular changes, plus larger visceral and bodily muscular patterns, provides information about the emotional state of another; they also activate physiological and emotional responses in the individual who is appraising others (Critchley, 2012). This reciprocal relational field operates at the implicit level, a reality that enables performers to gauge responses in fellow performers and the audience; it also allows implicit access for the audience to perceive and feel emotions that are expressed on stage.

Gaining skills to manage negative emotions is critical to enhance creativity, especially in children (Butcher & Niec, 2005). Ultimately when children have difficulty with emotion regulation they also have greater disruptive behavior problems such as increased aggression, attention deficit hyperactivity disorder, defiance and oppositional disorder. These problem behaviors do not allow children sufficient opportunities to problem-solve and tolerate the inherent frustrations that arise during problem solving (Butcher & Niec, 2005). Accessing emotion is important in creative play but equally important is emotion regulation during creative play activities (Hoffman & Russ, 2012). This same need is present in adults. Because creative people are open to experience, they require greater emotional regulation to support their creativity (Ivcevic & Brackett, 2015). Some of many optimal emotional regulating skills that performers implement during their practice sessions and before and after performances include muscle relaxation, deep breathing, positive mental practices, memory checking, and visualizations (Poczwardowski & Conroy, 2002). Other effective strategies include imagery work, mental practice, and self-talk; all tools to manage stressful situations (Tenenbaum, Edmonds, & Eccles, 2008). Certainly, emotion regulation does not increase creative responses but it does allow increased persistence and frustration tolerance and these features enable the generation of more creative ideas (Ivcevic & Brackett, 2015).

Artists, in general, experience more intense emotionality, yet they are also able to engage effective self-control to manage their emotionality (Sanchez-Ruiz, Perez-Gonzalez, & Petrides, 2010). Although performing artists may struggle with emotional distress they also have great self-control. For example, emotional distress is common in actors. While portraying characters, they also experience the emotional states of the characters. Consequently, they are subject to greater emotional distress, but their training helps them regulate intense emotional states (Burgoyne et al., 1999; Hetzler, 2008). Singers also express emotional distress, especially when singing the blues (Verhaeghen et al., 2005). Musicians empathically perceive emotional expression in music (Wollner, 2012), and dancers, singers, and musicians communicate emotions via similar neural pathways (Juslin & Laukka, 2003). Although performers are required to engage heightened emotional states during a performance, both generators (composers, choreographers, writers) and interpreters (actors, dancers, musicians, singers) modulate emotional biases (i.e., angry, playful, caring), with the exception of greater awareness of sadness and identification with spirituality (Thomson et al., 2009).

Collectively, performing artists possess a range of emotional expressivity, although some have greater difficulty regulating negative emotions (Thomson & Jaque, 2016b, 2016c). As performers they must activate emotions; emotional approach behavior is essential while performing (Lolich, Vazquez, Zapata, & Akiskal, 2015). All forms of performing arts have the potential to evoke emotional distress for the performers and their audiences (Lading & Schellenberg, 2012). Given these facts, performers need emotional support during their training years and throughout their careers (Nagel, 2009). They are intrinsically motivated to express emotions; the expression of these emotions is refined via training (St-Louis & Vallerand, 2015; Fig. 14.3).

Figure 14.3 "Integrated coordination." Performer: John Roscigno. Photographer: Lee Choo.

Resilience

Individuals who remain productive and healthy despite experiencing prolonged stress or trauma are considered "resilient." They are able to manage difficult circumstances for protracted periods of time without negative consequences (McGloin & Widom, 2001). In addition to overt trauma and/or significant loss experiences, low socioeconomic status (SES), low birth weight, premature birth, chronic illness, chronic poverty, and other life events are regarded as adversity and present as risk factors for further pathology. These "ordinary" events are commonly met with "ordinary" resilience; people are generally able to cope with these challenges and adapt (Masten, 2001).

Resilient individuals may actively resist adversity or they may suffer during the traumatic event but show no signs of distress after the event; they successfully adapt despite the stressors (McEwen et al., 2015; McGloin & Widom, 2001). It is irrefutable that chronic stress, especially in childhood, is deleterious (Evans & Kim, 2013), and yet some individuals escape the frequently reported increased morbidity and mortality. Studying resilience shifts the focus from examining increased risk markers related to chronic stress and trauma; resilience is measured as either the absence of psychopathology or the presence of adaptive functioning. Resilience studies gather information in one of two ways: (1) identifying variable factors such as attachment profiles, SES, IQ, social supports, developmental timing of risk experiences, and temperament or (2) the person-focused approach that compares adaptive and maladaptive individuals (measuring competence, self-esteem, talent) (Masten, 2001). Positive outcomes to adversity may result because individuals are fundamentally stable (good problem solving, strong cognitive functioning, greater neurophysiological resources, BAS temperament, secure attachment, academic achievement, prosocial attitudes), or they deploy positive adaptive strategies during and following the adverse event (capacity to tolerate and regulate pain, frustration, and fear, and mobilize effective solutions to resolve or address the adversity) (Metzl, 2009).

Unlike "pure" resilient individuals, many may suffer for a circumscribed period of time and then "recover." These individuals may manifest some form of resilience, but the fact that they manifested pathological symptoms, even if it was for a brief period of time, excludes them from the truly resilient grouping. These individuals demonstrate a mental toughness that can help them overcome adversity. It must be noted that normal human grief responses and transient trauma symptoms (i.e., sleeplessness, preoccupation about the traumatic event) are not considered pathology. Human existence involves trauma and loss experiences; individuals respond to these experiences while continuing to function in normal daily life (Bonanno, 2004). Although they are not "resilient" in the strictest sense of the term, their ability to grapple with difficulty also suggests that they have persistence. Despite frustration and failure they continue to function; many even successfully experience positive and rewarding flow states while engaging in their preferred activities (Crust & Swann, 2013; Thomson & Jaque, 2016c). Even individuals who are considered to have a personality trait that is neurotic, one that is biased toward negative emotion, can respond adaptively to adversity (Engelhard, van den Hout, & Lommen, 2009). This adaptive profile describes many performing artists who have encountered adversity (Thomson & Jaque, 2016c).

To determine resilience longitudinally, studies identify abused and neglected children and follow their development into adulthood (Rutter, 2006). To be regarded as resilient, this population must demonstrate a stable employment history, completion of educational programs, actively engage in social activities, maintain secure housing, have no criminal record, and have never received a psychiatric diagnosis or been treated for substance abuse (McGloin & Widom, 2001). According to McGloin and Widom (2001), approximately 22% of this population met criteria and were deemed resilient. More females were identified as resilient in this study, perhaps because maltreated males tend to be more vulnerable to morbidity and mortality risks in early development.

Experiencing trauma and loss in adulthood is much different than experiencing them in childhood. Adults draw upon more resources, including early attachment experiences, cognitive problem-solving skill development, interpersonal social networks, and coping strategies that have been trained and practiced (Mancini & Bonanno, 2006). Many pathways that promote resilience are possible, such as self-confidence and belief that the trauma/loss experience can be managed, viewing trauma and loss as opportunities for self-enhancement, and shifting emotional expression from negative to positive (Bonanno, 2004). These resources can effectively support resilience in adults, a reality that may be associated with lower pathology risks in adult trauma victims. On the other hand, adults with a history of childhood abuse and neglect possess insufficient resources to manage adversity and are more vulnerable to pathology.

Contributing factors that augment resilience include a genetic predisposition. There are genetic strengths to buffer deleterious effects of early childhood adversity (van Ryzin et al., 2015). Studies have demonstrated that long alleles found on 5-HTTLPR and genetic polymorphisms within the corticotropin-releasing hormone (CRH) receptors are related to decreased depression and posttraumatic stress disorder (PTSD) (Bradley, Davis, Kaye, & Wingo, 2014). As well, the capacity to experience childhood stress without significant neurobiological damage is partially related to endogenous levels of brain-derived neurotrophic factor (BDNF) within the hippocampus. BDNF helps modulate cortisol responses and facilitates hippocampal tonic inhibition of the HPA axis (stress axis) (Raglan & Schulkin, 2014). Increased neuromodulation from neuropeptide Y (NPY) also decreases depression and PTSD, in part because NPY is a cotransmitter in the sympathetic nervous system (SNS), a branch of the ANS that influences cardiovascular responsiveness and energy homeostasis (Enman, Sabban, McGonigle, & Van Bockstaele, 2015). Higher levels of dehydroepiandrosterone (DHEA) counteract the effects of cortisol secretion, consequently enhancing resilience during stressful situations. Dancers tend to possess increased serotonergic neurotransmission, a factor that augments positive social communication and spiritual connectivity (Bachner-Melman et al., 2006). These factors, and many more neurobiological factors, promote resilience. Future pharmacological agents may help those who lack a resilient genetic predisposition (Enman et al., 2015).

Neurophysiological factors also promote resilience. Greater heart rate variability in general, or a rapid parasympathetic respiratory sinus arrhythmia (RSA) response immediately following a stressor, ensures better self-regulation (Obradovic & Boyce, 2012; Porges, 2001). For performing artists who must cope with recurrent failed auditions and yet remain optimistically redirecting their efforts toward the pursuit of career goals, greater HRV supports these efforts. More variance in heart rate leads to greater inhibitory self-control, a feature that supports recovery from failure and maintenance of ongoing pursuit of goals (Geisler & Kubiak, 2009). Attenuated sympathetic (SNS) reactivity and cortisol response can also buffer the effects of stress (Gunnar & Adam, 2012; Obradovic & Boyce, 2012; Saxbe, Margolin, Shapiro, & Baucom, 2012).

Other factors that facilitate greater resilience, despite childhood adversity, include attachment security, authoritative parenting style (responsive and yet limit-setting), family resources (extended family, SES), peer and community influences and support, as well as, gender, racial, and cultural acceptance and appreciation (Bradley et al., 2014). The capacity to self-regulate (cognitive, emotional, and psychological) enhances resilience; whether it is employed to actively resist stressful effects or to quickly respond from the stressor (Bradley et al., 2014). Finding meaning promotes resilience; adults are able to draw meaning from their childhood adversity. Meaning can also be garnered during chronic stress endured during adulthood, and/or gleaned from spiritual beliefs or mindfulness practices; tools that help heal trauma and loss experiences (Park & Slattery, 2014). For performing artists, engaging in the creative process also promotes greater meaning, plus a sense of healing and transcendence (Nelson & Rawlings 2009; Thomson & Jaque, 2016d).

Music, either listening to or playing, is a powerful healing intervention; it enhances meaning and promotes optimism (Hodges, 2014). Exercise, such as the positive physical effects garnered from dancing, helps regulate the HPA axis and facilitates stress resilience in the hippocampus (dentate gyrus) via greater control of glucocorticoid secretion (Reul et al., 2015). Lastly, domain-general creative skills such as flexible thinking and originality can promote effective adaptive strategies during and after a traumatic event and foster resilience (Metzl, 2009).

Performing artists generally have increased empathy, whether that is a result of their genetic predisposition or introduced and augmented during their training (Jeffers, 2009; Kalliopuska, 1989; Verducci, 2000; Wollner, 2012). Empathy can operate as a facilitator of resilience; however, it can also intensify distress (Hein, 2014; Loggia et al., 2008). Performing artists are able to empathically connect to the characters they portray and to their fellow performers. When empathy works positively during performance, performers can freely express and connect with other artists, but when empathy intensifies personal distress they are less resilient during performance stress (Hein, 2014). Empathy can also increase parental sensitivity and enhance adult love relationships, factors that promote the nurturing of young performing artists and facilitate professional stability as adults (Collishaw et al., 2007).

Positive self-esteem is cultivated by interactions with supportive others, such as teachers, mentors, family, peers, or community. Successfully realizing talent and achieving competence directly enhances self-esteem. Not surprising, self-esteem also promotes greater physiological regulation, as evidenced by studies that investigated programs that promoted self-esteem. Results demonstrated that increased self-esteem augmented vagal tone (Martens et al., 2010). Self-esteem also protects against disease, another critical factor in resilience (O'Donnell, Brydon, Wright, & Steptoe, 2008). Increased self-esteem is associated with positive emotions; both self-esteem and positive emotions can be enhanced via greater compassion, including self-compassion (Neff et al., 2005; Zabelina & Robinson, 2010b). This can be augmented via regular compassion meditation practices (Lutz, Brefczynski-Lewis, Johnstone, & Davidson, 2008). Both self-esteem and compassion are powerful factors in regulating reactivity to stress and recovery from stress.

Multiple factors promote resilience, most particularly social support from caring family and community, adequate self-regulation skills, positive view of the self, and motivation to effect changes in self and the environment. Performing artists are well situated to possess resilience despite adversity. They work within a community of other artists, they are highly motivated, and they possess talents to effectively communicate with others. Assessing them for risk factors for pathology is important, but equally important is the need to assess their resilience factors. Adaptive systems such as attachment, exploration, and novelty seeking are powerful antidotes to the stress systems that are mobilized during times of adversity (Masten, 2001). Performing artists have typically cultivated these adaptive systems during countless years of training, practice, and performing. They have vast resources to draw upon and the majority of these resources nurture resilience (Thomson & Jaque, 2016b, 2016c).

Summary

- Self-regulation: definition, behavior and biology, executive functioning, metabolic resources.
- Organize–disorganize–reorganize: the role of self-regulation in the creative process: recursive process, open complex systems, tolerate shifting states.
- Temperament: definition, constructs and models, infant temperament and development, BAS and BIS and neurobiology.
- Temperament, attachment, and self-regulation: differences between temperament and attachment and interactions between the two constructs.
- Stimulus hunger, temperament, overexcitability, and flow: reactivity (sensory, emotional, physiological), engaging/disengaging in novelty.
- Coping strategies under stress: types (problem-focused, appraisal-focused, task-oriented, emotion-oriented, avoidant-oriented), fear of success/failure.
- Emotional regulation: definitions and types, appraisal, managing negative/positive emotions, intensity/valence.
- Resilience: definitions, neurophysiology, strategies to promote resilience, empathy, self-esteem.

Concluding remarks

Self-regulation, emotion regulation, and resilience are key factors in all human behaviors. Action is a fundamental element in the performing arts. Actors, dancers, singers, and musicians all move to give life to their art form. They must also engage emotionally to give expressive life to the performance. Self-regulatory strategies necessary to succeed in the performing arts require multidimensional skills and abilities. Performing requires intense metabolic demands and so does self-regulation. Through years of training, performers gain a greater ability to regulate physical, cognitive, and emotional fatigue. They learn to recognize early signs of dysregulation and they cultivate skills to promote regulation. Consequently, they employ diverse coping strategies that are operationalized during specific situational demands. For example, they are emotion-oriented while performing, task-oriented during training and performance, and they implement avoidance strategies if they need to delay responding to potentially difficult situations. They also learn to manage success and failure. Regardless of temperament—whether shy, introverted and inhibited, or extraverted and eagerly seeking novelty—performers learn to tolerate performance stress. Through constant exposure to performance situations, most performers habituate to performance anxiety and recover from frequent audition rejections. They learn to merge intention, action, and emotion as they flexibly adapt to the demands of a career in the performing arts.

Mood and creativity in the performing artist

15

Actors in any capacity, artists of any stripe, are inspired by their curiosity, by their desire to explore all quarters of life, in light and in dark, and reflect what they find in their work. Artists instinctively want to reflect humanity, their own and each other's, in all its intermittent virtue and vitality, frailty and fallibility.

Tom Hiddleston

Introduction

Emotional expression is at the heart of aesthetic experience in the performing arts. Understanding the effects of emotion on creativity runs deeper than the strategies necessary to regulate emotion as discussed in the previous Chapter 14, Self-Regulation, Emotion, and Resilience. For performing artists, emotion is often the driving force behind their creative inspiration and expression. In fact, they could be viewed as possessing high emotional intelligence (an ability that enhances convergent thinking, coupled with regulating negative emotions) and emotional creativity (involves divergent thinking and the ability to manipulate experiences to solve problems) (Ivcevic, Brackett, & Mayer, 2007). Mood, a state generally lasting much longer than an emotional state, can paradoxically augment or destroy creativity in the artist. Mood is also a critical factor in a creative product; for example, mood is consciously evoked in music, dance, theater, and film. This chapter will address mood and its direct effect on creativity in the performing arts.

Mood is continuous, is often not connected to an external object or event, and is frequently influenced by a particular emotion that is sustained over time (Damasio, 1999; Davis, 2009). According to Gross and Thompson (2007), moods influence cognitive processes more than action responses. They tend to be more diffuse and less intense compared to emotions and are categorized hedonically as positive (happy, relaxed, cheerful) or negative (sad, morose, hostile, contemptuous, guilty, irritable, anxious) (Baas, De Dreu, & Nijstad, 2008; Kring, Smith, & Neale, 1994). They alter physiological and neurological responses, especially because they are long lasting. Musculoskeletal changes alter posture and coordination in direct response to mood. For example, positive mood yields open posture and gestures that are expansive and quicker; whereas, negative mood provokes collapsed posture and smaller, slower gestures (van Dyck, Maes, Hargreaves, Lesaffre, & Leman, 2013). Other changes occur at the autonomic nervous system and endocrine levels as well as within different brain regions and neural networks (Damasio, 1999).

Creativity and the Performing Artist. DOI: http://dx.doi.org/10.1016/B978-0-12-804051-5.00015-9

Moods also alter emotional responses; they operate like a filter through which situations are viewed and emotions are provoked. When someone is in an irritable mood they may be easily frustrated and angry; whereas, if they are happy they may approach the same situation with curiosity or compassion. Likewise they will evaluate perceived recognitions and rewards through the same positive or negative mood filter (George & Zhou, 2002).

Although moods are generally experienced over longer periods of time compared to emotions, they can also be short in duration (Davis, 2009). Moods in this sense are still considered transitory states, unlike traits that are more dispositional in nature, such as trait-anxiety, internalized trait shame, or trait-flow. Earlier concepts such as the four humors, also known as the four temperaments, clearly captured the concept of trait moods. The four temperaments described individuals as sanguine (optimistic, social, and associated with the element of air), melancholic (analytical, quiet, earth), choleric (short-tempered, irritable, fire), and phlegmatic (relaxed, peaceful, water) (Buckingham, 2002). These temperaments have been the inspirational source for many performing arts works such as George Balanchine's ballet, *The Four Temperaments*, and Carl Nielson's Symphony No. 2. William Shakespeare directly built characters based on these temperaments.

If moods persist for protracted periods of time and they cause significant distress in daily life, then they are considered a disorder or psychopathology (Damasio, 1999). They negatively influence cognition and action and interfere in work and social settings. Marked alterations in neurobiological and physiological processes are evident in mood disorders (bipolar, anxiety, depression). Sleep, diet, personal hygiene, judgment, thought, and interpersonal relationships are all compromised. At the most severe end of the continuum, mood disorders can provoke psychotic episodes and at the milder end of the clinical spectrum, they significantly limit function (APA, 2013; Fig. 15.1).

Mood and creativity

Research investigating mood and creativity primarily explores the effects of hedonic tone or valence, with mixed results. Moods are induced (positive, negative, and neutral) in order to see the acute effects on problem solving and idea generation. Mixed results are reported: positive mood promotes creativity compared to neutral mood but not compared to negative mood (Davis, 2009). In contrast, negative mood is more preferential than positive mood to facilitate creative problem solving (Kaufmann & Vosburg, 1997). Positive mood facilitates more solution generation in creative problem solving, while negative mood biases response patterns that optimize strategies to discover one solution (Vosberg, 1998). When examining the effects of emotion on the timing of creative idea production a different set of patterns emerged, with positive mood promoting early idea production versus neutral and negative mood evoking late idea production (Kaufmann & Vosberg, 2002). This finding suggests that positive mood is most effective in unconstrained

Figure 15.1 "Shaping mood." Performers: Katie Alcazar, Elissa Brock, Jocelyn Brooke, Autumn Hooks, Savannah Hooks, Annie Sherman, Cheyenne Spencer. Photographer: Emma Rosenblatt.

situations, and surprisingly, negative mood is best in constrained situations. Another explanation is that, unlike negative mood, positive mood may evaporate if the task is demanding and prolonged.

Evaluating the contribution of hedonic mood to creativity also necessitates the consideration of how mood affects behavior. Metaanalysis studies suggest that approach behavior activated by positive mood (joy, pleasure) is ideal for creativity, but mostly when the creative task was framed as enjoyable and rewarding rather than serious and extrinsically rewarded (Baas et al., 2008). Unfortunately, positive relaxation mood states tend to be deactivating and are less helpful in creative work. When examining the effects of deactivating emotions such as sadness, fear, and anxiety, less creative problem solving and cognitive flexibility were evident. In a different study, approaching mood as either activating or deactivating altered the hedonic equation. Regardless if the mood was positive or negative (happy, angry, fearful), if it activated behavioral responses during a creative task more creative output resulted. Of note, positive activating moods directly enhance creative fluency, flexibility, and originality. Unlike positive mood, activating negative mood augmented persistence, a major factor in creativity (De Dreu, Baas, & Nijstad, 2008).

Mood intensity also directly influences creative output, with too little or too much intensity compromising creative productivity (Davis, 2009). This is especially true if negative mood is highly intense. Persistence will falter under the onslaught of elevated negative mood. Unfortunately, intensely elevated positive mood operates as a demotivating factor. These diverse findings indicate that the mood–creativity question is a complex equation (Davis, 2009; Kaufmann, 2003). Currently, the literature suggests a bias toward positive mood enhancing creative process engagement (To, Fisher, Ashkanasy, & Rowe, 2012).

Physiology, neurobiology, and mood

The dynamics of mood states directly influence physiological reactivity (Golland, Keissar, & Levit-Binnun, 2014). In particular, intense emotional responses tend to evoke similar autonomic and central nervous system activation patterns within individuals and across individuals; whereas, widespread individual differences occur during neutral moods. The findings of this study reinforce general knowledge that heightened mood states are experienced collectively as embodied states (Golland et al., 2014). When studying mood in isolation, without attempting to account for differences in personality, temperament, developmental history, and psychopathology, results may be conflicting. Although there are some overarching unifying patterns reported, the influence of mood on the response to creative tasks remains poorly understood. Below, some of the key findings are presented in an attempt to better understand this area of research, in particular, how it relates to creativity in the performing arts.

Mood influences judgment about task difficulty and colors motivational behavior (Silvestrini & Gendolla, 2009). Coupled with a specific hedonic mood (positive, negative, neutral), external demands influence both degree of mobilization efforts and the anticipation of success or failure (Richter & Knappe, 2014). Typically, when the task demand is high, negative mood increases cardiovascular reactivity simply because individuals in a negative mood amplify perceptions that the task is difficult, whereas, positive mood reverses this appraisal pattern (Silvestrini & Gendolla, 2009). Interestingly, when individuals focus on *task reward*, more cardiovascular effort (sympathetic nervous system) is exerted during positive moods compared to negative mood states. When attention is directed toward *task demand*, the opposite pattern is evident; less cardiovascular effort occurs during positive moods and more cardiovascular demands are mobilized during negative mood states (Richter & Knappe, 2014). These findings support the *motivational intensity theory*, a theory that describes matching effort to task difficulty during situations in which performance standards are fixed (Martin, 2001). This is a normal situation for all performing artists. On the other hand, the importance of success (rewards) tends to influence effort mobilization when performance standards are not prescribed, such as during improvisational situations. Based on the motivation intensity theory, task demand rather than task reward inversely influences cardiovascular effort relative to positive or negative mood states (Box 15.1).

A different study demonstrated how perceptions of a task (pleasant or unpleasant) further influence mood, motivational, and cardiovascular demands (Silvestrini & Gendolla, 2009). When a task was viewed as pleasant, even if it was difficult and the individual was in a negative mood, cardiovascular demands were lower. Individuals who were in a negative mood and were required to engage in unpleasant and difficult

Box 15.1 Motivation Intensity Theory, Task Reward, and Task Demand

High-intensity motivation	*Low-intensity motivation*
High approach, narrow attention	Low approach, broad attention
Task reward	***Task demand***
Focus on task reward increases cardiovascular exertion, especially during positive moods	High task demand increases cardiovascular reactivity, especially when negative mood colors the appraisal of task difficulty

tasks experienced higher cardiovascular demands. In this study, engaging in pleasant tasks also altered prior negative mood states. After completion of the task, the performers shifted into a positive mood; however, engaging in unpleasant tasks further amplified their negative mood and decreased prior positive mood states. This pattern is present in performing artists. They usually experience their talent field as pleasant; they are strongly motivated to participate and excel. Because their task is regarded as pleasant, regardless of prior mood state and task difficulty, they report a high frequency of positive flow experiences (Thomson & Jaque, 2016c). These findings suggest that positive hedonic tasks can enhance mood regulation (Silvestrini & Gendolla, 2009).

The performing arts require high physical exertion. For example, while conducting, conductors exerted work intensities considered "high" (Jaque et al., 2015). When these normal work effort demands are decreased, whether as a result of injury or career hiatus, exercise conditioning can decrease. This results in a decrease in fitness level that directly increases negative mood states and alters autonomic responses. Decreased regulation from the parasympathetic nervous system (PNS) can occur rapidly. In a study that investigated exercise withdrawal for a two-week period, decreased PNS tone and increased negative mood resulted (Weinstein, Deuster, & Kop, 2007). Understanding the effect of exercise withdrawal (detraining) on mood in performing artists, especially when they are injured or rehabilitating from a medical procedure, can help normalize resultant negative mood changes. Efforts can be taken to help injured or ill performers enhance positive moods; engaging in some form of remedial exercise directly increases positive mood and creativity. Regardless of injury or health status, dance, compared to aerobic activity, proved to be the best form of exercise to enhance positive mood and to increase creative cognitive fluency (Steinberg et al., 1997). This research finding reinforces the creative need for movement in performing artists.

Negative moods such as anger and sadness also increase sympathetic autonomic demands (Rochman & Diamond, 2008). For example, unresolved anger increases behaviors of blaming others, a desire to seek retribution or to avoid the situation completely. Unresolved sadness is usually related to loss experiences. This mood state can occur during bereavement of a loved one but it can also be evoked as a result of a debilitating injury or illness that deters performance. A pervasive mood of anger or sadness massively alters physiological regulation (Rochman & Diamond, 2008). Because these mood states are physiologically sympathetic-biased, more metabolic demands are placed on an already depleted system. Likewise, performers who experience prolonged periods of guilt have increased cardiac sympathetic activation (Fourie et al., 2011). To feel guilt, individuals must have the capacity to feel empathy for another. Performing artists have a high capacity to feel empathy (Jeffers, 2009), thus they are vulnerable to suffering the high arousal state of guilt. Especially in performers who are more behaviorally inhibited, the feeling of guilt evokes feelings of discomfort; consequently this mood can motivate performers to actively alter behavior and repair the situation in order for them to resume optimal performance (Fourie et al., 2011). Pride also engages more cardiac sympathetic tone but it does not provoke the feelings of distress that accompany guilt. Just like guilt, it is an activating mood. It encourages performers to seek positive social interactions and motivates them to strive toward goal mastery. The results of the Fourie et al. study (Fourie et al., 2011) reinforce the fact that highly activating moods such as guilt and pride are not necessarily equally psychologically depleting. The gain from experiencing positive pride rather than negative guilt moods, despite similar physiological energy expenditure, motivates approach behavior, a key ingredient in successfully achieving elite status in the performing arts.

Interestingly, artists who are more emotionally vulnerable, as evidenced by lower baseline DHEAS levels (adrenal stress-regulating steroid), are also more likely to respond with higher creativity. This was especially noted when they experienced social rejection rather than social approval (Akinola & Mendes, 2008). Negative evaluation in psychologically vulnerable artists may propel them to work harder to seek creative solutions, or they may use the creative task as a distraction from the painful experience of social rejection. Regardless of the reason, biological vulnerability to negative mood (depression) did provoke enhanced creativity.

With respect to the brain, activating a positive mood or introducing a cognitive stimulus via exposure to emotionally contagious music such as inducing cheerfulness increases alpha synchronization in the prefrontal region of the brain. Alpha wave coherence (synchronization) is strongly related to internal attentional processing; it is not an indicator of the brain idling (cortical deactivation) (Fink, Schwab, & Papousek, 2011). Increased alpha wave coherence activation in creative individuals, compared to less creative individuals, has been identified across multiple studies that examined performance patterns during creative tasks. It reinforces the effects of cognitive challenge or positive mood on internal attentional processing during creative task performance (Fink et al., 2011). In other studies that looked specifically at alpha-1 during creative activities (composing music, creating a story), greater alpha-1 coherence activations were observed across multiple regions and hemispheres of the brain (Petsche, Kaplan, von Stein, & Filz, 1997). These alpha-1 findings were also observed while experiencing elated or agitated moods; negative coherence patterns were found while experiencing drowsiness. Despite large individual differences in brain wave activity, compared to nonartists, artists had more pronounced coherence patterns that extended over large distances in the brain (Petsche et al., 1997; Fig. 15.2).

Figure 15.2 "Passion." Performers: CSUN Dancers. Photographer: Lee Choo.

Empathy, passion, rumination, and mood

Empathy augments both positive and negative mood, a process that can also influence performance in the same bidirectional manner. For example, if actors, dancers, or musicians perform a work that portrays intense human suffering and loss, the contagion effect of sadness can strengthen the mood of the performance (Loggia et al., 2008). Empathic understanding of mood is also facilitated when performers assume specific body postures, a process that dancers, opera singers, and actors engage in when performing (Critchley, 2012). Bodily empathy enhances aesthetic appreciation and understanding (Freedberg & Gallese, 2007), and is a powerful means of communication between performers while rehearsing and performing (Sevdalis & Raab, 2014). Audience members equally recognize mood states such as happiness and sadness based solely on decoding mood states derived from the performers' movement of the chest and torso (van Dyck, Vansteenkiste, Lenoir, Lesffre, & Leman, 2014). The viewers' ability to perceive bodily motion in performers (musicians) enables them to accurately determine musical expressive intention even when the auditory stimulus is removed. Body empathy powerfully informs and enriches aesthetic appreciation in both the audience members and the performers (Wollner, 2012).

Performing artists are passionate about their art form and careers, despite the reality that periodically they may doubt their abilities and career success. Passion propels them forward in their careers and offers meaning in their lives (Luh & Lu, 2012). The effect of mood on passion shares similar features to the mood–creativity equation. Positive and negative moods are operational during the creative process, although experienced in different ratios and intensities among artists who are classified as harmoniously passionate compared to obsessively passionate. All passionate performers tend to experience more frequent moderately positive moods; this pattern occurs during all four phases of the creative process (preparation, idea generation, production, and completion) (St-Louis & Vallerand, 2015). Further, artists identified as harmoniously passionate experience more positive mood states in general, and yet across all stages in the creative process they are more negative, even more so during the preparation phase. These findings indicate that all passionate artists experience moderate levels of positive and high levels of negative mood activation during the creative process. Harmoniously passionate artists tolerate even more intensity and frequency of positive and negative moods compared to obsessively passionate artists. In general, the intensity and valence of mood did not deter passionate artists from engaging in the creative process (St-Louis & Vallerand, 2015; see "Motivation" section in chapter: Personality and Motivation for further details).

Rumination is strongly linked to mood, whether positive or negative. It refers to internally generated thoughts that focus on a single topic despite a lack of external stimulus to activate the cluster of thoughts (Verhaeghen, Joorman, & Aikman, 2014). Rumination is a form of self-reflection, a process that can have positive or negative outcomes. Different hedonic moods can also influence rumination, especially when rumination is symptomatic of depression or bipolar disorder. Rumination is also present in passionate artists, regardless if they are harmonious or obsessive in their passion (Verhaeghen et al., 2005). Reflectiveness is regarded as active and "pondering" yet well-regulated emotionally; whereas, brooding is a passive perseveration or fixation on undesirable aspects of the self. Brooding is not well-regulated and increases emotional distress. Not surprisingly, self-reflective rumination enhances creative processes while brooding increases depressive moods (Verhaeghen et al., 2014). When immersed in a positive mood, rumination (both positive and negative), supports self-immersion while problem solving, especially when artists are creatively indecisive. Another strategy applied to rumination (self-reflective or ruminating) is "self-distancing"; literally the self or the activity is viewed from an imaginary distance. This strategy can be applied effectively across mood valence and intensity (Copeland, 2016).

Related to rumination, in particular the brooding form is the "critical internal judge." This is a common experience in the performing arts. Internal judgment, literally self-critical narratives that loop incessantly, can cripple performers. Some performers frequently and intensely experience these narratives, a clear indication that the rumination is a symptom of clinical depression; however, the majority of performers experience random negative intrusive thoughts, even when their prevailing mood is positive. Seemingly, for no apparent reason, a self-sabotaging idea will manifest. For example, an unwelcome internal "judge" voice will proclaim that they will forget a line, miss a note, or fall out of a turn. Although the self-reflective thoughts seem real, they are perceptions rooted within the background mood of the artist. Negative moods tend to increase the frequency and intensity of these ruminative thoughts. Whether artists term these ruminations as self-criticism, self-judgment, or even the dark/cruel/mean voice, they must learn to manage this form

Figure 15.3 "Joy." Performers: Amanda Loretta Mendez, Yuriko Mizushima, Steven Nunez, Rubina Vartanians, Matt Young. Photographer: Lee Choo.

of rumination (Taylor & Estanol, 2015; Wangh, 2000). Actors, in particular, suffer from this form of rumination because they are completely exposed. They have no music or choreography to hide behind (Wangh, 2000). As described in earlier chapters, training in the performing arts, including emotional regulation and coping strategies, can decrease the negative effects of this form of rumination. With experience, performers learn to tame self-criticism. If it is very intense and negative then they may seek professional help (Fig. 15.3).

Mood in music, dance, and acting

All cultures value their musical tradition; it is directly and indirectly embedded into daily life. Listening to music has direct effects on our health and wellbeing; it unites us socially and culturally (Morinville, Miranda, & Gaudreau, 2013; Unwin, Kenny, & Davis, 2002). Not surprising, music is the most researched area in the performing arts, including robust research that identifies complex effects of music on mood (Kantor-Martynuska & Horabik, 2015). It is undeniable that listening to music can alter mood and change physiological responses to match the evoked mood. It can also stimulate autobiographic memories, which then further amplify mood (Vuoskoski & Eerola, 2012). Listening to music, especially when it has relevance to the listener, can induce feelings, including sadness and joy (Morinville et al., 2013). For listeners high in empathy, even unfamiliar music can induce feelings of sadness or happiness (Vuoskoski & Eerola, 2012). When listeners select music they usually match their current mood with a particular mood-congruent musical work. This pattern of musical choice reinforces their mood; however, they may also elect to change their mood and select something that contrasts their current mood (De Marco et al., 2015).

Listeners who are generally open to experience appreciate more complex music (blues, jazz, classical, and folk music) (Renfrow & Gosling, 2003). They are also able to experience more awe states while listening to music (Silvia, Fayn, Nusbaum, & Beaty, 2015). This same pattern was evident in individuals who preferred heavy metal music; however, this open to experience group also endorsed negative attitudes to authority, lower self-esteem, greater need for uniqueness, and fewer religious beliefs (Swami et al., 2013). A combination of personality traits that includes open to experience and agreeableness indicates an ability to hear musical emotional expression with more intensity, especially sad music. Listeners with musical training respond positively to both sad and happy music, especially if they also have open to experience and agreeableness personality traits (Lading & Schellenberg, 2012).

Evoking mood through music is usually related to tempo and mode. Music played faster and in the major key is related to happy mood and slower, minor key music is associated with sad feelings. A combination of different tempos and modes elicits mixed emotions (Schellenberg & von Scheve, 2012); musical dissonance generally induces fear states (Unwin et al., 2002). Melodic structure, pitch amplitude, dynamics, and rhythm also influence the expression of mood. Different instruments add further musical color and emotional tone. In studies that manipulate these variables, listeners perceive music as happy or sad based primarily on whether it is played quickly and in a major key, or slowly and in a minor key, respectively. This finding persisted when the same Bach piece of music was transposed and performed to match tempo and mode typically associated with sad–happy moods (Hunter, Schellenberg, & Schimmack, 2010). Musical genres also have predictable mood-inducing effects. For example, grunge rock increases feelings of hostility, sadness, tension, and fatigue; whereas, listening to designer music (music composed specifically to induce relaxation responses) promotes feelings of caring, relaxation, mental clarity, and vigor (McCraty, Barrios-Choplin, Atkinson, & Tomasino, 1998). Consciously manipulating tempo, mode, and genre in music is a powerful tool to influence listener mood, a reality that can be applied in a myriad of settings (healing centers, political rallies, business offices, prisons, etc.).

Although much is known about the listener's response to music and the effect on mood, little is known about the performer's mood and the influence of music on it. Musical expertise clearly enhances self-perceived emotional responses to music that is heard or played. Fine-grained musical discrimination regarding emotion is better discerned by musical experts than by musically educated nonmusicians. They also perceive these affective states within themselves while listening to the music (Kantor-Martynuska & Horabik, 2015). Trained musicians are better able to appreciate the complexity of music that evokes mixed emotions; whereas, untrained listeners respond to music that induces either sad or happy feelings (Lading & Schellenberg, 2012). This capacity may be related to musical training and the resultant neurological changes in the auditory and limbic regions of the brain.

The actual experience of performing music generates a mixture of positive and negative moods. Musicians seldom self-select musical works to be performed. Practice and rehearsals enhance familiarity at the time of performance. Despite this familiarity, they often experience negative moods (anxiety) prior to performance but then derive pleasure and fulfillment (positive) during performance (Lamont, 2012). Singing, especially group singing, compared to listening to singing, increases mood states even more. One study demonstrated greater positive mood states endorsed by singers compared to listeners (Unwin et al., 2002). Despite these positive findings, singers report a decrease in emotional intensity and sense of purpose in life immediately after a performance (Bodner & Bensimon, 2008). The natural after-performance decrement in positive mood is a typical experience in all performing artists.

Sadly, minimal research has examined the effect of mood on dancers and actors. Creatively, performers need to engage in more self-reflection, simply because they are the expressive instrument. Because of the demands of their profession, they also experience fluctuations in mood states, partially as a result of changing demands in classes or lessons, rehearsals, and in performances (Quested et al., 2013). Actors are directly and immediately influenced by performance demands. Their feelings and moods shift in response to the circumstances of the character they are portraying (Hetzler, 2008). Although their training prepares them to express a wide range of emotional and hedonic toned moods, their own personal moods can color career choices. For example, an actor who seeks high-risk activities may audition and perform in more action-based works; whereas, a more introverted actor may gravitate to works that are dark and more emotionally rooted. Singers self-select to sing the blues and dancers may seek opportunities to perform highly athletic works. Each performer's personal mood preferences may lead them into careers that are mood-congruent (Verhaeghen et al., 2005).

Summary

- Introduction: definitions, emotional expression and aesthetic experience, duration.
- Mood and creativity: hedonic tone (valence) and intensity, influences on creativity.
- Physiology, neurobiology, and mood: complex interactive responses, judgment and appraisal, perception and attentional processes.
- Empathy, passion, rumination, and mood: influence on positive and negative mood, communication and motivation, internal judgment.
- Mood in music, dance, and acting: influence on process, product and audience, evoking mood.

Concluding remarks

Mood is a powerful factor in the performing arts, both within the artistic product and within the performers themselves. Despite countless studies that demonstrate a strong influence of negative mood on creativity, motivation, and perceptions of task difficulty, positive mood has an equally powerful effect. Performers demonstrate strong predilections to empathically resonate with the performance work and with their fellow performers. Artistic experiences have the potential to elicit emotions and moods that often transcend the normal range studied. For example, heightened positive mood states include aesthetic chills, awe, and spiritual transcendence (Keltner & Haidt, 2003). Performers endorse these experiences more frequently and at a higher intensity than the general public (Ayers et al., 1999; Thomson et al., 2009). Aesthetic chills are marked by physiological changes such as goosebumps and feelings of tingling or shuddering that move through the spine (McCrae, 2007). They are positive in nature, which differentiates them from chills experienced while in terror states, and thrills that are experienced during high risk-taking activities. Works of art that are either sad (negative) or joyous (positive) can evoke chills; the key ingredient is feeling "moved" emotionally by the artwork (i.e., viewing a film or dance) (Wassiliwizky, Wagner, Jacobsen, & Menninghaus, 2015). The likelihood of art-elicited chills is directly related to the heightened intensity of being moved, hence anxious mood states deter the potential to be moved and experience aesthetic chills (Nusbaum et al., 2014). Currently, research studies suggest that art-elicited chills are most effectively evoked by music, especially during the transitional moments when key changes occur, but chills can be equally activated while witnessing a powerful performance given by a dancer, actor, or singer (McCrae, 2007).

Feelings of awe are evoked when states of expansiveness and admiration are activated concurrently (Keltner & Haidt, 2003). A sense of connectedness and transcendence also accompany feelings of awe (Silvia et al., 2015). Positive variables such as great beauty, ability, and virtue can elicit awe, just as intense negative threats, whether by human devastation or natural disasters, evoke a more distressed feeling of awe. Awe is a mood state that most performers hope to elicit in their audiences, as is evident by great works that can be viewed in movie theaters, concert halls, or theaters. Works that push our perceptions beyond the familiar have the potential to inspire awe. Feelings of connecting to something beyond the mundane excite these awe-inspiring experiences, especially in those performers and audience members who have a predilection to openly experience events that nudge them beyond the familiar (Keltner & Haidt, 2003; Silvia et al., 2015).

Mood is a general state that colors experience and influences motivation. It encourages approach or withdrawal behavior and it colors perceptions of self, others, and the task. As part of the performing arts, experiencing mood states is a highly ritualized process (Nagy, 2015). Evoking a mood during the performance of any work is the performers' challenge. They induce the mood in themselves and within their audience. Mood influences their creativity and it also becomes part of the fabric of most works of art.

Unresolved mourning and posttraumatic stress disorder in performing artists

16

This will be our reply to violence: to make music more intensely, more beautifully, more devotedly than ever before.

Leonard Bernstein

Introduction

The word *trauma*, derived from the Greek word *titrōskein*, means *to wound* (Merriam Webster Dictionary). Trauma is the result of an external force, whether inflicted by nature or human. Examples of intentional human acts of trauma include violent sexual or nonsexual assault; nonviolent coercive sexual assault; torture; war; terrorism; witnessing or learning about a homicide, suicide, or extreme life-threatening event such as bombing or kidnapping; or being forced to commit violent atrocities. Nonintentional human acts include accidents (car, plane, train) or medical trauma that can occur when a medical intervention is particularly invasive or unexpected. Acts of omission, such as emotional and physical neglect, are equally traumatic, and include abandonment in orphanages, internment camps, or profound isolation. Nonhuman acts include hurricanes, floods, earthquakes, landslides, avalanches, tsunamis, fires, explosions, nuclear disasters, tornados, typhoons, drought, famine, animal attack, or serious illness. The effects of a traumatic event, regardless if it occurred in adulthood or childhood, can be long-lasting and life-changing (van der Kolk, 2014). The effects reach beyond the traumatized individual and are transmitted to the unborn fetus (Dubrowsky, 2015b).

To date, minimal investigative research explores the effects of trauma on performing artists. On the other hand, much research and media coverage examines the mental stability and instability of performers, with little regard to possible unhealed past trauma. A typical misconception is that emotionally unstable individuals perform as a way to express intense emotions. Sarah Benolken (2006) captures this in her statement: "My mother always told me I should go on the stage, because I cried so much. The reason I cried so much was that, as far back as I can remember, my father was hitting me in the face and yelling at me that I was too fat almost every day" (p. 55). Sarah did perform but her unresolved past trauma encroached on her sense of wellbeing. Performers are expected to portray human atrocities and give emotional life to these events; however, their audience is usually unaware and

Creativity and the Performing Artist. DOI: http://dx.doi.org/10.1016/B978-0-12-804051-5.00016-0

unconcerned about the performers' personal trauma histories. While performing, unbidden past traumatic memories can intrude and destabilize them at any time.

Bessel van der Kolk (2014) claims that traumatized individuals suffer from "a failure of imagination" (p. 17). Artists rely on imagination to give life to the works they perform. If they want to succeed, they must also imagine themselves in their future career. But unresolved trauma freezes creative thinking and constricts emotional exploration, consequently it foreshortens a sense of a future self. For example, traumatized individuals are frequently afraid to embody and give voice to their trauma, fearing that expressing their emotions will lead to a devastating loss of control. They fear they will go "crazy" or "break into a million pieces." Dysregulation sabotages curiosity, and ultimately, creativity (van der Kolk, 2014).

Ironically, performing arts are common therapeutic modalities used to address and heal trauma (Gobodo-Madikizela, 2006). When employed for therapeutic purposes, theater, music, and dance provide traumatized individuals an opportunity to express unspeakable traumatic experiences, but the healing resides in the relational field cocreated between the traumatized person and a trained therapist. What makes it healing is the reality that trauma therapists fully listen, without judgment, to the pain that was visited onto the individual (Felman & Laub, 1992; van der Kolk, 2014). Body movement, breath connection, and rhythmically organized expression become the medium to "tell" the stories. During this therapeutic process, traumatized individuals can learn to self-regulate and recover from their trauma (Dutton, 2001). Programs such as Theater of Witness or testimony theater give the survivors a voice; the "conspiracy of silence" that surrounds human atrocity is broken (Ehn, 200; Peleg, Lev-Wiesel, & Yaniv, 2014). The silence stops when traumatized people "perform" their stories, but empowerment and healing occur only when other people respond with empathy and full attention as they witness these testimonial performances (Caruth, 1996; Peleg et al., 2014). Sadly, for many traumatized performing artists, this attuned healing process is not the norm in training or rehearsal settings. Most performing artists learn to hide their personal struggles. They know better than to reveal psychological weakness in class or rehearsals; they are fully aware that their job includes expressing human suffering while simultaneously remaining in control of their technical and expressive skills (Kenny, Driscoll, & Ackerman, 2014).

What are posttraumatic stress disorder and unresolved mourning?

Posttraumatic stress disorder (PTSD) is a psychiatric disorder given when an overwhelming life-threatening event is experienced or witnessed. It can also occur when receiving news that a close family member or friend experienced a violent or accidental catastrophic event (APA, 2013). To receive a diagnosis of PTSD, intrusive, avoidant, and/or persistent negative thoughts and moods, plus intense arousal and reactive symptoms, must be present one month following the traumatic event, although these symptoms can emerge much later than one month postexposure and can persist for years or decades (see Box 16.1). The potential that individuals will be

Box 16.1 DSM-5 Trauma and Stress-Related Disorders—Posttraumatic Stress Disorder (APA, 2013, pp. 271–272).

To receive a diagnosis, all criteria must be met.	
Criteria A: Exposure to actual or threatened death, serious injury, or sexual violence in one (or more) of the following:	**Criteria A**: 1. Directly experiencing traumatic event(s) 2. Witnessing, in person, the traumatic event(s) that occurred to others 3. Learning that the traumatic event(s) occurred to a close family member or friend, actual or threatened death must be violent or accidental 4. Experiencing repeated or extreme exposure to aversive details of traumatic event(s)
Criteria B: Presence of one (or more) of the following intrusion symptoms associated with traumatic event(s), beginning after traumatic event(s) occurred:	**Criteria B:** 1. Recurrent involuntary, and intrusive distressing memories of the traumatic event(s) 2. Recurrent distressing dreams in which content and/or affect of dreams are related to traumatic event(s) 3. Dissociative reactions (e.g., flashbacks in which the individual feels or acts as if the traumatic event(s) were recurring (possible that dissociative reaction includes complete loss of awareness of present surroundings) 4. Intense or prolonged psychological distress at exposure to internal or external cues that symbolize or resemble an aspect of traumatic event(s) 5. Marked physiological reactions to internal or external cues that symbolize or resemble an aspect of traumatic event(s)
Criteria C: Persistent avoidance of stimuli associated with traumatic event(s), beginning after the traumatic event(s) occurred, as evidenced by one or both of the following:	**Criteria C:** 1. Avoidance of or efforts to avoid distressing memories, thoughts, or feelings about or closely associated with traumatic event(s) 2. Avoidance of or efforts to avoid external reminders (people, places, conversations, activities, objects, situations) that arouse distressing memories, thoughts, or feelings about or closely associated with traumatic event(s)

(Continued)

Box 16.1 (cont'd)

Criteria D: Negative alterations in cognitions and mood associated with traumatic event(s), beginning or worsening after the traumatic event(s) occurred, as evidenced by two (or more) of the following:	**Criteria D:** 1. Inability to remember an important aspect of traumatic event(s) (typically due to dissociative amnesia and not to other factors such as head injury, alcohol, or drugs) 2. Persistent and exaggerated negative beliefs or expectations about oneself, others, or the world (e.g., "I am bad," "No one can be trusted," "The world is completely dangerous," "My whole nervous system is permanently ruined") 3. Persistent, distorted cognitions about the cause or consequences of the traumatic event(s) that lead the individual to blame himself/herself or others 4. Persistent negative emotional states (e.g., fear, horror, anger, guilt, or shame) 5. Markedly diminished interest or participation in significant activities 6. Feelings of detachment or estrangement from others 7. Persistent inability to experience positive emotions (e.g., inability to experience happiness, satisfaction, or loving feelings)
Criteria E: Marked alterations in arousal and reactivity associated with the traumatic event(s), beginning or worsening after the traumatic event(s) occurred as evidenced by two (or more) of the following:	**Criteria E:** 1. Irritable behavior and angry outbursts (with little or no provocation) typically expressed as verbal or physical aggression toward people or objects 2. Reckless or self-destructive behavior 3. Hypervigilance 4. Exaggerated startle response 5. Sleep disturbance (e.g., difficulty falling or staying asleep or restless sleep)
Criteria F: Duration of the disturbance (Criteria B, C, D, and E) is more than one month	
Criteria G: The disturbance causes clinically significant distress or impairment in social, occupational, or other important areas of functioning	
Criteria H: The disturbance is not attributable to the physiological effects of a substance (e.g., medication, alcohol) or another medical condition	

Specify whether: With dissociative symptoms: The individual's symptoms meet the criteria for posttraumatic stress disorder, and in addition, in response to the stressor, the individual experiences persistent or recurrent symptoms of either of the following: ***Note***: To use this subtype, the dissociative symptoms must not be attributable to the physiological effects of a substance (e.g., blackouts, behavior during alcohol intoxication) or another medical condition (e.g., complex partial seizures).	**Specifier**: 1. **Depersonalization**: Persistent or recurrent experiences of feeling detached from, and as if one were an outside observer of one's mental processes or body (e.g., feeling as though one were in a dream; feeling a sense of unreality of self or body or of time moving slowly) 2. **Derealization**: Persistent or recurrent experiences of unreality of surroundings (e.g., the world around the individual is experienced as unreal, dreamlike, distant or distorted)
Specify if:	**With delayed expression**: If the full diagnostic criteria are not met until at least 6 months after the event (although the onset and expression of some symptoms may be immediate)

exposed to a traumatic event(s) during their lifetime is high, with 89.7% of the US population reporting experiencing at least one traumatic event (Kilpatrick et al., 2013). The percentage of individuals receiving a diagnosis of PTSD varies among studies; the range is between 8.9% (Kilpatrick et al., 2013) and 1.7% (Lukaschek et al., 2013). In a study that examined PTSD diagnosis over time, 28.8% of the PTSD sample received a diagnosis 1 month postexposure and 17% at 12 months postexposure (Santiago et al., 2013). This decreasing pattern of PTSD diagnosis over time is particularly relevant to *nonintentional traumatic events* such as natural disasters or traffic accidents. PTSD diagnosis after exposure to *intentional traumatic events* (i.e., sexual or nonsexual assault) actually increases over time, with 11.8% of the sample receiving a diagnosis 1 month postexposure, and 23.3% receiving a diagnosis 12 months after exposure. In this intentionally traumatized PTSD group, nearly 40% developed a chronic course of PTSD (Santiago et al., 2013).

These statistics demonstrate the frequency of traumatic exposure and the potential development of PTSD, but, sadly, few individuals in the general population are actually assessed and treated for PTSD. A large percentage of exposed individuals are misdiagnosed by primary care physicians, in part, because PTSD symptoms can resemble depression, chronic pain, sleep disorders, anxiety, substance abuse/dependence, or irritable bowel syndrome. Further, during medical appointments, primary care physicians commonly fail to inquire about past traumatic exposure in their patients (Liebschutz et al., 2007). Effectively treating PTSD ensures a decrease in symptoms. However, if left untreated, then PTSD symptoms convert into debilitating health problems, both physical (anemia, arthritis, asthma, back pain, diabetes, eczema, kidney disease, lung disease, sleep disorders, and ulcer) and psychiatric (anxiety, depression, dissociation, substance abuse) (Perez, Abrams, Lopez-Martinez, & Asmundson, 2012; Weisberg et al., 2002).

Determining the degree of resolution for past traumatic and loss events can add further information about the exposed individuals' psychological stability. Unresolved mourning is a classification given as part of the Adult Attachment Interview (AAI) procedure. It is not a psychiatric diagnosis; however, how a person speaks about trauma reveals how his or her mind is managing these experiences. During the coding of the AAI, psychological disorientation and disorganization are noted during passages in which participants speak about past loss of significant family members or close friends, past traumatic abuse in childhood/adolescence, or traumatic events that occurred in adulthood (Main et al., 2003). Marked lapses of monitoring of reason, discourse, or behavior indicate that the speaker remains unresolved for these events (see Box 16.2). Assessing whether individuals are aware of their brief moments of dysregulation adds to an understanding regarding their success or failure to psychologically integrate these events (trauma or loss). For example, some individuals speak about the dead person or traumatic event in the present tense, and then quickly add a statement that indicates they know it happened in the past. This self-correction suggests they are capable of monitoring their thoughts and feelings without getting lost. If these statements are not corrected, and the person fails to monitor disorganized thoughts, feelings or behaviors, then a state of unresolved mourning is identified.

Years after a traumatic event, absorbed sensory narrative discourse about the traumatic event(s) and a failure to monitor psychological processes are strongly linked to a diagnosis of PTSD (Crespo & Fernandez-Lansac, 2015; Ng,

Box 16.2 Markers for Unresolved Mourning in the Adult Attachment Interview (Main et al., 2003)

Identifying unresolved/disorganized attachment status within the AAI. Lapses in the monitoring of discourse and reason include:	Note: The events must meet criteria for loss (significant loved one) and trauma (physical abuse that left marks and hurt, sexual abuse, or bizarre and unusual punishment by a caregiver or close family member)
Lapses in the Monitoring of Discourse suggests that the topic has triggered a "state shift" indicative of considerable absorption, frequently appearing to involve entrance into peculiar, compartmentalized states of mind.	• Unusual attention to detail • Poetic phrasing with a memorialized quality • Prolonged silences (20–50 sec.) • Visual-sensory images related to experience intrusion • Words or phrases seeming to stand for the abuse or are used in an odd way, as though unable to name it

	• Weaker indices; unfinished sentences, sudden changes or move away from topic, other topics invaded by information regarding death, notable change in discourse style
More Detailed Explanation	• Sudden changes in speech register (e.g., shifting from normal speaking patterns into eulogistic/ funeral speech, as "She was beautiful, active, young, and she was taken from us by that most horrific of diseases, cancer") • Falling silent for 100 seconds midsentence, then continuing on in an unrelated topic • Giving extreme attention to details surrounding a loss or other potentially traumatic experiences inappropriate to the interview context (e.g., a 10-minute discussion involving minute details of a loss including time of day, furnishings of the room, and clothing worn to the funeral by each family member)
Lapses in the Monitoring of Reason suggest that the speaker is temporarily expressing ideas that violate our usual understanding of physical causality or time–space relationships	• Indications of disbelief that the person is dead • Sense of being causal where no material cause is present (loss or trauma) • Confusion between dead person and self • Time disorientation • Space disorientation • Psychologically confused statements • Unsuccessful denial of the occurrence, nature, or intensity of the abusive experience • Subject fears being taken over (possessed) by the abusive figure
More Detailed Explanation:	• Subtle indications that a deceased individual is believed simultaneously dead and alive in the physical (not religious or metaphysical) sense (e.g., "It was almost better when she died, because then she could get on with being dead and I could get on with raising my family") • Placement of the timing of a death at several separated periods (e.g., ages 9, 11, and 15 given for same loss experience at differing places in the interview) • Indications that self was responsible for the death where no material cause was present (e.g., death caused by having thought something negative about a person near the time of their death)

(Continued)

Box 16.2 (cont'd)

	• Claims to have been absent at the time of the death, juxtaposed with claims to having been present at a drowning, then later speaking as though the self had been present; "and we tried but none of us could swim to her")
Extreme Behavioral Responses	• Suicidal, unable to function, self-destructive behavior

Note: Lapses in monitoring such as monitoring of reason suggest that the speaker is temporarily expressing ideas that violate our usual understanding of physical causality or time–space relations and lapses in monitoring of discourse sometimes suggest that the topic has triggered a "state shift" indicative of considerable absorption, frequently appearing to involve entrance into peculiar, compartmentalized states of mind. These lapses in monitoring indicate an unresolved/disorganized (U/D) category and are included with a best-fitting alternate classification. When two separate attachment strategies exist as discrete classifications within the same transcript then a Cannot Classify descriptor is given.

Ahishakiye, Miller, & Meyerowitz, 2015). Sensory, perceptual, and emotional details invade trauma narratives, unlike descriptions of other life events. This narrative pattern demonstrates how trauma reminders automatically retrieve sensory–perceptual representations of the event, a pattern that typifies flashbacks (Crespo & Fernandez-Lansac, 2015). Multiple studies have demonstrated a statistically significant relationship between unresolved mourning and PTSD (Benoit, Bouthillier, Moss, Rousseau, & Brunet, 2010; Harari et al., 2009; Stovall-McClough & Cloitre, 2006). Further, a large metaanalysis assessing unresolved mourning distribution patterns revealed that unresolved mourning classifications varied across samples; clinical PTSD samples had higher distributions of the unresolved classification (68%) compared to nonclinical samples (23%) (Bakermans-Kranenburg & van IJzendoorn, 2009). Like PTSD, an unresolved classification indicates greater difficulty managing negative emotions (Creasey, 2002). Sadly, early developmental experiences of ineffective parenting also increase the likelihood of unresolved mourning (Fleming et al., 2002; Hesse & Main, 1999).

Unresolved mourning is also related to more negative interactions with healthcare providers (Green, Kaltman, Chung, Holt, Jackson, & Dozier, 2012). The nature of disorganized unresolved attachment behaviors indicates a pattern of seeking help and simultaneously rejecting it. Successfully receiving medical treatment is compromised by the very symptoms that need to be treated. For unresolved performing artists, this contradictory negative response pattern with healthcare providers can seriously compromise treatment when they are injured or ill. Unresolved mourning is also related to a grieving process that remains prolonged and disorganizing. The cause of loss, such as violent death, is more likely to complicate the grieving process (Anderson, Marwit, Vandenberg, & Chibnall, 2005; Currier, Holland, & Neimeyer, 2006; Davis, Wohl, & Verberg, 2007; Thomson, 2010). Unresolved trauma and depression share similar symptom patterns with unresolved loss, although they require treatment that addresses each as unique dimensions of distress (Boelen, 2013). Studies have demonstrated that performing artists, in general, and actors in particular, have a higher distribution of unresolved mourning compared to the general population (Thomson & Jaque, 2012a, 2012b), consequently they are at greater risk to experience PTSD symptoms and to have more negative interactions with healthcare providers. They also endorse more paranormal beliefs, which further compromises resolution for past traumatic events and prolonged grief (Thomson & Jaque, 2014).

Trauma, trauma reminders, and performance: effects on physiological and neurobiological systems

Physiological and neurobiological stress-related effects are clearly understood by scientists, including the long-term alterations resulting from chronic stress. Biological responses are mobilized during a traumatic event, a reality that is normative and expected. But for those with PTSD or unresolved mourning, these same biological states are reactivated when a strong reminder of the traumatic event is induced. On the other hand, those who are resolved and have no diagnosis of PTSD demonstrate minimal physiological reactivity when recalling trauma (Sledjeski & Delhanty, 2012). A "trigger" is any social/environmental or internal cue that serves as a reminder of the traumatic event. Triggered biological stress reactivations may be consciously or unconsciously activated. For example, many triggers are sparked by a particular smell, sound or place, a time of day, an anniversary date, or even an emotional state. The autonomic nervous system (ANS) itself can also be a trigger. For example, specific biological states can match past traumatic exposure states. An elevated heart rate prior to or during a performance can match the arousal level experienced during past physical or sexual assaults. The performer may be focusing on the performance but, unconsciously, increased heart rate triggers a memory of the rape or attack. The performer becomes flooded with terror, often including painful physical sensations and autobiographical memories of the rape. These cues are processed through a chain of associations causing activation in the ANS in preparation for fight, flight, freeze, or collapse responses. Once this pronounced activation has begun, performers are no longer able to meet the demands of performance (Swart, 2014). It is as if the event was happening again. Their performance is hijacked by the unbidden memories and physiological arousal of past trauma.

Activating the biological stress response is devastating to performers with unresolved trauma (Swart, 2014). As arousal increases beyond normal self-regulatory states, performers shift into a stress state called allostasis (Danese & McEwen,

2012). This allostatic survival state is mobilized when threatening environmental demands require greater neurobiological responses that exceed regular homeostatic needs, a condition that is pronounced in individuals with PTSD. Allostasis provides survival regulation that occurs during extreme change; it enables detection of threats to homeostasis, whether external (environment) or internal (i.e., high fevers, gastric distress, irregular heartbeats, etc.). Allostasis involves three major regulatory systems: the nervous, endocrine, and immune systems. These systems are highly integrated, thus they work interactively during stress events. One system can also activate the other two (Altemus, Dhabhar, & Yang, 2006). A return to homeostasis is induced by responses such as reduction in motor activities, calming or limiting social interactions, and increased sleep, all behaviors that manifest with any illness. But chronic stress, including PTSD, does not allow short-term "sick" behaviors to restore homeostasis.

Prolonged activation of the three allostatic systems results in structural and functional alterations, especially in stress-sensitive regions of the brain such as the prefrontal cortex, hippocampus, amygdala, hypothalamus–pituitary–adrenal (HPA) axis, and the ANS. Over time, chronic stress, and the resultant neurobiological adaptations, lead to age-related disorders such as cardiovascular disease, autoimmune disorders, and metabolic syndromes (Kendall-Tackett & Klest, 2009; Kibler, 2009). This trajectory of health-related problems can be transmitted across generations, as evidenced in increased health problems in the middle-aged offspring of Holocaust survivors (Shrira, Palgi, Ben-Ezra, & Shmotkin, 2011). The allostatic process is even more pronounced during childhood (Bornstein, Schuppenies, Wong, & Licinio, 2006; Danese & McEwen, 2012; Raikkonen, Matthews, & Salomon, 2003), especially increased coronary heart disease that is associated with early trauma exposure (Rooks et al., 2015). Complex adult psychopathology, including anxiety, depression, substance abuse, impulsivity, depression, mixed mood disorders, dissociation, and suicidality, also results from childhood adversity (Putnam, Harris, & Putnam, 2013).

PTSD flashbacks (triggers) provoke prolonged allostasis, ultimately leading to a loss of control of the major stress systems in the body (van der Kolk, 2014). During a PTSD trauma flashback, the eyes dilate in an effort to extract more peripheral information in the environment. If this happens during a performance it is devastating. For example, musicians will have extreme difficulty narrowing their focus on the conductor and processing musical cues that are given by the conductor. The diffuse focus of a flashback is biased to detect threat. They are now attempting to interpret, process, and respond to irrelevant peripheral visual information that may include small movements by audience members, performers, and others working in the wings offstage. Although these movements are irrelevant on stage, during a traumatic event, diffuse focus expands information-gathering regarding the nature of danger and the direction of the danger. The ability to process sound also shifts from narrow to diffuse attention. But onstage, this interferes with a performer's ability to orient to the sounds being played. For example, the dancer's inability to accurately detect sounds will destroy ensemble, pas de deux, or solo work, especially because sound and movement are tightly integrated neurobiologically.

This integration of sound and movement is highly trained in performers, and trauma triggers can derail this integration (Leube et al., 2003).

Decrease in saliva production, resulting from increased activation of the sympathetic nervous system (SNS), helps reduce metabolic demands during a traumatic event, but, for singers and actors, during a triggered trauma memory this can destroy performance. Coupled with decreased saliva production, multiple regions in the brain are altered during a trigger, such as transient deactivation in Broca's area (speech center) and the hippocampus (memory retrieval) (van der Kolk, 2014). Now singers and actors not only lack saliva to produce sound but they are also unable to speak or remember the words or notes to be performed. The same inability to retrieve memory, resulting from transient activation–deactivation fluctuations in the hippocampus, occurs for dancers and musicians during traumatic memories. Unfortunately, racing thoughts that often accompany this triggered state compound the painful awareness that they are unable to speak; these thoughts usually focus on terror and ineffective problem solving.

The higher cortical prefrontal region supports flexible thinking needed to solve problems, but during a trauma trigger this region is now hijacked by lower regions in the brain (van der Kolk, 2006). The amygdala, a subcortical structure, detects threat and sends inhibitory feedback loops between the prefrontal cortex and hippocampus, and excitatory loops to regions that will increase alertness and attention necessary to meet the environmental demands. Subcortical regions that release the alerting neurotransmitter norepinephrine, such as the hypothalamus, periaqueductal gray, and locus coeruleus, are strongly recruited. The hypothalamus increases activation in the SNS branch of the ANS, as well as the neuroendocrine system, which is managed by the pituitary and adrenal glands (Danese & McEwen, 2012). All these regions prepare for threat, a process that exponentially increases metabolic demands. Consequently, trauma triggers are energetically costly, especially when they occur during a performance.

Physiologically, efforts to respond to a potential threat during performance compromise breathing (shallower), heart rate (rapid), and blood pressure (elevated). Trauma triggers also increase a need to urinate and/or defecate, factors that add to performance distress. These physiological responses are understandable from a survival perspective. By decreasing metabolic energy needed to regulate these bodily functions, the body can redirect energy to motor striatal muscle tissues in order to effectively run away or fight. As well, during a traumatic event, greater secretion of analgesic neurohormones is released to reduce anticipated pain (van der Kolk, 2014), but this is problematic when it occurs during a performance. These stress responses reduce rapid neurobiological feedforward and feedback information that is essential for finely integrated elite performance. The risk for injury is increased during these flashback episodes because performers are not able to "feel" their bodies. These cascading physiological alterations mobilized during a trauma trigger are identical to stage fright or performance anxiety episodes (Kenny 2005; Kenny et al., 2014). Clinician assessment must determine if performers are suffering from PTSD/unresolved mourning or from stage fright/performance anxiety. These two conditions require different treatment protocols even though the physiological responses are similar (Fig. 16.1).

Figure 16.1 "Despair." Performers: Singer Annie Sherman and Dancers. Photographer: Lee Choo.

Effects of chronic stress

Obviously, chronic stress reactivity has deleterious consequences. Cardiorespiratory changes, coupled with a cascade of complex HPA axis alterations, directly influence the body's immune response (Danese & McEwen, 2012). Over time, greater inflammation is present as well as an inability to fight pathogens. With increased inflammation, healing tissue damage is further compromised (Danese & McEwen, 2012; McEwen et al., 2015). These neuroendocrine changes are very dangerous for performers, in part, because their work is naturally stress-inducing. They need to recover quickly and effectively between performances, as well as during intense rehearsal periods and the long hours of training. If performers are not able to manage stress caused by dysregulated PTSD triggers and unresolved mourning, then they are also at added risk for cardiovascular, metabolic, and inflammatory diseases (Danese & McEwen, 2012). For example, chronic pain, an inflammatory stress response, can be the precipitating factor for disorders such as substance abuse, depression, and anxiety. Managing chronic pain is debilitating for all, but when this occurs in performing artists, an inability to recover from injuries and to maintain normal performance-work demands influences career longevity (Otis, Keane, & Kerns, 2003; Figs. 16.2 and 16.3).

Study findings that report physiological alterations due to PTSD are sobering. For example, trauma-exposed individuals have elevated resting blood pressure and heart rate but they are not as high as those with PTSD. This pattern raises significant concerns about cardiovascular health in PTSD populations

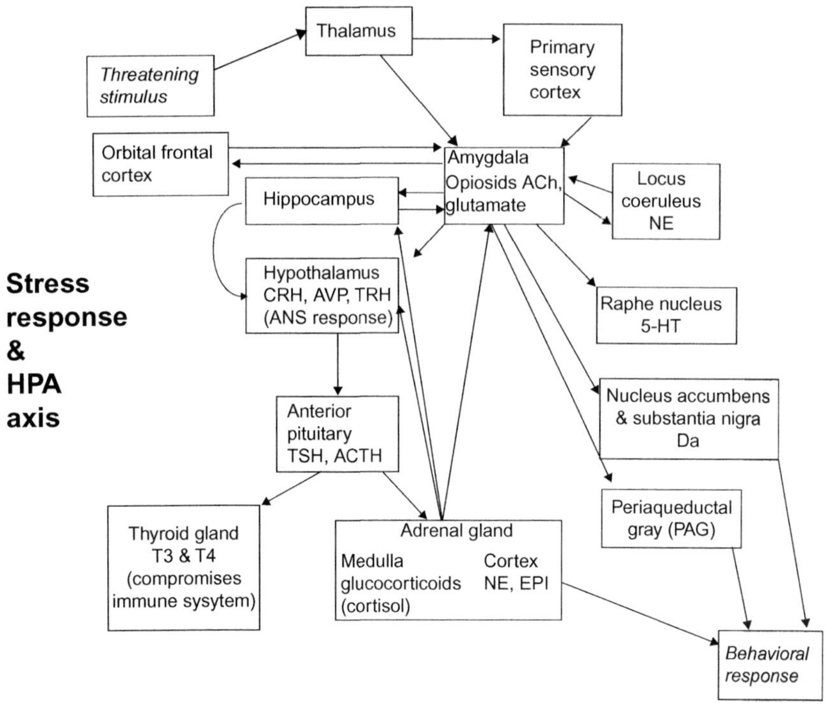

Figure 16.2 Hypothalamus–pituitary–adrenal axis and stress response.

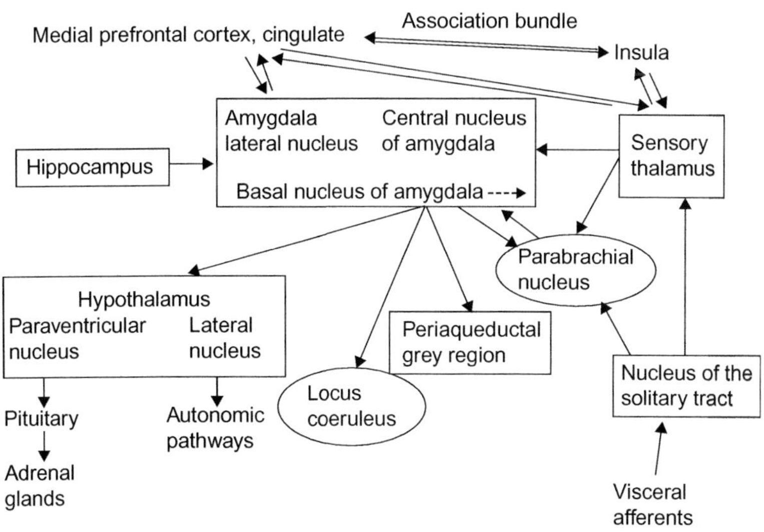

Figure 16.3 Cortical and subcortical pathways during hypothalamus–pituitary–adrenal axis activation.

(Paulus, Argo, & Egge, 2013). For example, in a study that examined PTSD veterans their physiological profile was marked by greater sympathetic activation and decreased parasympathetic tone. This ANS pattern predicted increased PTSD symptom severity and subsequent health problems in this sample (Brady et al., 2015). Heart rate variability was also decreased in PTSD individuals, indicating a dysregulated ANS response to chronic stress (Hauschildt, Peters, Moritz, & Jelinek, 2011). For adult performing artists with PTSD, this compromised ANS response will exacerbate the already stressful career demands that performers must manage. Contrary to these adult physiological responses, children from chronic domestic violence families have greater parasympathetic activation during peer provocation. This finding suggests that these children already learned to overrecruit parasympathetic tone as a way to self-regulate. Sadly, despite their atypical ANS efforts to manage stress, these children were more likely to be diagnosed with PTSD as well as conduct disorders; whereas, children from less aggressive families had a typical stress response, marked by greater physiological sympathetic reactivity when discussing conflict. They were also able to return to physiological balance when not stressed, a marker of greater heart rate variability (Hauschildt et al., 2011; Katz & Rigterink, 2012).

Traumatized children who are training to become performers may also manifest atypical stress reactions. Providing optimal environments to train young performers is critical; however, these training environments may be traumatizing. Verbally and physically aggressive training environments are not much different than adverse family environments; they lead to disruptions in the developing stress systems. For example, training children in environments that focus on intense evaluation and criticism compromises their ability to perform creatively (Byron, Khazanchi, & Nazarian, 2010). They will also respond with a decreased sense of control, including withdrawing from challenging tasks. These behaviors will limit the realization of their talent (Mittal & Griskevicius, 2014). Repeated exposure to stress, such as abusive training, also disrupts the natural ability to buffer anxiety (Abdollahi, Pyszczynski, Maxfield, & Luszczynska, 2011). Rather than habituating to stress, maltreated performers become more anxious. They either withdraw from performance training or are hypervigilant to potential training threats.

Cortisol reactivity, whether blunted or elevated, is a common pattern in individuals chronically exposed to trauma, a pattern that is also found in patients with recurrent depression and stress-related chronic pain disorders (Ehlert, Gaab, & Henirichs, 2001). When trauma occurs during infancy a delayed cortisol recovery pattern persists into adulthood, a clear marker that the HPA axis is profoundly altered and is less effective at regulating stress (Kuhlman, Vargas, Geiss, & Lopez-Duran, 2015). Trauma during infancy impairs HPA feedback responses, primarily because cortisol levels are not sufficient to effectively shut off the stress response in the hypothalamus. This alteration is significant given that during the first year of life HPA maturation is massively shaped by social/attachment relationships. It is a sensitive developmental time period for HPA functioning. Normal development of the HPA stress response is measured by increasing hyporesponsiveness of the HPA axis; trauma during this period makes the HPA axis more hyperresponsive (Kuhlman et al., 2015).

When traumatic events occur in childhood, males and females display different diurnal cortisol responses. Trauma-exposed males have high waking cortisol, lower cortisol during the day, and elevated cortisol in the evening; whereas, trauma-exposed females have high circulating cortisol throughout the day. Regardless of pattern, efficient physiological regulation during the day is compromised for both males and females. More total trauma (from infancy into adolescence) also negatively influences recovery time after stressful events (Kuhlman et al., 2015). Further, childhood maltreatment causes specific genetic changes in the glucocorticoid receptor genes (Romens et al., 2015). This was markedly evident in children from aggressive family backgrounds. While discussing aggression, they exhibited a dampened response to cortisol, suggesting that greater exposure to chronic aggression in childhood decreased physiological feedback loops in the HPA axis (Romens et al., 2015). In addition, a biological blunting of cortisol concentration reduces stress regulation, especially under conditions that are overtly threatening (Saxbe et al., 2012). Given this pattern of altered neuroendocrine function, performers raised in adverse environments will have altered cortisol responses, which will decrease their ability to manage stress; they will react more intensely to a stressor and will have greater difficulty recovering from it. Although performers are blessed with talent, their talent may not be sufficient to offset the epigenetic alterations that occur in their cortisol receptors; their resilience to stress will be significantly compromised (Romens et al., 2015).

Alterations in brain volume occur in direct relationship to early adversity, such as decreased volume in the hippocampus (Teicher, Anderson, & Polcari, 2012), and increased volume in the amygdala; whereas, an inverse pattern is found in children raised in families with parental support (Luby et al., 2012; Teicher, Tomoda, & Andersen, 2006). These changes are a direct result of epigenetics (gene–environment interaction) (Binder et al., 2008). Specifically, DNA methylation shapes brain formation during early development; the methylation process is influenced by environmental demands, including chronic states of allostasis (McEwen, Eiland, Hunter, & Miller, 2012). As stated earlier, many of these brain regions are integral components in the three allostatic stress systems (Meaney, 2001). Other regions that are critical to integrative processes, such as the cerebellum, corpus callosum, anterior cingulate, insula, and higher cortical somatosensory associative regions, are equally altered (DeBellis et al., 2002; Marusak, Etkin, & Thomason, 2015; Teicher et al., 2006). The cascade of brain changes, especially during childhood, increases the risk for later PTSD (Binder et al., 2008), although it is possible to have some genetic influences that can protect the developing integrative regions and stress systems in the brain (van Ryzin et al., 2015).

Childhood trauma, PTSD, and unresolved mourning ultimately disrupt connectivity within key brain networks. In particular, the salience and default mode networks are significantly compromised as a result of trauma exposure; consequently early adversity diminishes reward experiences and increases illness (Marusak et al., 2015). Alterations in interoception awareness and somatic marker sensitivity influence and amplify a negative sense of self, evidenced in both cognitive and affective misappraisals about self and others (van der Kolk, 2014). These changes profoundly harm a performer's sense of wellbeing, including engagement in complex activities such as creativity.

Trauma and self-harm

Self-harm behaviors are strongly associated with unresolved mourning, especially in younger individuals. Self-harm behaviors always involve tissue damage such as cutting, gauging, scratching, picking, or burning (Briere & Gil, 1998). Less overt but equally troubling behavior includes high-risk activities (sexual promiscuity, drug use, extreme sports). Related disorders may also be considered self-harm behaviors, such as eating disorders, substance abuse disorders, and impulse control disorders. Notably, these behaviors are not considered suicidal threats; however, prolonged or extreme self-harm behaviors may result in death. Ultimately, all self-harm behaviors indicate difficulty regulating emotional intensity. Robust research demonstrates self-harm is significantly associated with childhood adversity and trauma, in particular, physical abuse and emotion dysregulation (Gratz & Chapman, 2007). Studies also suggest that deliberate self-harm is not only related to childhood maltreatment, but it is heightened in maltreated individuals with personality traits such as sensation-seeking and openness to experience (Goldstein et al., 2009), traits that are common among performing artists. Histories of parental neglect and rejection also increase deliberate self-harm behaviors (Kaess et al., 2013). Interestingly, increased expression of emotional intensity and reactivity was associated with decreased self-harm behavior (Gratz & Chapman, 2007), a pattern that may protect some performers who are able to positively express negative feelings while practicing or performing (Fig. 16.4).

Figure 16.4 "Isolating shame." Performers: Michael Sean Florita, Darrell Hardaway.

Shame and performing artists

The devastating effects of shame, provoked by early adversity or traumatic events that occur during adulthood, are palpable in performers (Thomson & Jaque, 2013, 2015a). Performers are also vulnerable to intense shaming experiences during their training, rehearsals, and/or performances. For example, if they have been severely publically humiliated the actual reminders of these shaming events can trigger PTSD-like symptoms (Matos, Pinto-Gouveia, & Costa, 2013). Although exposure to adversity usually activates fear, shame can be equally dysregulating; it increases arousal in the three stress systems (Freed & D'Andrea, 2015). Remembering past events that are marked by violence and abuse not only elicit fear, they can also induce profound feelings of shame (Bash & Papa, 2014). Once shame is evoked, further physiological dysregulation ensues, leading to even more difficulty managing stress responses (Freed & D'Andrea, 2015).

Childhood emotional abuse and emotional neglect directly influence beliefs about self, in particular, self-schemas that are shame-based. They display greater beliefs that harm will befall them; they feel more vulnerable and perceive they lack adequate protection (Wright et al., 2009). Constantly being criticized, disapproved, rejected, or ignored is internalized into beliefs that the self is defective. When parents or significant mentor figures are emotionally abusive, children interpret these emotional assaults as validation that their core self is disappointing and not worthy of being loved and valued. These shame-based schemas frequently manifest in disorders such as anxiety, dissociation, or depression. Shame heightens a sense of catastrophic doom, the need to sacrifice the self, and feelings of helplessness; factors that further intensify shame-based self-schemas (Wright et al., 2009). Experiences of significant caregivers and mentors who are unavailable or unresponsive exacerbate feelings of being defective and unwanted. Repeated exposures to such emotional abuse and neglect become shame-based traumatic experiences (Matos et al., 2013).

Like PTSD triggers, reminders of past shaming events can become retraumatizing flashbacks. Performing artists are extremely vulnerable to these deleterious events. There is a long history of cruel (emotional and physical) treatment during training and rehearsal. Performers who grow up with maladaptive interpersonal exposure are at high risk for increased psychopathology, including poor body schemas, eating disorders, performance anxiety, and perfectionism (Arcelus, Witcomb, & Mitchell, 2014; Penniment & Egan, 2012). Internalizing a sense of being bad or unworthy intensifies withdrawal from social contacts and increases narcissistic rage episodes and addictions (Schimmenti, 2012). Shame-based schemas increase a need for perfectionism; this is employed to mediate a "defective" sense of self (Cumming & Duda, 2012; Eusanio et al., 2014). Shame also increases the frequency of physical aggression, especially in individuals with interpersonal trauma histories (Schoenleber, Sippel, Jakupcak, & Tull, 2015). These behaviors are extremely detrimental to performance success, and yet are frequently evident in performing artist populations (Kenny, Driscoll, & Ackerman, 2014).

Although feelings of shame increase as a result of past traumatic experiences, shame can provoke positive outcomes. It can serve as a warning sign that current actions are not optimal, which provides opportunities to redirect behavior in order to achieve the desired goal (Turner & Schallert, 2001). Performing artists tend to believe they are "no good" more than athletes do, thus they are vulnerable to greater feelings of shame; however, feelings of shame can propel them into actively working toward goal achievement (Conroy et al., 2001). Because training in the performing arts is contingent on complex relationships with teachers and mentors, performing artists believe their success is rooted in these relationships. Unfortunately, compared to athletes, they also experience greater pressure not to fail and disappoint their teachers or mentors, and with failure, feelings of shame increase for performers (Conroy et al., 2001).

Performing artists: PTSD, unresolved mourning, and shame

According to Perry (1999), performing arts can capture the emotional memory of trauma. Performers recreate the events through music, film, theater, or dance, and by giving expression, society can begin to comprehend the devastating effects of trauma and the resilience that is possible in many. Performers who continue to work at optimal levels manage or compensate for their traumatic and loss experiences (Swart, 2014). This pattern is evident in countless musicians who suffered life-changing illnesses, or survived wars, political censorship, and exile. In all cases, these events occurred after their talent was identified and significant skill level had been achieved (Swart, 2014). However, familial trauma exposure can compromise talent identification and skill development, including decreased emotional resilience and increased performance stress (Swart, 2014). Trauma experiences, coupled with trait-like anxiety, exacerbate intrusive recollections of traumatic events (Laposa & Alden, 2008); consequently anxious performers are extremely vulnerable to intrusive memories. For example, many traumatized performers suffer terror that is less related to performance anxiety and more to do with the flooding of past traumatic memories. Rachmaninoff frequently suffered recurrent depressive episodes in response to professional criticism; this form of criticism matched his childhood adversity and frightening attachment experiences (Swart, 2014). Beethoven is commonly described as one of many artists who suffered from bipolar disorder; however, few scholars examine the consequences of his early traumatic experiences. He was frequently physically abused by his alcoholic father, experiences that certainly altered his developing brain; hence the manifestation of traumatic symptoms that may have mimicked bipolar disorder (Swart, 2014). Thematic patterns that recur in the performing arts also reflect the traumatized artist's effort to rework, or perhaps, reenact, past unresolved trauma or loss. Wagner is a strong example of someone who struggled with a fatherless childhood, a situation that is reflected in his great operatic themes (Swart, 2014).

PTSD trauma triggers impair consistent hippocampal functioning leading to memory deficits. This is massively challenging for performers, especially when attempting to recall negative affective memories (Swart, van Nieker, & Hartman, 2010). The nature of unresolved mourning is evidenced in the failure to access coherent trauma memories; it is a dissociative strategy to compartmentalize memories of trauma. This defensive strategy exacerbates dissociative amnesia, including pervasive problems with memory recall (Stovall-McClough & Cloitre, 2006). For example, in a study on musicians with a trauma history, they had greater problems with memory and concentration (Swart et al., 2010). Dancers and actors with unresolved mourning also had greater dissociative processes (Thomson & Jaque, 2011a, 2012a, 2012b, 2013). Increased internalized shame was also found in these performing artists (Thomson & Jaque, 2013, 2015a). Further, a greater proportion of dancers had a diagnosis of PTSD compared to the general population, and they also had greater internalized shame and dissociation (Thomson & Jaque, 2015b). A significant proportion of singers (26%) and dancers (15%) also endorsed toxic levels of shame; in fact the proportion was much higher than in an athlete group (6%) (Thomson & Jaque, 2016b). Beyond difficulties with memory and dissociation, interpersonal trauma exacerbates negative self-assumptions (Platt & Freyd, 2012). As evidenced in Rachmaninoff, these responses all reduce an ability to accept negative criticism. Without an ability to metabolize criticism, performers will not be as successful in skill acquisition; they will have greater difficulty mobilizing efforts to improve performance (Beghetto, 2014).

Trauma and resilience

Although neurobiological changes alter stress reactions, performing artists may be somewhat protected from these deleterious effects. Many of the stress regions in the brain are highly interactive; they are also operational in complex tasks such as creativity and elite performance. Training effects in the performing arts increase neural integration (Bar & DeSouza, 2016; Kleber et al., 2013), and this training effect may buffer some of the negative effects of early adversity. For example, the vestibular system is a powerful regulating system; it is strengthened directly by repeated movements of rocking and swinging, movements that are inherent when dancing and playing a musical instrument. The vestibular system enhances hippocampal and cerebellar interconnections, which augments regulatory function (Teicher et al., 2006). Because performing artists regularly engage the vestibular system, along with the hippocampal and cerebellar regions of the brain, they may be actively enhancing their ability to regulate past traumatic events. Research studies demonstrate that participating in regular vestibular-enhancing movement mediates the severity of PTSD symptoms (Rutter, Weatherill, Krill, Orazem, & Taft, 2013). Consequently, performers who suffer from PTSD may be less distressed while engaging in movements that promote physiological stress regulation.

Figure 16.5 "Traumatic loss." Performers: CSUN Dancers. Photographer: Lee Choo.

More life satisfaction also decreases distress after a traumatic event (Gunty et al., 2011). Despite all the obstacles in a performing arts career, most performers describe greater career satisfaction, a fact that may promote resilience (Bennett, 2009; Kogan, 2002). Performers with PTSD, while creating, derive meaning and a sense of integration and healing, a finding that points toward cognitive and emotional regulation (Thomson & Jaque, 2016d). These factors suggest that post-traumatic growth may be possible (Zoellner & Maercker, 2006). Further, some life-time adversity actually operates as a source of resilience and increases wellbeing, perhaps because learning how to cope with adversity promotes a sense of strength (Seery et al., 2010). In a study that investigated resilience after a natural disaster (Hurricane Katrina), cognitive flexibility, tolerance for frustration, and engaging in problem solving all fostered resilience (Metzl, 2009). Although natural disaster trauma is extremely distressing, it is unknown if these same creative skills would operate as buffers for interpersonal traumatic recovery; however, the cross-sectional study findings investigating creativity in performers with PTSD suggests that creative experiences are associated with positive flow states and wellbeing (Thomson & Jaque, 2016d; Fig. 16.5).

Recall of trauma: testimony theater/theater of witness

Testimonial art has great healing potential; especially for large-scale social trauma associated with gross human rights abuses. In this artistic medium, artist and observer frequently share the complexity of being both victim and witness (Gobodo-Madikizela, 2006). The act of creating a testimonial artistic work provides an opportunity to claim the traumatic events while simultaneously distancing from them (Kaplan, 2005). The process of creating an artistic work and sharing it with a community offers perspective-taking, without the insistence of silencing the victim and denying the atrocity. The very act of expressing these events through art opens the possibility of resolution and healing (Felman & Laub, 1992; Malpede, 1999). At its best, the process of creating testimony theater is one of healing. It facilitates recognition of strengths and increases self-worth in the performers (Dutton, 2001). Testimony theater demands that the performer present horrific events and that the performer and audience assume responsibility to bear witness (Culbertson, 1995; Felman, 1995); however, they also cocreate a community that embraces human existence (Weine, 1996).

Working with traumatic material can provide opportunities to heal, particularly when the creative process is dependent on clear direction, trust, freedom, respect, challenge, collaboration, and unity with the performers, director, and audience (Nemiro, 1997). However, revealing past trauma can also increase distress and pathological dissociation (Malpede, 1996; Reddy, Seligowski, Rabenhorst, & Orcutt, 2015), as was evident in studies that investigated actors who created testimony theater (Weine, 1996; Thomson & Jaque, 2011a). Successfully exposing trauma through narrative may require implementing more positive emotional words to offset the bias toward negative wording. Performers also need to maintain higher cognitive processing and actively decrease focusing on themselves (Jaeger, Lindblom, Parker-Guilbert, & Zoellner, 2014). Emotional distancing, although not generally encouraged in the performing arts, is critical when working with traumatic material (Burgoyne et al., 1999; Kaplan, 2005). Including these strategies may counter the negative effects of dissociation and mobilize positive dissociation (Thomson et al., 2009). As a performer of testimony theater, adopting a positive and meaningful stance can be empowering and can restore a sense of self-efficacy, wellbeing, and resilience (Harris, 2007; Peleg et al., 2014).

Working with performing artists with PTSD

When treating performing artists with a history of unresolved mourning and PTSD, it is essential to recognize internalized beliefs, such as that their self is worthless, the world is threatening and dangerous, and others are abusive. Because performers generally have greater proneness to fantasize, they are also at increased risk for higher levels of internalized shame and anxiety (Thomson et al., 2009; Thomson & Jaque, 2011–2012, 2013, 2015a). Also, many performers have a higher frequency of mental illness in their families and express greater difficulty speaking about their trauma (Thomson & Jaque, 2015b). Their innate emotional and imaginational overexcitability traits also elevate shame and anxiety. Treatment strategies should incorporate ways to positively direct and manage emotional and imaginational processes (Thomson & Jaque, 2016b), especially when they perform works that activate past trauma memories. Because actors, singers, and dancers assume body postures that express trauma, they are at increased risk for retrieving their own autobiographical memories of trauma (Dijkstra et al., 2007). Educating performers about this reality, including accompanying feelings of terror, anxiety, and shame (Chatard et al., 2012), can help normalize distressing experiences during rehearsals and performances. Distancing strategies can then be deployed prior to rehearsing or performing works with traumatic content (Burgoyne et al., 1999; Kaplan, 2005).

Creating a standardized psychological screener when young performers enter conservatory training programs may help identify performers with trauma histories, as well as other psychopathological disorders (Hamilton, Solomon, & Solomon, 2006). Therapeutic interventions and education can be directed at helping traumatized performers trust the arousal cycle again (Swart, 2014). They can once again derive pleasure from heightened sympathetic physiological states and rediscover that the stage is a safe place to express intense emotional states (Thomson & Jaque, 2011a). Increased exposure to performance settings can also decrease performance arousal and performance anxiety, a reality that will enhance a sense of safety (Boucher & Ryan, 2011). Performers with PTSD can achieve positive flow states and derive meaning from their creative efforts, all components that potentially heal trauma (Thomson & Jaque, 2016c, 2016d). Trauma-related dissociation can also be decreased via trauma-focused therapeutic interventions. Along with decreased trauma-related dissociation, the ability to memorize increases (Swart et al., 2010).

Increasing awareness of survival skills cultivates a positive balance to the trauma-related work (Howard, 2008). Helping performers reconnect with the healing power of music and performance ritual can decrease symptoms while rekindling an appreciation for the artistic medium that they perform (McClary, 2007). Clinicians and educators need to hold awareness that talented young performers who are exposed to conditions of war trauma may reject their creative talents. They experience conflicting emotions that dysregulate them, hence they withdraw from the expression of their talent. They need help to reinvest and cultivate their talents (Knezevic & Ovsenik, 2002). Encouraging companionship among performers who share similar trauma histories may also reduce PTSD symptoms (Gregory & Embrey, 2009). Helping young (as well as older) traumatized performers rediscover play, including psychodrama, can help heal trauma and reinstall hope (Carbonell & Parteleno-Barehmi, 1999; Lantz & Raiz, 2003; Reyes & Asbrand, 2005).

Encouraging musicians, singers, and actors to engage in meaningful physical exercise, and dancers to continue with their dance practices, can positively augment standard trauma treatment protocols (Powers et al., 2015). They require encouragement because many individuals suffering from PTSD decrease participation in physical activity due to diminished motivation (de Assis et al., 2008). Specific best practice treatment approaches to manage complex PTSD include emotion regulation strategies, narration of trauma memories, cognitive restructuring, anxiety and stress management skills, and interpersonal skills. Meditation and mindfulness are considered second-line intervention strategies that positively enhance trauma treatment (Cloitre et al., 2011). Mindfulness-based stress reduction protocols also decrease inflammatory processes in trauma-exposed individuals (Gallegos, Lytle, Moynihan, & Talbot, 2015). Learning to manage unresolved mourning, whether from loss or trauma, can promote greater physiological and psychological regulation. These positive results occur when trained clinicians effectively manage emotional disclosure while empathically attuning to the performer (O'Connor, Allen, & Kaszniak, 2005; Peleg et al., 2014). One positive by-product that may emerge is wisdom; performers who work on healing past trauma can recognize and accept human limitations while integrating emotional and cognitive responses (Linley, 2003).

Summary

- Introduction: definition, intentional / nonintentional trauma, imagination and expression of trauma.
- What are PTSD and unresolved mourning?: risk rate, long-term consequences, criteria and symptoms for diagnosis, markers to determine resolution for past traumatic or loss events.
- Trauma, trauma reminders, and performance: effects on physiological and neurobiological systems: physiological and neurological changes, triggers, flashbacks, allostatic systems.
- Effects of chronic stress: cardiovascular changes, autoimmune and inflammatory responses, HPA axis, cortisol reactivity, alterations in brain volume.
- Trauma and self-harm: childhood maltreatment, dysregulation.
- Shame and performing artists: abuse (emotional, physical, sexual), shame triggers and physiological stress, negative and positive outcomes.
- Performing artists: PTSD, unresolved mourning, and shame: trait anxiety and intrusive recollections, dissociation.
- Trauma and resilience: neural integration, vestibular system, life satisfaction after trauma.
- Recall of trauma: testimony theater/theater of witness: healing potential, narrative creations, positive dissociation.
- Working with performing artists with PTSD: internalized beliefs of self, other and the world; psychological screener; supportive healing protocols.

Concluding remarks

A high percentage of performing artists suffer from PTSD. In a study by Kenny, Driscoll, and Ackermann (2014), 22% of musicians had PTSD. A similar distribution (20.2%) was found in a dancer sample (Thomson & Jaque, 2015b). As well, a greater proportion of actors had unresolved mourning compared to a healthy control sample (Thomson & Jaque, 2012b). Although these statistical distributions are high when compared to the general population (8.9%) (Kilpatrick et al., 2013), all the samples studied were derived from successful and active performers. Helping performers with PTSD manage difficult negative emotions can promote optimal performance. They need to learn how to shift away from emotion-oriented coping strategies to more task-oriented coping behaviors during high-pressure rehearsal and performance experiences (Thomson & Jaque, 2016d). This is critically important given the reality that regulating difficult negative emotions during daily activities is even more problematic when PTSD symptoms are severe (O'Bryan, McLeish, Kraemer, & Fleming, 2015). Encouraging evidence suggests that engaging in the performing arts may help manage symptoms while creating, although the higher distribution of performers with PTSD and unresolved mourning indicates that they continue to struggle in their daily functioning. Although unknown, without this creative outlet, performers may be more distressed (Kaufman & Sexton, 2006). Performers with PTSD can experience transcendence, integration, and healing while creating (Thomson & Jaque, 2016d).

Performing artists and psychopathology

17

Inspiration is an awakening, a quickening of all man's faculties, and it is manifested in all high artistic achievements. ... Art is a kind of illness.

Puccini

Introduction

The association between creativity and madness persists. Plato described how artists are "bereft of their senses." They are God's vehicles to communicate to the populace (Plato, c.338BCE/1976). With the emergence of the formal disciplines of psychiatry and psychology, this association has been either strengthened or challenged (Rothenberg & Hausman, 1976; Schlesinger, 2009). The mad genius debate has primarily examined poets, writers, visual artists, and scientists (Neihart, 1998), with minimal studies examining performing artists. One consistent method to address this question has been historiometric research; literally researchers select eminent artists from the past and determine a psychiatric diagnosis based on biographical literature. They then conduct statistical analyses to determine patterns and trends (Simonton, 1999b, 2014). Other study methods are based on directly sampling artists, administering self-reports and diagnostic assessments, and then reporting statistical findings (Akiskal et al., 2005; Preti & Vellante, 2007; Strong et al., 2007). These studies demonstrate that a significant proportion of artists experience mental illness, but not all artists are vulnerable to psychopathology. Findings also indicate an overrepresentation of mental illness in families of creative professionals (Kyaga et al., 2011). They often have relatives diagnosed with schizophrenia, bipolar disorder, anorexia nervosa, or siblings with autism (Kyaga et al., 2013). To date, there is a strong link between creativity and mental illness; however, it may be a more curvilinear pathway, with mild-to-moderate psychopathology related to creativity but not severe psychopathology (Simonton, 2014). As well, common factors between creativity and psychopathology may coexist, including genetic roots (Power et al., 2015), but that does not mean that creativity is the cause of psychopathology (Schuldberg, 2000–01).

This chapter examines major psychiatric disorders that are frequently reported in performing artist samples. Determining the nature of these disorders, how they influence the careers of performing artists, and some treatment approaches to manage these disorders is the primary focus of this chapter. Offering answers to prove or disprove the mad genius debate is not the goal of this chapter.

Creativity and the Performing Artist. DOI: http://dx.doi.org/10.1016/B978-0-12-804051-5.00017-2

The intention of this chapter is to illuminate some of the common struggles that performing artists encounter and how psychiatric disorders may or may not enhance their creativity. There is a large body of research that directly examines the relationship between mental illness and the creative process, such as divergent thinking and schizotypy disorders. Some of these studies will be included but the overarching goal is to shed light on mental illness. Rather than examining domain-specific performing arts, this chapter investigates performing artists and specific disorders related to them.

Vulnerability, resilience, and stereotypes of mental illness

The stereotype that creative people are "mad geniuses" is endorsed by many, including the general population, patients with mental illness, and creative healthy artists. For example, mentally ill people may minimize their disorder and claim they are creative, when in fact they may be minimally creative but want to present as a "mad genius." On the other hand, highly creative people may fear that if they present as "normal" they will be regarded as boring and so they amplify any mental instability to justify being an interesting artist. In a study by Kaufman, Bromley, and Cole (2006–07), highly creative individuals tended to self-identify as a "mad genius," regardless of any inherent mental illness. This study demonstrated that creative individuals perpetuate the stereotype in an effort to appear unique. However, adhering to this stereotype can deter treatment for those who do suffer from mental illness; many fear that treatment will remove their creativity along with their "mad genius" status and so they avoid seeking help.

Studies have also indicated that engaging in professions that demand emotional expressivity increases vulnerability to mental illness (alcoholism, drug use, and depression) (Ludwig, 1998). Highly creative individuals, such as performing artists, are sensitive to the effects of negative content and negative emotions; they are biased toward negative perceptions and responses (Drus, Kozbelt, & Hughes, 2014), thus increasing the risk for psychopathology (Strong et al., 2007). Decreased risk for mental illness is found in careers rooted in objectivity; they require more analytic, objective, and formal forms of expression. The careers of performing artists are dominated by subjectivity and increased psychological instability (Ludwig, 1998). Some theorists suggest that creative individuals engage in creative expression as a defensive strategy to ward off psychological distress and psychopathology (Drus et al., 2014). This line of research unfortunately denies the reality that performers are disciplined and highly motivated, markers of mental stability and health.

Performers actively approach novel creative experiences. Motivational behaviors such as novelty seeking and appetitive approach behaviors are essential in successful performance careers; consequently disorders that are inherently approach-avoidant, such as depression and anxiety, are detrimental to performers (Silvia & Kimbrel, 2010). If they suffer from these disorders, they still have to engage in approach behaviors to continue working in their career. On the other hand, psychotic disorders such as bipolar disorder, and to some degree, schizophrenia, can enhance novelty-seeking behaviors. Individuals with these disorders, or with subthreshold disorders such as cyclothymia or schizotypy, may actually benefit from the associated appetitive behaviors. Looser associations and unusual subjective beliefs that are inherent in these conditions may also enhance creative processing (Akiskal et al., 2005; Kyaga et al., 2011; Preti & Vellante, 2007; Box 17.1).

Box 17.1 Common Psychiatric Disorders Related to Performing Artists (APA, 2013)

Psychosis	"Abnormalities in one or more of the five following domains: delusions, hallucinations, disorganized thinking (speech), grossly disorganized or abnormal motor behavior (including catatonia) and negative symptoms" (p. 87).
Bipolar disorder	Originally termed "manic depression," this disorder is marked by manic or hypomanic episodes and major depression. Cyclothymic disorder is marked by less extreme emotional fluctuations. Manic or hypomanic episodes include inflated self-esteem or grandiosity, decreased need for sleep, more verbose, flights of ideas, distractible, increased goal-directed activities that may be harmful or damaging (p. 124).
Depressive disorders	Disorders include *disruptive mood dysregulation* (severe recurrent temper outbursts); *major depression* (marked diminished interest or pleasure, feelings of worthlessness, diminished ability to concentrate, suicidal thoughts, insomnia or hypersomnia, weight loss or gain); *persistent depressive disorder* (dysthymia) (depressed mood for at least two years); *premenstrual dysphoric disorder* (marked emotional lability, irritability or anger, depressed mood, anxiety, difficulty concentrating, lethargy, change in appetite, change in sleep patterns, sense of being overwhelmed or out of control).
Anxiety disorders	Anxiety disorders share features of excessive *fear* (emotional response to real or perceived threat), *anxiety* (anticipation of future threat), and related behavioral disturbances (p. 189). *Panic attacks* may accompany anxiety disorders as well as other disorders; they are marked by abrupt and extreme cognitive and/or physiological arousal such as fear of losing control, feelings of dizziness, choking, shortness of breath, sweating, palpitations, shaking, numbness, derealization/depersonalization, fear of dying. *Panic disorder* involves recurrent panic attacks and fear of panic attacks. *Social anxiety* is marked by fear of negative evaluation and avoidance of social interactions and situations (stage fright or performance anxiety is considered social anxiety). *Generalized anxiety disorder* includes prolonged excessive worry and anxiety that are not easily controlled, feeling restless and on edge, easily fatigued, difficulty concentrating or mind going blank, irritability, muscle tension, sleep disturbances.

(Continued)

Box 17.1 (cont'd)

Obsessive-compulsive disorders	"*Obsessions* are recurrent and persistent thoughts, urges, or images that are experienced as intrusive and unwanted. *Compulsions* are repetitive behaviors or mental acts that an individual feels driven to perform in response to an obsession or according to rules that must be applied rigidly" (p. 235). *Obsessive-compulsive disorder* is marked by significant distress and extreme time-consuming behavior. *Body dysmorphic disorder* is a preoccupation with one or more perceived defects or flaws in physical appearance that involves excessive repetitive behaviors or mental acts. *Trichotillomania* (hair-pulling disorder) and *excoriation* (skin-picking disorder) are repetitive behaviors that are difficult to stop and cause significant distress.
Dissociative disorders	These disorders are "characterized by disruptions and/or discontinuity in normal integration of consciousness, memory, identity, emotion, perceptions, body representation, motor control, and behavior" (p. 291). Dissociative disorders frequently occur after a traumatic event or under anxiety-provoking stressful situations. Symptoms include losses of continuity of a subjective experience or an inability to access information or control mental functioning. *Dissociative identity disorder* includes disruption of identity with two or more distinct personality states and recurrent gaps of recall. *Dissociative amnesia* involves an inability to recall important autobiographical information. Perceptual dissociative disorders: *Depersonalization* (perceptual alterations of a subjective sense of self such as feeling unreal, detached, outside observer) and *derealization* (perceptual alterations of environment such as feeling dreamlike, foggy, or visual distortions).
Eating disorders	"Feeding or eating disorders are characterized by persistent disturbances of eating or eating-related behavior that results in altered consumption or absorption of food and that significantly impairs physical health or psychosocial functioning" (p. 329). *Avoidant/restrictive food intake disorder* is marked by an apparent lack of interest in eating or avoiding food based on sensory characteristics or concern about aversive consequences. *Anorexia nervosa* is extreme restriction of food leading to significant low body weight coupled with intense fear of gaining weight. Intense preoccupation about body weight and shape is experienced, often with a lack of recognition of the dangers of these behaviors. *Bulimia nervosa* is recurrent episodes of binge eating and compensatory behavior (vomiting, excessive exercise, laxatives, fasting) to prevent weight gain. A sense of lack of control accompanies this disorder, along with excessive self-evaluation. *Binge-eating disorder* involves recurrent episodes of binge eating without the compensatory behaviors to manage weight.

Somatic symptoms and related disorders	These disorders are marked by somatic symptoms associated with significant distress and impairment. The somatic symptoms are always accompanied by abnormal thoughts, feelings, and behaviors in response to the symptoms. *Somatic symptom disorder* includes disproportionate and persistent thoughts about the symptoms including anxiety and excessive time and energy devoted to the symptoms or health concerns. Pain is a major symptom related to this disorder. *Illness anxiety disorder* is a preoccupation with having or acquiring a serious illness and involves elevated anxiety and excessive health-related behaviors and preoccupations.
Sleep–wake disorders	Sleep–wake disorders cause marked daytime distress and impairment. *Insomnia disorder* is marked by dissatisfaction with sleep quality or quantity such as difficulty falling asleep, staying asleep, or early morning awakening and inability to return to sleep. *Hypersomnolence disorder* is excessive sleepiness despite sleeping at least 7 h per night.
Substance-related and addictive disorders	Persistent use of a substance or engagement with a behavior that causes harm cognitively, behaviorally, and/or physiologically. Impaired control, social impairment, risky use, strong craving, and increased tolerance for the substance coupled with intoxication and withdrawal symptoms are hallmarks of these disorders. Substances that lead to substance use disorders include alcohol, caffeine, cannabis, hallucinogens, inhalants, opioids, sedatives, hypnotics, anxiolytics, stimulants, tobacco, other known or unknown substances, gambling (including Internet).
Personality disorders	"Personality disorder is an enduring pattern of inner experience and behavior that deviates markedly from the expectations of the individual's culture, is pervasive and inflexible ... and is stable over time" (p. 645). Personality disorders include *paranoid* (distrust and suspiciousness and interpret motives as malevolent), *schizoid* (detachment from social relationships, restricted range of emotional expression), *schizotypal* (acute discomfort with close relationships, cognitive and perceptual distortions, and eccentric behavior), *antisocial* (disregard for the rights of others), *borderline* (instability in interpersonal relationships, self-image, emotions, and impulsivity), *histrionic* (excessive emotionality and attention seeking), *narcissistic* (grandiosity, need for admiration, lack of empathy), *avoidant* (social inhibition, feeling inadequate, hypersensitive to negative evaluation), *dependent* (submissive and clinging behavior, excessive need to be taken care of), *obsessive-compulsive* (preoccupation with orderliness, perfectionism and control).

Psychotic disorders: schizophrenia, schizotypy, and bipolar disorder

Psychotic disorders are defined by positive and negative symptoms. Positive symptoms include (1) delusions (fixed beliefs that are bizarre), (2) hallucinations (perceptual experiences that manifest without external stimuli), (3) disorganized thinking (thought disorders and linguistic disorganization), and (4) grossly disorganized behavior. Negative symptoms comprise (1) diminished emotional expression, (2) avolition (decrease in motivated and purposeful activities), (3) alogia (diminished speech), and (4) anhedonia (decreased ability to experience pleasure from positive stimuli) (APA, 2013). Psychosis is often associated with overinclusive thinking, remote associations, and decreased latent inhibition (i.e., diminished or defective filtering of external and internal stimuli). According to the latent inhibition theory, the generation of new ideas or associations increases when stimuli are less vigorously screened out, literally, the gating mechanism in the brain is less active thus more information reaches the associational regions of the brain (Nelson & Rawlings, 2010). Overinclusive thoughts, loose associations, and latent inhibition are associated with psychosis and promote creative cognitive processes such as divergent thinking. However, the negative psychotic symptoms limit creative achievement; they compromise motivation, cognitive flexibility, and openness to experience (Abraham, Windmann, Daum, & Gunturkun, 2005; Crabtree & Green, 2016). In general, psychoticism and creativity typically share features of associative cognitive processes but they are dissimilar when evaluating productive goal-related thinking and behavior.

Latent inhibition theory accounts for freer flow of information during creative thought; whereas, severe psychosis is measured by a significant decrease in sensory gating of stimuli leading to extremely disorganized behavior and thought. Psychotic patients are flooded with information that is not effectively filtered for organized cognitive or perceptual processing. The timing of the gating process also indicates creative wellbeing. For example, highly trained performers, such as musicians, display significant decreased sensory gating, but only during the early phase of receiving sensory stimuli. This enables them to preattentively assimilate information about pitch and tempo. They are then able to filter stimuli in a rapid and efficient manner, which enables precise musical performance decisions (Kizkin, Karlidag, Ozcan, & Ozisik, 2006). Research findings suggest that timing and discrimination during creativity is different when compared to sensory gating response patterns in severely psychotic individuals (Kizkin et al., 2006; Park et al., 2015).

Schizotypy personality, although not considered a florid psychotic disorder, indicates a vulnerability to schizophrenia, which is a psychotic disorder (Nelson & Rawlings, 2010). Schizotypy personality is marked by impulsive nonconformity and odd or unusual beliefs and thoughts, traits that are often associated with heightened creativity. These traits decrease anxiety and promote greater novelty-seeking behavior (Acar & Sen, 2013). Positive schizotypy traits are also associated with deeper absorption, heightened experiences of creative pleasure, and a greater sense of empowerment (Nelson & Rawlings, 2010). The negative schizotypy traits, such as inappropriate or constricted emotions coupled with diminished social and physical engagement, do not increase creativity (Acar & Sen, 2013; Fisher et al., 2004). When schizotypy is examined on a continuum, the positive features that are associated with high openness to experience plus high neuroticism enhance divergent thinking and heightened emotionality; whereas, negative features, marked by low openness to experience and agreeableness plus high neuroticism, do not enhance creativity (Fisher et al., 2004).

Creative thinking and schizotypy share similar activation and deactivation patterns in multiple neurocognitive networks, such as regions in the prefrontal cortex, inferior temporal gyri, insula, angular gyrus, precuneus, and parietal lobes (Fink et al., 2014; Park et al., 2015). These neurological studies demonstrate unique differences between severe psychotic patients and creative/schizotypy individuals. They also reinforce the reality that highly creative individuals are not all diagnosed with schizotypy psychopathology. Similarly, schizotypy individuals are not all creative (Armstrong, 2012); however, individuals who are high fantasy prone or have schizotypy collectively engage in magical thinking and cognitive–perceptual eccentricities (Sanchez-Bernardos & Avia, 2006). They are described as "happy schizotypes" because they do not manifest the negative features of schizotypy (Parra, 2006).

Both high fantasy proneness and schizotypy are predictive factors for supernatural beliefs and paranormal experiences; e.g., seeing and feeling ghosts or out-of-body experiences. Paranormal phenomena may result due to decreased latent inhibition or disturbed self-processing in the temporoparietal association regions in the brain (Blanke & Arzy, 2005). Failure to integrate an embodied self may also drive these experiences. It is important to note that a high proportion of performers claim that they experienced paranormal events (Ayers et al., 1999). These experiences are described and valued across all cultures, suggesting they are not necessarily markers for psychopathology (Parra, 2006). However, heightened paranormal experiences, although culturally normative, may indicate psychological difficulties resolving past trauma or loss experiences (Thomson & Jaque, 2014).

Bipolar disorder and creativity are frequently linked; this association may be related to behavioral similarities found in creative individuals with and without bipolar disorder (Johnson, Tharp, & Holmes, 2015). It is suggested that creative achievement is augmented by manic episodes (increased ambitiousness, loose associations, increased idea generation, and positive affectivity) (Goodwin & Jamison, 2007). Also, highly successful creative individuals are at greater risk for bipolar disorder (Andreasen, 2005). Although these study findings offer compelling evidence linking creativity with bipolar disorder, the major failing in these studies is related to how a clinical diagnosis of bipolar disorder was determined; most diagnoses were given retrospectively based on biographical writings. Studies that actually assess individuals for bipolar disorder before data gathering reveal a different picture. For example, when studying children with bipolar parents, increased creativity was evident in these children compared to children without bipolar parents (Simeonova, Chang, Strong, & Ketter, 2005). It is hypothesized that a genetic risk for this disorder may be associated with increased creativity (Power et al., 2015). Similarly, other studies demonstrate that creative individuals have more mentally ill family members (Kyaga et al., 2011, 2013), a pattern that has been demonstrated in samples of actors (Ayers et al., 1999) and dancers (Thomson & Jaque, 2015b). Creativity studies that investigate individuals with cyclothymia, a less extreme form of bipolar disorder, indicate that this population is no different than healthy

Figure 17.1 "Dreaming." Performers: Sara Adler, Arianna Douglas, Hilary Gereaux, Cameron Shim. Photographer: Lee Choo.

creatives (Johnson et al., 2015). Indeed, cyclothymia has clearly been linked with creative output (Akiskal et al., 2005; Strong et al., 2007; Vellante et al., 2011), as well as creative intuition (Srivastava et al., 2010). These studies demonstrate that psychopathology may exist in families of highly creative individuals, but when performers suffer severe psychopathology they experience limitations in their career and their creativity. However, less severe forms of psychopathology may actually augment creative performances, as evidenced by studies that compared individuals with bipolar disorder versus cyclothymia, and schizophrenia versus schizotypy (Kyaga et al., 2011; Simonton, 2014; Fig. 17.1).

Dissociation, maladaptive daydreaming, and excessive fantasy proneness

The hallmark of pathological dissociation is a failure to integrate perceptions, memories, and/or identity (APA, 2013). Dissociation also operates on a continuum ranging from normal dissociative absorption to pathological dissociative processes that severely disrupt interpersonal, cognitive, and social functioning (Butler, 2006). Elevated levels of dissociation have been found in performing artists (Thomson et al., 2009; Thomson & Jaque, 2011a, 2012c), artists (Perez-Fabello & Campos, 2011), and high fantasy-prone individuals (Lynn & Rhue, 1988; Rauschenberger & Lynn, 2002–03; Rhue & Lynn, 1987).

When dissociation reaches pathological levels, memory retrieval is compromised and memory errors increase (Giesbrecht, Geraerts, & Merckelbach, 2007)—a pattern that is often exacerbated by posttraumatic stress disorder (PTSD) in performing artists (Swart et al., 2010). Dissociation provokes greater emotional dysregulation, heightened sympathetic activation (Farina, Speranza, Imperatori, Quintiliani, & Marca, 2015; Giesbrecht et al., 2007), and decreased flow experiences (Thomson & Jaque, 2012c). Performers with pathological levels of dissociation such as depersonalization (disruptions in a sense of self) and derealization (disruptions in perceptual sense of the world) suffer more negative emotions and negative imaginative content (Levin, Simeon, & Guralnick, 2004). Even more problematic, pathological dissociation is related to coping behaviors that include social isolation, self-blame, and rumination, behaviors that are detrimental to performing artists (Wolfradt & Engelmann, 2003).

Dissociation can provoke a disenfranchisement from body ownership. For example, compared to rhythmic gymnasts, dancers with higher levels of dissociation believed they lost ownership of their body (Thomson, Kibarska, & Jaque, 2011). Dissociative individuals with a trauma history also disconnect from an embodied sense of self (Haven, 2009). Not only do dissociative individuals experience a subjective disconnection from their body, they have blunted autonomic responses and decreased resilience (Schmahl, Elzinga, & Bremner, 2002). As well, they feel greater alienation and betrayal (Bernstein, Delker, Knight, & Freyd, 2015; DePrince, Huntjens, & Dorahy, 2015).

High fantasy-prone individuals and compulsive daydreamers share similar features; in particular, the quantity of their fantasizing interferes with daily functioning, increases shame, and decreases self-esteem (Bigelsen & Schupak, 2011; Lynn & Rhue, 1988; Somer, 2002; Somer et al., 2016). Excessive daydreaming is also associated with obsessive-compulsive disorders, dissociative disorders (Somer et al., 2016), and increased nightmares (Simor, Csoka, & Bodizs, 2010). High fantasy-prone individuals often report childhood histories of intense loneliness and/or severe physical punishment. Engaging in compulsive daydreams or fantasies can serve to block pain, evoke revenge scenarios, or offer an escape into worlds that are more inviting (Rhue & Lynn, 1987). Unlike excessive daydreamers or high fantasy-prone individuals, performing artists tend to manifest moderate fantasy proneness. This is not surprising because their work must exist in reality and not reside only within the internal world of their imagination. However, some performers do struggle with excessive fantasy, which exacerbates anxiety, shame, and dissociation, and decreases optimal performance and flow experiences (Thomson & Jaque, 2011a, 2011–2012, 2012c, 2013, 2015a).

Normative dissociative processes typify creative performance work, especially during character development or musical interpretation (Thomson et al., 2009; Thomson & Jaque, 2011a, 2012a). Surprisingly, absorption, fantasy proneness, and dissociative processes are also found in celebrity worshippers (Maltby, Day, McCutcheon, Houran, & Ashe, 2006). Like performers, they construct elaborate fantasy relationships with celebrities and lose a sense of self (Maltby et al., 2006). Whether performer or celebrity worshipper, less severe forms of dissociation can enrich meaning-making efforts, whereas, dissociative psychopathology limits career success (Klinger et al., 2009).

Anxiety and performers

Highly creative individuals tend to have increased anxiety (Carlsson, 2002). Performance anxiety, often known as "stage fright," is clinically diagnosed as social anxiety or social phobia (APA, 2013). It is severe situational anxiety that arises when performers believe they are not able to control their behavior in front of an audience. They are afraid they will make a catastrophic mistake (Cox & Kenardy, 1993). Both negative cognitive appraisal and increased physiological sympathetic activation operate during performance anxiety. If performers are unable to manage their frightening thoughts and somatic reactions, then the feedback loop between the two will continue to increase until performance anxiety reaches the level of panic attacks. Clearly these responses are deleterious to performance. After experiencing a panic attack, the apprehension that future performances will be sabotaged by performance anxiety grows. Performers become sensitized (increased response) to the somatic and cognitive features of anxiety, rather than habituating (decreased response) to performance situations (Barnard, Broman-Fulks, Michael, Webb, & Zawilinski, 2011; Taylor & Cox, 1998). For example, performers can become anxiously sensitized to respiratory symptoms. They fear shortness of breath, a response that may be associated with a previous anxiety episode or with an event in which they witnessed someone else struggling to get sufficient oxygen. Anxious sensitivities can also gravitate around cardiac sensations that evoke memories of witnessing someone having a heart attack or experiencing their own arrhythmic heartbeats. Cognitive dyscontrol such as thoughts experienced during flashbacks or dissociative events can also increase anxiety sensitivities (Kilic, Kilic, & Yilmaz, 2008; Taylor & Cox, 1998).

Although performance anxiety may occur in actors, dancers, and singers, this disorder has been extensively studied in musicians. Specifically, compared to musicians who perform in ensembles or orchestras, solo performers are at greater risk, and musicians with high trait anxiety (generalized anxiety disorder) are more likely to suffer from performance anxiety (Cox & Kenardy, 1993). Many professional musicians (58%) who work in symphonic orchestras suffer from this disorder. They report anticipatory anxiety hours, days, weeks, and even months before a performance, causing increased distress in their professional and personal lives (van Kemenade, van Son, & van Heesch, 1995). Many of these musicians report experiencing performance anxiety as children, which persisted throughout their career. Effective education to manage anxiety during childhood training may ameliorate performance anxiety in adulthood; however, even with psychological skill management, this problem continues to trouble many musicians (Kenny et al., 2014). Choral singers also report high performance anxiety but less than other musicians, perhaps because they work in an ensemble and receive indirect social support from other singers (Ryan, 2009). Evidence also indicates that they have a different genetic profile from nonsingers; they have more serotonin and arginine vasopressin receptors, which may help manage anxiety, even during intense performance pressures (Morley et al., 2012).

As described by musicians, anxiety often emerges in early development. For young performers, public exposure may be experienced as a terrifying, even traumatizing, event. Whether due to a genetic predisposition for anxiety disorders or an inhibited temperament (i.e., shyness), critical appraisal provokes early shame-based and anxiety-ridden schemas. Their anxiety may also be driven by perfectionistic beliefs (Kenny & Osborne, 2006). What is unknown is whether young performers who develop anxiety resemble the general population of children. Do they share the same etiology for anxiety disorders, including greater difficulty regulating negative emotional reactivity and effectively reappraising situations (Carthy, Horesh, Apter, Edge, & Gross, 2010)? In general, trait anxiety is marked by elevated baseline anticipatory anxiety (Nitschke et al., 2009). Hence, early trait anxiety can increase the risk for PTSD after a traumatic event, and then PTSD stress reactions further increase anxiety (Kilic et al., 2008). With repeated exposure to anxiety-provoking situations any ability to buffer anxiety is diminished. Consequently, anxiety compromises appropriate situational appraisal and effective emotional regulation (Abdollahi et al., 2011; Carthy et al., 2010).

Anxiety increases dissociative depersonalization disorders, an experience that alienates performers from their sense of self. This limits optimal performance (Simeon, Riggio-Rosen, Guralnik, Knutelska, & Nelson, 2003; Thomson & Jaque, 2012c). For children, warm and positive parental support decreases dissociative stress responses; however, authoritarian parenting (and teachers) increases depersonalization and anxiety (Wolfradt, Hempel, & Miles, 2003). Another factor that can increase or decrease performance anxiety is fantasy (Thomson & Jaque, 2011–2012). The creative ability to imagine situational outcomes can exacerbate performance anxiety or enhance performance skills. For example, despite clinical levels of anxiety, dancers still achieved high flow states and endorsed elevated fantasy proneness, even though they also experienced more past traumatic events (Thomson & Jaque, 2011–2012).

Anxiety disorders are marked by heightened anticipatory activation in the amygdala and anterior cingulate. In anxious individuals, these brain regions, and the neural networks associated with them, rapidly respond to both aversive and neutral stimuli. Because these regions directly influence pain perceptions and autonomic activation, treatments that target reducing anticipatory arousal will also alter brain reactivity (Nitschke et al., 2009). Performance anxiety also increases cocontraction of antagonistic and agonistic muscle groups. It compromises implicit motor network integration, hence poorer motor control and diminished performance (Yoshie, Kudo, & Ohtsuki, 2008). Increased heart rate and sympathetic activation is a natural response during performance (Spahn et al., 2010; Thomson & Jaque, 2011b). For anxiety-sensitive performers, these somatic changes may trigger more performance anxiety, rather than performance excitement (LeBlanc, Jin, Ober, & Sivola, 1997; Jaque et al., 2015).

Mild performance anxiety lessens positive rewards, but at moderate-to-severe levels it disrupts performance skill and dampens all performance rewards (Sataloff et al., 1999). Both somatic and cognitive manifestations of anxiety can cripple performance. For example, in a group of ballet dancers, cognitive anxiety, rather than somatic anxiety, disrupts performance; the cognitive appraisal of being physically out of control increases anxiety (Walker & Nordin-Bates, 2010). Dancers with trait anxiety also experience poorer health, which results in decreased time practicing and performing. This negative cascade propels dancers to actively avoid setting performance goals (Lench, Levine, & Roe, 2010). Unlike dancers, singers tend to experience more somatic anxiety, especially concerns related to vocal health (Sandgren, 2009). Orchestra players often experience a negative feedback loop, one that involves anxiety, musculoskeletal problems, sleep disorders, and cardiovascular disorders; each variable interacts with the other variables (Langendorfer et al., 2006). This recursive pattern compromises performance and career satisfaction. Sadly, these negative interactive factors tend to increase with performance experience and age. For all performers, genetic traits such as hypermobility, alterations in brain networks, and daily personal stressors further increase performance anxiety.

Learning to manage anxiety is essential for performers. Coping strategies include addressing both positive (diet, exercise, negative emotional regulation) and negative responses (rumination, learned helplessness, substance abuse, resignation and hopelessness, intensified perfectionism). The most effective way to manage performance anxiety is via integrative treatment models that focus on biological and psychological factors that are directly related to performance situations (Quentzel & Loewy, 2010). Many performers implement these strategies, indicating their desire to alleviate somatic and cognitive markers of anxiety. They want to enjoy their performance experiences (Langendorfer et al., 2006). For young children, integrative treatment models must include play; in particular, helping them experience performance settings as arenas for pretend play (Russ, 2007).

A specific anxiety reduction skill is engaging in physical activity. For example, less physically active musicians tend to experience greater performance anxiety (Rocha, Marocolo, Correa, Morato, & da Mota, 2014). Another skill is journaling, accompanied by repeated self-assessments prior to and following a performance. This practice can reduce performance anxiety over time, as was evident in a group of musicians (Sadler & Miller, 2010). Both physiological reactivity and cognitive apprehension decreased over time, in part due to heightened subjective awareness of performance anxiety, and the efficacy engendered from implementing anxiety-reducing skills (Kusserow et al., 2012). Clinical treatment can explore the unconscious psychodynamic factors that drive severe anxiety. This approach examines unresolved complex emotions and psychological defense strategies designed to protect early maltreatment (Kenny et al., 2014). The success of this approach indicates that severe anxiety may stem from hidden factors that extend beyond performance settings. Thus it is prudent for performing artists with severe anxiety to seek professional care (Barbar, Crippa, & Osoiro, 2014; Fig. 17.2).

Depression

Clinical depression is a devastating disorder. Individuals can experience one major depressive episode, recurrent episodes, or persistent depression (APA, 2013). One of the features of clinical depression is an excessive sensitivity to negative environmental cues that intensify a feeling of failure and unworthiness, coupled with decreased ability to accept positive results (rewards, recognition, success) (Chui & Deldin, 2007). Neural circuit abnormalities such as increased activation in the

Figure 17.2 "Despair." Performers: Janice Dulak, Hugh Murphy. Photographer: Cylla von Tiedemann.

anterior cingulate and paralimbic structures, with decreased activation in the prefrontal cortex, typify neurological profiles of depressed patients (Chui & Deldin, 2007). Dysregulation in the hypothalamic–pituitary–adrenal (HPA) axis is also present; this is evidenced in decreased adrenal steroids and increased cortisol (Akinola & Mendes, 2008; Nestler et al., 2002). This neurological profile influences appraisal errors such as amplified detection of negative cues and misappraisal of positive stimuli (Chui & Deldin, 2007). Depressed individuals also have difficulty integrating contextual information, as well as a deficit in understanding the "mind" of others (Wolkenstein et al., 2011). Consequently, performing artists with a history of depression are at great risk for appraisal errors, cognitive distortions, and decreased confidence in achieving their goals (Lane, Whyte, Terry, & Nevill, 2005; Thomas & Duke, 2007), a reality that negatively influences training, rehearsals, and performances.

Depression has a negative effect on discriminating sound. This is a significant problem for performing artists; they must be exquisitely attuned to sound with heightened ability to perceive and interpret auditory clues (Young, Parsons, Stein, & Kringelbach, 2012). Not only does this disruption influence performance, it can also influence social interactions and parenting. The ability to discriminate distress facilitates timely and appropriate interpersonal interactions (Young et al., 2012). Without this ability, depressed performing artists may falter in a myriad of situations such as pitch discrimination when singing or playing an instrument or emotionally modulating responses while acting. The ability to discriminate facial expression is also compromised in depressed individuals. Depression hampers an ability to decode facial movement and integrate these movements into accurate emotional expressions. Depressed individuals are biased to overly appraise facial expression as angry, an experience that increases a belief that the environment is threatening and unpleasant (Mendlewicz, Linkowski, Bazelmans, & Philippot, 2005).

Depression and anxiety are often provoked by psychological trauma, a phenomenon related to early adversity and complex alterations of neurobiological stress systems. Activation of gene expression during early developmental stress also increases a genetic vulnerability for depression (Heim et al., 2004). For example, childhood abuse has been strongly associated with adult clinical depression, with increased severity of abuse directly related to severity of depression and suicidal behaviors (Bifulco, Moran, Baines, Bunn, & Stanford, 2002). Treatment of trauma and depression should address these comorbid symptoms, including the complex alterations in the neurobiological stress systems (Laugharne, Lilee, & Janca, 2010). Treating all these factors in performing artists is critical, in part because of the high frequency of early adversity and a genetic vulnerability for depression (Heim et al., 2004).

A significant factor that increases depression is chronic pain; in fact, assessing depression severity can be an excellent indicator of pain severity (Kenny & Ackermann, 2015). The majority of dancers and musicians frequently experience pain (Kenny & Ackermann, 2013; Kelman, 2000). More injuries and pain result from pronounced muscular imbalances due to excessive tension (Yoshie et al., 2008). Career stress in the performing arts also compromises health, e.g., being a musician can increase stress-related skin diseases. Dermatological lesions can limit performance, which then increases depression (Onder, Cosar, Oztas, & Candansayar, 2000). Depression is also related to somatoform disorders; however, performers with chronic pain and other somatoform symptoms disavow any symptoms of depression. They are diagnosed with a somatoform disorder; treatment will also focus on chronic pain as well as depressive and anxiety symptoms (Kenny & Ackermann, 2015).

Despite the obvious difficulties experienced by depressed individuals, mild-to-moderate levels of depression can result in more creative productivity (Akinola & Mendes, 2008). Cognitive traits inherent in depression may also propel individuals to seek creative enterprises to find an effective medium to express their internal experiences (Young, Winner, & Cordes, 2013). Attaining an elite level is certainly possible for depressed performers; however, they may be susceptible to increased depression if their mood is influenced by failure-based appraisals (Hammond, Gialloreto, Kubas, & Davis, 2013). Dancing has a powerful effect on mood, and can reverse mild states of depression immediately following a dance class (Lane, Hewston, Redding, & Whyte, 2003), but this is not the case when dancers experience moderate-to-severe levels of depression. Sadly, many professional performing artists may function with clinical depressive disorders (and anxiety) that remain underdiagnosed and untreated (Barbar et al., 2014).

Joint hypermobility and psychopathology

Joint hypermobility is a common benign connective tissue condition. It is genetically inherited and affects 10–25% of the population, with Asian populations and females experiencing a higher proportion of joint hypermobility (Eccles et al., 2012; Garcia-Campayo, Asso, & Alda, 2011). It is a condition marked by exceptional joint mobility, assessed by either actively moving through a range of motion or passively being moved through an extreme range of motion (Bulbena et al., 2011). Individuals with this condition often present with recurrent joint dislocations or subluxations, traumatic or overuse soft-tissue lesions, low-grade inflammation or degenerative arthritis (Garcia-Campayo et al., 2011). Not surprising, a high proportion of dancers are considered hypermobile, although musicians, singers, and actors may be equally hypermobile. For the purpose of this chapter, joint hypermobility will be examined in relationship to psychopathology; however, more detail related to physical injuries and illnesses will be discussed later (see chapter: Injuries and Rehabilitation, and chapter: Health Concerns and Burnout).

Individuals who suffer panic and anxiety disorders are more likely to have joint hypermobility conditions. This association may be related to laxity in the heart and mitral valve prolapse (Garcia-Campayo et al., 2011). This heart condition is also related to increased dysautonomia, and manifests as an inability to maintain homeostasis during environmental fluctuations such as rapid temperature changes (Durante, Durante, & Furiasse, 2002). Longitudinal studies demonstrate that individuals with a hypermobile syndrome are more likely to develop anxiety disorders (panic, agoraphobia, social and simple phobias) and be classified with trait anxiety (Bulbena et al., 2011). Depression, although it tends to be higher in performers, is weakly related to hypermobility (Hakim & Grahame, 2004). This weak association suggests that anxiety and panic disorders may have etiological roots in hypermobile syndrome; whereas, depression seems to have a different etiology (Scheper et al., 2013).

The neurological differences between hypermobile and nonhypermobile groups include greater amygdala volume, hence greater fear and anxiety levels (Eccles et al., 2012). They are more neurologically reactive to negative faces and emotions, a response that indicates greater insula activation (Mallorqui-Bague et al. (2014). These individuals have high interoceptive sensitivity (Eccles et al., 2012; Mallorqui-Bague et al., 2014); a personality trait that heightens internal awareness of physiological responses such as heart rate and increases accuracy identifying emotional states (Mallorqui-Bague et al., 2014). Consequently, greater physiological and emotional awareness (interoception) intensifies awareness of markers (somatic and cognitive) that elevate anxiety sensitivity. The relationship between hypermobility, interoception, and anxiety reinforces the undeniable association between the mind and the body; literally, psychiatric disorders are rooted within a body and these symptoms are often related to physiological dysregulation (Fig. 17.3).

Obsessive-compulsive disorders, eating disorders, body image, and perfectionism

Compulsive behavior often emerges as an attempt to reduce anxious thoughts that are obsessional in nature. Obsessive-compulsive disorder (OCD) is a common

Figure 17.3 "Extending." Performers: Ann Baltz, Lilia Kibarska. Photographer: Lee Choo.

mental disorder and is ranked as the fourth most common in the world (Maltby, Tolin, Worhunsky, O'Keefe, & Kiehl, 2005). Whether severe or mild, OCD is considered a chronic anxiety disorder. Some of the OCD symptoms can be found in the panic, phobia, social, and general anxiety disorders (APA, 2013). Many performers often exhibit less extreme forms of OCD behavior; e.g., they have highly ritualized behaviors before performances and during practice sessions. They recognize that these behaviors may be excessive; however, they are not extreme enough to deter successful functioning. However, performing artists with obsessional passions may actually have an OCD disorder. Such behaviors increase injury and decrease creativity (Bonneville-Roussy et al., 2010; Mageau et al., 2009; Rip et al., 2006; St-Louis & Vallerand, 2015).

It is believed that obsessive maladaptive thoughts and compulsive behaviors are rooted in difficulties processing rewards. Fear of punishment when engaging in goal-directed activities can also drive obsessional thoughts and compulsive behaviors (Jung et al., 2013). It is generally understood that OCD behaviors are triggered when errors are detected; the anterior cingulate is hyperresponsive to any minor error and initiates signals to correct them. Contrary to the *error-detecting perfectionistic theory, the high conflict rejection model* claims OCD is related to deficits in impulse inhibition. During high conflict situations, the OCD individual actually rejects the successful behavior, such as hand washing, and repeats this pattern again and again. Neuroimaging studies support this high conflict rejection theory. Excessive action monitoring takes place in the rostral and caudal anterior and posterior cingulate, left prefrontal cortex, orbitofrontal cortex, caudate nucleus, and thalamus. Inhibitory activation is thwarted in these high conflict states; OCD individuals are not able to reject repetition impulses (Maltby et al., 2005). Their high anxiety-driven behaviors are associated with alterations in functional connectivity in corticostriatal–limbic networks; consequently they struggle to modulate emotional and motivational states (Jung et al., 2013).

Eating disorders and body dysmorphic disorders share similar OCD symptomatology. Clearly identified conditions that produce eating disorders include a critical family environment, coercive parental control, and hyperfocus on weight (Haworth-Hoeppner, 2000). Families tend to mediate the culture of thinness for eating disordered patients. Parental mental illness also increases the risk for eating disorders (Rosenvinge & Pettersen, 2015). These family conditions are similar to many performing arts training environments; the teachers and coaches reinforce the aesthetic culture of these art forms, which can result in greater discomfort regarding body image. Inappropriate eating attitudes and behaviors are often the result of these environmental conditions (Ravaldi et al., 2003). Disordered eating can take many forms, although specific performing arts domains have higher distributions of one or two of the disorders. For example, more dancers may experience anorexia nervosa or bulimia nervosa (Thomas, Keel, & Heatherton, 2005). One unique disorder, orthorexia nervosa, is a pathological obsessive need to eat healthy food and avoid unhealthy food. In a study on performing artists, a high prevalence of opera singers (82%) had orthorexia nervosa compared to ballet dancers (32%) and orchestra musicians (36%) (Aksoydan & Camci, 2009).

Within the dance population, awareness of food, weight, and body image increases with age. Studies have demonstrated that dancers in all age groups consume too few calories to meet the demands of their physical workload (Dotti et al., 2002). Comparing dancers to nondancers reveals a higher prevalence of disordered eating, as well as actual eating disorder diagnoses (anorexia nervosa, bulimia nervosa, eating disorder not otherwise specified) (Friesen et al., 2011). In general, dancers internalize a norm of thinness, which can drive disordered eating (Scoffier-Meriaux et al., 2015). This pattern exists in Asian countries as well as Western countries; unfortunately, the aesthetic culture of dance is the common denominator. Dancers strive for lean bodies and are excessively aware of body image and weight concerns (Tseng et al., 2007). The dance culture extends beyond female dancers, with male dancers internalizing a similar aesthetic body image of leanness. Concern regarding the male dancer's ability to acquire a larger muscular body indicates that male dancers may selectively turn away from traditional male gender-role body images. The reality is that the majority of male and female dancers must fulfill the traditional body stereotype, including hyperfocused attention on physical performance and body aesthetic (Ravaldi et al., 2006).

Like aesthetic sports, participating in aesthetic performance careers (actors, singers, dancers, pop musicians) increases awareness and preoccupation regarding body image and weight concerns (Ackard, Henderson, & Wonderlich, 2004). An emphasis on appearance increases these concerns, especially when engaging in aesthetic activities that begin at an early age. For example, 5-year-old girls who participated in aesthetic activities were more aware of weight compared to the same-aged girls who did not participate in these activities (Davison, Earnest, & Birch, 2002). In a study that examined individuals who participated in childhood dance, they manifested increased eating disordered behavior as an adult, including a drive for thinness, perfectionism, and preferred lean body size. This pattern persisted even when they no longer participated in dance (Ackard et al., 2004).

Eating disorders are extremely dangerous, and sometimes life threatening (APA, 2013). Disruptions are evident in neurobiological systems, cognition, and emotional processing. For example, when anorexia nervosa symptoms are present, patients' interpersonal interactions are limited. They inaccurately interpret social actions, are biased to attend to angry threatening cues, and have difficulty recognizing emotional states in self and other (Harrison, Tchanturia, & Treasure, 2010). Anorexic individuals are limited in emotional awareness and expression, and slower to respond to facial recognition and to recall self-referential emotional words (Jansch, Harmer, & Cooper, 2009). They also have difficulties discerning emotional cues in vocal sounds; like depressed individuals, they are not able to discriminate pitch changes and respond to emotional coloration (Kucharska-Pietura, Nikolaou, Masiak, & Treasure, 2004). These deficits are challenging for performers who must accurately recognize, recall, perceive, and respond effectively to visual and auditory emotional cues. These deficits also constrain empathic awareness (Kucharska-Pietura et al., 2004), a feature that is critical for effective performance within ensembles (Sevdalis & Raab, 2014).

Disordered eating (problem eating) and the more severe eating disorder diagnoses are associated with increased rate of lifetime injuries. Many eating disordered patients also abuse exercise (Calogero & Pedrotty, 2004), which adds to the risk of repetitive or overuse injuries. This pattern is especially pronounced in ballet dancers (Thomas, Keel, & Heatherton, 2011). Because these disorders are regarded as one of the most treatment-resistant forms of psychopathology (Riva, 2014; Rosenvinge & Pettersen, 2015); eating disorder-related illnesses, injuries, and poor performance increase the likelihood of a shortened career. Sadly, performing artists frequently view their disordered eating as a positive attribute that will promote their careers; maintaining their eating disorder is congruent with their beliefs, especially the notion that they have greater self-control and are functioning at a higher level (Riva, 2014).

Adding to this negative profile, individuals with eating disorders acquire a belief that their body is an object that is rigid and fixed, rather than one that is subject to continual perceptual fluctuations (Riva, 2014). Once the body is objectified, it becomes the object of pride or shame. It is a tool that can be controlled and manipulated by self or others. For example, dancers, unlike rhythmic gymnasts, who also perform in an aesthetic art form, believe that they lack ownership of their body; they dissociate their body from a subjective sense of self (Thomson et al., 2011). Dancers' self-objectification is maintained by intense self-surveillance, and to a lesser degree, by body shame (Tiggemann & Slater, 2001). Dancers' body attire (leotard and tights) during years of training exacerbates these objectification beliefs (Price & Pettijohn, 2006). The consequence of this behavior is decreased self-esteem and body image (Bettle et al., 2001; Eusanio et al., 2014; Heiland et al., 2008). Body objectification is perpetuated in female actors and singers, in part, due to recurrent thin-ideal media images that infiltrate our public consciousness (Hargreaves & Tiggemann, 2002, 2003). Unlike females, males generally identify with muscular-ideal media images (Hargreaves & Tiggemann, 2009), although some male dancers identify with thinness (Ravaldi et al., 2009); their self-esteem and body image is positively or negatively influenced by these male media images (Hargreaves & Tiggemann, 2009).

Compared to the other performing arts, actors are directly affected by popular media-ideal images. Often their employment is contingent on meeting these ideal images, in fact, they are frequently the "objects" that the general public consumes. Performers are the objects that audiences "look at." Performers must not only understand how their image translates to the general population but also encourage acceptance of the image they are promoting. The great risk is that performers objectify themselves and become disenfranchised from an embodied sense of self. When performers internalize these cultural body standards and attach a sense of self to their professional image, they become vulnerable to disordered eating and distorted body image (McKinley & Hyde, 1996).

Perfectionistic tendencies exacerbate this objectification process. Perfectionism is generally regarded as multidimensional, with both personal and social components influencing perfectionistic drives (Mor, Day, Flett, & Hewitt, 1995). Perfectionism is an internalized standard. Individuals can direct perfectionistic standards toward themselves or others, or they believe that others expect perfectionism from them (socially prescribed perfectionism) (Hewitt & Flett, 2004). Studies have demonstrated a strong association between perfectionism and anxiety, in particular, socially prescribed perfectionism that involves believing others hold expectations of perfect performances/appearances. Curiously, anxiety was not increased when individuals maintained high perfectionistic standards for themselves (self-oriented perfectionism) or when they held high performance expectations of others (other-oriented perfectionism) (Mor et al., 1995). However, when performers have both elevated self-oriented and socially prescribed perfectionism, performance anxiety increases, goal satisfaction diminishes, and self-control decreases (Mor et al., 1995). Sadly, perfectionism tends to decrease a sense of personal control and elevate guilt-inducing and shaming experiences (Eusanio et al., 2014; Stoeber, Harris, & Moon, 2007). Repeated perfectionistic-driven motivation can lead to increased injury (Krasnow, Mainwaring, & Kerr, 1999), dysfunctional goal setting, burnout, and career dissatisfaction (Hall & Hill, 2012).

Despite perfectionistic efforts, if self-efficacy is preserved, then striving for perfection may actually enhance performance and goal satisfaction (Mor et al., 1995). Perfectionism may form as a coping strategy to manage fear of failure; it is an insurance policy to maintain favor from a judging other (Sagar & Stoeber, 2009). Helping perfectionistic performing artists cultivate skills to cognitively restructure their beliefs regarding audience and teacher expectations (socially prescribed perfectionism) and develop problem-solving skills to enhance self-efficacy will decrease more rigid perfectionistic tendencies (Cumming & Duda, 2012; Mor et al., 1995; Thomson, Jaque, & Baltz, 2017). Training in a supportive environment can also decrease the drive for debilitating perfectionism, a factor that intensifies performance anxiety (Thomson et al., 2017). When these skills are implemented, perfectionism can operate as a positive motivator to improve performance (Stoeber & Otto, 2006).

Chronic pain, insomnia, substance abuse, and suicidality

Chronic pain plagues performing artists. Managing pain is not only physically exhausting but also emotionally debilitating. Sadly, pain is associated with increased anxiety, depression, and somatoform disorders. This pain profile is found consistently in performing artists (Kenny & Ackermann, 2015). For example, approximately 84% of musicians experience performance-impairing pain (Kenny & Ackermann, 2015). A majority of these musicians also report performance anxiety and depression. In a different study, 86% of musicians reported musculoskeletal pain in the last 12 months, with risk increasing for women and those suffering depression and/or psychosomatic disorders. In contrast to other studies, this study demonstrated that anxiety and career stress were only weakly related to pain (Leaver, Harris, & Palmer, 2011).

In a cluster analysis, musicians who report the most extreme pain deny depression; however, they score highest for a psychosomatic disorder (Kenny & Ackermann, 2013). Many musicians are at elevated risk for psychosomatic disorders (i.e., chronic pain) and these disorders are exacerbated by anxiety (Leaver et al., 2011). Psychosomatic symptoms complicate the treatment of physical disorders, such as hand injuries that require consultation with a hand surgeon (Spahn, Ell, & Seidenglanz, 2001). The psychosomatic component interferes with effective diagnosis and treatment. PTSD symptoms also increase chronic pain disorders, in part related to stress system alterations (Otis et al., 2003). Childhood abuse, in particular, sexual abuse, profoundly predicts adult sleep disorders and chronic pain (Lind, Aggen, Kendler, York, & Amstadter, 2016). These psychosomatically rooted pain disorders profoundly compromise performance and treatment efficacy.

Studies suggest that a predisposition for psychopathology is evident long before elite status in the performing arts occurs (Wills, 2003). Family mental illness and early markers of stress-related disorders are often evident in many performers. How these psychological disorders are managed becomes the challenging problem, especially for musicians who self-identify with substance abuse problems. For example, many eminent jazz musicians are addicted to heroin, cocaine, and alcohol. Whether musician substance abuse is related to a predisposition for psychopathology, sensation-seeking personality trait, or creative disposition remains undetermined (Tolson & Cuyjet, 2007; Wills, 2003). Substance abuse is a significant problem for many performing artists, despite the fact that it is often normalized as part of the artistic culture, especially in the popular music industry.

Disruption in sleep patterns can also increase substance use, pain, depression, and suicidality (Cowley, 2015). Many performing artists struggle with sleep deprivation due to alterations in traditional working hours (most performances are in the evening) or psychopathology. Prolonged sleep deprivation alters brain structures, most notably, the anterior insula and anterior cingulate, regions that also monitor somatic processes and pain (Dubrovsky, 2015a). These alterations distort threat appraisal; in fact, recognition of all social signals is impaired and distorted in sleep-deprived individuals. These distortions and excessive responses to threat are extremely problematic for performing artists; they rely on their ability to discriminate between subtle emotional cues, an ability that must operate during practice, rehearsal, and performance.

Sleep deprivation, substance abuse, pain, psychopathology, and career stress can increase suicidal thoughts and suicidal risk (Roy-Byrne, 2015a). In a report in the *Victoria News* (2015), an Australian National Survey indicated that "10% of professional singers attempted suicide, 59.5% of entertainment industry workers sought help for mental health issues, 40% of performers were diagnosed with a mental illness, 36% of roadies (stage crew) reported 'suicide ideation' in their lifetime." There is a strong pattern of suicidality in highly creative individuals (Drapeau & DeBrule, 2013). In an earlier study, compared to the general population, more artists committed suicide, with contributing factors that included mental illness, adverse financial circumstances, and stress attributed to high rejection rates (Preti & Miotto, 1999). Perhaps because of the transient lifestyle that most musicians experience, or the increased use of substances coupled with depression, musicians, especially young musicians, tend to have a higher rate of suicide (Carton, Carton, & Vazquez, 2010). Despite these high numbers, musician suicide rates were lower than writers and visual artists, perhaps because playing and listening to music provides some psychological relief (Preti, de Biasi, & Miotto, 2001).

Suicide risk is related to a gene–environment interaction that includes alterations in serotonin genes that may be associated with increased psychotic, anxiety, and depressive disorders (Byrne, 2013). Risk also rises markedly when traumatic experiences increase, including more childhood adversity. Individuals who already have a diagnosis of PTSD are at even higher risk (LeBouthillier, McMillan, Thibodeau, & Asmundson, 2015). In a study that examined dancers with a diagnosis of PTSD, 17% endorsed frequent suicidal ideation; they also had more childhood adversity and traumatic events in adulthood (Thomson & Jaque, 2015b).

Recovery from the loss of someone who committed suicide is intensely difficult. These individuals are at high risk to develop major depression, PTSD, as well as suicidal behaviors (Byrne, 2015). They also carry intense feelings of guilt, confusion, shame, and anger (Young et al., 2012). Unfortunately, many performing artists experience the loss of someone as a result of suicide, a circumstance that compounds their efforts to manage their career as a performing artist. For example, high rates of death are common in popular musicians. Life expectancy is shortened in this group of artists, especially among musicians under 25 years of age, in part because of increased suicides, homicides, accidental deaths, and liver-related diseases (Kenny & Asher, 2016; Fig. 17.4).

Treating psychopathology in performing arts

Despite the fact that the psychiatric disorders discussed above often cause incredible suffering, many performers are reluctant to receive treatment. They are often highly skilled

Figure 17.4 "Falling." Performers: Elissa Brock, Jocelyn Brooke, Cheyenne Spencer, Christian Vidaure. Photographer: Lee Choo.

at masking their psychological distress (Fraser, 2010), especially because their training reinforces practices that include "leaving personal troubles outside the door" of rehearsal halls, studios, or performance settings. If they seek treatment, they usually fear that their emotional responses will be blunted and their creative output will decrease. For example, when psychopharmacological medication was the only form of treatment, a mixed-response picture emerged, with some performers enhancing their abilities, some returning to preillness states, and others performing worse (Andreasen, 2008).

Although research findings are robust regarding elevated psychopathology in performing artists, their talents are also expressed via their elevated overexcitabilities (Thomson & Jaque, 2016b). According to Dabrowski, heightened sensitivity does not equal illness. Learning to tolerate the challenges of talent should be part of the training of performing artists; clinical treatment must also include skills to regulate overexcitabilities (Mika, 2005). Further, many performers experience painful loneliness and intense soul-searching during their training and careers. They must grapple with the tension of sensitivity and strength as they cultivate resilience (Fraser, 2010). Through counseling, exercise, diet, meditation, and nurturing a sense of purpose and creative expression, wellbeing can be enhanced. Psychotherapy must include the creative process; engaging in the creative process furthers performers' innate talents and nurtures their intrinsic passion to perform. Gaining an understanding and acceptance of their overexcitabilities will promote self-understanding, as well as opportunities to heal a sense of wellbeing (Fraser, 2010; Mendaglio & Tillier, 2006). Performers live with paradoxical realities; they are passionate yet objective, imaginative and pragmatic, playful and disciplined, ambitious and humble, self-directed and generous, rebellious and conforming. Although managing these opposing forces can be exhausting, they also experience great joy and fulfillment. Treatment that is oriented toward positive psychology modalities is ideal. The focus of treatment should include helping performers manage creative performance tensions and provide skills to integrate them into a cohesive sense of self (Howard, 2008). This approach reinforces the reality that adversity can serve to increase resilience and hardiness (Crust & Swann, 2013; Seery et al., 2010).

During university training, performing artists are at increased risk for mental disorders such as substance abuse, depression, anxiety, suicidality, self-harm, and eating disorders (Tosevski, Milovancevic, & Gajic, 2010). Intense high-level training during this developmental stage also intensifies fatigue, depression, and anxiety (Hildebrandt, Nubling, & Candia, 2012). Educators and counselors have an opportunity to assist young artists during this stage of development. Providing sufficient resources is critical for students during their training, and as a preventive measure for future mental and physical illness. Sadly, many performers express frustration because clinical treatment does not address their performance needs (Hamilton & Robson, 2006). Training programs can help these performers; skills can be taught to improve mental attention (learning and memorizing), positive self-talk, and effective recovery after a poor performance. These skills can be directed to help manage anxiety and increase resilience (Osborne et al., 2014).

Addressing childhood adversity, as discussed in Chapter 16, Unresolved Mourning and Posttraumatic Stress Disorder in Performing Artists, is vital, particularly because these experiences are often the root of psychopathology. For example, increased beliefs in paranormal experiences are not only related to creativity but also to psychological coping mechanisms to manage a perceived sense of loss of control (Watt, Watson, & Wilson, 2007). This coping mechanism is insufficient to prevent later adult anxiety disorders and difficulties managing negative emotions, hence clinical focus should include both paranormal beliefs and childhood adversity (Thomson & Jaque, 2014). Encouraging the cultivation of positive social support can nurture renewed trust and safety (Schneider & Chesky, 2011); the performing artist community has the potential to be a positive "family" rather than another system that inflicts abuse.

Performing artists are well aware that they are embodied performers. Clinical attention that addresses the relationship between physical and psychological responses provides inclusive treatment that addresses the complex demands of performers. For example, understanding the relationship between joint hypermobility and psychiatric disorders can destigmatize mental illness and normalize the psychophysiological responses that emerge under stress. Encouraging engagement in physical activities, especially dancing, can reverse symptoms of depression, anxiety, and PTSD, plus it decreases social isolation (Powers et al., 2015; Rutter et al., 2013). Physical exercise has a profound positive influence on managing clinical levels of psychopathology, including decreasing inflammatory processes (Roy-Byrne, 2013a, 2013b). Unfortunately, performers with psychopathology tend to withdraw from physical exercise due to increased fatigue and lack of motivation (hallmark symptoms of depression) (de Assis et al., 2008; Glass et al., 2004).

A novel treatment approach, as outlined by Raeburn (2000), involves intervening at the band "family" level, as well as at the popular music industry culture. These approaches do not preclude individual treatment; rather they address a more systemic problem that besets performers. EEG-neurofeedback and biofeedback have also been validated as a reliable treatment approach to increase performance and creativity, as well as reduce performance anxiety (Gruzelier, 2014; Wells et al., 2012). Ultimately, a combination of therapeutic approaches (medication, psychotherapy, adjunctive therapies) provides the best evidence-based treatment protocol to reduce pain, sleep disorders, and psychopathology (Manger & Motta, 2005; Strohle, 2009; Vaag, Bjorngaard, & Bjerkeset, 2016).

Summary

- Introduction: creativity and madness.
- Vulnerability, resilience, and stereotypes of mental illness: mad genius, approach behavior, and novelty.
- Psychotic disorders: schizophrenia, schizotypy, and bipolar disorder: definitions, positive and negative symptoms, latent inhibition, high fantasy proneness, and psychosis.
- Dissociation, maladaptive daydreaming, and excessive fantasy proneness: definitions, absorption and memory, normative and pathological forms.
- Anxiety and performers: state/trait, stage fright, panic, influence on other pathologies, managing anxiety.
- Depression: definition and types; effects on perception, memory, attention, and performance.
- Joint hypermobility and psychopathology: definition, association with psychopathology (anxiety), dysautonomia.
- Obsessive-compulsive disorder, eating disorders, body image, and perfectionism: definitions, influences on aesthetics and expressivity, objectification.
- Chronic pain, insomnia, substance abuse, and suicidality: definitions, risk rate, vulnerability.
- Treating psychopathology in performing artists: diagnosing and misdiagnosing, screening recommendations, novel treatment approaches.

Concluding remarks

There is a high incidence of psychopathology in performing artists. Although studies vary somewhat, approximately 20% of dancers have clinical levels of anxiety (Thomson & Jaque, 2011–2012) and 15% have clinical levels of depression. Singers report more anxiety, with 32% reaching clinical levels. Like dancers, a similar percentage of singers reached clinical levels of depression (15%) (Thomson & Jaque, 2016a). About 30% of musicians report severe performance anxiety, with a similar percentage reporting clinical depression (Barbar et al., 2014; Fehm & Schmidt, 2006; Kenny, Driscoll, & Ackermann, 2014). Alcohol abuse is evident in 22% of an orchestral sample, with anxiety somewhat driving alcohol use (Kenny, Driscoll, & Ackermann, 2014).

During training and rehearsals, performers can be encouraged to employ coping strategies such as task-oriented rather than emotion-oriented attention (Brugues, 2011; Thomson & Jaque, 2016d; Yoshie et al., 2008). Task-oriented coping is advantageous, in part, because it directs attention toward performance preparation, including increased concentration on the task and awareness of sensory–perceptual feedback that informs goal achievement (Papageorgi, Hallam, & Welch, 2007). Individuals who cope emotionally focus on fear, anger, or despair; this coping strategy resembles perseverating thoughts and feelings in anxiety, depression, OCD, and eating disorders. Although performers embrace emotional expression while performing, this expressivity is very different than emotional coping while under stress. Performers learn to authentically give expression to works of art when practicing, rehearsing, and performing. This is far different than the emotional hijacking that grips individuals who suffer mental illness. Despite consistent findings that indicate mental illness is associated with increased creativity (Fink, Slamar-Halbedi, Unterrainer, & Weiss, 2012), mentally ill performers who remain untreated are at risk for more frequent and severe injuries and shortened careers.

It could be argued that performers reside at the border between stability and instability. Embracing their talent and cultivating skills to realize their talent situates performers in careers that require active movement toward novelty; they must openly explore subjectively evocative content. Performers tolerate physiological and psychological tension as they give life to works that express the extremes of our humanity; their performance work inhabits madness and sanity.

Section V

Professional Reality of a Performing Artist

Careers in the performing arts

18

I believe in professionalism, but playing is not like a job. You have to be grateful to have the opportunity to play.

Wynton Marsalis

Protean careers and job satisfaction

The performing arts are professions that generally require performers to manage multiple career enterprises beyond performing. Performers need to be self-sufficient; the majority cobble together a career that includes skills in teaching, business, and technology (Bennett, 2009). A small percentage of performers are employed full-time, e.g., employed by rock bands, jazz ensembles, or major dance, orchestra, or theater companies. However, the majority of performers work in multiple settings with short-term contracts. This *portfolio* career enables performers to self-manage their careers and allows them to rapidly respond to opportunities. At the extreme end of portfolio careers is the *protean* career (Bennett, 2009). This career type requires continual skill development to meet the demands of the job market. Personal identity and professional image can be managed directly by the performer; employability becomes the target goal, rather than steady employment (Bridgstock, 2005). Because of this autonomy, greater options for work satisfaction are available, including sustaining professional commitments in a field that is fundamentally unstable. A protean career necessitates that performers engage in multiple skill sets. A typical workweek for most performers includes performances, rehearsals/practice, teaching, social media promotion, project management, and technology (editing, digital media, etc.). Many also work in fields that are tangential to their profession such as working in the restaurant industry, health professions, or finance (Montgomery & Robinson, 2003). A protean career consists of high mobility and low job security, multiple occupational roles, multiple sources of income, subjective measures of success, and personal responsibility for career development and professional networking (Bridgstock, 2005).

Job satisfaction is driven, in part, by an ability to regulate emotional responses. Although distress is common during many performance engagements, working performers learn to enhance interpersonal interactions during the creation of a production. When a job is particularly unpleasant, positive relationship connections decrease intentions to quit a job. Performers learn to amplify pleasant emotions in order to sustain engagement in a difficult job contract (Cote & Morgan, 2002). In general, working as a performing artist provides opportunities to engage highly developed skills and knowledge, with some degree of determining working conditions. The rewards gained while performing tend to balance the efforts required to

Creativity and the Performing Artist. DOI: http://dx.doi.org/10.1016/B978-0-12-804051-5.00018-4

present an inspired performance. As a result, performing arts is ranked higher than most professions for job satisfaction, despite the fact that performers experience high levels of job stress (Wilhelm et al., 2004).

Artists' job satisfaction is directly related to work attitude. They value opportunities to explore within their talent domain and decide when and where they will work (Quested & Duda, 2009). Taking the initiative to learn new works and to accept or reject jobs all add to feeling happy in their career. Because the majority of performers are self-employed, there is more overlap between professional and personal activities; they have more flexibility and autonomy in their careers. These positive features counter the higher risk of unemployment (Bille, Fjaellegaard, Frey, & Steiner, 2013). Given these positive findings regarding artists' job satisfaction, established companies that employ performers for temporary project-driven productions should provide positive working conditions and promote artists' sense of self-determination and autonomy. Under these conditions performers accept the charismatic requests of directors, conductors, or choreographers. While working in these short-term contracts, performers lose a sense of self-determination, but both parties maintain an awareness of the agreed-upon working conditions and contract stipulations (Mangset, Kleppe, & Royseng, 2012).

Working within rigid hierarchical organizations such as national theater companies or orchestras, performers may be constrained by the charismatic leaders at the top. Often the leaders of these large companies shape the identity of the company via a combination of their personality, aesthetics, and values. Despite job security, an increased potential for job dissatisfaction coexists. Ideally, performers maintain a sense of autonomy as they adapt to the rhythms of daily working conditions. Unlike their freelance colleagues, company performers must manage long hours rehearsing, including waiting for the start of rehearsals or performances. However, the hours spent in the theater or concert hall also help facilitate the emergence of an artistic family unit, one that is headed by a charismatic leader who cultivates an expressive voice and an artistic aesthetic that identifies a company, and by association, the performer employed by the company (Mangset et al., 2012; Box 18.1).

Box 18.1 Psychosocial Model to Sustain Careers in the Performing Arts

Motivation	Competency and Expertise	Regulation
Intrinsic (i.e., flow)	Self-efficacy (learning, problem solving)	Recognize and manage positive and negative emotions
Extrinsic (i.e., rewards, jobs)	Clear goal setting	Ability to derive meaning from work
Harmonious passion	Ability to assess situation and redirect	Present-moment attention as a way to manage career unpredictability
Persistence	Building and maintaining social support	Physical caregiving (diet, exercise, sleep)

Resilience and career success

Performers who are resilient engage more positive emotions to recover from negative emotional experiences (Tugade & Fredrickson, 2004). This is vital in the performing arts, simply because the nature of live performance includes unexpected mishaps or "off" performances. The ability to reframe the experience and focus on the next performance is critical to job satisfaction. Performers must learn to infuse their creative projects with awareness that other performers share in the creation; it is a collective process that requires continuous adaptation. This collective awareness promotes resilience and increases a sense of stability, purpose, and fulfillment in performing artists, including those who may be struggling with negative emotional experiences (Simon, 2006).

For example, rock bands that remain together tend to be successful. They have collectively created an identity, one that enhances their visibility and promotes greater creative productivity (Phillips & Strachan, 2016). Despite the inherent resources provided in collective performing groups, when an individual moves into a solo career, such as rock musicians who leave their band and venture out as solo artists, they tend to do better financially (Phillips & Strachan, 2016). They have more self-determination and less financial overhead cost. Although they are no longer working within a band or ensemble, they can maintain social support and preserve a positive reputation (Dobson, 2011). Cultivating resources, such as support from family, other freelance musicians and band members, audiences, and professional networks, can offset the unpredictable and highly pressured lifestyle of a self-employed artist. Their dedication to making and performing music reinforces their career choices and adds meaning and satisfaction to their lives (Vaag, Giaever, & Bjerkeset, 2014).

Both creating new performance works and performing traditional repertory offer meaningful opportunities for performers. A performance ritual emerges that promotes cognitive and emotional self-realization in the performer (Nagy, 2015). The pre- and postperformance rituals created by performers forge stronger connections between fellow performers and their audience (Geeves, McIlwain, & Sutton, 2016). Performers prepare with the full awareness that their audience will help shape the performance, for example, through their spontaneous laughter, applause, and vocal "noises" (coughing, throat clearing) (Broth, 2011). These ritual practices and experiences are deeply related to flow states in the audience and the performers. They are profoundly regulating and meaningful, and provide enhanced resilience under adversity (Keller & Bless, 2008). Although these experiences have been directly studied during musical creativity (performance), they also occur within all the performing arts. Resilience is evident in performers' longitudinal success patterns. Regardless of periodic career lulls, creative output tends to persist in resilient performers, but only after the performing artist has successfully emerged and become established in the profession (Hass & Weisberg, 2009). At a certain point in career development, this level of career success and productivity may be related more to psychosocial support and less to creative potential (Helson, 1999). This is certainly true for performers who develop protean career skills essential to manage career fluctuations.

Serendipity in a performing arts career

Performing artists are well aware that they must cultivate and maintain their artistic skills; they are always prepared for any "chance" opportunity that may cross their path. Many performers experience serendipitous moments that changed the trajectory of their career, whether beneficial or detrimental. For example, an unexpected injury may remove one dancer from performing while another dancer may simultaneously arrive at the theater, and not only know the role but also fit into the costume. This is a serendipitous moment for both dancers; sadly the injured dancer loses an opportunity to be recognized within the well-rehearsed role while the other dancer may emerge as a new rising star. Serendipity, a term frequently employed in science discoveries, is the accidental finding of something unsought and regarded as valuable (de Chumaceiro, 2004). When successful it is regarded as luck or good fortune. Pseudoserendipity is similar to serendipity with the exception that something is accidentally found but it was also being sought. In the above example, a dancer was needed to fulfill the role; hence it was a pseudoserendipitous act.

In a study that investigated successful female conductors, pseudoserendipity facilitated career advancement. A chance encounter opened up a career path that was previously closed; however, the female conductors were internally and externally prepared and ready for the opportunity. Under pressure, these successful conductors were able to take risks, work flexibly, maintain a sense of optimism and confidence, effectively plan and problem solve, and maintain feelings of security and safety. They had previously received support from family and friends to master the skills needed to be accomplished musicians and conductors, and they were given support from family and friends when the serendipitous moment arrived. In the study that investigated four female conductors, these career breaks resulted in career level changes as a result of an unexpected substitution, such as moving from a choral to an orchestral conductor. Another female conductor took a course in opera conducting, which resulted in a request that she accept a contract to conduct an opera. A third career shift occurred in response to a serendipitous moment when an artistic director unexpectedly attended a music competition, noticed one of the conductors, and then launched her career. Many of these opportunities do not afford rehearsal time; if the performer is able to meet the challenge more opportunities can arise. Although it is more difficult for females to achieve these successes, in part because of the highly competitive environment and the inherent stereotype that dominates the performing arts, serendipity can help facilitate careers of performers who may not be officially positioned for these opportunities (Fig. 18.1).

Unions, guilds, and working conditions

Professional performing arts organizations work with membership-driven unions or guild guidelines. Performers who are members are offered opportunities to work within union (or guild) standards and guidelines that are created and approved by

Figure 18.1 "Classical standards." Performers: Monisha Jayakumar, Anjana Pathmarajah, Tharini Shanmugarajah. Photographer: Lee Choo.

the membership. Contract negotiations are based on union criteria, including pay scale and working conditions (hours, facilities). These unions typically have international agreement partnerships, thus performers can work internationally and remain protected by their union agreements. One of the largest organizations in the world is the American Federation of Musicians of the United States and Canada (AFM/AFofM). It negotiates fair agreements, protects ownership of recorded music, secures benefits such as health care and pension, and lobbies legislators (www.afm.org). The merger of the Screen Actors Guild (SAG) and American Federation of Television and Radio Arts (AFTRA) created one of the largest unions for professional actors. Like the musicians' union, SAG-AFTRA negotiates agreements that represent minors and adults who work within all media formats, including royalties, residual payments, and safety protocols (https://www.sagaftra.org). Actors' Equity Association (AEA), a union that represents performers working in live theater productions, represents actors, singers, and dancers. The American Guild of Musical Artists (AGMA) and the American Guild of Variety Artists (AGVA) (www.dansetrack.com) represent dancers and musical theater artists.

Many highly successful self-employed performers employ managers, talent agents, attorneys, and publicity agents to manage and promote their careers. Because performers continually interact, whether meeting at auditions, in classes, or while working in productions, they are well informed about the specific organizations operating in their community. For example, networking in regional communities requires different efforts to manage a career compared to those required when working in large centers such as New York, Los Angeles, Chicago, London, Paris, Toronto, Sydney, New Delhi, or Tokyo. The performance unions negotiate different agreements for large and small communities; these agreements reflect the needs of each community and the available resources to support performers. The subsidiary infrastructure also reflects the demands of each region. For example, an actor working in Los Angeles must have an agent in order to gain entry into most casting agent auditions; whereas, an actor working in a small community may learn about auditions from other actors or from the director/producer who will mount the show.

Union negotiations also influence copyright policy-making. This is particularly challenging because access to digital media is readily available to all. For example, pirated films and music are regular occurrences. Music is commonly sampled, altered, and embedded into other musical compositions. Although copyright is designed to protect the artists, many performing artists hold mixed feelings about these policies (Phillips & Street, 2015). Performers who work at the margins of the performing arts industry tend to justify copyright infringement because they are not able to afford securing rights to use digitized media. They express conflicting views regarding copyright; mainly their views serve their own self-interests. Given the fluid delivery service of digital media, establishing copyright regulations includes a host of complex factors; the landscape is ever changing and evolving (Phillips & Street, 2015).

Many performers commonly strive to work in large organizations such as orchestras and national theater or dance companies. Achieving this status means that they have successfully auditioned into one of the most competitive organizations in their profession. These companies generate billions of dollars and support a thriving city economy; they also receive substantial funding from private and public sectors (Brodsky, 2006). Once accepted into these organizations, performers are guaranteed constant professional and personal interactions with others who share their passion for music, dance, or acting. However, they often lose a sense of autonomy and freedom due to the highly competitive hierarchical structure of these organizations. The long, often tedious, hours of rehearsals, low wages, and the loss of a normal lifestyle further compromise a sense of autonomy (Brodsky, 2006). Added to the loss of autonomy, long hours place intense physical demands and excessive energy expenditure on performers. Even though company performers receive the legally mandated breaks, many performers use the break to continue practicing. These cumulative energy demands increase injury rates and exhaustion (Twitchett et al., 2010). This pattern is consistent for all company members, with slight differences in work demands between soloists and ensemble performers (Twitchett et al., 2010).

Unlike other professions, company performers are constantly evaluated while performing. Not only does the audience evaluate their abilities but so do the other performers, conductor/director/choreographer, and the company artistic team. Even though they successfully gained employment in a large company, they are offered time-limited contracts and these contracts will be renewed only if they maintain or improve their performance skills. Company performers frequently experience feelings of animosity, resentment, jealousy, and bitterness; the loss of autonomy means they are dependent on the artistic leaders to determine their career progression. Many performers surrender hopes of performing as soloists simply because of the stereotypes and biases that dominate the culture of the company (Brodsky, 2006).

While working in rehearsals, many performers are confronted with internal and external tensions, in large part stemming from evaluation. How they handle tension will influence how they are perceived professionally and socially. Although not a routine occurrence, many performing artists are scapegoated and publicly humiliated during a rehearsal process. They are emotionally bullied and shamed, and for some, the emotional pain is too great; they leave their career as a performing artist. For others, they are praised and adored; the rehearsal process is one of joy and satisfaction. Literally, a sense of identity is reshaped during intense rehearsal experiences; these identity renarrations positively or negatively influence relationships between self and others and between self and self (Beech et al., 2012). These "renarrations" persist postrehearsal and postperformance, often requiring other outside influences to alter them (Bodner & Bensimon, 2008).

Companies that elicit positive emotions and optimism will enhance creativity in their company employees (Amabile, Barsade, Mueller, & Staw, 2005). Performers have choices; they may remain in the company and accept company limitations, seek employment in other companies, or work as freelance artists. Regardless of where performers work, tenacity, discipline, self-direction, and challenging career stereotypes can empower performers (Stremikis, 2002). These skills, coupled with opportunities to creatively explore, express, and realize artistry, promote success within the performing arts (Watson et al., 2012; Box 18.2).

Box 18.2 Occupation Satisfaction Models and Career Versatility Models (Bennett, 2009; Wilhelm et al., 2004)

Model	Description
Demand–control model	Two dimensions are factored: decision latitude facilitates the use of individual skills and knowledge, and decision authority determines how much a worker can define working conditions.
Effort–reward imbalance model	Acknowledges an individual's need for control. A need for high control can suggest inflexibility when situations require high effort and low rewards. This situation provokes more stress and diseases related to stress. Individuals with a need for low control are not vulnerable to these consequences.
Person–environment fit model	The balance between the worker's abilities and job demands, and between goals and the supplies/opportunities to meet these goals, determines satisfaction. When there is incongruence between these factors then more stress is provoked.
Social cognitive career theory	The balance of social and economic factors influences career choice and job satisfaction. It focuses on cultivating self-efficacy that is derived from expected outcomes or compensations, and the ability to achieve goals.
Portfolio careers	Careers that are determined and evaluated by the individual and are based on the ability to shape self-identity and intrinsic success. It is a movement away from traditionally prescribed and defined careers. It involves multiple skills and maintaining multiple concurrent roles.

(Continued)

Box 18.2 (cont'd)

Protean careers	The extreme form of portfolio careers, one in which individuals change skills to meet the changing job market. It requires continual preparation as well as actively creating opportunities. The individual career is completely self-managed, hence an ability to rapidly respond to labor force trends. This is a typical career for most of today's performing artists.

Environmental challenges in the workplace

Fostering creativity in the performing arts occurs at many levels, beginning with the individual artist, followed by the interactions between artists, and finally, the aesthetic goals of the artistic and production teams. The quality of the physical environment in which the work is rehearsed and performed also directly influences creative output. Research findings suggest that physical environments conducive to greater creative productivity include complex visual detail, view of natural environment, use of natural materials in the rehearsal or concert halls, and inclusion of warm colors in the interior design (McCoy & Evans, 2002). Attention to these environmental details enhances job satisfaction, positively influences performance success, and decreases mood disorders and suicidal ideation (Woo & Postoiache, 2008).

The quality of the environment is also critical to the health and safety of performers. Many venues have toxic substances that can harm performers, e.g., medical treatment centers specializing in laryngology determined that opera singers were exposed to five toxic substances in their workplace: "(1) aromatic diisocyanates, (2) *penicillium frequentans* in cork granulate, (3) formaldehyde in cork granulate, (4) cobalt and aluminum (pigment components), and (5) quartz sand capable of entering the alveolae" (Richter et al., 2002, p. 72). Floor surfaces are also a major factor in supporting dancer health or increasing hazardous conditions (Wanke, Mill, Wanke, Davenport, & Koch, 2012). Raked (inclined) floors present significant hazards for all performers, contributing to increased injury rates in actors, singers, and dancers, especially in female performers (Pappas, Kremenic, Liederbach, Orishimo, & Hagins, 2011). For musicians, ergonomic challenges of playing particular instruments (static awkward positions) increase performance-related injuries and illnesses (Chan et al., 2000). Touring poses even more challenges for performer wellbeing. Conditions vary drastically, such as stage size, temperature and humidity balance, size of the backstage area, extra space in the wing area, lighting, acoustic resonance, stage surface, and floor resilience (Wanke et al., 2012). Crew expertise and temperament of theater/tour personnel add to the enjoyment or stress of a touring location. Hotel accommodations and transportation also influence performing conditions (Fig. 18.2).

Figure 18.2 "Concert stage." Performers: Paula Thomson & CSUN Dancers. Photographer: Lee Choo.

Performing arts entrepreneurship

The performing arts is recognized as an industry marked by a substantial rise in not-for-profit companies. It is estimated that $5.4 billion was generated by not-for-profit performing arts organizations in California. It supported "160,000 jobs and generated almost $3 million in state and local taxes" (California Arts Council, 2004). These not-for-profit performing arts organizations have shaped the concept of "creative cities" and their mission is to offer performances that serve the community (Preece, 2011). The creation of a not-for-profit organization requires the establishment of a board of directors who oversee the administrative and financial wellbeing of the company; fundraising and outreach programs; and the creation, promotion, and performance of the product. The uniting force in these organizations is the commitment to artistic expression. The reason to form a not-for-profit company is to maximize efficiency in all these domains, while facilitating the creative visions, passions, and desires of the performing artists (Preece, 2011). Establishing not-for-profit performing arts companies is an essential means to promote the arts and to provide work alternatives to the administrators and artists who wish to enter into a performing arts entrepreneurship (Preece, 2011).

The performing arts organizations now encompass diverse settings and present a myriad of performance intentions and missions. For example, on the margin of traditional performing arts organizations is the creation of not-for-profit companies whose mission is to stage research data. This is considered a unique way to analyze and present research. Health research comes alive when it is performed. Public health information can be rapidly communicated as a work of theater or as a digitized medium. Performing research findings necessitates the employment of performing artists. Performers are also engaged within research studies such as testing the ability to generate emotions, sounds, and movements. Actors are employed in medical and business schools; they assume the role of patients or clients in training scenarios (Rossiter et al., 2008). These nontraditional projects are just some of a diverse range of performing arts groups that are emerging in this rapidly changing economy.

Careers and anxiety

Performers are subjected to countless auditions and adjudications. These are high-pressure situations; they are not only being judged for their talent but also for their attractiveness (Ryan, Wapnick, Lacalle, & Darrow, 2006). Attractiveness, dress, and behavior of the performer, as well as the gender of the adjudicator, all influence ratings. While this information is disconcerting, especially for musicians and singers, professional performers are aware that their visual appearance is often equally important to, if not more important than, the auditory quality of the performer. Career aspirations increase a desire for success, which makes the knowledge that audition selection is biased toward physical attraction even more stressful, and sometimes discouraging (Kenny et al., 2004).

Many performers physically and emotionally strain to manage the work demands in their career and this effort tends to increase anxiety (Grove et al., 2013). When career aspirations are high and the work environment is harsh and judgmental, performer anxiety also grows. Their effort to be perfect intensifies along with hopes of receiving praise and promotions (Kenny et al., 2004). As demands increase, a sense of self-efficacy decreases; this is often a recipe for performer demotivation, including decreased interest in engaging in physical activities (Dishman et al., 2005). The perpetual pressure of a performing arts career accounts for substantial situational anxiety. Pressure stems from the uncertainty of work and the high physical demands to maintain elite-level performance (Marchant-Haycox & Wilson, 1992).

To counter these anxiety-provoking factors, many performers cultivate practices that promote wellbeing. For example, they strive to eat a well-balanced diet, get adequate sleep, maintain fitness levels to support performance demands, develop relaxation techniques, and nurture personal support from colleagues, family, and friends (Kenny et al., 2004). Anxiety is troublesome for performers; however, it can also motivate them to seek solutions to manage anxiety. For examples, musicians who experienced high trait anxiety also engaged in behaviors that enhanced wellbeing (Kenny et al., 2004). Redirecting attention toward personal activities that elicit positive emotions also aids wellbeing and creativity (Dalebroux, Goldstein, & Winner, 2008). The creative process of performing improves mood for the majority of performers (De Petrillo & Winner, 2005), unless they are debilitated by severe depression or anxiety. Successful performers are able to acknowledge their anxiety, take active measures to manage it, continue pursuing a career as a performer, and derive career satisfaction.

Diversity and stereotypes

Critical race theory (CRT) analyzes social disparities between dominant and marginalized groups (Hiraldo, 2010). It examines race, racism, and power, and how they are covertly and overtly reinforced. CRT is comprised of five tenets and each tenet examines and exposes racism. First, counter-storytelling is a collection of stories regarding marginalization that are shared by nonwhite, heterosexual male individuals. Because unconscious cultural bias accepts the dominant perspective, counter-storytelling challenges these assumptions and reveals patterns of exclusion. The permanence of dominant racial control must be challenged to effect change. These beliefs, if left unexamined, will continue to dominate political, social, and economic structures. Because racism is deeply entrenched in the cultural psyche, the dominant group of white heterosexual males is regarded as a property right. They are the primary beneficiary of civil rights and financial privilege; efforts to disenfranchise this group are experienced as revoking their civil and property rights. Lastly, liberalism tends to approach racism with the notion of color-blindness; literally, all people are part of the human race and should be treated equally. This liberal stance may be well intended but it fails to expose the micro- and macro-biases experienced within all diversity groups (Hiraldo, 2010).

In the performing arts, the phrase color-blind casting was coined to counter disparities; unfortunately this strategy does not work. Without consciously addressing discrimination, both conscious and unconscious, minority groupings are vulnerable to stereotype threats. They are at high risk to be negatively judged based on stereotyped beliefs about social identity group membership rather than individual contributions. These stereotype threats perpetuate significant decreases in wages and constrain career opportunities (Block, Koch, Liberman, Merriweather, & Roberson, 2011). Individuals in stereotyped groups tend to counter these racial and gender biases. They may overcompensate (increase work to prove competence) and/or they may engage in internal locus of control (internalizing a belief that the individual failed because of personal shortcomings rather than acknowledging that the locus of control resides outside talent and effort) (Block et al., 2011). Many psychologically distance themselves from identifying with their race or gender, or make overt efforts to become assimilated within the dominant group; they hope they will be freed of negative stereotyping and be afforded opportunities offered to the dominant group (Block et al., 2011). When efforts falter at challenging stereotyped threats, many individuals withdraw or disengage from their careers. This pattern of withdrawal is devastating to the performing arts; highly talented minority artists may forsake their goals and leave the profession. Their departure perpetuates the dominance of a nonmarginalized group and removes the rich texture of cultural diversity (Block et al., 2011).

Despite these diversity challenges, performers can succeed; many possess a resilience that helps them persevere (Block et al., 2011). They challenge the stereotype threats without disowning their own diversity. For example, Misty Copeland, the first African American ballerina in American Ballet Theatre to be cast as a soloist in ballets such as *Romeo and Juliet* and *Swan Lake* (Stahl, 2016), demonstrates resilience while openly acknowledging racial stereotypes. She successfully challenged the ballerina stereotype. Sadly, countless talented African American ballerinas before Misty Copeland were thwarted. Despite Copeland's successful efforts to expose racism, it remains a significant problem in the performing arts, as evidenced in the nominees for the 2016 Academy Awards; all were Caucasian actors. This event publicly exposed the racial bias that drives Hollywood.

Like all organizations, issues of inequality and the propagation of stereotyped images are pervasive in the performing arts. For example, increased concern has been expressed regarding rap music video productions and gender exploitation. The ongoing "booty video" reinforces the objectification of the female body, in particular black sexuality, power, and gender/sexual orientation; an alienation from embodiment and social support is provoked (Fitts, 2007). These concerns are a reality in the production process of hip hop music. The hierarchy is male dominated and many "video girls" are selected from a roster of nonunion performers. They are vulnerable to the whims of the production company, including working hours, wages, and performance content. The female performers are subjugated to harsh physical scrutiny, including the humiliation of being treated as a sexual commodity, and the implicit knowledge that they must trade sexual "favors" to succeed in this performance medium (Fitts, 2007). Many "video girls" are unable to separate their media image from their personal lives. This renders them vulnerable to physical and psychological illness. The efforts to "keep it real" when creating and producing rap music unfortunately reifies stereotypes; unfortunately, this industry is inadvertently compromising its own efforts to challenge inequalities within gender and race politics (Fitts, 2007).

Sexual orientation, although generally more accepted within the performing arts community, still involves painful experiences of sexual discrimination and marginalization. For example, young male dancers are exposed to heterocentric bias, such as adhering to gender norms, including masculine body images. They frequently experience intense marginalization (Risner, 2007). Moving beyond race and gender, talented disabled performers remain underserved, although some companies do perform works that integrate able and disabled performers. Because the performing arts are dependent on physical ability, ageism is equally problematic. Many highly talented and experienced performers are overlooked or replaced by the new "shiny" young talent that has just emerged in the field.

Despite the myriad of diversity offenses, the performing arts offer a universal voice that can transcend race, gender, and disability. For example, regardless of whether it is as a listener or performer, interacting with music alleviates a sense of alienation and isolation and enables identity formation (Aronoff & Gilboa, 2015). Music can challenge or reify cultural norms and assumptions. Today, easy access to digital media enables those who feel marginalized to find a sense of belonging; they can find musical expression that speaks directly to them. This is powerfully evident in "gay music" and hip hop music; these subcultures strongly identify a sense of uniqueness and belonging via the musical medium that expresses their particular values and sensibilities (Aronoff & Gilboa, 2015).

Figure 18.3 "Shadows." Performer: Jocelyn Brooke. Photographer: Lee Choo.

Directly addressing differences and challenging the myriad microaggressions that transpire consciously and unconsciously is the primary tool to effect change and achieve equity and diversity (Block et al., 2011; Sue et al., 2007). Microaggressions manifest as unconscious assaults, insults, or invalidations (Sue et al., 2007). These microaggressions can transpire between student and teacher, director/conductor and performer, performer and performer, or performer and audience. They ultimately lead to further mistrust and alienation unless they are exposed and acknowledged (Sue et al., 2007). Employing a "token" ethnic or disabled performer to represent diversity will not remedy the problem. The performing arts, like all organizations, must reflect the culture it resides within. Promoting and creating opportunities for diverse performers should include talented performers from all minority groups. Performers rely on mentors to guide them to levels of expertise and the performers themselves become mentor figures to the larger public. Performing artists are situated to model and educate others about microaggressions. Cultural awareness, including revealing and addressing diversity, can enrich understanding and promote greater equality among minority groups within the performing arts and society at large (Osula & Irvin, 2009; Fig. 18.3).

Challenges and abuse in a performing arts career

Unemployment, a worldwide stressor, has profound negative consequences, including suicide (Nordt, Warnke, Seifritz, & Kawohl, 2015). In the performing arts, a majority of professional performers are unemployed at any given time, and all experience bouts of unemployment (Kogan, 2002). Whether working or seeking work, many performers network while socializing, including consuming alcohol. This entrepreneurial networking behavior is critical for self-employed performers, although it increases the possibility that they will develop a substance abuse disorder (Dobson, 2011). Abuse of alcohol or other substances increases risk for suicide; they are key variables indicating career difficulty and personal distress (Nordt et al., 2015). Financial strain, loss of physical assets (home ownership or car), relational discord, chronic illnesses, or death in the family, as well as a preexisting anxiety disorder, increases risk for substance abuse and suicidality (Wilhelm et al., 2004).

Within some performing arts genres, young performers are exposed to alcohol and drugs with the caveat that this usage will enhance creativity and guarantee success. Many pop icons and jazz musicians believe this common misconception. Sadly, this behavior is frequently related to significant decreased life expectancy (Kenny & Asher, 2016). Performers such as Jim Morrison, Janis Joplin, and Jimi Hendrix actually lost the motivation to create as their addictions worsened. Their creativity was sacrificed, in part, by the chaotic personae they offered to their audience, personae that ultimately led to their demise (Holm-Hadulla & Bertolino, 2014). The wild pop icon persona may also be driven by a personality trait that is high sensation-seeking. These musicians are often drawn to intense/rebellious or energetic/rhythmic musical genres as a means to actualize their self-expression; they are intensely embodying their personality trait, often at the cost of their life (Miller & Quigley, 2012).

Another factor that increases physical and emotional harm to performers is the demand to express violence. For example, singers and actors, and sometimes dancers, are commonly asked to grunt, sob, shout, or groan during performances. Producing these sounds, especially within the context of emotionally driven work, can physically damage the larynx, with potential long-term consequences. Vocal training will help reduce vocal damage; however, adequate and appropriate rehearsal and performance conditions must be offered to the performers to ensure sufficient vocal health (Roy, Ryker, & Bless, 2000). Even when these harsh vocal demands are not required in violent work, the physical and emotional demands are intense. Performers are not recreating emotions. Performers, especially actors, are genuinely reacting to circumstances that evoke genuine emotional responses. Highly dramatic roles, including violent roles, can deplete performers emotionally and physically. Again, rehearsal and performance accommodations are invaluable to prevent stress-related injuries (Hetzler, 2008).

Some performers experience emotional, physical and/or sexual abuse by their teachers, directors, casting agents, or fellow performers (Ledoux et al., 2009). These experiences are painful and humiliating. Many performers fear saying no because they want approval from the teacher or they want to be employed by the arts organization. The effort to survive as an elite performer necessitates acceptance and approval. These compromising situations (or overt assault), although not commonly studied in academic research, are well known among performers. The "casting couch" remains a reality for many performers. Performing arts teachers can easily breach student–teacher boundaries, especially when they are training artists to perform sexual or sensual material. Emotional and verbal abuse can be a common experience between elite performers and their directors, or teachers/coaches (MacArthur, 2011). Studies investigating sexual abuse between athletes and coaches are now examined; similar investigations need to take place in the performing arts (Leahy, Pretty, & Tenenbaum, 2002; Parent & Demers, 2011). Once these emotional, physical, and sexual abuse practices are exposed, it is hoped that performing artists can seek treatment (see Box 14.2 for emotion regulation strategies in stressful situations). Therapeutic healing can help them make meaning of their abuse and regain freedom and trust in the creative process (Grossman, Sorsoli, & Kia-Keating, 2006). They deserve opportunities to thrive in the performing arts and to realize their talents.

Aging, retirement, and career endings

Aging is challenging for everyone, but perhaps even more challenging for performers because they are dependent on their body as their expressive instrument. Beverly Sills, a world-renowned opera singer, stated, "My voice had a long, non-stop career. It deserves to be put to bed with quiet and dignity, not yanked out every once in a while to see if it can still do what it used to do. It can't." Every performer recognizes that they must adjust to the aging process. For example, singers often develop more vibratos due to weakening vocal muscular support and musicians are not able to play as quickly as they once did in their youth. Actors and singers age out of ingénue roles and must prove that they can adequately perform older roles. Dancers gradually lose speed and power; consequently they can no longer perform repertory that was created for dancers who possess physical prowess. Across all the performing arts, most aging performers lose confidence in their memory (Csikszentmihalyi, 1996). The speed of rapid memorization of scripts or scores becomes a slower process, and often an uncertain process. Although physical prowess diminishes, elite older performers have cultivated a wealth of knowledge and experience. If they were employed in other professions, they would be heralded as leaders in their company; however, physical decline necessitates "early" retirement for most performers. Performers generally experience aging as a process of change and loss (Roncaglia, 2006).

The performing arts unions have branches that specialize in career transition and aging performer needs; however, many aging performers struggle with a loss of identity and a loss of community. Many also confront the need to establish a second career, a daunting task for those who have reached elite levels within their professional domain (Roncaglia, 2006). Despite these career realities, continuing to engage in the performing arts slows the aging process. Studies have demonstrated that dancing enhances mobility and cognitive flexibility (Coubard et al., 2011; Verghese, 2006). Continued dancing improves postural stability and control, and cognitive performance (Kattenstroth et al., 2013). Because performers' sense of identity is tied to their talent domain, many retired performers also lose a sense of identity, a painful and often depressing process. Continuing to engage in performance, even if it is modified, improves mood, promotes wellbeing (Alpert et al., 2009), and strengthens memory (Franklin et al., 2008; Jakobson et al., 2003). This continuation is possible for many musicians and actors; they are able to age positively and remain active members of a symphony orchestra or theater company (Brodsky, 2011).

Other complications for retired performers include high risk for significant chronic pain and decreased musculoskeletal health. A long career in the performing arts is physically demanding; the cost extends into retirement years, as evidenced in retired professional dancers who had significantly greater difficulty walking due to knee and hip pain (Ronkko et al., 2007). Compounding these aging challenges is altered self-perception. The performer's instrument is the body; how they view aging stems from their highly trained kinesthetic perceptual appraisal process. For many performers, aging disenfranchises them from their embodied identity. Gradually, aches and pains dominate their attention as confidence to perform with speed and agility diminishes (Wainwright & Turner, 2006). Unfortunately, neurological motor pathways still send signals to execute complex movement patterns, but physical prowess has eroded over time and the movement can no longer be performed with the same refined skill as younger performers. There is a sensorimotor and cognitive dissonance between their highly trained neurological pathways and their physiological abilities. The challenge for aging performers is that they also lose their younger able-bodied self-identity; retirement must include modifications to their acceptance of a different embodied self and a different means to express their creativity.

The natural course of aging is ultimately an accumulation of cellular dysfunction and damage that culminates in death (Izaks, & Westendorp, 2003). Unlike chronic disease, which usually begins earlier in development, aging is the natural progression of physical alterations. Both chronic illness and the aging process are deteriorating physiological processes, even when one is described as disease and the other is regarded as a normal process of aging. For aging performing artists, the ability to recover from illness, injury, and workload demands decreases simply because the biological process of aging includes decreased metabolic resources vital for tissue repair and systemic rebalancing (Hamerman, 1999). Receiving appropriate medical care for the aging population is an essential aspect of medical care; performing artists often require more care as a consequence of longstanding physical demands placed on their physical bodies, including the chronic stress that is inherent in a performing career. Although aging is not a disease per se, many performers are actually aware that the aging process significantly constricts the expression of their talent. Elderly performers know they lack the metabolic resources of younger performers; they feel more infirmed (Izaks & Westendorp, 2003). However, for older performers, the expanse of their creative career can promote a sense of wisdom and integrity; they continue to love their art form and derive satisfaction from it (Gembris & Heye, 2014). They can become a source of inspiration for younger performers who follow in their footsteps (Brodsky, 2011; Csikszentmihalyi, 1996), especially for performers who have a portfolio career (Gembris & Heye, 2014).

Summary

- Protean careers and job satisfaction: definitions, work attitude and efficacy, hierarchical systems.
- Resilience and career success: self-determination, daily practices, productivity.

- Serendipity in a performing arts career: chance opportunities.
- Unions, guilds, and working conditions: definitions and examples, negotiations, large and small organizations, emotions and outcomes.
- Environmental challenges in the workplace: fostering creativity, toxins and physical conditions, touring.
- Performing arts entrepreneurship: industry economy, arts organizations.
- Careers and anxiety: unstable career and stress, anxiety-provoking factors, and wellbeing.
- Diversity and stereotypes: critical race theory, color-blind casting, objectification and exploitation, addressing diversity (microaggressions).
- Challenges and abuse in a performing arts career: unemployment, self-employment, exposure to drugs, abuse (emotional, physical, sexual).
- Aging, retirement, and career endings: definitions, career transitions, skill maintenance and loss.

Concluding remarks

Performing artists need to understand the nature of protean careers. Many are trained within institutions that prepare them for a career in a traditional company. They are given the illusion they will gain career stability and security once they leave the training institution. The growth of higher education training programs has intensified the dissonance between an imagined stable career in a performance company and the realities of a protean career. In fact, studies indicate that too many performers are being trained for too few jobs. Education programs need to include tools to successfully manage a protean career. Performers must learn to accept a realistic understanding that multiple sources of income are the norm; it is not a failure to incorporate working in nonperforming arts jobs while still maintaining the status of a performer (Bennett, 2009; Wilkie, 2015). Performers must acknowledge and accept that their protean career requires management skills and ongoing cultivation of new skills to meet changing artistic trends (Bridgstock, 2005). They also need to learn to cultivate networking skills to successfully maintain an entrepreneurial career (Coulson, 2012).

Assessing a career in the performing arts can be organized around three key factors as outlined in the demands–control–support model (Karasek & Theorell, 1990). This model identifies *work demands* as working hours, physical demands, and work intensity. *Control* references skill ability and decision-making, while *support* includes physical, economic, and emotional support from family, colleagues, and leaders. These three factors interact to promote a positive work environment. If poorly managed they can also lead to poor health, increase job stress, and decrease productivity. Applying this model to freelance performing artists confirms the overall positive factors in this career (Vaag et al., 2014). Performers describe considerable work demands but the demands are balanced by a sense of control and support. For example, unlike permanently employed musicians, freelance musicians have greater challenges managing the interplay between demands and resources. Strong entrepreneurial skills, emotional/cognitive flexibility, and a robust passion for musicmaking offsets job strains. In general, these performing artists tolerate the challenges of their freelance career and positively endorse their choice to work as a performing artist (Vaag et al., 2014).

Multiple jobs, although often within related artistic fields, require similar creative problem-solving skills. Performers learn to effectively manage performance schedules and nonperformance work-related demands such as taxi driving or waitering. Performing artists must challenge the shared myth that only artists who earn their full income as a performer are viewed as professional artists. Today's climate necessitates that performers maintain their identity as artists, even while working in occupations outside the performing arts (Bain, 2005). Despite lower financial rewards, job satisfaction is high; performers derive meaning and purpose within their self-determined protean career (Throsby & Zednik, 2011).

Shaping a career in the performing arts is a dynamic process that includes multiple variables that are challenging and/or pleasurable (Frade, 2005). Despite precarious employment, success in the performing arts is measured by meaning making. The performer and audience share an experience that is enriching and aesthetically rewarding (Foreman-Wernet & Dervin, 2011). A career in the performing arts occurs within a network of social and professional connections. These networks are often at the heart of what makes a career in the performing arts rich and satisfying.

Injuries and rehabilitation

19

I have actual acting scars.

Benedict Cumberbatch

When you're a dancer who is injured, you are at the bottom of the food chain. We are so replaceable.

Bebe Neuwirth

Introduction

Deliberate focused training over many years is one of the major factors in preparing a performing artist to secure a professional career as a performer. The physical demands of most performing artists match demands experienced by elite athletes (Bird, 2009). While training or working professionally, performers are at high risk for injury, whether from acute traumatic incidents or chronic repetitive overuse that results in microtrauma (Gamboa, Hagins, & Manal, 2005). Medical professionals are the critical experts who assess, diagnosis, and treat performers; they recognize the severity of the injury and the potential limits to a performer's career (Macintyre & Joy, 2000). Like sports medicine practitioners, performing arts medical specialists are experts in understanding the demands of a performing arts career, including training, rehearsal, and performance practices (Bird, 2009). They can specify treatment protocols appropriate for each performing artist's professional demands. Treating performers requires the inclusion of specific physical and aesthetic factors, e.g., dancers need to maintain flexibility and lean body mass, singers and actors need to continually engage striatal muscles of the larynx for vocal production, and musicians' hand and upper body dexterity requirements are dictated by their instrument and the musical repertory they are to perform.

Because the performing arts profession is often societally devalued, performers tend to overcompensate in order to be included in their profession (Barker, Soklaridis, Waters, Herr, & Cassidy, 2009). They are forced into heightened competition to secure performance opportunities; consequently many performers place unrealistic demands on themselves while minimizing the stigma of injury, including hiding current injuries (Barker et al., 2009). They accept contracts in shows that place high psychological and physical demands on them. For example, Broadway performers will perform in a minimum of eight shows per week, dancers in the Rockettes will perform three shows each day, and performers in Las Vegas will often perform two shows per day. High-intensity performance schedules inevitably lead to overuse injuries and compromise career longevity (Miller, 2006). Unfortunately, many performers are reticent to seek treatment, fearing they will be forced to take time off, which increases loss of employment opportunities

Creativity and the Performing Artist. DOI: http://dx.doi.org/10.1016/B978-0-12-804051-5.00019-6

Figure 19.1 "Risk-taking." Performers: Jesse Ricaldi, Courtney Ryan & CSUN Performers. Photographer: Lee Choo.

(Bird, 2009). If they do seek treatment, in most communities, there are limited or no performing arts medical specialists available who can treat their unique performance-related medical needs. Performers remain an underserved population, although efforts are being made to create specializations in this field (Dick et al., 2013).

High injury rates persist among performers, in particular, musicians and dancers (Lee et al., 2013; Manchester, 2006). Most injuries are soft-tissue in nature (regional muscle pain syndromes or tendon/ligament damage), caused by factors that are either intrinsic (structural and functional characteristics of the performers' anatomy or technical usage) or extrinsic (environmental influences). These factors may be modifiable (change length of time practicing, increase strength, etc.) or nonmodifiable (gender, genetics, hypermobility syndrome). Some domain-specific causes may be somewhat modifiable, such as changing the construction of pointe shoes, whereas, others are more difficult to modify, for instance the key width and resistance on standard concert piano keyboards (Manchester, 2006). Although efforts are directed at preventing injury, most performers experience at least one injury in their career, and many need to seek medical help to resume their performance career (Rietveld, Macfarlane & de Haas, 2013). Despite these harsh conditions, performers still maintain a strong belief in their work (Barker et al., 2009; Fig. 19.1).

Injury and the effects on creativity

Performing artists are creative. For many performing artists, risk of injury can compromise creative physical risk-taking, although they may willingly take aesthetic risks when expressing novel ideas or challenging cultural norms (Eisenman, 2001). However, when performers are confronted with injuries that lead to loss of work, increased despair and isolation, and decreased creative desire often coexist during injury recovery (Rietveld, 2013). Initially, injuries that disrupt performance work schedules are disorienting, and often frightening. Daily routines can no longer be practiced, regular interactions with fellow performers decrease, and the vehicle for self-expression is curtailed. The performer's instrument can no longer be expressive. This is generally a temporary loss of the performer's creative voice; however, most performers fear it may be a permanent loss. This fear constricts creative flow experiences (Rietveld, 2013).

The experience of injury is certainly challenging, but it can also force introspection. If performers accept their injury, especially when caused by misuse or overuse, they can then address injury-provoking behaviors, such as poor technique, denial of the injury, and minimization of pain signals. To preserve creative aspiration and maintain career goals, they can implement more effective coping strategies (Csikszentmihalyi, 1990). They can attend to the task of healing, manage their emotional responses to the injury, cultivate a realistic balanced perspective, and seek appropriate treatment. Learning to transform adversity into a positive opportunity can instill greater self-confidence and self-awareness. Performers can learn about the causal factors that lead to the injury and effective strategies to avoid future injuries. Efforts to focus beyond their performance domain may enrich their creative expression. The time required to heal is an opportunity to redirect attention away from the self and into the world. Although injury can be a time of isolation from a small community of performing artists, it is also a time to connect with a much larger and diverse community (Csikszentmihalyi, 1990). It is an opportunity to cultivate an autotelic self, one that can flexibly set and adjust goals.

Deriving joy and meaning can reside within the injury healing process. Performers who are passionate about their art are frequently able to direct their passion toward novel healing approaches. They can flexibly explore other ways to stay engaged in their talent domain; e.g., performers can observe rehearsals or performances and mentally apply corrections or performance choices. They can learn more about other aspects of the profession, such as stage management, lighting, and marketing. When passion is harmonious, performers heal faster and gain more skills while healing; whereas, passion that is obsessional limits healing speed and decreases flexible coping strategies (Rip et al., 2006). Encouraging performers to direct their passion toward open exploration regarding the nature of their injury, the recommended healing protocols to restore health, and novel ways to practice their art form will not only promote healing, but also creativity (Csikszentmihalyi, 1990; Rip et al., 2006). When the rehabilitation environment is positive, performers can engage their creative talents. Ongoing communication with fellow performers diminishes isolation and aids healing (Rietveld, 2013), and confidence in job security greatly helps performers heal and nurtures future creative goals (MacEachen, Clarke, Franche, & Irvin, 2006).

Injury safety policies and workers' compensation

Safety policies are instituted in training programs and within professional working environments. Union negotiations stipulate specific safety standards to ensure that working environments are safe for performers. Stringent policies are also required in training programs, with detailed descriptions and procedures outlined in manuals designed to protect performers, designers, and stage crew. This is demonstrated in the Performing Arts Safety Manual prepared by the University of California (UCLA, 2014). Despite awareness of safety needs, many producers are reluctant to enforce stipulated policies, fearing they will compromise artistic choices. They often defer to the expertise of the performing artists and follow their requests (Ledoux et al., 2009). Reporting accidents is often an equally vague practice, with many "minor" injuries left unreported. This is complicated by the reality that approximately 17% of injuries result from external causes such as floor surfaces; whereas, the majority of injuries are related to the actions of the performer (e.g., landing incorrectly from a jump). Fatigue is the greatest factor leading to muscle strains; cognitive fatigue (e.g., concentration lapses) adds to the increased pattern of injuries and accidents (Ledoux et al., 2009).

Underreporting of injury is also reflected in the underutilization of workers' compensation. For example, 93% of injured professional musicians self-reported their injury but only 23% filed a workers' compensation claim (Chimenti et al., 2013). Many musicians believed their injury was not severe enough to file a claim; however, 47% reported that their pain persisted long after performing, with many (16%) indicating their symptoms interfered with daily activities. A similar pattern was found in a group of modern dancers; 61% of injuries reported were considered severe and yet only 12% were covered by workers' compensation insurance (Shah, Weiss, & Bouchette, 2012). In this same study, modern dancers had higher rates of injuries (81%) compared to American football players (80%); they were aware that the demands of performing were high and the majority performed despite pain and injury. A similar pattern was found in orchestral musicians, with 86% of musicians reporting regional pain (Leaver et al., 2011). The lack of insurance coverage and underreporting highlights the reality that performer perception of injury severity is skewed, including the right to apply for workers' compensation insurance. Although performers can purchase entertainment disability insurance (www.allaboardbenefits.com/disability-insurance/entertainment), many lack sufficient funds to purchase the insurance. Long-term disability due to performance-related injuries may be career ending, and yet performers continue to tolerate pain while performing; they fail to seek necessary medical help.

Artistic production personnel are integral partners in reducing injury risk; they can educate performers about safety policies and make informed decisions regarding rehearsal organization and safety procedures. For example, compression forces when rehearsing and performing partnered lifts usually exceed recommended loads outlined by the National Institute of Occupational Safety and Health. Artistic teams can plan rehearsals so that repeated lift practice is minimized (Aldersen, Hopper, Elliott, & Ackland, 2009). Moving beyond performers, designers, managers, and stage crew are at increased risk for injury while executing their skills. They work with heavy equipment and ensure that the theatrical setting meets the artistic demands of the production, while remaining safe for performers. Fatigue is a problem for this group of workers; they often work long hours, with limited breaks and tight schedules (Ledoux et al., 2009). Effective organizational schedules created by the artistic and technical production teams can alleviate some stress; however, limited budgets often dictate reduced hours to accomplish challenging tasks.

An optimal work environment for injury prevention includes effective communication among all participants, clear planning, access to professional equipment, and employment of experts who are experienced and skilled. Educating performers about occupational health risks, and providing them with information regarding occupational-related injuries, as well as legal compensation for time lost from work, will help performers advocate for their rights (Raymond, Romeo, & Kumke, 2012). Lastly, increasing legislation policies and engaging policy-makers who understand and support the performing arts will help ensure safe environments and conditions; they can protect highly trained and talented performing artists (Ledoux et al., 2009).

Injury prevalence and severity

According to estimates from the Survey of Occupational Injuries and Illnesses (SOII) conducted by the U.S. Bureau of Labor Statistics, performing arts organizations reported nonfatal injuries as days away from work, or days of job transfer or restriction (2015). The performing arts field is divided into two sectors, *Information* (motion picture and sound recording industries) and *Leisure and Hospitality* (performing arts, spectator sports, and related industries). The total recordable injury rate is a representation of the number of injuries and illnesses per 100 full-time workers in 2014. The motion picture and recording industry sector reported 1.6 total recordable injuries and the performing arts sector reported 4.7 total recordable injuries.

Studies that investigate groups of performers report alarming rates of injuries; however, most of the injured performers continue working. For example, a South African study reported that 84.4% of musicians were injured (Barnes et al., 2011). A similar number were reported in Australia (Ackermann, Driscoll, & Kenny, 2012), Spain (Roset-Liobet, Rosines-Cubellis, & Salo-Orfila, 2000), and Scotland (Berque, Gray, & Angus, 2016). Prevalence of injuries in dancers is equally high, ranging from 50% to 97% (Air, 2013; Toledo, Akuthota, Drake, Nadler, & Chou, 2004). This same proportion of injured performers is found in performing arts schools, with many students missing an average of 4.5 weeks of classes (Miller & Moa, 1998). At least half of those injured never fully recover; they learn to perform despite musculoskeletal pain (Ackerman et al., 2012). In research that investigates injury rate, an *injury rate equivalent scale* is calculated to determine the number of injuries occurring in a group of performers over 1000 hours of activity exposure. Activity participation and risk of injury are classified on a four-point scale ranging from low to high risk (Liederbach, Schanfein, & Kremenic, 2013). In studies that implemented the injury rate equivalent scale, compared to other groups, performing artists have a high rate (risk) of injury.

Physical injuries manifest as loss of range of motion, endurance, and fine motor control accompanied by pain (Sternbach, June 15, 2016, retrieved from www.gmu/centers/artswellness/overuseinjury.html). They are classified in levels of severity, with Grade I related to overuse syndromes with low-grade inflammation, which resolve with rest, ice, stretching and massage. Grade II is determined when pain exists in multiple sites and strength is reduced (some tissue damage); however, no interference in normal activities results. Like Grade I injuries, this classification of injury will resolve with a similar treatment protocol. Grade III, IV, and V injuries require assessment from a medical specialist. These forms of injury include multiple sites of pain that persist, with varying degrees of impairment, from minimal loss in functioning to disabling pain and full loss of function (Mueller-Wohlfahrt et al., 2012). All five grades of injury as well as muscle tear grading (type 1A to 4) are provoked by intrinsic factors (structural insufficiency) or extrinsic (environmental and workload changes) (Motta-Valencia, 2006; Mueller-Wohlfahrt et al., 2012). Illnesses and stress can also operate as extrinsic factors (Dersh, Gatchel, Polatin & Mayer, 2002; Boxes 19.1 and 19.2).

Box 19.1 Common Injuries in the Performing Arts

Performance Domain	Injury Type	Locations
Musicians	Microtrauma, nerve entrapment, compartment syndrome, focal dystonia, overuse	Upper limb (fingers, wrist, forearm, elbow, shoulder), neck, back (lumbar, sacral), skin
Singers	Traumatic injuries (fractures, sprains, muscle contractures), vocal overuse	Neck, throat, shoulders
Actors	Traumatic injuries (fractures, sprains, muscle contractures), overuse	Lower extremities, neck, throat, shoulders, skin
Dancers	Traumatic injuries (muscle contractures, sprains, fractures, luxations, subluxations), overuse	Ankle, foot, knee, thigh and leg, hip and pelvis, shoulder, elbow/forearm, wrist, hand, lumbar, thorax, cervical, abdomen, face, skin

Box 19.2 Mechanisms of Injury

Performance Domain	Mechanism of Injury
Musicians	Overuse, poor posture and muscular imbalance, ergonomic imbalances related to instrument and chairs, lighting
Singers	Overuse, vocal strain due to acoustic difficulties, poor posture and technique
Actors	Overuse, ergonomic difficulties with set design, stunts and stage combat, vocal strain
Dancers	Overuse, recurrence of injury, falls, landings from jumps, pirouettes and rotational movements, lifts and partnering

Pain appraisal and coping

Most performing artists have difficulty discerning performance pain from injury pain. Pain is a physical and psychological experience that is interpreted by the brain (nociception) (Damasio, 1999). It serves as a warning signal to move away from the source of pain or to restrict movements that provoke pain. Pain operates as a motivating factor to alter behavior; it is a subjective experience that is difficult to empirically measure. In general, performers learn to tolerate intense levels of pain during performance; however, this increased tolerance level is also deployed when injured. They deny injury and avoid seeking treatment.

Appraising pain ranges from a catastrophic threatening experience to an opportunity to gain control over pain (Anderson & Hanrahan, 2008). Although performers tend to tolerate higher levels of pain, some magnify, catastrophize, or become helpless when confronted with pain. These responses actually intensify distress and increase subjective pain appraisals (Paparizos, Tripp, Sullivan, & Rubenstein, 2005). When performers cognitively appraise pain as threatening, they cope with the pain by avoidance or histrionic terror; strategies that are ineffectual for optimal healing (Anderson & Hanrahan, 2008). Regardless of pain appraisal as either threatening or positive opportunities, performers tend to continue performing despite experiencing severe pain. In a study on Brazilian orchestra musicians, some musicians reported experiencing ongoing pain for up to 30 years. Unfortunately, neglecting severe pain increases sleep disorders and personal distress (Kaneko, Lianzo, & Dawson, 2005). Sadly, many performers hide their pain symptoms, often believing it is a normal part of being a performer.

Hypermobility condition and injury vulnerability

Dancers, in particular, and performers in general, may be diagnosed with joint hypermobility syndrome, a connective tissue genetic condition that is marked by joint and tissue laxity. This condition increases risk for injuries that impede

Figure 19.2 "Mobility." Performer: Janice Dulak. Photographer: Cylla van Tiedemann.

performance (Scheper, de Vries, Juul-Kristesen, Nollet, & Engelbert, 2014). For example, hypermobility in dancers is considered one of the factors associated with talent; however, it also increases physiological and psychological vulnerability (Scheper et al., 2013). One of the symptoms of joint hypermobility is decreased physical fitness and muscle strength, despite similar training programs and time invested in training. Performers with this condition also experience more physical fatigue (Scheper et al., 2013). This condition is significantly associated with joint dislocation, musculoskeletal complaints, bone fragility (osteopenia or osteoporosis), and overuse injuries (Scheper et al., 2013). This risk is related to the structural reality that ligamentous laxity increases the need for more joint stabilization, a demand that is often difficult to meet (Miller, 2006).

Rehabilitation efforts with performers who have hypermobile joint conditions should include extra efforts to build muscular strength and cardiovascular fitness. Core stability is paramount for all performers but even more so for those with hypermobility. For example, core stability enables dancers to freely and independently move their limbs without risk of injury (Rietveld, 2013). Because they also experience more fatigue, achieving increased physical conditioning may be more challenging. Hypermobile performers also experience greater pain (Scheper et al., 2013). Along with pain and fatigue, hypermobile performers experience increased panic and anxiety symptoms, factors that may complicate injury risk and rehabilitation (Rietveld et al., 2013; Fig. 19.2).

Overuse, overtraining, and fatigue

Overuse or overtraining injuries plague all performing artists. This pattern exists among classical and nonclassical performers (Buckley & Manchester, 2006).

Typically, overuse injuries and stress-related injuries increase when working in abusive environments. Abusive leaders demand high-level performance at the expense of performer safety. This type of artistic leadership decreases creativity, even though performers struggle to meet the leader's demands. Consequently, more injuries occur in performers and the artistic support team; all are driven by the behaviors of an abusive artistic director (Liu, Liao, & Loi, 2012). Injury rates can also be increased via a combination of perfectionism and stress (effort to please the artistic leader) (Krasnow et al., 1999). Perfectionism intensifies when performers fear they will fail or they attempt to avoid shaming experiences (Eusanio et al., 2014; Sagar & Stoeber, 2009).

Researchers and performers commonly attribute the cause of injury to overwork (overuse) conditions. Fundamentally, they are describing fatigue and its effect on performance. Fatigue is marked by a transient reduction in work capacity, hence a reduced ability to perform effectively (Murgia, 2013b). Physiologically, fatigue is either decreased muscular activation due to a shortage of metabolic substrates such as glycogen depletion or failure in the signaling processes of the central nervous system. Muscular fatigue, also known as peripheral fatigue, occurs when muscular demands include holding a position for prolonged periods of time, or repetition of isometric or isotonic muscular contractions. Central neurological fatigue manifests as an unexpected drop in performance that is not related to illness or injury. This form of fatigue results when a performer has insufficient rest and recovery periods. It is often described as "burnout," "staleness," or "overtraining" and is usually accompanied by mood disturbances and diminished level of achievement (Liederbach et al., 2013). Insufficient metabolic energy not only compromises motor functioning, it also reduces cognitive processing. Sensorimotor functioning falters when peripheral and central neurological feedback is diminished. Fatigue impairs both the stabilization of postural control and the dexterity of fine motor performance; injury risk increases as a result.

The physiological and biomechanical nature of fatigue consists of a complex array of interacting factors. In general, fatigue is associated with insufficient nutrients and rest to support the metabolic demands of recovery. This energy input/output equation is difficult for many performing artists, especially if they are performing multiple shows per week and if their art form requires more lean body mass (Murgia, 2013b). Many performers are undernourished and sleep deprived—physiological factors that intensify the deleterious effects of fatigue. Repeated physical activity, lack of rest, and deficiencies in calcium and vitamin D impair bone remodeling (Raisz, 2005). This pattern is intensified in eating disordered performers, often described as the "female athlete triad"; they develop osteoporosis in response to malnutrition coupled with decreased estrogen levels (Prisk, O'Loughlin, & Kennedy, 2008). Also, constant compressive force from landings induces bone fatigue. This culminates in micro and/or macro fractures, particularly when the strength of bone tissue is compromised in osteoporotic bone (Motta-Valencia, 2006). Bone remodeling and resorption requires substantial recovery time; e.g., when activity levels change drastically bone adaptation may take approximately 3–9 months (Murgia, 2013a). Similar healing timeframes are required to regenerate tendons (sturdy, stiff collagen tissue that connects muscle to bone), cartilage (covers and protects bony surfaces, acts as a shock absorber), ligaments (flexible but non-elastic collagen tissue that joins bone to bone), and muscle tissue (Hodgkins, Kennedy, & O'Loughlin, 2008; Murgia, 2013a). Repetitive strain and fatigue increases muscle soreness, inflammation, and pain. Fatigue in muscle tissue is related to insufficient ATP (energy), depletion of acetylcholine, phosphagen, and glycogen, and accumulation of lactate and other substrates such as hydrogen, ammonia, and phosphates (Kelly, 1996).

Insufficient rest and nutrition, coupled with repetitive strain, lead to deterioration of joint structures, which contributes to degenerative joint disease. In particular, cartilage tissue frays, tendons and ligaments become inflamed or rupture, and muscle fibers tear (Murgia, 2013a). In most performing arts injury rate studies, injuries tend to occur after performers have been working more than five hours per day and/or later in the performing season (Bird, 2009, 2013; Liederbach et al., 2013). This injury pattern reflects accumulated fatigue; it is also related to increased injury severity such as fractures or ruptured ligamentous or tendon tissue (Liederbach, Dilgen, & Rose, 2008; Murgia, 2013a). The combination of increased fatigue and intrinsic/extrinsic factors profoundly heightens injury rates, a reality that exists for actors, singers, musicians, and dancers (Heming, 2004; Ledoux et al., 2009; Liederbach et al., 2013).

Unique injury profiles in performing artists

All performers are vulnerable to overuse injuries, and many also experience acute traumatic injuries. Sadly, there is a pervasive cultural belief regarding injury that is negative and shame-based. Performers tend to view injury as a sign of personal weakness; they believe injuries are caused by poor technique, laziness, or clumsiness (Rickert, Barrett, & Ackermann, 2014a). Many also view the psychosocial stress of a performance career, especially the lack of control regarding repertory choices, as a contributor to injury risk (Noh, Morris, & Andersen, 2011; Rickert, Barrett, & Ackermann, 2014b). This stressed, shame-based culture encourages performers to conceal, ignore, or underreport injuries; performers tend to perform despite pain and do not realize the long-term consequences of these choices (Askling, Lund, Saartok, & Thorstensson, 2002; Rickert et al., 2014a). The reality is that every dancer, musician, actor, and singer will likely experience at least one injury in their career (Rietveld, 2013). Research studies actually reveal that company dancers will experience at least one injury per year (Ramkumar, Farber, Arnouk, Varner, & McCulloch, 2016).

Performing arts medicine research tends to focus on biomechanical movement analysis, including deficits in muscle strength and incorrect motor patterning (Kaufman-Cohen & Ratzon, 2011). Most orthopedic specialists suggest that improper technique and poor motor control are the primary causes of injury (Roussel et al., 2013). Physical limitations also predispose performers to injuries, such as tight hip sockets inhibiting turn out in dancers, or small hand span leading to hand strain in pianists. Structural abnormalities such as spinal scoliosis or fatigue-related elevated lactic acidosis may further intensify injury severity. Even when performers have impeccable technique, fatigue and overtraining can lead to a transient collapse in technical execution; stress and unexpected accidents also contribute to injury risk, despite excellent technique (Liederbach, Kremenic, Orishimo, Pappas, & Hagins, 2014; Rietveld, 2013). Another significant factor is that expert performers have the ability to execute skills with more power, speed, and agility, but this means they must also physically absorb the force exerted during the execution of these skills. For example, professional dancers jump higher than nonprofessional dancers, which mean they must sustain greater ground impact (Peng, Chen, Kernozek, Kim, & Song, 2015). Collectively, these factors need to be considered when assessing injury risk (Fig. 19.3).

Injury vulnerability is exacerbated by insufficient sleep, a reality that most performing artists experience. Their peak time to perform is usually in the evening, which significantly delays sleep onset. Most performers must attend morning classes or other work-related activities (teaching, administrative tasks, etc.), which further compromises sleep (Adam, Brassington, Steiner, & Mattheson, 2004; Scialom, Goncalves & Padovani, 2006). Negative mood and perceived stress adds to sleep loss and increases injury risk (Adam et al., 2004; Kenny & Ackermann, 2015). For example, a stressed and sleep-deprived performer who continually overworks can develop a joint compartment syndrome injury. If the performer's behavior remains

Figure 19.3 "Resistance." Performers: Dana Fukagawa, Logan Schyvynck. Photographer: Chuck Valerio.

unchanged, the compartment syndrome will progress to muscle necrosis or microfractures; the severity of this form of injury can potentially destroy a performance career (Bird, 2013).

Young performers are at increased risk for injuries simply because of their physiological immaturity; this is even more pronounced during the adolescent years (Bowerman, Whatman, Harris, Bradshaw, & Karin, 2014). The long hours required to train a performer, even if the training is anatomically and biomechanically correct, place increased physical demands on a developing system. High levels of overuse injuries occur in the younger population; consequently educators must provide adequate recovery time, including many brief breaks (Ekegren, Quested, & Broderick, 2014). Careful attention must be directed toward vitamin deficiencies, in part, because the training demands on young performers can rapidly deplete their nutritional stores (Ducher et al., 2011). Muscular pain in multiple regions of the body is normative during the training of young artists; unfortunately, this pain persists into adulthood, often becoming musculoskeletal disorders. Young girls are more vulnerable to muscular pain, although in some studies young boys are at greater risk (Luke et al., 2002; Nawrocka, Mynarski, Powerska-Didowska, Grabara, & Garbaciak, 2014). Fractures, including stress fractures related to decreased bone mineral content, are associated with the greatest amount of time loss during training, a pattern that is elevated in young dancers who strive for thinness (Ekegren et al., 2014). Adding to this vulnerability pattern, younger injured performers experience greater psychological distress, such as worsening somatization and depression symptoms, coupled with brooding and perseverating behaviors. Young performers lack adequate coping strategies; as a result, medical practitioners need to provide other types of care while treating the musculoskeletal injuries of young performers (Air, 2013).

Similarly, older performers, especially over 60 years of age, have increased musculoskeletal pain that limits performance or increases discomfort while performing (Abreu-Ramos & Micheo, 2007; Heredia, Hinkamp, Brodsky, & Llapur, 2014). For example, musicians and singers from the Orquesta Buena Vista Social Club describe greater weakness, numbing, swelling, cramping, and pain. They recognize the need for interventions that can prevent injury, such as warm up, more frequent breaks, postural alignment and support, and musculoskeletal care. These musicians described a lack of health care support, including health care practitioners who specialize in performing arts medicine (Heredia et al., 2014).

Injury patterns in specific performing arts domains

Actors and singers

Investigative studies have examined specific needs of most instrumentalists and dancers, as well as musicians and dancers from different genres (jazz, rock, folk, orchestra, ballet, tap, modern, hip hop, breakdancing); however, minimal studies examine specific injury patterns in actors and singers. This research pattern may be related to the reality that substantial differences in injury rate exist among these performing groups. For example, in a student sample, dancers had the greatest frequency of injuries followed by musicians, and then actors (Miller & Moa, 1998). The few studies that examine actors generally report lower limb, back, and neck problems; actors frequently share similar injuries to singers' (Ledoux et al., 2009). Vocal injury, in both actors and singers, is related to environmental conditions, such as toxins or humidity levels. Hearing loss is associated with acoustic levels in concert halls or orchestral pits (Ledoux et al., 2009). Actors also share similar injury patterns with dancers; both commonly experience lower limb injuries (Ledoux et al., 2009).

Musicians

In general, musicians suffer more neck/shoulder, and hand/wrist pain and injury. However, when discussing injury patterns in musicians it is essential to include analysis related to the instrument they play. Each instrument requires a specific body position and hand usage (Nolet, 2013); in fact, research study designs cluster performers by instrument type when discussing injury patterns. For example, marching band musicians must support their instrument while marching, a reality that directly influences injury rates (Moffit, Russ, & Mansell, 2015). The repertory to be performed also influences injury patterns among musicians, because specific genres, composers, and overall tempo provoke musculoskeletal strains. For example, pianists may be required to initiate 72 bimanual finger movements per second when playing something like Mendelssohn's piano fantasy (lasting approximately 5 minutes), whereas, string players must actively and precisely press on the strings to influence vibrational quality when playing a classical or bluegrass piece of music (Nolet, 2013). Gender is a heightened risk factor for musicians. Females tend to experience more musculoskeletal symptoms and the symptoms persist longer (Paarup, Baelum, Holm, Manniche, & Wedderkopp, 2011). Regardless of gender or instrument played, all musicians are at risk for noise-induced hearing loss. This occupational hazard is massively damaging to their careers (Raymond et al., 2012).

Dancers

Repetitive movements contribute to repetitive strain injuries (Bird, 2013). For example, dancers experience higher incidence of ankle sprains (67.6%), and in male dancers, higher incidence of muscle strains (54.5%) (Costa, Ferreira, Silva, & Felicio, 2016). The most common injury locations in ballet dancers are foot and ankle, lumbar and cervical spines, followed by hip (Ramkumar et al., 2016). Most injuries occur during rehearsals, possibly related to the fact that the majority of time is spent in rehearsal compared to performance (Scialom et al., 2006). Younger performers (under age 35) account for approximately 78% of injuries even though they represent 42% of the performing arts workforce. Injuries are disproportionately caused by the body's reaction to exertion, falls, or contact with objects or equipment. Injury hazards for many dancers are related to poor floor conditions,

Box 19.3 Managing Low-Grade Injuries

- ✓ Warm up: general full body mobilization and then domain-specific warm up (with instrument, dance style, singing genre, character demands)
- ✓ Rest periods scheduled during the day and the week
- ✓ Physical conditioning (aerobic/cardiovascular, strength, endurance, core stability, alignment)
- ✓ Include restorative practices (massage, yoga, meditation)
- ✓ Recognize pain and overtraining symptoms
- ✓ Diet and daily life practices (e.g., avoid smoking)
- ✓ Personal hygiene (skin care, sleep)
- ✓ After a long period of vacation, rest or injury, gradually recondition before resuming a full performance schedule

partnering mishaps (Wanke et al., 2014), and unsafe stages (Cahalan & O'Sullivan, 2013). Other risk factors are related to specific dance form requirements, although the pattern of injury sites remains the same (lower extremities, hips, back, shoulders) (Noon, Hoch, McNamara, & Schimke, 2010; Wanke et al., 2014). This intrinsic and extrinsic injury pattern is evident in rehearsals and on stage, although in one study, more injuries occurred during performance (79%), with 22% related to partnering and 21% related to floor surfaces (Wanke et al., 2014). Despite the fact that dancers have the highest rate of injuries, very few dancers thoroughly investigate or report the cause of their injury (Scialom et al., 2006). Perhaps this lack of personal investigation is related to the reality that dancers are a medically underserved population (Bronner, Ojofeitimi, & Spriggs, 2003; Box 19.3).

Rehabilitation practices

Performing artist medicine is a growing field of specialization (Bird, 2009). Most performing arts medical specialists draw upon the established practices in sport medicine; many of the treatment approaches are identical to the needs of athletes (Dick et al., 2013). Medical specialists in multiple fields such as orthopedics, chiropractic, physical therapy, and athletic training directly attend to the specific needs of performers, with the goal of rehabilitation and return to active career performance. Treatment recommendations begin with traditional conservative methods such as ice/heat, compression, elevation, rest, and nonsteroidal antiinflammatory drugs. In more severe injuries, physical therapy modalities are introduced such as ultrasound, electrical stimulation therapies, prolonged stretching, trigger point massage, and myofascial massage, as well as therapeutic exercises that are prescribed by trained practitioners. Severe injuries may require surgical repair followed by physical therapy treatment (Miller, 2006).

Screening performers prior to entering intensive training programs may identify musculoskeletal imbalances, diet and sleep health status, and psychosocial stressors. A screening process when performers are young can also address anatomical anomalies such as scoliosis and joint hypermobility. Early detection can reduce the progression of such disorders and maintenance programs can be designed to decrease potential injuries (Steinberg, Hershkovitz, et al., 2013; Steinberg, Siev-Ner, et al., 2013). Early determination of anatomical injury risk factors can influence the construction of individualized preventative programs (Gamboa, Roberts, Maring, & Fergus, 2008).

Ensuring physical fitness can decrease injury rates; if a performer sustains a traumatic injury, prior physical fitness decreases the time lost during the recovery period. Aerobic fitness along with lower proportions of body fat significantly decrease injury incidence (Twitchett et al., 2010). Adding aerobic, endurance, strength, and motor control exercises to regular training for dancers (as well as musicians, singers, and actors) increases physical fitness, decreases musculoskeletal injuries, and yet, does not alter aesthetic expression. Such programs are ideal approaches to prevent many mild-to-moderate injuries (Mistiaen et al., 2012). Supportive training programs can include Pilates, yoga, Alexander Technique, or traditional cross-training fitness programs (Wilke, Priebus, Biallas, & Frobose, 2011). Even working with exercise DVDs can decrease musculoskeletal disorders (Chan, Driscoll, & Ackermann, 2014). All have positive benefits for performers (Kava, Larson, Stiller, & Maher, 2010; Rietveld et al., 2013).

Injury prevention can fall into three categories: (1) implementing intervention protocols prior to an injury, (2) preventing an injury from getting worse, or (3) learning from an injury for future optimal training and performance situations (Rietveld et al., 2013). Prior to an injury performers can implement the following practices: (1) walking the stage and checking floor surfaces, as well as the security and condition of props and sets; (2) accurate assessment of personal skills required to meet the demands of the performance; (3) staying fit and properly warming up and cooling down; (4) proper nutrition and adequate sleep; (5) proper selection of footwear (including orthotics if necessary), instrument sizing and/or including ergonomically designed supportive devices; and (6) seeking preventive treatments such as massage, acupuncture, chiropractic, Alexander Technique, or other somatic practices (Ledoux et al., 2009; Rietveld et al., 2013; Russell, 2013). During formative training, performers endorse courses in injury and injury prevention as invaluable in their education. These courses are both educational and practical. Performers gain greater awareness and tend to respond with more self-efficacy; they actively assume responsibility for assessing and correcting alignment and technical accuracy (Fuhrmann, Brayer, Andrus, & McIntosh, 2010; Fig. 19.4).

Understanding the pattern of injury risk can help avert injuries. For example, performers can anticipate that injuries may occur as a result of changing teachers, instruments, studios, theaters, or cities, as well as fatigue during practicing, training, and rehearsals. Providing brief breaks near the end of a rehearsal session or including rest periods within a long intensive program can offset the effects of fatigue (Liederbach et al., 2013). Implementing strategies to manage changes in routine can address performer vulnerability during these stressful periods

Figure 19.4 "Strain." Performers: Janice Dulak, Hugh Murphy. Photographer: Cylla von Tiedemann.

(Rietveld et al., 2013). Performing arts is similar to the long seasons of professional sports programming. In sports training, periodization is implemented to reduce fatigue-related injuries. Periodization requires alternating periods of training load and recovery periods to avoid overuse and excessive fatigue (Smith, 2003). Although this is not formally incorporated into performing arts programs, periodization can be built into class lessons or rehearsal sessions. This approach may improve career longevity and optimize performance.

Direct information about compensatory patterns and how to relearn biomechanically correct motor patterning is a highly effective approach for injured performers (Rietveld, 2013). Compliance rates are higher when performers in rehabilitation programs are given corrective exercises that will enhance their performance skills (Shah et al., 2012). Ideally, this should be done under the supervision of trained performing arts medical specialists. For example, physical therapists and movement specialists can train dynamic stabilization in performers. Performers require core stability that is developed via increased strength in trunk flexor and extensor muscles; they must also engage stabilization while dynamically moving as they perform. Through specific sequencing of exercises an optimal neuromuscular kinetic chain can be nurtured. One approach is dynamic neuromuscular stabilization, a holistic treatment approach that increases efficient central nervous system control over movements that require both flexibility and stability. The goal is to achieve dynamic coordination over antagonistic and synergistic neuromuscular motor patterns in the core, while supporting flexible movement in the limbs (Frank, Kobesova, & Kolar, 2013). It is important to note that postural stability is often poorer in injured performers; whether this is a result of being injured or if it is a predisposing factor for injury is unknown. Regardless, attention to postural stability in injured performers is critical; e.g., enhancing sensorimotor feedback during medial-lateral directional balance training helps offset this imbalance (Lin, Lee, Liao, Wu, & Su, 2011). For musicians, attention directed toward trunk stabilization and muscular endurance offsets the demands of maintaining postural support for sustained periods of time. For all performers with back injuries, a similar stabilization and muscular endurance approach, along with movement reeducation, leads to a pain-free return to performance (Hagins, 2011). The incorporation of endurance, stabilization, and performance-specific exercises is vital for all performing artists (Kava et al., 2010).

Many injured performers tend to not "stop," even when medically advised (Rietveld et al., 2013). Given this reality, performers require training in nonmovement activities such as visualization (mental practice) to help maintain technical skills. Mental imagery is a process of internally visualizing and/or kinesthetically imagining the execution of a movement. This enables motor practice without straining the injured region (Milton et al., 2008; Nordin-Bates et al., 2011). Injured performers derive great hope from study findings that demonstrate that accurate biomechanical and kinesthetic mental practice enhances movement efficiency (Couillandre, Lewton-Brain, & Portero, 2008). Effective mental imagery practice can also improve cognitive and motivational functions, as well as physiological regulation. Mental practice (a.k.a mental imagery, visualization) is further strengthened when an audio recording of relaxation music plus a voice-enhanced audio narration is played during the mental practice session (Karageorghis, Smith, & Priest, 2012).

Broad-based coping skills can be integrated during the healing process and can be maintained posthealing to prevent future injuries. This psychologically rooted intervention protocol consists of autogenic/self-hypnosis (focus on breathing, slowing heart rate, and feeling warmth/coolness while in a state of self-induced relaxation), imagery (visualizing successfully managing pain, stress, technical demands, etc.), and self-talk (challenging negative thoughts and redirecting thoughts toward optimal thoughts and behaviors) (Noh, Morris, & Andersen, 2007). Helping performers develop a daily practice of "body mapping," an internally directed somatic meditation that focuses on sensory–motor integration, can promote technical and expressive performances while allowing performers to assess muscle tension or compensatory motor patterning (Thompson, Ruthmann, Anttila, & Doan, 2014). These psychological skills are trained concurrently with traditional rehabilitation programs that address musculoskeletal pain and misuse.

The relationship between a performer and health care practitioner is a powerful component in optimal healing. For example, when occupational stress is low for physical therapists they are able to individualize treatment for their injured performers while demonstrating genuine care during the healing process. However, when working in stressful environments, they tend to distance from their patients and are more irritable and aggressive in their interactions, which inevitably limits the healing of their patients (Alesli, Damiani, & Pernice, 2005). Medical practitioners' distancing strategies are often subtle and manifest in nonverbal facial expression (less smiling and more frowning). Although seemingly unimportant, there is a direct negative effect on long-term healing outcomes in their patients (Ambady, Koo, Rosenthal, & Winograd, 2002). Also, medical practitioners who value their own health care and wellbeing and who openly interact with their patients positively affect long-term healing outcomes in their patients (Alesli et al., 2005; Sabo, 2013). To further facilitate optimal rehabilitation, practitioners must understand the specific performance needs of their patients. Understanding the terminology used by dancers, musicians, singers, and actors promotes more effective communication and facilitates a stronger working relationship between the performer and the practitioner. When these factors are present, performers tend to be highly motivated, respond positively to ongoing evaluation, and willingly implement healing strategies (Sabo, 2013).

Summary

- Introduction: definitions, deliberate training and compensation, high injury rates.
- Injury and the effects on creativity: positive and negative influences of injury, transform adversity, isolation and job security.
- Injury safety policies and workers' compensation: union negotiations, underreporting, reduction of injury, fatigue.
- Injury prevalence and severity: statistical reports, Grades I–V injuries.
- Pain appraisal and coping: catastrophic or minimizing pain, strategies to manage pain.
- Hypermobility conditions and injury vulnerability: hypermobility syndrome and injury risk, rehabilitation efforts.
- Overuse, overtraining, and fatigue: definitions, causes of fatigue, central and peripheral fatigue, nutritional factors, repetitive strain, rest and recovery.
- Unique injury profiles in performing artists: general causes, young and old performers, gender and injury.
- Injury patterns in specific performing arts domains: actors, singers, musicians, dancers.
- Rehabilitation practices: screening, fitness levels, injury prevention, compensatory patterns, physical therapy, imagery, broad-based coping skills, relationships between performers and health care practitioners.

Concluding remarks

Most performing artists suffer some form of musculoskeletal pain, injury, and/or disability during their career. The field of performing arts medicine is growing (Barker et al., 2009), but it still lags far behind the field of sports medicine (Bird & Macdonald, 2013). Because of the complex needs of performing artists, and the reality that they are an underserved population, a team of health care professionals is ideally suited to address the complex biopsychosocial requirements of injured artists (Bird & Macdonald, 2013; Dersh et al., 2002). Rehabilitation should include therapeutic exercises that are performed with proper biomechanical motor functioning. This is essential for performers who suffer sudden onset traumatic injuries, as well as common injuries associated with overuse (Heming, 2004; Wilke et al., 2011). Medical specialists must fully understand the technical demands of specific performance domains so that they can target specific movement exercises designed to strengthen weakened areas, repattern inefficient motor movement, and increase performer confidence while executing movement efficiently (Costa et al., 2016; de Greef, van Wijck, Reynders, Tousaint, & Hesseling, 2003).

A sense of self-identity is deeply entwined with talent and career. When performers are injured their identity is often equally injured. While healing from an injury, performers require psychological support in order to maintain self-efficacy and self-esteem (Wainwright, Williams, & Turner, 2005). Performers are keenly aware of the psychosocial factors that potentiate injury risk. As professionals, they learn to acknowledge and manage performance stress, the complications of interpersonal relationships, the hierarchical organizational structure of most companies, and lack of control over scheduling and programming decisions (Rickert et al., 2014b). They are also aware of the ever-changing demands of composers, writers, and choreographers, as well as the audience that has grown to expect performers who will push the physical limits of their body. Despite risk of injury, performers tend to dissociate their subjective embodiment; as artists they extend the boundaries of materiality while giving life to ever more demanding expressive feats (Aalten, 2007).

Most performers believe that pain and injury are an integral part of being a performer; it shapes their sense of identity and infiltrates how they subjectively experience flow states (Turner & Wainwright, 2003). However, once they are seriously injured, many performers avoid flow experiences, fearing that a lack of awareness during flow states may promote a lack of awareness of their body. They use strategies to disrupt flow in an effort to continue their career and avoid subsequent injuries (Guptill, 2012). Helping performers learn to trust their healing recovery may allow them to once again experience the joy of flow-like performances. Health care professionals must understand these underlying factors in injured performers. They are vulnerable and require care that is specifically directed toward the complex requirements of being a professional performer (Guptill, 2011).

It is recommended that future research studies address how injuries are determined and diagnosed. Because of the physical nature of all performing arts, the degree of injury (classification level) should be standardized across performing arts medicine (Hincapie, Morton, & Cassidy, 2008; Liederbach, Hagins, Gamboa, & Welsh, 2012). Sharing standardized criteria with student performers during their formative training will encourage greater awareness of injury types, causes, and prevention; ideally, it will cultivate a culture of performer wellness (Wilson & Deckert, 2009). Despite this lack of standardization, consistent findings demonstrate that injury risk is associated with fatigue, poor coping strategies under stress, and previous history of injury with the resultant strain on the contralateral limb (Jacobs, Hincapie, & Cassidy, 2012). Performers are generally impatient with healing and rehabilitation. The sociopolitical climate in the performing arts generally discourages actively seeking professional help for musculoskeletal injuries, mental health disorders, and low health conditions (Raymond et al., 2012). Performers want to practice, rehearse, and perform without hindrance (Scialom et al., 2006). Most of the time they are able to perform, despite minor musculoskeletal symptoms; only when symptoms become chronic or moderate to severe is performance limited (Paarup, Baelum, Manniche, Holm, & Wedderkopp, 2012). Encouraging performers to reveal injury status will ensure implementation of greater personal, organizational, and legislative practices and policies (Foxman & Brugel, 2006). Hopefully these practices will decrease injury rates and increase the care of performers.

Health concerns and burnout

20

Medical science has proven time and again that when the resources are provided, great progress in the treatment, cure, and prevention of disease can occur.

Michael J. Fox

Introduction

Maintaining a healthy body ensures the ability to perform. Performers are acutely aware of the dynamic nature of health and wellbeing; they must sustain a delicate relationship between their body and their art form (Schoeb & Zosso, 2012). Severe illnesses can provoke changes in career focus (Zausner, 1998); for example, performers may transition from performing careers and become writers, composers, or choreographers. Other performers may gain training and work as a member of a medical team specializing in treating performing artists (massage therapist, Pilates instructor, physical therapist, nutritionist, Alexander teacher, medical doctor). Crippling diseases may lead to permanent disability, a possibility that many performers fear. For some disabled performers, they continue to perform with modification. Others shift their focus and become researchers in the field of disability or work as political activists for the disabled (Lubet, 2002). To become an expert in the performing arts requires years of training and experience. The sad reality is that performance-related diseases are not always transformative. The performing arts suffer when talented performers' lives are cut short (Adams, 2009; Fig. 20.1).

The etymological meaning of *disease* is without (*dis*) ease (rest, peace, comfort). The earlier term, illness, has an etymological meaning that includes evil or bad moral quality (see www.etymonline.com). Because the term *illness* contains negative connotations, disease became the primary term. Today, disease generally includes a convergence of factors that are often cumulative in nature, resulting in a disease process. Ultimately, aberrant functioning in an organ, structure, or system is based on genetic or developmental errors, nutritional deficiency, toxicity, poisons, or infections. Risk factors for expressing specific diseases must preexist within the individual (Izaks & Westendorp, 2003). Disease is fundamentally an indication that the body, or body part, is no longer working effectively.

Although the frequency of musculoskeletal disorders is high in performing artists, other nonmusculoskeletal conditions are equally troubling. For example, in a sample of popular musicians, 50% of the musicians had a nonmusculoskeletal disorder that affected their ability to perform, with 42% reporting noise-related

Creativity and the Performing Artist. DOI: http://dx.doi.org/10.1016/B978-0-12-804051-5.00020-2

Figure 20.1 "Able and disabled." Performers: Allen Chung, Crystal Duran. Photographer: Lee Choo.

hearing loss (Raeburn, Hipple, Delaney, & Chesky, 2003). Like performers with musculoskeletal injuries, performers who grapple with illness frequently continue to work and are often unwilling to seek medical help; this lack of treatment decreases quality of life and exacerbates depression and pain sensitivity (Raymond et al., 2012). Unlike the high distribution of ill professional adult performers, in a young student sample, most performers remained healthy, and if they acquired an illness it usually resolved quickly. However, approximately 10% of student performers suffered severe health issues. This group of students also demonstrated very poor coping skills; they were not well equipped to manage their significant health concerns (Spahn, Nusseck, & Zander, 2014). These young performers are at high risk for long-term health complications that require significant medical care. They also need psychoeducational skill training, in order to effectively cope with the collective demands of their health and performing career (Spahn et al., 2014).

At some point, health concerns play a role in every individual's life. In this chapter, specific performance-related disorders will be discussed. A detailed account of each disorder is beyond the scope of this book or chapter. The goal of this chapter is to highlight some of the health problems that may befall performers, along with general suggestions to manage these concerns. Seeking medical attention is critical, especially from medical practitioners who are well informed about the professional demands of the performing arts, and the unique health concerns that manifest in the performers (Dommerholt, 2009). Promoting knowledge in this field not only enhances quality of life for performers but it can also cultivate training environments that address performer wellbeing (Clark, Gupta, & Ho, 2014; Clark, Williamon, & Redding, 2013; Hamilton & Robson, 2006; Hays, 2002; Zander, Voltmer, & Spahn, 2010).

Hypermobility condition, dysautonomia, and health concerns

Joint hypermobility syndrome is common in dancers, although it can be found in approximately 10–30% of the general population (Day, Koutedakis, & Wyon, 2011; Hakim & Sahota, 2006). It is strongly associated with stress-sensitive medical disorders, such as irritable bowel, fibromyalgia, chronic fatigue, and chronic pain (Eccles et al., 2012; Scheper et al., 2013). Dysregulation in the autonomic nervous system is also associated with joint hypermobility syndrome. The most common autonomic abnormalities include mitral valve prolapse (palpitations, atypical chest pain), postural orthostatic intolerance syndrome or postural tachycardia syndrome, which are disorders marked by difficulty regulating blood pressure when changing positions (Hakim & Sahota, 2006; Mallorqui-Bague et al., 2014). Many performers comment on feeling dizzy, light headed, or faint (presyncope or syncope) when suddenly changing levels such as moving from sitting to standing. They also experience more cardiovascular reactivity during activities such as increased heart rate or palpitations (Eccles et al., 2012). As well as the associated joint hypermobility phenomenon of weakened vascular tissue, these cardiovascular impairments are likely related to neuronal abnormalities that compromise vasoregulation (Hakim & Grahame, 2004). Other symptoms include increased joint pain, nausea, migraine, allergies, rashes, night sweats, fevers, poor sleep, and lymph gland pain (Hakim & Grahame, 2004). Collectively, these symptoms suggest dysfunction in the autonomic nervous system (dysautonomia), rather than disorders within a single organ or system (Hakim & Grahame, 2004; Hakim & Sahota, 2006).

Hypermobility is directly associated with alterations in connective tissue (collagen, elastin, fibrillin, and extracellular matrix proteins). Several severe pathologies are associated with hypermobility, including Marfan syndrome, Ehlers–Danlos syndrome, osteogenesis imperfect, and Stickler's syndrome (Hakim & Sahota, 2006). Ruptures in connective tissue within the heart and colon are profoundly life threatening. Less life-threatening connective tissue disorders affect the dermis layer of skin. This layer of skin not only provides a physical barrier between the external world and internal environment of the body, it is also involved in sensory perception and thermoregulation (Hakim & Sahota, 2006). The associated skin symptoms related to joint hypermobility include skin laxity and increased bruising, stretch marks (striae), and scarring. Increased early onset osteoarthritis, dislocations, scoliosis, dental, ocular, and headaches are also linked with joint hypermobility condition.

Hypermobility conditions may be due to different brain volumes in the amygdala and insula, regions that influence pain responses and physiological monitoring (Eccles et al., 2012; Mallorqui-Bague et al., 2014). As well, brain scans reveal differences in inferior parietal regions; thus proprioceptive awareness is also compromised, a factor that may increase injury frequency. Other neural regions related to joint hypermobility include alterations in the brainstem, caudate nucleus, thalamus, anterior cingulate, inferolateral prefrontal cortex, and inferotemporal cortex—regions implicated in autonomic nervous system regulation, body mapping, and

Figure 20.2 "Supportive hypermobility." Performers: Elissa Brock, Cheyenne Spencer, Christian Vidaure. Photographer: Lee Choo.

interoception awareness (Mallorqui-Bague et al., 2014). Neuroendocrine alterations also contribute to joint hypermobility, including progesterone-related disorders such as polycystic ovary syndrome and endometriosis (Foley & Bird, 2013). This cluster of variant brain regions, along with a complex grouping of symptoms and disorders, suggests a neurovisceral phenotype that underlies the joint hypermobility syndrome. Further research is necessary to confirm this phenotype. Joint hypermobility is ultimately a disorder that manifests in multiple regions; it is a disorder that involves neurological, psychological, and physiological symptoms (Eccles et al., 2012; Fig. 20.2).

Despite the substantial health risks related to joint hypermobility syndrome, this disorder may facilitate increased success in dancers, in part because of the extreme flexibility required in dance. It is important to remember that most dancers achieve their flexibility from training and not because they have a hypermobile syndrome (Day et al., 2011). If dancers maintain adequate fitness, nutrition, and effectively regulate stress, they are less likely to sustain injuries or acquire illness, compared to their hypermobile colleagues who do not follow this regime. Because proprioception is compromised by joint laxity, they will also need to increase kinesthetic awareness to optimize sensorimotor feedback. Carefully selecting an appropriate dance style and teachers who are well informed about joint hypermobility will help dancers achieve stabilization, strength, and endurance required to perform safely (Foley & Bird, 2013). Actors, singers, and musicians may also have joint hypermobility, although their hypermobility may be less advantageous to their careers. They also require a similar health care regime as hypermobile dancers to manage the myriad symptoms associated with this syndrome. Performing arts medical specialists must understand the many factors that contribute to this condition; hypermobile performers require specialized attention and support during training and this need will continue throughout their career (Foley & Bird, 2013; Boxes 20.1 and 20.2).

Box 20.1 Joint Hypermobility Indicators (Hakim & Sahota, 2006, p. 523)

A 5-part questionnaire for identifying hypermobility:

1. Can you now (or could you ever) place your hands flat on the floor without bending your knees?
2. Can you now (or could you ever) bend your thumb to touch your forearm?
3. As a child, did you amuse your friends by contorting your body into strange shapes or could you do the splits?
4. As a child or teenager, did your shoulder or kneecap dislocate on more than one occasion?
5. Do you consider yourself double jointed?

Answer in the affirmative to two or more questions suggests hypermobility with sensitivity 80–85% and specificity 80–90%.

Box 20.2 Maladies Associated With Joint Hypermobility (Eccles, Owens, Mathias, Umeda, & Critchley, 2015)

- ✓ Anxiety
- ✓ Asthma
- ✓ Carpel tunnel syndrome
- ✓ Chronic constipation
- ✓ Chronic regional pain syndrome
- ✓ Crohn's disease
- ✓ Dysautonomia
- ✓ Fecal incontinence
- ✓ Fibromyalgia
- ✓ Functional gastrointestinal disorder
- ✓ Headaches
- ✓ Hiatal hernia
- ✓ Mitral valve prolapse
- ✓ Migraine
- ✓ Pelvic organic prolapse
- ✓ Postural tachycardia syndrome
- ✓ Psychological distress (depersonalization disorder)
- ✓ Rectal evacuatory dysfunction
- ✓ Somatosensory amplification
- ✓ Urinary stress incontinence
- ✓ Vasovagal syncope

Cumulative wear and tear, hearing loss, and health problems

Most performing artists do not manifest any health-related illness until at least 10 years after they begin seriously training as a performer (Ledoux et al., 2009).

Repetitive musculoskeletal overuse propels performers toward a path of gradual onset pain and musculoskeletal disorders. Over time, performers begin to reevaluate the physical (and vocal) demands of a role, dance, or musical repertory. These chronic overuse states increase psychological distress, especially anxiety, depression, substance use, eating disorders, and psychosomatic disorders (Dersh et al., 2002; Ledoux, 2009). Chronic pain, associated with musculoskeletal disorders, initiates an anxiety-pain feedback loop; the other psychiatric disorders frequently follow the onset of this feedback loop. For example, eating disorders and substance use may serve as attempts to alleviate pain. Depression, psychosomatic disorders, and anxiety may reflect feelings of helplessness related to chronic pain (Dersh et al., 2002; Kenny & Ackermann, 2015; Kenny, Driscoll, & Ackermann, 2014).

Other cumulative wear and tear conditions experienced by performers, especially musicians, include headaches, eyestrain, and hearing loss (Dick et al., 2013; Fjellman-Wiklund & Chesky, 2006). Performing artists typically experience *noise-induced hearing loss* due to continuous or intermittent noise exposure. This disorder affects the cochlear hair cells in the inner ear, and if not addressed immediately, will lead to permanent hearing loss. Most musicians and singers work in environments where the decibel level exceeds occupational recommended health standards (Dick et al., 2013). Other hearing disorders associated with performance distress include tinnitus (neurological inability to inhibit inner ear sound), hyperacusis (hypersensitivity to sound levels), distortion (music perceived as out of tune), and displacusis (inability to inhibit pitch and frequency matching) (Kahari, Zachau, Eklof, & Moller, 2004; Kahari, Zachau, Eklof, Sandsjo, & Moller, 2003). Musicians in general have greater incidence of hearing disorders compared to the general population, although jazz/rock musicians have more hearing disorders compared to classical musicians (Kahari et al., 2004). Generally, female musicians tend to have better hearing, but when they suffer hearing problems, they tend to have increased rates of hyperacusis and tinnitus; whereas, male musicians have more hearing loss and tinnitus (Kahari et al., 2003). Hearing disorders are intensely problematic for musicians; they are dependent on their ability to discriminate sounds and produce accuracy in pitch, and yet constant sound exposure frequently compromises auditory ability (Schuele & Lederman, 2004; Walter, 2009). When hearing loss and noise-induced hearing disorders become severe, they are career ending for musicians (Zuskin et al., 2005).

Scoliosis, dystonia, and neurological disorders

Scoliosis is a disorder of the spine; it curves in either an S or C shape rather than in a straight line; sometimes the spine can also corkscrew or twist (Bird & Pinto, 2013). Although mild scoliosis is usually asymptomatic, because of the increased physical demands of performance, scoliosis in performing artists frequently becomes symptomatic and requires medical attention (Bird & Pinto, 2013). Ideally, scoliosis should be identified early in development; training and medical attention can then be directed toward keeping scoliosis asymptomatic. Performing artists tend to compensate for their scoliosis, such as by tilting or twisting the hips and/or shoulders to accommodate both the curvature of the spine and to meet the physical demands of performance. These compensations often exacerbate musculoskeletal injuries, and with severe scoliosis, respiratory function is also affected. Careful monitoring and specific physical therapy regimes can address compensatory patterns and provide the performer added strength and stability (Bird & Pinto, 2013).

Focal dystonia (TFD) is a condition that involves small muscle cramping and weakness leading to an inability to play music. Despite an inability to perform, there is no pain involved and daily functioning is often not disrupted (Rietveld et al., 2013). It is believed that TFD is a neurological disorder (Rietveld & Leijnse, 2013; Ruiz, Strubing, Jabusch, & Altenmuller, 2011); however, sometimes it is caused by structural anatomical limitations (Rietveld et al., 2013), limitations that may be genetically based (Schmidt et al., 2011). Generally, focal task-specific dystonia is related to occupational hazards, due to excess repetition of highly skilled movements, for example pianists and string players commonly develop TFD of the hand, singers develop dystonia in the vocal cords, and brass- and reed-playing musicians develop it in the facial musculature. The early symptoms of dystonia often evoke compensatory motor patterns, often occurring outside the awareness of the performer (Leijnse, Hallett, & Sonneveld, 2015). Gradually, overcompensation leads to more severe entrapment neuropathies, such as carpal tunnel syndrome, ulnar neuropathy, or thoracic outlet syndrome; literally the nerves are compressed (Sakai, 2002).

Entrapment neuropathies and severe TFD are very painful and severely restrict function; they require significant rest to recover. Sometimes they lead to long-term disability and career termination (Schuele & Lederman, 2004). Two contributing factors to TFD are anxiety and perfectionism, psychological components that compromise accurate neurovisceral awareness of fatigue or discomfort (Jabusch, Muller, & Altenmuller, 2004). A combination of chronic pain, chronic anxiety, and TFD leads to degradations in sensorimotor representations in the cortex. Consequently, a loss of sensorimotor cortical organization alters feedback loops between subcortical limbic and basal ganglia structures that regulate movement and emotion, thus intensifying anxiety, pain, and despair (Jabusch et al., 2004; Jankovic & Ashoori, 2008). These interconnected disorders have led researchers to speculate that TFD and anxiety may actually share common pathophysiological mechanisms (Enders et al., 2011; Jabusch & Altenmuller, 2004).

The most important element in treating TFD is to receive an accurate diagnosis. Many physicians are unfamiliar with this disorder in performing artists, although it is a relatively common phenomenon, especially in musicians (Rietveld & Leijnse, 2013). Recommendations to standardize an evaluative method to determine TFD include subjective (patient history and self-reports), objective (physical examination), and automated (EMG) assessments (Spector & Brandfonbrener, 2007). Once diagnosed, TFD treatment modalities include medication, biofeedback, relaxation techniques, and relearning new motor pathways to circumvent faltering neural signals (Rietveld & Leijnse, 2013). Changing equipment, Botox injections, splinting, and surgery are sometimes implemented when the less-invasive protocols fail (Dick et al., 2013). Psychological correlates during treatment should include reducing the dysregulating effects of perfectionism and anxiety, as well as dispelling perceptions that TFDs and musculoskeletal pain are associated with serious illness (Kok, Vlieland, Fiocco, Kaptein, Nelissen, 2013).

Other *neurological motor disorders* that affect performers include alterations in sensorimotor timing, muscular force, excessive involuntary movements, or an inability to perform coordinated movements (ataxia and apraxia) (Jankovic & Ashoori, 2008). Some motor disorders may result from a stroke, either hemorrhagic (bleeding in the brain) or ischemic (blood clot). Strokes are the fifth leading cause of death and the leading cause of adult disability (Thom et al., 2006). Strokes may compromise sensory, motor, or sensorimotor signaling, which alters the execution of intricate and specific movement. Efficient neurological organization of movement is required to speak, dance, or play an instrument. Other neurological motor disorders, such as tremors, Tourette syndrome, and Parkinsonism, can severally alter performance. Even though these disorders may be associated with enhanced creativity, they decrease performance (Acosta, 2014; Jankovic & Ashoori, 2008). Medical diagnosis and treatment are essential for these more severe neurological motor disorders. A performance arts medical team must coordinate an integrated multipronged treatment approach to help neurologically impaired performers maintain a career in the arts, although many performers will be forced to terminate their careers. Certainly, incorporating music, dance, and singing in the healing process has powerful effects (Hackney et al., 2007; Jankovic & Ashoori, 2008); however, these expressive arts healing modalities do not resemble the rigorous demands required of performers.

Respiratory and cardiovascular disease

Respiratory problems affect all performers. For example, singers and musicians learn to glottal stop to produce specific sounds; they are highly trained to produce these effects, but it can also increase respiratory problems. Further studies need to be conducted to determine if this vocal training is associated with a higher occurrence of asthma in these two populations (Zuskin et al., 2005). Occupational hazards that directly influence respiratory disorders are commonly found in musicians who play wind instruments. Because of the force of air being expelled, these musicians are at high risk to rupture blood vessels in their mouth and lungs. When a rupture occurs in the lungs, the risk of death is high (Zuskin et al., 2005). Some performers acquire *chronic pulmonary disease*, a disorder that is marked by a constriction of expiratory flow, leading to compensatory breathing patterns such as hyperinflation of the lungs. Unfortunately, rapid breathing patterns reduce expiration time and alter efficient respiratory patterns. Because singing requires increased expiratory flow, it is an ideal treatment for patients with chronic pulmonary disease. For singers who acquire this disease, continuing to sing will help improve quality of life, including reducing insufficient breathing-related anxiety (Lord et al., 2010). Helping singers (and nonsingers) train efficient respiratory patterning, such as acquiring concepts based on the Alexander Technique, will greatly enhance postural support and sensory awareness, factors that promote respiratory health (Hudson, 2002).

Happily, performing artists are ranked in the lowest group for *metabolic disorders* (Davila et al., 2010). Metabolic syndrome is commonly viewed as a precursor for cardiovascular disease. It is rated as a major disorder in the general population, with one-third of the American population diagnosed with this syndrome. This condition is diagnosed based on a cluster of symptoms. They include central adiposity (excess fat in the midriff as measured by waist circumference), high levels of triglycerides, low levels of HDL cholesterol, high blood pressure, and high fasting blood glucose levels (Davila et al., 2010). Most performing artists are physically active, which is an essential component in decreasing risk for metabolic syndromes; however, some performers are obese. In a study that investigated entrained heart rate responsiveness to music genre and tempo, professional musicians with a higher body mass index had elevated heart rate responses, regardless of fitness level, tempo, and musical genre (rock, metal, western, Christian) (Vellers, Irwin, & Lightfoot, 2015). Of note, several studies suggest that performing or conducting repertory that is demanding (musical genre, rapid tempo) requires high physiological workloads (Jaque et al., 2015; Vellers et al., 2015). Consequently, performers with a high body mass index are at increased risk for metabolic-related cardiovascular disease, especially given the physical rigor and stamina required to perform.

Cardiovocal disorder is a symptom associated with cardiovascular disease (Mulpuru, Vasavada, Punukollu, & Patel, 2008). The vagal nerve that enervates the larynx also enervates the heart. When the nerve is partially impinged it can cause hoarseness, dyspnoea, increased vocal effort, and palsy. Clinical treatment requires medical specialists in both cardiology and laryngology to determine the interacting complications, most importantly to determine the respiratory complications related to cardiovocal disorders. Aneurysms of the aorta and the pulmonary artery are directly related to cardiovocal palsy. Surgery to repair aneurysms may also lead to cardiovocal hoarseness, a condition that is often irreversible. This condition is extraordinarily debilitating to singers and actors. Performers who present with vocal palsy should also be checked for cardiovascular disease and vice versa (Mulpuru et al., 2008).

Skin disorders

Specific dermatological conditions typify occupational hazards in the performing arts. These conditions are caused by long hours of playing an instrument, dancing on inconsistent floor surfaces, or performing in poorly fitting footwear (Gambichler, Boms, & Freitag, 2004). Classical and nonclassical musicians commonly experience skin conditions associated with instrument positioning, persistent practice, and strenuous performances. Dancers suffer blisters, corns, and split skin on their feet. The skin is an organ that is highly responsive to external and internal stressors. Although many skin disorders are provoked by external factors, psychological states such as anxiety and depression are often comorbid with stress-related skin disorders. These dermatological disorders are often classified as stress-related inflammatory diseases (Crnivec, 2004). In fact, the neuroendocrine immune system directly influences skin receptivity. Severe skin disorders, caused by external or internal factors or an interaction of the two, include abrasions, allergic reactions, and autoimmune conditions such as plaque psoriasis (inflammatory skin condition). Other common stress-related skin disorders are pruritus (unpleasant sensation activating a desire to scratch), lichen planus (a rash that is triggered directly from the immune system), urticarial (hives), hyperhidrosis (excessive sweating), or atopic dermatitis (eczema) (Onder et al., 2000). Although most skin disorders are highly treatable, they significantly constrict performance and are extremely uncomfortable. For example, musicians rely on intricate feedback from contact surfaces in the skin; without this tactile information, creative expression is compromised, and worse, the ability to play may be temporarily suspended (Zuskin et al., 2005).

Performers who seek treatment should establish a trusting relationship with a dermatologist. Currently there are concerns regarding unethical practices. The promotion of cosmetic services during in-office visits may add to the cost of a visit and increase risk for unnecessary treatments. Ethical guidelines are provided by the Academy of Dermatology; however, coercive practices with patients are pervasive in this industry (Baumann, 2012). Educating performing artists about these ethical conflicts is beneficial. Most performing arts communities are tight-knit; performers are usually well informed about medical practitioners and their understanding of performer needs. Despite the ethical guidelines, some performers suffer harmful cosmetic dermatological care, and the Hollywood tabloids voraciously reveal these failures (Fig. 20.3).

Figure 20.3 "Musical integration." Performers: David Aks, Jayme Allilaw, Carol De Leon, James Dorsa, Ramon Osa. Photographer: Lee Choo.

Vocal disorders

Singers and actors acquire highly trained voices. Vocally, they give life to the characters they portray, reveal the emotional state of the character, and for singers, capture the unique tonal quality of the musical genre (Bjorkner, 2008). Long hours of practice and prolonged periods of vocal performance increase vocal fatigue, a common factor contributing to vocal pathology (Caraty & Montacie, 2014). These demands, and the powerful dependence on a fragile vocal instrument, can intensify performance anxiety and increase preoccupation with bodily sensations and functioning (Sandgren, 2002). Vocal insecurities manifest in a typical set of behaviors. For example, professional opera singers commonly test their voice throughout the day; they vocalize regularly to check the condition and availability of their voice. They are acutely aware of the judgment of others and fear rejection based on the quality of their voice (Sandgren, 2002). The vocal quality of actors is less harshly judged but they are also powerfully dependent on their vocal production.

Assessing and managing vocal disorders has led to a highly specialized medical field. Laryngologists have a thorough understanding of professional voice-users (singers, actors, broadcasters, public speakers) (Klein & Johns, 2007). Sophisticated technological evaluations aid in more precise diagnosis and treatment recommendations. Many professional singers and actors seek multiple physicians before an accurate diagnosis is provided, a phenomenon that increases their anxiety and distrust (Franco & Andrus, 2007). To determine diagnoses, medical specialists must get a detailed history (vocal demands such as new production, rehearsal schedules, vocal repertory, and presence of toxins). Physicians also need to consider interacting diagnoses that may complicate vocal problems (i.e., anxiety, sleep disorder, diet, smoking, asthma) (Franco & Andrus, 2007). Vocal problems may emerge gradually, with complaints ranging from losing vocal range, voice "breaks," increased vocal fatigue, or loss of vocal clarity. Persistent anterior neck muscle tightness may exacerbate vocal discomfort; however, it may also indicate the presence of other vocal problems. For example, one of the major causes of vocal symptoms is "silent" laryngopharyngeal reflux, a reflux disorder that does not manifest as the typical heartburn that occurs when lying down. This form of reflux occurs in upright positions and affects the upper esophageal sphincter (Franco & Andrus, 2007). It is a common disorder that is often misdiagnosed in professional voice-users.

Following any vocal disorder, professional voice-users may experience vocal fold scarring. This is challenging to treat, with only fair results, hence the need to educate singers and actors about good vocal hygiene in order to avoid vocal scarring (Franco & Andrus, 2007). Vocal hygiene and vocal health include consistent vocal practice with adequate rest between practice periods, as well as maintaining healthy diet and lifestyle behaviors. There is a diverse range of treatment recommendations for less-severe vocal problems. For example, singers and actors with laryngopharyngeal reflux should avoid singing, exercise, and lifting or bending over for 3–4 hours following a meal (Franco & Andrus, 2007). Persistent laryngitis may be related to other disorders, including patients with a history of diabetes, nutritional deficits, or patients undergoing immunosuppressive therapy. These conditions need to be factored into any vocal treatment recommendations. The sensation of persistent burning in the throat may indicate fungal laryngitis, a disorder that may be associated with more severe diseases such as malignancies (Franco & Andrus, 2007). Determining the cause of fungal laryngitis will greatly determine treatment approaches and healing outcomes. Female estrogen plays a vital role in laryngeal health. Decreased levels resulting from premenstrual changes, menopause, or medically induced menopause, can increase vocal disorders. Many female singers and actors are encouraged to use transdermal estrogen replacement therapy to support vocal production (Caruso et al., 2000). The normal changes due to aging also lead to decreased vocal efficiency, a condition that can be minimized with vocal hygiene practices (Franco & Andrus, 2007). Lastly, muscle tension can lead to dysphonia, which is a disorder that includes abnormally low-pitched speaking voice, poor control of breath stream, glottal stopping (or gaps), and excess muscle tension in the neck, tongue, and jaw. To effectively treat vocal dysphonia, it is essential to rule out organic disorders or vocal fold paresis that affect the laryngeal complex (Franco & Andrus, 2007).

Vocal emergencies for singers and actors must be immediately assessed and treated by skilled physicians; failure to accurately evaluate a vocal emergency can terminate a professional career. There are several disorders that are considered vocal emergencies. Vocal fold hemorrhage usually occurs suddenly during a strenuous vocal task. The use of antiinflammatory drugs and hormonal changes increase the risk for sudden vocal fold hemorrhages. Absolute rest and corticosteroids can resolve this disorder (Klein & Johns, 2007). Vocal fold mucosal tears (i.e., nodules, polyps, cysts) result from yelling, coughing, retching, or harmful singing (Franco & Andrus, 2007). Symptoms include hoarseness, loss of vocal range, and pain. Like vocal fold hemorrhage, complete vocal rest is prescribed (Klein & Johns, 2007). In severe cases, intracordal vocal fold cysts or lesions may require surgery; however, this is a delicate procedure and is avoided unless absolutely necessary (Burns, Stadelman-Cohen, & Zeitels, 2009). Phonotrauma leads to increased effort to produce sound; it usually results in acute edema, cysts, and fibrovascular changes and requires complete vocal rest. Upper respiratory infections can cause acute laryngitis. Vocal rest and avoiding antiinflammatory medication is advised. Allergies, laryngopharyngeal reflux, acute asthma, and functional dystonia can all compromise vocal production; if severe they are also regarded as vocal emergencies.

In large companies, laryngologists may work onsite (theater, concert hall, stadium, film set), although it is best for performers to be treated at the medical office of the physician. Vocal emergencies are stressful for the performer, the company, and the physician. The production is contingent on the vocal health of the performer; an inaccurate diagnosis and treatment can terminate professional careers for singers and actors, and yet the decision to cancel a high-profile show has significant negative consequences (Klein & Johns, 2007). These are difficult situations for all involved. Given these high-risk situations, vocal hygiene is critical, especially for performers who have higher vocal demands (i.e., pop singers, repertory that is poorly suited to the performer). Hydration and vocal rest, along with ongoing vocal coaching to ensure efficient physical execution, can alleviate many vocal disorders, including high-pressure vocal emergencies (Roy et al., 2000; Timmermans et al., 2005).

Genitourinary conditions

Urinary incontinence is associated with involuntary leakage of urine, usually caused by intraabdominal pressure such as coughing, sneezing, lifting heavy loads, or jumping. The loss of bladder–sphincter balance occurs when the forces of expulsion are greater than retention. Urinary incontinence sometimes triggers urinary urgency, a condition that is more common in women; the initial involuntary loss of some urine is accompanied by a sudden and uncontrollable urge to urinate (Bo & Borgen, 2001). Anatomical predispositions increase the likelihood of this condition, such as a thinner pelvic diaphragm, less muscular activity in the elevator muscles of the anus (levator ani), and a greater proportion of the urethra extending below the pubis. Sometimes urinary incontinence is associated with failed retention; however, it is more frequently related to increased expulsion forces (Wallace, 1994). Excess flexibility in the arch of the foot further compounds the effects of expulsion forces simply because a highly flexible foot is less able to absorb the landing forces generated during jumping or lifting (Pozo-Municio, 2007).

Urinary incontinence poses problems socially and hygienically and it affects men and women equally. Dancers are at high risk for urinary incontinence, with approximately 70% of dancers experiencing this condition and 30% claiming it is significantly problematic (Pozo-Municio, 2007). Dancers also suffer from urinary incontinence at a much higher rate than the general population, due to the combination of greater foot flexibility and increased forces when landing from jumps. This disorder is more pronounced in ballet dancers and flamenco dancers; it is directly linked to repetitive jumps and vibrations from foot tapping (Pozo-Municio, 2007). Despite this high frequency, this health condition is seldom investigated in other performing artists. Although no empirical evidence exists to date, musicians, singers, and actors most likely share a similar profile as dancers. They must also exert greater intraabdominal pressure when playing many musical instruments or singing demanding vocal repertory. Ways to manage urinary incontinence include increasing the muscular strength of the foot as well as muscles of the pelvic floor. These pelvic floor muscles need to be specifically trained to handle the exertion forces demanded in performances. The ability to contract the pelvic floor muscles is not an innate skill; rather it is learned and practiced (Thyssen, Clevin, Olesen, & Lose, 2002).

Pregnancy and endocrine control related to menstrual cycles

Changes in hormonal levels related to menstrual cycles, pregnancy, and menopause profoundly influence vocal production and vocal quality (Caruso et al., 2000; Franco & Andrus, 2007; Sandgren, 2002). Dancers face similar challenges related to hormonal fluctuations but their hormonal shifts manifest in joint laxity (Foley & Bird, 2013). Further, dancers share similar hormonal alterations and menstrual cycle cessation as female athletes, a phenomenon associated with the female athlete triad (Otis, Drinkwater, Johnson, Loucks, & Wilmore, 1997). The physical demands of dance alter the functioning of the hypothalamus and pituitary as evidenced by increased levels of gonadotropin and testosterone and decreased levels of estrogen. These hormonal changes influence menstrual cycle length and regularity (Buchanan et al., 1991). As a result, a higher rate of secondary amenorrhea is found in dancers (and athletes). Secondary amenorrhea is diagnosed when cessation of the menstrual cycle has persisted for 3–6 months or more. One of the leading causes of secondary amenorrhea is a ratio imbalance of energy intake versus energy output. Dancers commonly display this energy imbalance, in part because their efforts to maintain low body weight to meet the demands of an "ideal" body image require decreased caloric intake. Low body weight and intense physical exertion not only increases risk for secondary amenorrhea but also osteoporosis (Chartrand & Chatfield, 2005). Unlike secondary amenorrhea, primary amenorrhea is diagnosed when the onset of menarche is delayed. Young dancers (and athletes) may initially receive a diagnosis of primary amenorrhea that is then changed to secondary amenorrhea once menarche begins (Chartrand & Chatfield, 2005).

One of the roles estrogen plays is regulation of vascular function; consequently decreased estrogen levels compromise coronary and peripheral vessel efficiency. The female athlete triad response is associated with a steroid hormonal profile similar to menopause; this hormonal alteration impairs endothelial cell function and arterial dilation (Hoch, Papanek, Szabo, Widlansky, & Gutterman, 2011). Performers with secondary amenorrhea frequently suffer endothelial dysfunction; the combination of these two disorders increases risk for cardiovascular disease, which is the number one cause of death in females in the United States. Unfortunately, dancers experience a higher rate of endothelial dysfunction (64%), leaving them at greater risk for future cardiovascular disease. Treating endothelial dysfunction with dietary supplements is proving effective. For example, dancers' hormonal imbalances were substantially decreased by folic acid supplementation (10 mg/day); this positive outcome only occurred when treatment lasted at least four weeks (Hoch et al., 2011). This hopeful noninvasive treatment approach may also decrease risk of future cardiovascular disease and reduce symptoms of secondary amenorrhea.

Pregnancy presents challenges for many performers. Because dance is a high-exercise profession, fetal development in pregnant dancers may be compromised. Exercise-induced challenges to fetal development are common, such as fetal malnutrition, hypoxia, hyperthermia, and risk of preterm labor (Sanders, 2008). For pregnant performers, increased risk of fall is associated with alterations in balance. During pregnancy, there are significant changes in maternal metabolic resources, including oxygen availability. Pregnant performers may experience a decrease in performance endurance and increase in joint laxity. Musculoskeletal injuries and pain may increase due to these hormonal changes. For performers who are dependent on exercise to manage depression and anxiety, decreasing physical activity may increase psychopathology symptoms, including elevated risk for postpartum depression (Weinstein et al., 2007). However, under medical supervision, pregnant performers can modify activity level and continue to perform for a portion of their pregnancy. Performers may be able to continue (with modifications) without negative consequences to themselves or their babies (Sanders, 2008).

Malignancies and cancer

Studies clearly indicate the main cause of death in jazz musicians is related to malignant neoplasms (tumors) (Herer, 2000). Malignancies are often not readily observed or diagnosed in performers; attention is usually directed toward treating musculoskeletal disorders. Without ruling out underlying malignancies in performers, valuable time is lost to initiate cancer treatment (Macfarlane & Rietveld, 2013). Performers generally do not overestimate pain; they are notorious underreporters of pain (Guptill, 2012; Raeburn et al., 2003). Given this pattern, when conservative treatment and manual therapy does not alleviate physical distress, further testing should follow.

For performing artists, continuing to practice and perform during cancer treatment may enhance their healing outcome, as well as decrease cancer-related fatigue. Generally, increased patient fatigue and decreased quality of life are associated with the process of receiving a cancer diagnosis, determining a treatment approach, and undergoing treatment. Studies suggest that engaging in dance can decrease fatigue and promote wellbeing (Sturm, Baak, Storek, Traore, & Thuss-Patience, 2014). As well, adding body–mind approaches conducted by trained psychotherapy movement specialists can reduce secondary cancer treatment symptoms and increase self-efficacy (Payne, 2009). This adjunctive treatment approach may be particularly appealing to performing artists (Fig. 20.4).

Figure 20.4 "Vulnerability." Performers: Emeka Nwoye, Kenneth Miles & CSUN Dancers. Photographer: Lee Choo.

Liver-related diseases, hepatitis, and infectious diseases

Performers are vulnerable to liver-related diseases such as alcohol-induced cirrhosis of the liver. A high mortality rate is associated with these diseases, and performing artists, as a group, struggle at a disproportionately higher rate compared to other professions. In Great Britain, actors and other performing artists were considered the third-highest group to die from this disorder. Factors that promote alcohol-related mortality include easy access to substances, separation from family and home, very high or low annual income, colleague collusion, and stress from job insecurity (Baker, 2008). Obviously, all these factors are endemic in the performing arts, and even more so for performers on tour. Even though working conditions and medical advancements are improving, many performers still suffer from stress-related diseases (Adams, 2009). For example, great jazz artists are often marginalized within the music industry; they struggle financially and physically and are at high risk for cirrhosis and the associated hepatocellular carcinoma (liver cancer).

Infectious diseases related to neglect or overindulgence include HIV/AIDS, hepatitis C, tuberculosis, syphilis, and liver-related diseases; performers are at increased risk for these diseases as well (Sartin, 2010). Musicians, dancers, and singers who work in popular media (rock, jazz, hip hop, pop, and music video) are exposed to risky lifestyles, hence greater infectious diseases. A long career of neglect and indulgence, along with aging, also elevate risk of developing a cardiovascular disease. Interestingly, these conditions can decrease after a performer has worked for over 25 years in a highly successful career, although this pattern is more prevalent in European artists compared to North American artists. Sadly, the performing arts community is cheated by the loss of so many talented artists; however, for those who reach celebrity status, they are in a powerful position to educate young performers and encourage more health-informed behaviors vital to sustaining a long and successful performance career (Bellis et al., 2007).

Inflammatory and autoimmune diseases

Autoimmune and inflammatory diseases often result from a combination of genetic predisposition and heightened stress. Performing artists, with a genetic predisposition to both autoimmune disorders and anxiety, are at increased risk (Crnivec, 2004; Pilger et al., 2014). There are multiple disorders that are considered autoimmune-based such as rheumatoid arthritis, inflammatory bowel disease, systemic lupus, pernicious anemia (decreased red blood cells due to inability to absorb vitamin B12), psoriasis (skin condition), Hashimoto's disease (inflammation in thyroid gland), Addison's disease (adrenal hormone insufficiency), and type-1 diabetes (retrieved from https://www.nlm.nih.gov/medlineplus/autoimmunediseases.html). Inflammatory processes also increase skin disorders, which are highly prevalent in musicians and other performing artists (Gambichler et al., 2004; Onder et al., 2000). Although there are a myriad of symptoms associated with these disorders, common symptoms include increased fatigue, fever, and general malaise (feeling ill).

Autoimmune and inflammatory research is relatively unexamined in performing artists; however, some studies have revealed sobering findings. For example, professional musicians report greater inflammatory diseases compared to the general population (Crnivec, 2004). In general, performance stress increases oxidative stress, which elevates inflammatory markers (Pilger et al., 2014). Preliminary evidence suggests that increased inflammatory disease may also be associated with long-term side effects caused by beta-blocker usage, a common treatment for performance anxiety in musicians (Patston & Loughlan, 2014). Elevation in musician anxiety and inflammatory disease are distressing. Participating in music should actually decrease anxiety and inflammatory responses. Musical engagement generally strengthens the immune system via increasing immunoglobulin levels (Kuhn, 2002). The fact that music is not having this effect on many performers indicates other factors that influence inflammatory responses (and anxiety). For example, childhood adverse experiences are associated with greater autoimmune and pain-related problems (Sachs-Ericsson, Cromer, Hernandez, & Kendall-Tackett, 2009). Performing arts medical specialists should assess performers' early adverse history along with appraising their current stress levels. The combination of early stress and professional stress may provoke a performer's genetic predisposition for acquiring an autoimmune disorder. Given these risk factors, when performers present with increased fatigue, fever, and malaise, medical practitioners should consider autoimmune disorders; performers may be at greater risk (Crnivec, 2004).

Somatoform disorders and health care

Somatoform disorders are challenging to diagnose because patients with these disorders are generally poor historians regarding previous health conditions. They present with vague symptoms that fluctuate rapidly, they respond poorly to treatment, and they are biased to believe their symptoms are related to physical health rather than stress-related psychological distress (Zinn, McCain, & Zinn, 2000). Some performing artists respond to career stress via the manifestation of somatoform disorders. They often complain of feeling ill and do not recover from their illness with traditional treatment. They have a pattern of excessively seeking and utilizing medical care; they have a cluster of stress-related physiological symptoms, and have marked negative affect (Zinn et al., 2000). In general, patients with somatoform disorders have greater threat perceptions, which exacerbates the disorder (Kenny & Ackermann, 2015). Over time, somatoform disorders can convert to stress-related diseases; the wear and tear manifests on body organs and stress systems.

Hypochondriasis is considered a somatoform disorder; it is defined as excessive worry about health issues. For example, a high proportion of singers display increased vocal health concerns. A combination of behaviors reveals their

Figure 20.5 "Looping." Performers: Bethana Rosenblatt, Evan Rosenblatt. Photographer: Lee Choo.

preoccupation, such as more vocal checking to determine vocal health and more visits to medical providers (Sandgren, 2009). Their health anxieties are usually rooted in very high career expectations. Audiences, producers, and other performers demand vocal perfection from the singer and these performances typically occur in stressful situations. These pressures intensify performance anxiety, which is translated into vocal health concerns (Sandgren, 2009; Fig. 20.5).

Burnout

Burnout is a syndrome that shares features with overtraining and staleness, and yet, it is also a distinct disorder. In some countries it is formally classified as a disease; in all countries it is accepted as a debilitating syndrome with negative consequences on health and job performance (Schaufeli, Leiter, & Maslach, 2009). It has a gradual onset with features that include fatigue, exhaustion, performance decrement, demotivation, feeling distanced from self and others (depersonalization), and devaluation of a once-desired profession or activity (Grove et al., 2013; Maslach & Jackson, 1981). Burnout is a complex condition with an indeterminate cause, although the symptom profile is consistent. It can be classified as either acute, which has a shorter duration (one month), or chronic, which persists for months and sometimes years (Koutedakis, 2000). Poor concentration and decreased curiosity are hallmark cognitive features of burnout. Psychological markers include depersonalization, anxiety, and depression, along with a higher risk for substance abuse behaviors and suicidal ideation (Maslach, Schaufeli, & Leiter, 2001; Weber & Jaekel-Reinhart, 2000). A cluster of physical symptoms manifest to varying degrees, such as excessive fatigue, an inability to recover from muscular exertion, chronic muscle weakness, poor coordination, sleep disturbances (including nightmares and excessive dreaming), menstrual disturbances, loss of appetite, and difficulty recovering from infections (especially respiratory and skin infections) (Koutedakis, 2000). Other somatic symptoms include headaches, gastrointestinal disorders (irritable stomach and bowel), and cardiovascular symptoms (arrhythmia, tachycardia, hypertonia) (Weber & Jaekel-Reinhart, 2000). Burnout shares many features with chronic stress; some argue that it is actually a psychosocial–environmental stress disorder (Koutedakis, 2000; Weber & Jaekel-Reinhart, 2000).

Physical exhaustion is directly related to long hours required to achieve skill mastery, a prerequisite for a career in the performing arts. It is often stated that mastery requires at least 10,000 hours or 10 years of deliberate practice to realize talent (Coakley, 2009). Success in the performing arts is based on meeting high expectations; performers and their audience internalize these expectations. The long hours, combined with intense pressure, often result in the fact that many young talented performers never reach adulthood as a performer; they "burnout" (Coakley, 2009; Dick et al., 2013). Performing artists, like athletes and health care professionals, are at high risk to experience burnout (Hernandez, 2012; Maslach & Jackson, 1981).

Burnout is intensified by chronic emotional demands and interpersonal career stress; it leads to performance inefficiency and despair (Maslach et al., 2001). For example, emotional expressivity in the performing arts may lead to emotional exhaustion, a key ingredient in burnout. This emotional exhaustion is exacerbated when performers implement emotional coping strategies under stress (see chapter: Self-Regulation, Emotion, and Resilience for further details) (Sears et al., 2000). As well, when obsessional passion, unlike harmonious passion, operates as a driving motivator the likelihood of burnout increases (Trepanier, Fernet, Austin, Forest, & Vallerand, 2014). Work overload and boredom are stress factors that may also provoke emotional exhaustion (Parasuraman & Purohit, 2000). These conditions are experienced by most performers at some point in their training or professional careers.

One of the features of emotional exhaustion is increased cynicism and callousness (Maslach & Jackson, 1981). These are dangerous emotional states for all performers; such intense negative attitudes constrain creative exploration. Burned out cynical teachers or coaches may actually intensify burnout states in the performers they are training (Dick et al., 2013). This same phenomenon can occur in medical practitioners who may be treating performing artists. Working in a state of burnout compromises compassion and may promote dehumanizing perceptions and negative attitudes toward their patients (performers) (Maslach & Jackson, 1981). Burned out performers (including students), educators, and/or medical providers also begin to view themselves negatively; they feel dissatisfied with their performance, which increases absenteeism in their work setting, or worse, leads to a termination of their career (Dick et al., 2013; Koutedakis, 2000; Maslach & Jackson, 1981; Maslach et al., 2001).

A decreased sense of autonomy is a major factor in burnout (Quested & Duda, 2011). During the long years of training to become a performing artist, or the protracted hours working in performing arts companies, a sense of autonomy often diminishes. With decreased autonomy, dissatisfaction intensifies, including diminished self-esteem and efficacy (Quested & Duda, 2011). Having a sense of autonomy involves the capacity to shape goals and influence working conditions. Optimal interpersonal interactions, including a sense of belonging within the performing arts community, enrich a sense of relatedness and self-esteem. A sense of competence is established by mastery of skills coupled by experiences of accomplishment. These three psychological factors (autonomy, relatedness, and competence) powerfully enhance feelings of self-determination; they promote career satisfaction, enhance behavioral regulation, and prevent burnout in performers (Quested & Duda, 2011).

Managing burnout is a complicated process, especially for performing artists. Some approaches to decreasing burnout include creating supportive arts organizations that foster wellbeing. When performers feel valued their negative self-appraisal diminishes (Asad & Khan, 2003). Another approach to shifting burnout is to address stressful situations with a task-oriented coping strategy. Some highly trained professionals will actually seek another avocation in order to gain a sense of self-determination. When tasks are effectively managed, whether within an avocational activity or career setting, performers experience a sense of accomplishment. Small markers of success renew optimism, and hopefully fatigue levels will decrease as interest is restored (Sears et al., 2000). Another stressor that increases burnout is boredom. When performers are given opportunities to determine their own repertory, including level of difficulty, boredom decreases along with burnout (Parasuraman & Purohit, 2000). Unfortunately, creativity, in and of itself, is not an antidote to burnout. Individual creativity does not affect the level of stress experienced at work or the potential to experience burnout (Asad & Khan, 2003). Despite the lack of protection from being creative, when creativity is nurtured in a supportive environment burnout is reduced and performance improves (de Jonge, Spoon, Sonnentag, Dormann, & van den Tooren, 2012; McManus, Jonvik, Richards, & Paice, 2011).

Although performers feel called to their profession, the risk of burnout is high. Like other professions, the performing arts have a high incidence of compassion fatigue (emotional exhaustion), stress, and diminished autonomy. The original metaphor of burnout captures the experience of many performers. They begin with a burning passion to pursue their talent; they have vigor, tenacity, and creative curiosity. Without adequate metabolic and psychosocial resources to fuel their fire, it gradually burns out. This is the cumulative effect of burnout (Schaufeli et al., 2009). Performers with burnout lack sufficient fuel to maintain their passion. Helping performers counter burnout is vital. For example, engaging in outside avocation/leisure activities can serve as psychological "fuel" (McManus et al., 2011). Diversifying interests may also reduce obsessional behaviors and perfectionistic tendencies, factors that constrain expressivity and increase intrinsic stress (Hall & Hill, 2012; Trepanier et al., 2014). Most importantly, increased awareness of burnout may reduce this hazard, especially among performing arts administrators, educators, parents, medical providers, and the performers themselves (Coakley 2009). Medication and psychotherapy may be necessary to heal severe burnout, but professional care must be rooted in a deep understanding of performer needs and the reality that

Box 20.3 Causes of Burnout Syndrome

1. Result of stress (chronic overwork and emotional demands)
2. Insufficient metabolic resources to manage cognitive and physical workload
3. Person–environment misfit (performer desiring more creative freedom and yet is only employed in a hierarchically rigid organization)
4. Lack of fulfillment of basic needs (autonomy, competence, and relational satisfaction)

burnout is a performance-related disorder (Koutedakis, 2000; Weber & Jaekel-Reinhart, 2000; Box 20.3).

Treatment and wellbeing

Creativity enhances health and healing (Flaherty, 2011). Participating creatively in the performing arts, regardless of whether the individual is a professional or amateur, promotes health, especially in an aging population (Cohen, 2006; Turiano, Spiro, & Mroczek, 2012). Over the course of a lifespan, individuals who remain open to experience and continually explore creative activities have a 12% decrease in mortality risk, significantly slower cognitive decline, and greater adaptive resilience (Turiano et al., 2012). Physical exercise and REM sleep also enhance creativity and aid in recovery from illness (Flaherty, 2011). For example, aerobic fitness promotes cardiorespiratory efficiency, a major factor in decreasing risk for diseases such as cancer, diabetes, and heart disease (Ward, 2008); it also increases creative output (Blanchette et al., 2005).

A well-balanced diet is also vital to support healing and prevent illness (Sousa, Carvalho, Moreira, & Teixeira, 2013). Because of the physical demands, performers should be screened for vitamin insufficiency, such as that of vitamin D (Constantini, Arieli, Chodick, & Dubnov-Raz, 2010). Performers need to be well informed about the importance of carbohydrate intake following training, rehearsal, or performances that exceed one hour in duration. Insufficient carbohydrate levels have deleterious effects on musculoskeletal and cognitive fatigue (Lieberman, Falco, & Slade, 2002; Sousa et al., 2013). A major problem in the performing arts is the challenge to maintain a healthy diet while also meeting the physical demands of rehearsals and performances. Despite high awareness and adequate nutrition knowledge, many performers indicate difficulty maintaining a healthy diet, especially while on tour. Complicating factors that limit nutritional support include cost, availability, and tour schedule (Cizek, Kelly, Kress, & Mattfeldt-Beman, 2016).

Sleep disorders are commonly associated with disease and disease risk. Educating performers about sleep hygiene and ways to ensure sufficient sleep greatly improves disease prevention and healing. Consistently, studies on musicians report that more than 50% of musicians struggle with some form of sleep disorder (Kaneko et al., 2005; Paarup et al., 2011; Spence, 2001). Although studies do not directly investigate singers, actors, and dancers, it is highly likely that they share similar difficulties. One factor that may negatively influence sleep in performers is their excessive ingestion of energy drinks (caffeinated drinks). Although these drinks do not predict substance use disorders, there is a statistical association, which raises concerns about performer health and wellbeing (Miller & Quigley, 2011). Further concerns are associated with the fact that a majority of performers use dietary supplements and sleep aids to help them meet performance demands; however, most performers lack sufficient medical information to determine accurate dosage and dietary needs (Brown & Wyon, 2014).

Receiving a correct diagnosis can greatly influence health outcome. Effective treatment for performing artists should also include patient education. Most performers thrive when given concrete suggestions to promote the success of their careers, including how to prevent and heal from occupational-related diseases (Dommerholt, 2010b). This information fosters an internal locus of control; performers can become their own health care advocates (Spahn, Burger, Hildebrandt, & Seidenglanz, 2005; Fig. 20.6).

Figure 20.6 "Exuberant focus." Performers: COSMIC Dance Crew. Photographer: Chuck Valerio.

Summary

- Introduction: definitions, disorders, seeking medical care.
- Hypermobility syndrome, dysautonomia, and health concerns: stress-sensitive disorders, alterations in connective tissues, challenges and advantages.
- Cumulative wear and tear, hearing loss, and health problems: overuse and fatigue, multiple symptoms and disorders (headaches, eyestrain, hearing disorders).
- Scoliosis, dystonia, and neurological disorders: definitions, compensations and symptom increase, neuropathies, treating TFD.
- Respiratory and cardiovascular diseases: definitions and symptoms, occupational hazards, metabolic disorders, cardiovocal disorder.
- Skin disorders: overuse, irritations, treatment.
- Vocal disorders: performance demands, vocal insecurities, assessing and managing, vocal emergencies and treatment.
- Genitourinary conditions: incontinence and physical strain.
- Pregnancy and endocrine control related to menstrual cycles: training and performance stress, role of estrogen and health.
- Malignancies and cancer: performing during treatment.
- Liver-related diseases, hepatitis, and infectious diseases: increased risk, infectious diseases due to neglect or overindulgence.
- Inflammatory and autoimmune diseases: definitions and examples of autoimmune disorders, stress and inflammatory disease.
- Somatoform disorders and health care: definitions, diagnosis and treatment, health anxieties.
- Burnout: distinct from overtraining, symptoms, acute/chronic, emotional demands, decreased sense of autonomy, managing burnout.
- Treatment and wellbeing: creativity enhances health and healing, diet, sleep, correct diagnosis.

Concluding remarks

The attention of researchers and performance-related health care practitioners is predominantly directed toward musculoskeletal disorders, with less focus on performance-related disorders and diseases. This field of investigation remains an underrepresented area (Guptill, 2008). Increased knowledge in this area of performance-related health conditions will potentially improve work environments and provide guidelines for optimal training programs. For example, when provided with knowledge about the standard recommended decibel levels to prevent hearing loss, performers can responsibly choose to rehearse and perform in optimal acoustical environments (Dick et al., 2013). Likewise, performers who exert physical force while performing can receive training to strengthen their feet and pelvic floor muscles to decrease urinary incontinence (Pozo-Municio, 2007).

Performing arts medicine specialists need to conduct an accurate extended history (including childhood adversity) to determine disease processes. Knowledge of the performers' daily demands, as well as their training, employment, and repertoire can help identify underlying performance-related disorders (Dommerholt, 2009). This information will assist in diagnoses, as well as prescribing appropriate treatment interventions (Dommerholt, 2010a, 2010b). One of the most important risks in diagnosing performance-related illnesses is identifying underlying causes of musculoskeletal pain. Neurodegenerative diseases or malignancies may be missed in early diagnostic evaluations. It is paramount that medical practitioners understand the behavioral patterns of performing artists; they underreport pain, tolerate years of performance pain without complaint, and are reticent to seek medical assessment (Geyer & Bressman, 2006; Guptill, 2012; Raeburn et al., 2003).

Implementing wellness programs into conservatory training institutions will hopefully change the current underreporting of health concerns and diminish fear regarding seeking medical help. Wellness programs can also provide education about proper nutrition and fitness regimes (Clark et al., 2014). Including regular health and injury screening can promote optimal performance and wellness, and perhaps more importantly, improve early detection of disease processes (Clark et al., 2013; Zander et al., 2010). Helping performers internalize a locus of control will offset their belief that all control resides outside of them. This is especially important when addressing their health and injury concerns (Spahn, Burger et al., 2005). Concurrent to the implementation of wellness programs in training institutions, professional performing arts health and wellness centers should be established in all major cities that sponsor performing arts organizations. Performing arts medicine is a growing field that requires a diverse range of highly specialized health care professionals who understand the unique demands placed on performers (Clark et al., 2014).

Conclusion/Epilogue

> *The medium of creativity granted me was the dance, always and ever the dance. Therein I could invent and create. Therein I have found my poetry, have given shape and profile to my visions, have molded and built, toiled and worked on the human being, with the human being, and for the human being.*
>
> *Mary Wigman (1966, p. 8)*

Performing artists are highly talented, trained, and passionate. Despite uncertain careers, they derive meaning and job satisfaction. They give life to scripts, scores, and choreographic works. Without performers there are no performing arts, and without the performing arts, the economic and cultural health of our global society would falter. Understanding performing artists, going behind the mask, sheds insight on their tremendous strengths and their vulnerabilities. For example, research studies demonstrate an ambiguous relationship between creativity, health, and mental illness. Compared to the general population, performers suffer more musculoskeletal injuries, performance-related diseases, and psychopathology; however, they also have much higher career satisfaction and better quality of life. Many of the performance-related disorders are considered career hazards; consequently they are not necessarily dispositional markers of physical or psychological weakness in performers.

Creativity research, although growing in range and complexity, is still developing (Sawyer, 2006a). One of the domains that is poorly studied is the performing arts; however, musicians as a group are better researched compared to the other performing arts domains. Surprisingly, actors are the least studied, despite their high visibility in popular media. Investigating creativity in the performing arts requires an expansion of methodological approaches. Performers operate within a collective art form; a confluence of factors intersect when identifying, training, and promoting a performer. Researching performing artists necessitates an interdisciplinary approach. Psychological, biological, developmental, sociological, cultural, epidemiological, and economic fields are some of many required to understand creativity in the performing arts. Fundamentally, performing arts is an emergent phenomenon; it must be analyzed at multiple levels to appreciate the creativity that resides within these domains (Hennessey & Amabile, 2010; Sawyer, 2006a).

Creativity and performing arts: constraints and future recommendations

1. Measuring creativity in the performing arts is challenging. Do you measure the creative novelty and value of the work performed? Is the performer the creative product? Do you

evaluate the collaborative interplay among performers, audience, and the work being performed?

2. Performing artists are the embodied expressive medium through which creativity is revealed. Most creative measurements evaluate one small aspect, such as personality, cognition, motivation, or emotion. Currently, ambulatory neurobiological and physiological measures lack the sophistication to fully capture the creative dynamic of a performance (Dietrich, 2008). For example, fMRI brain images are all static images. Dancers, singers, actors, and musicians are not able to perform at their fullest capacity while being scanned; hence the images do not portray an accurate creative brain at work in a performing artist.
3. Measurements to evaluate kinesthetic or motor creativity are still underdeveloped (Bournelli et al., 2009; Stinson, 2004). Operationalizing the concept "kinesthetic creative knowing" is in its pioneering stages, and yet most performers rely on this "knowing" to determine if they experienced creative flow states (optimal performance).
4. Within the performing arts there are a disproportionate number of studies investigating musicians compared to other performance domains. A myriad of studies have examined the influence of music, musical creativity, the brains of musicians, and a host of other topics on music. Although dancers are receiving more research attention, singers, followed by actors, remain understudied. This imbalance needs to be addressed to get a fuller perspective on all the performing arts domains.
5. Examining the effect of stress on creative performance remains critical. Performers face a multitude of stressors and each stressor influences creative output differently. These same stressors are also experienced differently at different points in development and during a professional career (Byron et al., 2010).
6. Understanding the relationship between performance stress and immune responses will help clarify why performers suffer more inflammatory-related diseases, despite the proven benefits gained from the direct or indirect use of expressive arts as a healing tool to reduce inflammatory processes (Fancourt, Ockelford, & Belai, 2014).
7. Investigation into the long-term consequences of performing a single role for a protracted period of time will help educators, medical practitioners, producers, and performers. For example, what happens to actors when they must portray a violent character for several years during the run of television series or while on Broadway? What happens to an opera singer who travels the world portraying one or two roles for their entire career?
8. Examining developmental history, especially childhood adversity, may offer explanations about physiological and psychological effects on performers. Modeling from current sport studies may help address these questions, as evidenced in the findings offered by a British research group. They discovered that elite Olympic athletes who regularly medal had childhood adversity compared to the other nonmedaling Olympic athletes (Nuwer, 2016). These medal winners claimed that their childhood adversity served to motivate them toward success. They wanted to find meaning and purpose in life to counter their early childhood maltreatment. Future studies in the performing arts, in particular, and creativity, in general, may add to this investigative direction.
9. Greater understanding about the role injury and illness plays on creativity will help educators and medical practitioners. Performers need to understand the nature of their creativity as it relates to their talent and the embodied expression of their talent.
10. Performers, as a group, tend to underreport illness, injury, and abuse (Spahn et al., 2005). Further research is needed, especially determining the frequency, severity, and consequences of workplace abuse (emotional, physical, sexual).

11. Performance-related medical specialties are required to diagnosis and treat performers (Dick et al., 2013). The higher morbidity and mortality rate of performers is disturbing (Gajewski & Poznanska, 2008). The performing arts are better served when highly talented and trained artists continue to contribute to their respective domain. Creating integrative research centers, such as the British Centre for Performance Science located at the Royal College of Music and the Imperial College, has the potential to further insight into performance itself. Including a wide range of disciplines (engineering, medicine, business, science, and aesthetic philosophy) allows investigation into the core processes and products of performance (see www.PerformanceScience.ac.uk).
12. The popular belief that performers are eccentric, flaky, crazy, sensitive, etc. needs to be challenged. They may be all of these things but they are also highly disciplined, principled, and persistent. Reporting results on performing artists requires the full picture of these contradictory factors. They may be highly anxious and yet they are courageous. They may be dreamers and yet they are grounded by the reality of daily practice. The paradoxical nature of a performing artist is a more accurate picture. They perform with mind and body, psyche and soma; they are never one-dimensional or isolated creators.

Epilogue

Performers offer society tremendous gifts. Gregory Peck shared an observation. He stated, "I like actors. I think they're good, fine, intelligent, openhearted, generous souls, every last one of them ... It's courageous to try to become someone else" (Brown, 1999, p. 69). His perspective about actors is applicable to all performers; throughout this book performers are portrayed as talented and courageous. They embrace Shakespeare's request of performers: "... the purpose of playing, whose end, both at the first and now, was and is, to hold, as 'twere, the mirror up to nature; to show virtue her own feature, scorn her own image, and the very age and body of the time his form and pressure" (Hamlet, Act III, Scene II, lines 21–25). Performers who accept the rigors of a performing arts career are dedicated, and they are also incredibly vulnerable. Their career is filled with stress, both physically and psychologically. It takes a toll on them, and yet they also experience the creative rewards of finding meaning and purpose. They have the opportunity to live a creative life. They share the experience of Rabindranath Tagore, who stated (1928, p. 270):

When death comes and whispers to me,
"Thy days are ended,"
let me say to him, "I have lived in love
and not in mere time."
He will ask, "Will thy songs remain?"
I shall say, "I know not, but this I know
That often when I sang I found my eternity."

References

Aalten, A. (2007). Listening to the dancer's body. *The Sociological Review*, *55*(1), 109–125. Available from http://dx.doi.org/10.1111/j.1467-954X.2007.00696.X.

Abdollahi, A., Pyszczynski, T., Maxfield, M., & Luszczynska, A. (2011). Posttraumatic stress reactions as a disruption in anxiety-buffer functioning: Dissociation and responses to mortality salience as predictors of severity of posttraumatic symptoms. *Psychological Trauma: Theory, Research, Practice, and Policy*, *3*(4), 329–341. Available from http://dx.doi.org/10.1037/a0021084.

Abdul-Kareem, I. A., Stancak, A. J., Parkes, L. M., Al-Ameen, M., AlGhamdi, J., Aldhafeeri, F. M., … Sluming, V. (2011). Plasticity of the superior and middle cerebellar peduncles in musicians revealed by quantitative analysis of volume and number of streamlines based on diffusion tensor tractography. *Cerebellum*, *10*, 611–623. Available from http://dx.doi.org/10.1007/s12311-011-0274-1.

Abraham, A., & Windmann, S. (2007). Creative cognition: The diverse operations and the prospect of applying a cognitive neuroscience perspective. *Methods*, *42*, 38–48. Available from http://dx.doi.org/10.1016/j.ymeth.2006.12.007.

Abraham, A., Windmann, S., Daum, I., & Gunturkun, O. (2005). Conceptual expansion and creative imagery as a function of psychoticism. *Consciousness and Cognition*, *14*, 520–534. Available from http://dx.doi.org/10.1016/j.concog.2004.12.003.

Abreu-Ramos, A., & Micheo, W. F. (2007). Lifetime prevalence of upper-body musculoskeletal problems in a professional level symphony orchestra: Age, gender, and instrument-specific results. *Medical Problems of Performing Artists*, *22*(3), 97–104.

Abuhamdeh, S., & Csikszentmihalyi, M. (2012). The importance of challenge for the enjoyment of intrinsically motivated, goal-directed activities. *Personality and Social Psychology*, *38* (3), 317–330. Available from http://dx.doi.org/10.1177/0146167211427147.

Acar, S., & Sen, S. (2013). A multilevel meta-analysis of the relationship between creativity and schizotypy. *Psychology of Aesthetics, Creativity, and the Arts*, *7*(3), 214–228. Available from http://dx.doi.org/10.1037/a0031975.

Acharya, U. R., Joseph, K. P., Kannathal, N., Lim, C. M., & Suri, J. S. (2006). Heart rate variability: A review. *Medical Biology, Engineering, and Computers*, *44*, 1031–1051. Available from http://dx.doi.org/10.1007/s11517-006-0119-0.

Ackard, D. M., Henderson, J. B., & Wonderlich, A. L. (2004). The associations between childhood dance participation and adult disordered eating and related psychopathology. *Journal of Psychosomatic Research*, *57*, 485–490. Available from http://dx.doi.org/10.1016/j.jpsychores.2004.03.004.

Ackerman, C. M. (2009). The essential elements of Dabrowski's Theory of Positive Disintegration and how they are connected. *Roeper Review*, *31*(2), 81–95. Available from http://dx.doi.org/10.1080/02783190902737657.

Ackermann, B., Driscoll, T., & Kenny, D. T. (2012). Musculoskeletal pain and injury in professional orchestral musicians in Australia. *Medical Problems of Performing Artists*, *27* (4), 181–187.

Acosta, L. M. (2014). Creativity and neurological disease. *Current Neurological Neuroscience Report*, *14*(464), 1–6. Available from http://dx.doi.org/10.1007/s11910-014-04064-6.

Adam, M. U., Brassington, G. S., Steiner, H., & Matheson, G. O. (2004). Psychological factors associated with performance-limiting injuries in professional dancers. *Journal of Dance Medicine and Science*, *8*(2), 43–46.

Adams, P. (2009). The lost years: The impact of cirrhosis on the history of jazz. *Canadian Journal of Gastroenterology*, *23*(6), 405–406.

Adler, S. Retrieved from www.goodreads.com/quotes/tag/acting.

Air, M. (2013). Psychological distress among dancers seeking outpatient treatment for musculoskeletal injury. *Journal of Dance Medicine and Science*, *17*(3), 115–125. Available from http://dx.doi.org/10.12678/1089.313X.17.3.115.

Akinola, M., & Mendes, W. B. (2008). The dark side of creativity: Biological vulnerable and negative emotions lead to greater artistic creativity. *Personality and Social Psychology Bulletin*, *34*(12), 1677–1686. Available from http://dx.doi.org/10.1177/0146167208323933.

Akiskal, K. K., & Akiskal, H. S. (2005). The theoretical underpinnings of affective temperaments: Implications for evolutionary foundations of bipolar disorder and human nature. *Journal of Affective Disorders*, *85*, 231–239. Available from http://dx.doi.org/10.1016/j.jad.2004.08.002.

Akiskal, K. K., Savino, M., & Akiskal, H. S. (2005). Temperament profiles in physicians, lawyers, managers, industrialists, architects, journalists, and artists: A study in psychiatric outpatients. *Journal of Affective Disorders*, *85*, 201–206. Available from http://dx.doi.org/10.1016/j.jad.2004.08.003.

Aksoydan, E., & Camci, N. (2009). Prevalence of orthorexia nervosa among Turkish performance artists. *Eating Weight Disorder*, *14*, 33–37.

Aldersen, J., Hopper, L., Elliott, B., & Ackland, T. (2009). Risk factors for lower back injury in male dancers performing ballet lifts. *Journal of Dance Medicine and Science*, *13*(3), 83–89.

Aleman, A., Nieuwenstein, M. R., Bocker, K. B. E., & de Haan, E. H. F. (2000). Music training and mental imagery ability. *Neuropsychologia*, *38*, 1664–1668.

Alesli, A., Damiani, C., & Permice, D. (2005). The physical therapist–patient relationship. Does physical therapists' occupational stress affect patients' quality of life. *Functional Neurology*, *20*(3), 121–126.

Alias, A., Rahman, S., Majid, R. A., & Yassin, S. F. M. (2013). Dabrowski's overexcitabilities profile among gifted students. *Asian Social Science*, *9*(16), 120. Available from http://dx.doi.org/10.5539/ass.v9n16p120.

Alpert, P. T., Miller, S. K., Wallman, H., Havey, R., Cross, C., Chevalia, T., ... Kodandapari, K. (2009). The effect of modified jazz dance on balance, cognition, and mood in older adults. *Journal of the American Academy of Nurse Practitioners*, *21*, 108–115. Available from http://dx.doi.org/10.1111/j.1745-7599.2008.00392.x.

Altemus, M., Dhabhar, F. S., & Yang, R. (2006). Immune function in PTSD. *Annals of the New York Academy of Science*, *1071*, 167–183.

Alter, J. B. (1984). Creativity profile of university and conservatory dance students. *Journal of Personality Assessment*, *48*(2), 153–158. Available from http://dx.doi.org/10.1207/s15327752jpa4802_8.

Alvarez, J. A., Emory, J. A., & Emory, E. (2006). Executive function and the frontal lobes: A meta-analytic review. *Neuropsychology Review*, *16*(1), 17–42. Available from http://dx.doi.org/10.1007/s11065-006-9002-x, PMID 16794878.

Amabile, T. M. (1996). *Creativity in context*. Boulder, CO: Westview Press.

Amabile, T. M., Barsade, S. G., Mueller, J. S., & Staw, B. M. (2005). Affect and creativity at work. *Administrative Science Quarterly*, *50*, 367–403.

Amano, M., Kanda, T., Ue, H., & Moritani, T. (2001). Exercise training and autonomic nervous system activity in obese individuals. *Medicine and Science in Sports and Exercise*, *33*(8), 1287–1291.

Ambady, N., Koo, J., Rosenthal, R., & Winograd, C. H. (2002). Physical therapists' nonverbal communication predicts geriatric patients' health outcomes. *Psychology and Aging*, *17*(3), 443–452. Available from http://dx.doi.org/10.1037//0882-7974.17.3.443.

American heritage® dictionary of the English language. (5th ed.). (2011). New York: Houghton Mifflin Harcourt Publishing Company.

American Psychiatric Association (APA) (2013). *Diagnostic and statistical manual of mental disorders* (5th ed.). Washington, DC: American Psychiatric Publishing.

Anderson, L. M. (2011). Myself or someone like me: A review of the literature on the psychological well-being of child actors. *Medical Problems of Performing Artists*, *26*(3), 146–149.

Anderson, M. J., Marwit, S. J., Vandenberg, B., & Chibnall, J. T. (2005). Psychological and religious coping strategies of mothers bereaved by the sudden death of a child. *Death Studies*, *29*, 811–826.

Anderson, R., & Hanrahan, S. J. (2008). Dancing in pain: Pain appraisal and coping in dancers. *Journal of Dance Medicine and Science*, *12*(1), 9–16.

Andreasen, N. C. (2005). *The creating brain: The neuroscience of genius*. New York: Dana Press.

Andreasen, N. C. (2008). The relationship between creativity and mood disorders. *Dialogues in Clinical Neuroscience*, *10*, 251–255.

Andrews-Hanna, J. R., Smallwood, J., & Spreng, R. N. (2014). The default network and self-generated thought: Component processes, dynamic control, and clinical relevance. *Annals of the New York Academy of Sciences*, *1316*(1), 29–52. Available from http://dx.doi.org/10.1111/nyas.12360.

Angioi, M., Metsios, G. S., Twitchett, E., Koutedakis, Y., & Wyon, M. (2009). Association between selected physical fitness parameters and aesthetic competence in contemporary dancers. *Journal of Dance Medicine and Science*, *13*(4), 115–123.

Arcelus, J., Witcomb, G. L., & Mitchell, A. (2014). Prevalence of eating disorders amongst dancers: A systematic review and meta-analysis of eating disorders and dance. *European Eating Disorders Review*, *22*(2), 92–101.

Ardeleanu, D., & Ardeleanu, S. (2014). The voice of music resounding on HRV. *Romanian Journal of Artistic Creativity*, *4*, 245–254.

Arden, R., Chavez, R. S., Grazioplene, R., & Jung, R. E. (2010). Neuroimaging creativity: A psychometric view. *Behavioral Brain Research*, *214*, 143–156. Available from http://dx.doi.org/10.1016/j.bbr.2010.05.015.

Aristotle (1958). On poetry and style *(G. M. A. Grube, Trans. & Intro.)*. New York: The Bobbs-Merrill Company, Inc.

Armstrong, D. (2012). The contributions of creative cognition and schizotypal symptoms to creative achievement. *Creativity Research Journal*, *24*(3), 177–190. Available from http://dx.doi.org/10.1080/10400419.2012.677329.

Aronoff, U., & Gilboa, A. (2015). Music and the closet: The roles music plays for gay men in the "coming out" process. *Psychology of Music*, *43*(3), 423–437. Available from http://dx.doi.org/10.1177/0305735613515943.

Arrighi, G. (2015). Circus studies: Where to next? *Popular Entertainment Studies*, *6*(1), 62–65.

Artaud, A. (1958). *The theatre and its double*. New York: Grove Press, Inc.

Asad, N., & Khan, S. (2003). Relationship between job-stress and burnout: Organizational support and creativity as predictor variables. *Pakistan Journal of Psychological Research, 18*(3–4), 139–149.

Asakawa, A. (2010). Flow experience, culture, and well-being: How do autotelic Japanese college students feel, behave, and think in their daily lives? *Journal of Happiness, 11* (2), 205–233. Available from http://dx.doi.org/10.1007/s10902-008-9152-3.

Asakawa, K. (2004). Flow experience and autotelic personality in Japanese college students: How do they experience challenges in daily life? *Journal of Happiness Studies, 5*, 123–154.

Askling, C., Lund, H., Saartok, T., & Thorstensson, A. (2002). Self-reported hamstring injuries in student-dancers. *Scandinavian Journal of Medical Science in Sports, 12*, 230–235.

Aston-Jones, G., & Cohen, J. D. (2005). An integrative theory of locus coeruleus-norepinephrine function: Adaptive gain and optimal performance. *Annual Reviews in Neuroscience, 28*, 403–450. Available from http://dx.doi.org/10.1146/annurev.neuro.28.061604.135709.

Atkinson, A. P., Turnstall, M. L., & Dittrich, W. H. (2007). Evidence for distinct contributions of form and motion information to the recognition of emotions from body gestures. *Cognition, 104*, 59–72. Available from http://dx.doi.org/10.1016/j.cognition.2006.05.005.

Ayers, L., Beaton, S., & Hunt, H. (1999). The significance of transpersonal experiences, emotional conflict, and cognitive abilities in creativity. *Empirical Studies of the Arts, 17*(1), 73–82.

Ayman-Nolley, S. (1992). Vygotsky's perspective on the development of imagination and creativity. *Creativity Research Journal, 5*(1), 77–85.

Baas, M., De Dreu, C. K. W., & Nijstad, B. A. (2008). A meta-analysis of 25 years of mood-creativity research: Hedonic tone, activation, or regulatory focus. *Psychological Bulletin, 134*(6), 779–806. Available from http://dx.doi.org/10.1037/a0012815.

Baas, M., De Dreu, C. K. W., & Nijstad, B. A. (2011). When prevention promotes creativity: The role of mood, regulation, focus and regulatory closure. *Journal of Personality and Social Psychology, 100*(5), 794–809. Available from http://dx.doi.org/10.1037/a0022981.

Bachner-Melman, R., Dina, C., Zohar, A. H., Constantini, N., Lerer, E., Sella, S., ... Ebstein, R. P. (2005). AVPR1a and SLC6A4 gene polymorphisms are associated with creative dance performance. *PLoS Genetics, 1*(3), e42. Available from http://dx.doi.org/10.1371/journal.pgen.0010042.

Bachner-Melman, R., Zohar, A. H., Epstein, R. P., Elizur, Y., & Constantini, N. (2006). How anorexic-like are the symptoms and personality profiles of aesthetic athletes. *Medicine and Science in Sports and Exercise, 38*(4), 628–636. Available from http://dx.doi.org/10.1249/01.miss.0000210188.70295.c0.

Baer, J. (2010a). The strength-of-weak-ties perspective on creativity: A comprehensive examination and extension. *Journal of Applied Psychology, 95*(3), 592–601. Available from http://dx.doi.org/10.1037/a0018761.

Baer, J. (2010b). Is creativity domain specific? In J. C. Kaufman, & R. J. Sternberg (Eds.), *The Cambridge handbook of creativity* (pp. 321–341). New York: Cambridge University Press. Available from http://dx.doi.org/10.1017/CBO9780511763205.021.

Baer, J. (2015). The importance of domain-specific expertise in creativity. *Roeper Review, 37*, 165–178. Available from http://dx.doi.org/10.1080/02783193.2015.1047480.

Baert, S., Casier, A., & De Raedt, R. (2012). The effects of attentional training on physiological stress recovery after induced social threat. *Anxiety, Stress and Coping, 25*(4), 365–379. Available from http://dx.doi.org/10.1080/10615806.2011.605122.

Bagley, C. (2008). Educational ethnography as performance art: Towards a sensuous feeling and knowing. *Qualitative Research*, *8*(1), 53–72. Available from http://dx.doi.org/10.1177/1468794107085296.

Baillie, Y., Wyon, M., & Head, A. (2007). Highland dance: Heart-rate and blood lactate differences between competition and class. *International Journal of Sports Physiology and Performance*, *2*, 371–376.

Bain, A. (2005). Constructing an artistic identity. *Work, Employment and Society*, *19*(1), 25–46. Available from http://dx.doi.org/10.1177/0950017005051280.

Baker, A. (2008). Alcohol-related deaths by occupation: What do data for England and Wales in 2001–2005 tell us about doctors' mortality? *Alcohol and Alcoholism*, *43*(2), 121–122. Available from http://dx.doi.org/10.1093/alcalc/agm170.

Bakermans-Kranenburg, M. J., & van IJzendoorn, M. H. (2009). The first 10,000 adult attachment interviews: Distributions of adult attachment representations in clinical and non-clinical groups. *Attachment and Human Development*, *11*(3), 223–263. Available from http://dx.doi.org/10.1080/14616730902814762.

Bakker, A. B. (2005). Flow among music teachers and their students: The crossover of peak experiences. *Journal of Vocational Behavior*, *66*, 26–44. Available from http://dx.doi.org/10.1016/j.jvb.2003.11.001.

Bakker, F. C. (1991). Development of personality in dancers: A longitudinal study. *Personality and Individual Differences*, *12*(7), 671–681.

Balanchine, G. Retrieved from www.goodreads.com/quotes/tag/dancing.

Balconi, M., & Bortolotti, A. (2012). Empathy in cooperative versus non-cooperative situations: The contribution of self-report measures and autonomic responses. *Applied Psychophysiology Biofeedback*, *37*, 161–169. Available from http://dx.doi.org/10.1007/s10484-012-9188-z.

Banfield, J. F., Wyland, C. L., MacRae, C. N., Munte, R. F., & Heatherton, T. F. (2004). The cognitive neuroscience of self-regulation. In R. F. Baumeister, & K. D. Vohs (Eds.), *Handbook of self-regulation: Research, theory, and application* (pp. 62–83). New York: The Guilford Press.

Bar, R. J., & DeSouza, J. F. X. (2016). Tracking plasticity: Effects of long-term rehearsal in expert dancers encoding music to movement. *PLoS ONE*, *11*(1), e0147731. Available from http://dx.doi.org/10.1371/journal.pone.0147731.

Barba, E., & Savarese, N. (2005). *A dictionary of theatre anthropology: The secret art of the performer*. New York: Routledge.

Barbar, A. E. M., Crippa, J. A. S., & Osorio, F. L. (2014). Performance anxiety in Brazilian musicians: Prevalence and association with psychopathology indicators. *Journal of Affective Disorders*, *152–154*, 381–386. Available from http://dx.doi.org/10.1016/j.jad.2013.09.041.

Barker, K. K., Soklaridis, S., Waters, I., Her, G., & Cassidy, J. D. (2009). Occupational strain and professional artists: A qualitative study of an underemployed group. *Arts and Health: An International Journal for Research, Policy and Practice*, *1*(2), 136–150. Available from http://dx.doi.org/10.1080/17533010903031390.

Barnard, K. E., Broman-Fulks, J. J., Michael, K. D., Webb, R. M., & Zawilinski, L. L. (2011). The effects of physiological arousal on cognitive and psychomotor performance among individuals with high and love anxiety sensitivity. *Anxiety, Stress, and Coping*, *24*(2), 201–216. Available from http://dx.doi.org/10.1080/10615806.2010.494328.

Barnes, R., Attwood, H., Blom, J., Jankielsohn, S., van Rensburg, J., Smith, T., ... Nel, M. (2011). Injury profile of musicians in the Bloemfontein-based free state symphony orchestra: A short report. *South African Journal of Physiotherapy*, *67*(2), 41–44.

Barnes, T. D., Kubota, Y., Hu, D., Jin, D. Z., & Graybiel, A. M. (2005). Activity of striatal neurons reflects dynamic encoding and recoding of procedural memories. *Nature*, *437*, 1158–1161. Available from http://dx.doi.org/10.1038/nature0453.

Barron, F. (1955). The disposition towards originality. *Journal of Abnormal and Social Psychology*, *51*, 478–485.

Bartholomew, J. B., Moore, J., Todd, J., Todd, T., & Elrod, C. C. (2001). Psychological states following resistance exercise of different workloads. *Journal of Applied Sport Psychology*, *13*, 399–410.

Bartle, G. A. R. (1990). How 1,250,000,000 people train their performing musicians: A microscopic view of the training of pianists, opera singers, orchestral musicians, conductors, organists, and sound recording technicians in six countries. *International Journal of Music Education*, *15*, 31–37.

Baryshnikov, M. Retrieved from www.goodreads.com/quotes/tag/dancing.

Bash, H. L., & Papa, A. (2014). Shame and PTSD symptoms. *Psychological Trauma: Theory, Research, Practice, and Policy*, *6*(2), 159–166. Available from http://dx.doi.org/10.1037/a0032637.

Bateman, C. (2014). What are we playing with? Role-taking, role-playing, and story-play with Tolkien's legendarium. *International Journal of Play*, *3*(2), 107–118. Available from http://dx.doi.org/10.1080/21594937.2013.877639.

Batey, M. (2012). The measurement of creativity: From definitional consensus to the introduction of a new heuristic framework. *Creativity Research Journal*, *24*(1), 55–65. Available from http://dx.doi.org/10.1080/10400419.2012.649181.

Batson, G. (2013). Exercise-induced central fatigue: A review of the literature with implications for dance science research. *Journal of Dance Medicine and Science*, *17*(2), 53–62. Available from http://dx.doi.org/10.12678/1089-313X.17.2.53.

Baumann, L. (2012). Ethics in cosmetic dermatology. *Clinics in Dermatology*, *30*, 522–527. Available from http://dx.doi.org/10.1016/j.clindermatol.2011.06.023.

Baumann, N., & Turpin, J.-C. (2010). Neurochemistry of stress: An overview. *Neurochemistry Research*, *35*, 1875–1879. Available from http://dx.doi.org/10.1007/s11064-010-0298-9.

Baumrind, D. (1991). Parenting styles and adolescent development. In R. M. Lerner, A. C. Petersen, & J. Brooks-Gunn (Eds.), *Encyclopedia of adolescence* (Vol. 2, pp. 746–758). New York: Garland Publishing.

Beaty, R. E. (2015). The neuroscience of musical improvisation. *Neuroscience and Biobehavioral Reviews*, *51*, 108–117. Available from http://dx.doi.org/10.1016/j/neruobiorev.2015.01.004.

Bechara, A., & Damasio, A. (2005). The somatic marker hypothesis: A neural theory of economic decision. *Games and Economic Behavior*, *52*, 336–372. Available from http://dx.doi.org/10.1016/j.geb.2004.06.010.

Beck, S., Redding, E., & Wyon, M. A. (2015). Methodological considerations for documenting the energy demand of dance activity: A review. *Frontiers in Psychology*, *6*(568), 1–14. Available from http://dx.doi.org/10.3389/fpsyg.2015.00568.

Beech, N., Gilmore, C., Cochrane, C., & Greig, G. (2012). Identity work as a response to tensions: A re-narration in opera rehearsals. *Scandinavian Journal of Management*, *28*(1), 39–47. Available from http://dx.doi.org/10.1016/j.scaman.2011.12.005.

Beekman, C., Neiderhiser, J. M., Buss, K. A., Loken, E., Moore, G. A., Leve, L. D., … Reiss, D. (2015). The development of early profiles of temperament: Characterizations, continuity, and etiology. *Child Development*, *86*(6), 1794–1811. Available from http://dx.doi.org/10.1111/cdev.12417.

Beghetto, R. A. (2014). Creative mortification: An initial exploration. *Psychology of Aesthetics, Creativity, and the Arts*, *8*(3), 266–276. Available from http://dx.doi.org/10.1037/a0036618.

Bekhtereva, N. P., Starchenko, M. G., Klyucharev, V. A., Vorob'ev, V. A., Pakhomov, S. V., & Medvedev, S. V. (2000). Study of the brain organization of creativity: II. Positron-emission tomography data. *Human Physiology*, *26*(5), 12–18.

Belanger, A.-E., Bernier, A., Simard, V., Bordeleau, S., & Carrier, J. (2015). Attachment and sleep among toddlers: Disentangling attachment security and dependency. *Monographs of the Society for Research in Child Development*, *80*(1), 125–140. Available from http://dx.doi.org/10.1111/mono.v80.1/issuetoc.

Bellis, M. A., Hennell, T., Lushey, C., Hughes, K., Tocque, K., & Ashton, J. R. (2007). Elivs to Eminem: Quantifying the price of fame through mortality of European and North American rock and pop stars. *Journal of Epidemiology and Community Health*, *61*, 896–901. Available from http://dx.doi.org/10.1136/jech.2007.059915.

Belsky, J., & Rovine, M. (1987). Temperament and attachment security in the strange situation: An empirical rapprochement. *Child Development*, *58*, 787–795.

Bender, T. S., & Hancock, C. B. (2010). The effect of conductor intensity and ensemble performance quality on musicians' evaluations of conductor effectiveness. *Journal of Band Research*, *46*(1), 13–22.

Benedek, M., Borovnjak, B., Neubauer, A. C., & Kruse-Weber, S. (2014). Creativity and personality in classical, jazz and folk musicians. *Personality and Individual Differences*, *63*, 117–121. Available from http://dx.doi.org/10.1016/j.paid.2014.01.064.

Benedek, M., Franz, F., Heene, M., & Neubauer, A. C. (2012). Differential effects of cognitive inhibition and intelligence on creativity. *Personality and Individual Differences*, *53*, 480–485. Available from http://dx.doi.org/10.1016/j.paid.2012.04.014.

Bennett, D. (2009). Academy and the real world: Developing realistic notions of career in the performing arts. *Arts and Humanities in Higher Education*, *8*(3), 309–327. Available from http://dx.doi.org/10.1177/1474022209339953.

Benoit, M., Bouthillier, D., Moss, E., Rousseau, C., & Brunet, A. (2010). Emotion regulation strategies as mediators of the association between level of attachment security and PTSD symptoms following trauma in adulthood. *Anxiety, Stress and Coping*, *23*(1), 101–118. Available from http://dx.doi.org/10.1080/10615800802638279.

Benolken, S. (2006). My so-called life in the art: Why I ran away from the theatre of cruelty to join the social sciences. *Psychology of Aesthetics, Creativity, and the Arts*, *5*(1), 55–58. Available from http://dx.doi.org/10.1037/1931-3896.S.1.55.

Berk, L. E., & Meyers, A. B. (2013). The role of make-believe play in the development of executive function: Status of research and future directions. *American Journal of Play*, *6* (1), 98–110.

Bernard, R. (2009). Music making, transcendence, flow, and music education. *International Journal of Education and the Arts*, *10*(4), 1–21.

Bernardis, P., & Gentilucci, M. (2006). Speech and gesture share the same communication system. *Neuropsychologia*, *44*, 178–190.

Bernstein, L. Retrieved from www.goodreads.com/quotes/tag/music.

Bernstein, R. E., Delker, B. C., Knight, J. A., & Freyd, J. J. (2015). Hypervigilance in college students: Associations with betrayal and dissociation and psychometric properties in a brief hypervigilance scale. *Psychological Trauma: Theory, Research, Practice, and Policy*, *7*(5), 448–455. Available from http://dx.doi.org/10.1037/tra000070.

Berntson, G. G., & Cacioppo, J. T. (2007). Integrative physiology: Homeostasis, allostasis and the orchestration of systemic physiology. In J. T. Cacioppo, L. G. Tassinary, &

G. G. Berntson (Eds.), *Handbook of psychophysiology* (3rd ed., pp. 433–452). New York: Cambridge University Press.

Berntson, G. G., Norman, G. J., Hawkley, L. C., & Cacioppo, J. T. (2008a). Cardiac autonomic balance versus cardiac regulatory capacity. *Psychophysiology*, *45*, 643–652. Available from http://dx.doi.org/10.1111/j.1469-8986.2008.00652.x.

Berntson, G. G., Norman, G. J., Hawkley, L. C., & Cacioppo, J. T. (2008b). Spirituality and autonomic cardiac control. *Annals of Behavioral Medicine: A Publication of the Society of Behavioral Medicine*, *35*(2), 198–208. Available from http://dx.doi.org/10.1007/s12160-008-9027-x.

Berque, P., Gray, H., & McFadyen, A. (2016). Playing-related musculoskeletal problems among professional orchestra musicians in Scotland: A prevalence study using a validated instrument, The Musculoskeletal Pain Intensity and Interference Questionnaire for Musicians (MPIIQM). *Medical Problems of Performing Artists*, *31*(2), 78–86. Available from http://dx.doi.org/10.2109/mppa.2016.2015.

Bettle, N., Bettle, O., Neumarker, U., & Neumarker, K.-J. (2001). Body image and self-esteem in adolescent ballet dancers. *Perceptual and Motor Skills*, *93*, 297–309.

Biasi, V., Bonaiuto, P., Giannini, M. A., & Chiappero, E. (1999). Personological studies on dancers: Motivations, conflicts, and defense mechanisms. *Empirical Studies of the Arts*, *17*(2), 171–186.

Biasutti, M., & Frezza, L. (2009). Dimensions of music improvisation. *Creativity Research Journal*, *21*(2–3), 232–242. Available from http://dx.doi.org/10.1080/10400410902861240.

Bifulco, A., Moran, P. M., Baines, R., Bunn, A., & Stanford, K. (2002). Exploring psychological abuse in childhood: II. Association with other abuse and adult clinical depression. *Bulletin of the Menninger Clinic*, *66*(3), 241–258.

Bigelsen, J., & Schupak, C. (2011). Compulsive fantasy: Proposed evidence of an under-reported syndrome through a systematic study of 90 self-identified non-normative fantasizers. *Consciousness and Cognition*, *20*, 1634–1648. Available from http://dx.doi.org/10.1016/j.concog.2011.08.013.

Bille, T., Fjaellegaard, C. B., Frey, B. S., & Steiner, L. (2013). Happiness in the arts—International evidence on artists' job satisfaction. *Economics Letters*, *121*, 15–18. Available from http://dx.doi.org/10.1016/j.enconlet.2013.06.016.

Binder, E. B., Bradley, R. G., Liu, W., Epstein, M. P., Deveau, T. C., Mercer, K. B., ... Ressler, K. J. (2008). Association of FKBP5 polymorphisms and childhood abuse with risk of posttraumatic stress disorder symptoms in adults. *Journal of American Medical Association*, *299*(11), 1291–1305.

Bird, H. A. (2009). The performing artist as an elite athlete. *Rheumatology*, *48*, 1469–1470. Available from http://dx.doi.org/10.093/rheumatology/kep257.

Bird, H. A. (2013). Overuse syndrome in musicians. *Clinical Rheumatology*, *32*, 475–479. Available from http://dx.doi.org/10.1007/s10067-013-2198-2.

Bird, H. A., & Macdonald, I. (2013). Expert care of the performing artist. *Clinical Rheumatology*, *32*, 421–423. Available from http://dx.doi.org/10.1007/s10067-013-2194-6.

Bird, H. A., & Pinto, S. O. (2013). Scoliosis in musicians and dancers. *Clinical Rheumatology*, *32*, 515–521. Available from http://dx.doi.org/10.1007/s10067-013-2190-x.

Bjorkner, E. (2009). Musical theater and opera singing—why so different? A study of subglottal pressure, voice sources, and format frequency characteristics. *Journal of Voice*, *22*(5), 533–540. Available from http://dx.doi.org/10.1016/j.voice.2006.12.007.

Black, J., & Barnes, J. L. (2015). Fiction and social cognition: The effect of viewing award-winning television dramas on theory of mind. *Psychology of Aesthetics, Creativity, and the Arts*, *9*(4), 423–429. Available from http://dx.doi.org/10.1037/aca000031.

Blakeslee, S., & Blakeslee, M. (2007). *The body has a mind of its own*. New York: Random House.

Blanchette, D. M., Ramocki, S. P., O'del, J. N., & Casey, M. S. (2005). Aerobic exercise and creative potential: Immediate and residual effects. *Creativity Research Journal, 17* (2&3), 257–264.

Blanke, O., & Arzy, S. (2005). The out-of-body experience: Disturbed self-processing at the temporal-parietal junction. *Neuroscientist, 11*(1), 16–24. Available from http://dx.doi.org/10.1177/1073858404270885.

Blasing, B., Calvo-Merino, B., Cross, E. S., Jola, C., Honisch, J., & Stevens, C. J. (2012). Neurocognitive control in dance perception and performance. *Acta Psychologica, 139*, 300–308. Available from http://dx.doi.org/10.1016/j.actpsy.2011.12.005.

Blasing, B., Tenenbaum, G., & Schack, T. (2009). The cognitive structure of movements in classical dance. *Psychology of Sport and Exercise, 10*, 350–360. Available from http://dx.doi.org/10.1016/j.psychsport.2008.10.001.

Block, C. J., Koch, S. M., Liberman, B. E., Merriweather, T. J., & Roberson, L. (2011). Contending with stereotype threat at work: A model of long-term responses. *The Counseling Psychologist, 39*(4), 570–600. Available from http://dx.doi.org/10.1177/0011000010382459.

Blood, A. J., & Zatorre, R. J. (2001). Intensely pleasurable responses to music correlate with activity in brain regions implicated in reward and emotion. *Proceedings of the National Academy of Science, 98*, 11818–11823. Available from http://dx.doi.org/10.1037/pnas.191355898.

Bo, K., & Borgen, J. S. (2001). Prevalence of stress and urge urinary incontinence in elite athletes and controls. *Medical Science Sports Exercise, 33*, 1797–1802.

Bodner, E., & Bensimon, M. (2008). After the curtain falls: On the post-performance adjustment of solo singers. *Medical Problems of Performing Artists, 23*(4), 172–178.

Bodrova, E., Germeroth, C., & Leong, E. J. (2013). Play and self-regulation. *American Journal of Play, 6*(1), 111–123.

Bodrova, E., & Leong, D. J. (2015). Vygotskian and post-Vygotskian views on children's play. *American Journal of Play, 7*(3), 371–388.

Boelen, P. A. (2013). Symptoms of prolonged grief, depression, and adult separation anxiety: Distinctiveness and correlates. *Psychiatry Research, 207*, 68–72. Available from http://dx.doi.org/10.1016/j.psycres.2012.09.021.

Boerner, S., & Jobst, J. (2013). Enjoying theatre: The role of visitor's response to the performance. *Psychology of Aesthetics, Creativity, and the Arts, 7*(4), 391–408. Available from http://dx.doi.org/10.1037/a0034570.

Boerner, S., Jobst, J., & Wieman, M. (2010). Exploring the theatrical experience: Results from an empirical investigation. *Psychology of Aesthetics, Creativity, and the Arts, 4*(3), 173–180. Available from http://dx.doi.org/10.1037/a0018460.

Bogart, A. (2001). *A director prepares: Seven essays on arts and theatre*. New York: Routledge.

Bonanno, G. A. (2004). Loss, trauma, and human resilience: Have we underestimated the human capacity to thrive after extremely aversive events? *American Psychologist, 59*(1), 20–28. Available from http://dx.doi.org/10.1037/0003-066X.59.1.20.

Bonanno, G. A., & Burton, C. L. (2013). Regulatory flexibility: An individual differences perspective on coping and emotion regulation. *Perspectives on Psychological Science, 86*(6), 591–612. Available from http://dx.doi.org/10.1177/1745691613504116.

Bonanno, G. A., Papa, A., Lalande, K., Westphal, M., & Coifman, K. (2004). The importance of being flexible: The ability to both enhance and suppress emotional expression predicts long-term adjustment. *Psychological Science, 15*(7), 482–487.

Bonnemeier, H., Wiegand, U. K. H., Brandes, A., Kluge, N., Katus, H. A., Richardt, G., & Potratz, J. (2003). Circadian profile of cardiac autonomic nervous modulation in healthy subjects: Differing effects of aging and gender on heart rate variability. *Journal of Cardiovascular Electrophysiology*, *14*(8), 791–799.

Bonneville-Roussy, A., Lavigne, G. L., & Vallerand, R. J. (2010). When passion leads to excellence: The case of musicians. *Psychology of Music*, *39*(1), 123–138. Available from http://dx.doi.org/10.1177/0305735609352441.

Bornstein, M. H., & Lamb, M. E. (2005). *Developmental science: An advanced textbook* (5th ed.). Mahwah, NJ: Lawrence Erlbaum Associates, Publishers.

Bornstein, S. R., Schuppenies, A., Wong, M.-L., & Licinio, J. (2006). Approaching the shared biology of obesity and depression: The stress axis as the locus of gene-environment interactions. *Molecular Psychiatry*, *11*, 892–902.

Bouchard, M.-A., Target, M., Lecours, S., Fonagy, P., Tremblay, L.-M., Schacter, A., & Stein, H. (2008). Mentalization in adult attachment narratives: Reflective functioning, mental states, and affect elaboration compared. *Psychoanalytic Psychology*, *25*(1), 47–66. Available from http://dx.doi.org/10.1037/0736-9735.25.1.47.

Boucher, H., & Ryan, C. A. (2011). Performance stress and the very young musicians. *Journal of Research in Music Education*, *58*(4), 329–345. Available from http://dx.doi.org/10.1177/0022429410386965.

Bourgeois-Bougrine, S., Botella, M., Glaveau, V., Guillou, K., De Biasi, P. M., & Lubart, T. (2014). The creativity maze: Exploring creativity in screenplay writing. *Psychology of Aesthetics, Creativity, and the Arts*, *8*(4), 384–399. Available from http://dx.doi.org/10.1037/a0037839.

Bournelli, P., Makri, A., & Mylonas, K. (2009). Motor creativity and self-concept. *Creativity Research Journal*, *21*(1), 104–110. Available from http://dx.doi.org/10.1080/10400410802633657.

Bowerman, E., Whatman, C., Harris, N., Bradshaw, E., & Karin, J. (2014). Are maturation, growth, and lower extremity alignment associated with overuse injury in elite adolescent ballet dancers? *Physical Therapy in Sport*, *15*, 234–241. Available from http://dx.doi.org/10.1016/j.ptsp.2013.12.014.

Bowers, M. T., Green, B. C., Hemme, F., & Chalip, L. (2014). Assessing the relationship between youth sport participation settings and creativity in adulthood. *Creativity Research Journal*, *26* (3), 314–327. Available from http://dx.doi.org/10.1080/10400419.2014.929420.

Bowie, D. Retrieved from www.goodreads.com/quotes/tag/acting.

Bowlby, J. (1988). *A secure base: Parent-child attachment and healthy human development*. New York: Basic Books, Inc.

Bradley, B., Davis, T. A., Kaye, J., & Wingo, A. (2014). Developmental social factors as promoters of resilience in childhood and adolescence. In M. Kent, M. C. Davis, & J. W. Reich (Eds.), *The resilience handbook: Approaches to stress and trauma* (pp. 197–208). New York: Routledge.

Brady, R. E., Constans, J. I., Marx, B. P., Spira, J. L., Kimbrell, T. A., & Kramer, T. L. (2015). Effect of symptom over-reporting on heart rate variability in veterans with post-traumatic stress disorder. *Journal of Trauma and Dissociation*, *16*, 551–562. Available from http://dx.doi.org/10.1080/15299732.2015.1021505.

Brattico, E., & Pearce, M. (2013). The neuroaesthetics of music. *Psychology of Aesthetics, Creativity, and the Arts*, *7*(1), 48–61. Available from http://dx.doi.org/10.1037/a0031624.

Bremner, J. D. (2005). *Brain imaging handbook*. New York: W. W. Norton and Company.

Bridges, L. J., Denham, S. A., & Ganiban, J. M. (2004). Definitional issues in emotion regulation research. *Child Development*, *75*(2), 340–345.

Bridgstock, R. (2005). Australian artists, starving and well-nourished: What can we learn from the prototypical protean career. *Australian Journal of Career Development*, *14*(3), 40–48.

Briere, J., & Gil, E. (1998). Self-mutilation in clinical and general population samples: Prevalence, correlates, and functions. *American Journal of Orthopsychiatry*, *68*, 609–620.

Brockett, O. G., & Hildy, F. (2007). *History of the theatre* (10th ed.). New York: Holt, Rinehart and Winston, Inc.

Brodsky, W. (2006). In the wings of British orchestras: A multi-episode interview study among symphony players. *Journal of Occupational and Organizational Psychology*, *79*, 673–690. Available from http://dx.doi.org/10.1348/096317905X68213.

Brodsky, W. (2011). Rational behind investigating positive aging among symphony orchestra musicians: A call for a new arena of empirical study. *Musicae Scientiae*, *15*(1), 3–15. Available from http://dx.doi.org/10.1177/1029864910393425.

Bronner, S., Ojofeitimi, S., Lora, J. B., Southwick, H., Kulak, M. C., & Gamboa, J. (2014). A preseason cardiorespiratory profile of dancers in nine professional ballet and modern companies. *Journal of Dance Medicine and Science*, *18*(2), 74–85. Available from http://dx.doi.org/10.12678/1089-313X.18.2.74.

Bronner, S., Ojofeitimi, S., & Spriggs, J. (2003). Occupational musculoskeletal disorders in dancers. *Physical Therapy Review*, *8*, 57–68. Available from http://dx.doi.org/10.1179/1083319032250024.

Bronner, S., & Rakov, S. (2014). An accelerated step test to assess dancer pre-season aerobic fitness. *Journal of Dance Medicine and Science*, *18*(1), 12–21. Available from http://dx.doi.org/10.12678/1089-313X.18.1.12.

Broth, M. (2011). The theatre performance as interaction between actors and their audience. *Nottingham French Studies*, *50*(2), 113–133.

Brown, A. C., Wells, T. J., Schade, M. L., Smith, D. L., & Fehling, P. C. (2007). Effects of plyometric training versus traditional weight training on strength, power, and aesthetic jumping ability in female collegiate dancers. *Journal of Dance Medicine and Science*, *11*(2), 38–44.

Brown, D., & Wyon, M. (2014). An international study on dietary supplementation use in dancers. *Medical Problems of Performing Artists*, *29*(4), 229–234.

Brown, S., & Martinez, M. J. (2007). Activation of premotor vocal areas during musical discrimination. *Brain and Cognition*, *63*, 59–69. Available from http://dx.doi.org/10.1016/j.bandc.2006.08.006.

Brown, S., Martinez, M. J., Hodges, D. A., Fox, P. T., & Parsons, L. M. (2004). The song system of the human brain. *Cognitive Brain Research*, *20*, 363–375. Available from http://dx.doi.org/10.1016/j.cogbrainres.2004.03.016.

Brown, S., Martinez, M. J., & Parsons, L. M. (2006). The neural basis of human dance. *Cerebral Cortex*, *16*, 1157–1167. Available from http://dx.doi.org/10.1093/cercor/bhj057.

Brugues, A. O. (2011). Music performance anxiety—Part I: A review of its epidemiology. *Medical Problems of Performing Artists*, *26*(2), 102–105.

Bruya, B. (2010). Introduction: Toward a theory of attention that includes effortless attention and action. In B. Bruya (Ed.), *Effortless attention: A new perspective in the cognitive science of attention and action* (pp. 1–28). Cambridge, MA: MIT Press.

Buchanan, A., White, S. W., Walters, W. A., Redman, S., Quail, A. W., Coffee, D. B. F., & Hennessy, E. J. (1991). Teenage ballet dancers as a model of the female athlete sensitivity of endocrine control of the menstrual cycle to exercise. *The Australian Journal of Science and Medicine in Sport*, *24*, 63–67.

Buckingham, R. M. (2002). Extraversion, neuroticism and the four temperaments of antiquity: An investigation of physiological reactivity. *Personality and Individual Differences*, *32*, 225–246.

Buckley, T., & Manchester, R. (2006). Overuse injuries in non-classical recreational instruments. *Medical Problems of Performing Artists*, *21*(2), 80–87.

Budden, A. (2009). The role of shame in posttraumatic stress disorder: A proposal for a socio-emotional model for DSM-V. *Social Science and Medicine*, *69*, 1032–1039. Available from http://dx.doi.org/10.1016/j.socscimed.2009.07.032.

Bulbena, A., Gago, J., Pailhez, G., Sperry, L., Fullana, M. A., & Vilarroya, O. (2011). Joint hypermobility syndrome is a risk factor for trait anxiety disorders: A 15-year follow-up cohort study. *General Hospital Psychiatry*, *33*, 363–370. Available from http://dx.doi.org/10.1016/j.genhosppsych.2011.03.004.

Burgoyne, S., Poulin, K., & Rearden, A. (1999). The impact of acting on student actors: Boundary blurring, growth, and emotional distress. *Theatre Topics*, *9*(2), 157–179.

Burns, J. A., Stadelman-Cohen, T., & Zeitels, S. M. (2009). Phonomicrosurgical treatment of intracordal vocal-fold cysts in singers. *The Laryngoscope*, *119*, 419–422. Available from http://dx.doi.org/10.1002/lary.20001.

Busch, H., Hofer, J., Chasiotis, A., & Campos, D. (2013). The achievement flow motive as an element of the autotelic personality: Predicting educational attainment in three cultures. *European Journal of Psychology of Education*, *28*(2), 239–254. Available from http://dx.doi.org/10.1007/s10212-012-0112-y.

Butcher, J. L., & Niec, L. N. (2005). Disruptive behaviors in creativity in childhood: The importance of affect regulation. *Creativity Research Journal*, *17*(2 & 3), 181–193.

Butler, L. D. (2006). Normative dissociation. *Psychiatric Clinics of North America*, *29*, 45–62.

Buttsworth, L. M., & Smith, G. A. (1995). Personality of Australian performing musicians by gender and by instrument. *Personality and Individual Differences*, *18*(5), 595–603.

Byng-Hall, J. (2002). Relieving parentified children's burdens in families with insecure attachment patterns. *Family Process*, *41*(3), 375–388.

Byron, K., Khazanchi, S., & Nazarian, D. (2010). The relationship between stressors and creativity: A meta-analysis examining competing theoretical models. *Journal of Applied Psychology*, *95*(1), 201–212. Available from http://dx.doi.org/10.1037/a0017868.

Cacioppo, J. T., Tassinary, L. G., & Berntson, G. G. (2007). *Handbook of psychophysiology* (3rd ed.). Cambridge, UK: Cambridge University Press. Available from http://dx.doi.org/10.1017/CBO9780511546396.

Cahalan, R., & O'Sullivan, K. (2013). Injury in professional Irish dancers. *Journal of Dance Medicine and Science*, *17*(4), 150–158. Available from http://dx.doi.org/10.12678/1089-313X.17.4.150.

Calkins, S. D. (2004). Early attachment processes and the development of emotional self-regulation. In R. F. Baumeister, & K. D. Vohs (Eds.), *Handbook of self-regulation: Research, theory, and application* (pp. 324–339). New York: The Guilford Press.

Callas, M. Retrieved from www.goodreads.com/quotes/tag/opera.

Calogero, R. M., & Pedrotty, K. N. (2004). The practice and process of healthy exercise: An investigation of the treatment of exercise abuse in women with eating disorders. *Eating Disorders: The Journal of Treatment and Prevention*, *12*(4), 273–291. Available from http://dx.doi.org/10.1080/10640260490521352.

Calvo-Merino, B., Glaser, D. E., Grezes, J., Passingham, R. E., & Haggard, P. (2005). Action observation and acquired motor skills: An fMRI study with expert dancers. *Cerebral Cortex*, *15*, 1243–1249. Available from http://dx.doi.org/10.1093/cercor/bhi007.

Calvo-Merino, B., Grezes, J., Glaser, D. E., Passingham, R. E., & Haggard, P. (2006). Seeing or doing? Influence of visual and motor familiarity in action observation. *Current Biology*, *16*, 1905–1910. Available from http://dx.doi.org/10.1016/j.cub.2006.10.065.

Calvo-Merino, B., Jola, C., Glaser, D. E., & Haggard, P. (2008). Towards a sensorimotor aesthetics of performing art. *Consciousness and Cognition*, *17*, 911–922. Available from http://dx.doi.org/10.1016/j.concog.2007.11.003.

Cameron, J. E., Duffy, M., & Glenwright, B. (2015). Singers take center stage! Personality traits and stereotypes of popular musicians. *Psychology of Music*, *43*(6), 818–830. Available from http://dx.doi.org/10.1177/0305735614543217.

Camurri, A., Lagerlof, I., & Volpe, G. (2003). Recognizing emotion from dance movement: Comparison of spectator recognition and automated techniques. *International Journal of Human-Computer Studies*, *59*, 213–225. Available from http://dx.doi.org/10.1016/S1071-5819(03)00050-8.

Cantero, M.-J., Alfonso-Benlliure, V., & Melero, R. (2016). Creativity in middle childhood: Influence of perceived maternal sensitivity, self-esteem, and shyness. *Creativity Research Journal*, *28*(1), 105–113. Available from http://dx.doi.org/10.1080/10400419.2016.1125246.

Caraty, M.-J., & Montacie, C. (2014). Vocal fatigue induced by prolonged oral reading: Analysis and detection. *Computer Speech and Language*, *28*, 453–466. Available from http://dx.doi.org/10.1016/j.csl.2012.12.003.

Carbonell, D. M., & Parteleno-Barehmi, C. (1999). Psychodrama groups for girls coping with trauma. *International Journal of Group Psychotherapy*, *49*(3), 285–306.

Card, P. J., Leviit, P., Gluhovsky, M., & Rinaman, L. (2005). Early experience modifies the postnatal assembly of autonomic emotional motor circuits in rats. *Journal of Neuroscience*, *25*(40), 9102–9111.

Cardena, E., Lynn, S. J., & Krippner, S. (Eds.), (2000). *Varieties of anomalous experience: Examining the scientific evidence.* Washington, DC: American Psychological Association.

Career Transition for Dancers. www.careertransition.org.

Carlson, E. A., Sroufe, L. A., & Egeland, B. (2004). The construction of experience: A longitudinal study of representation and behavior. *Child Development*, *75*(1), 66–83.

Carlsson, I. (2002). Anxiety and flexibility of defense related to high or low creativity. *Creativity Research Journal*, *14*(3 & 4), 341–349.

Carlsson, I., Wendt, P. E., & Risberg, J. (2000). On the neurobiology of creativity: Differences in frontal activity between high and low creative subjects. *Neuropsychologia*, *38*(6), 873–885.

Carmeli, Y. (2001). Circus play, circus talk, and the nostalgia for a total order. *The Journal of Popular Culture*, *35*(3), 157–164. Available from http://dx.doi.org/10.1111/j.0022-3840.2001.3503_157.x.

Carney, R. M., Freedland, K. E., & Veith, R. C. (2005). Depression, the autonomic nervous system, and coronary disease. *Psychosomatic Medicine*, *67*(S1), S29–S33.

Carnicke, S. M. (2000). Stanislavsky's system: Pathway for the actor. In A. Hodge (Ed.), *Twentieth century actor training* (pp. 11–36). New York: Routledge.

Carson, S. H., Peterson, J. B., & Higgins, D. M. (2003). Decreased latent inhibition is associated with increased creative achievement in high-functioning individuals. *Journal of Personality and Social Psychology*, *85*(3), 499–506. Available from http://dx.doi.org/10.1037/0022.3514.85.3.499.

Carter, L. A. (2005). Temperament as a unifying basis for personality and psychopathology. *Journal of Abnormal Psychology*, *114*(4), 505–521. Available from http://dx.doi.org/10.1037/0021-843X.114.4.505.

Carthy, T., Horesh, N., Apter, A., Edge, M. D., & Gross, J. J. (2010). Emotional reactivity and cognitive regulation in anxious children. *Behavior Research and Therapy*, *48*(5), 384–393. Available from http://dx.doi.org/10.1016/j.brat.2009.12.013.

Carton, M. M. S., Carton, C. S., & Vazquez, M. S. (2010). P01-342—suicide in musicians in the last forty years: A creative mind condition or a psychopathologic expression. *European Psychiatry*, *25*(1), 555. Available from http://dx.doi.org/10.1016/S0924-9338(10)70550-2.

Caruso, S., Roccasalva, L., Sapienza, G., Zappalá, M., Nuciforo, G., & Biondi, S. (2000). Laryngeal cytological aspects in women with surgically induced menopause who were treated with transdermal estrogen replacement therapy. *Fertility and Sterility*, *74*(6), 1073–1079. Available from http://dx.doi.org/10.1016/S0015-0282(00)01582-X.

Caruth, C. (1996). *Unclaimed experience: Trauma, narrative, and history*. Baltimore, MD: The John Hopkins University Press.

Carver, C. S., & White, T. L. (1994). Behavioral inhibition, behavioral activation, and affective responses to impending reward and punishment: The BIS/BAS scales. *Journal of Personality and Social Psychology*, *67*(2), 319–333.

Casey, B. J., & Jones, R. M. (2010). Neurobiology of the adolescent brain and behavior: Implications for substance use disorders. *Journal of American Academy of Child and Adolescent Psychiatry*, *49*(12), 1189–1201.

Casey, B. J., Jones, R. M., & Hare, T. A. (2008). The adolescent brain. *Annals of the New York Academy of Science*, *1124*, 111–126. Available from http://dx.doi.org/10.1196/annals.1440.010.

Casey, B. J., Jones, R. M., & Somerville, L. H. (2011). Braking and accelerating of the adolescent brain. *Journal of Research on Adolescence*, *21*(1), 21–33. Available from http://dx.doi.org/10.1111/j.1532-7795.2010.00712x.

Catterall, J. S., Dumais, S. A., & Hampden-Thompson, G. (2012). *The arts and achievement in at-risk youth: Findings from four longitudinal studies (Research Report # 55)*. Washington, DC: National Endowment for the Arts.

Cervellin, G., & Lippi, G. (2011). From the music-beat to heart-beat: A journey in the complex interactions between music, brain and heart. *European Journal of Internal Medicine*, *22*, 371–374. Available from http://dx.doi.org/10.1016/j.ejim.2011.02.019.

Chamorro-Premuzic, T., Furnham, A., Goma-i-Freixanet, M., & Muro, A. (2009). Personality, self-estimated intelligence, and uses of music: A Spanish replication and extension using structural equation modeling. *Psychology of Aesthetics, Creativity, and the Arts*, *3*(3), 149–155. Available from http://dx.doi.org/10.1037/a0015342.

Chan, C., Driscoll, T., & Ackermann, B. (2014). Exercise DVD effect on musculoskeletal disorders in professional orchestra musicians. *Occupational Medicine*, *64*, 23–30. Available from http://dx.doi.org/10.1093/occmed.kqt117.

Chan, R. R. M., Chow, C.-Y., Lee, G. P. S., To, L.-K., Tsang, X. Y. S., Yeung, S. S., & Yeung, E. W. (2000). Self-perceived exertion level and objective evaluation of neuromuscular fatigue in a training session of orchestral violin players. *Applied Ergonomics*, *31*, 335–341.

Chang, H. J., & Kuo, C. C. (2009). Overexcitabilities of gifted and talented students and its related research in Taiwan. *Asia-Pacific Journal of Gifted and Talented Education*, *1*(1), 41–74.

Chaplin, L. N., & Norton, M. I. (2014). Why we think we can't dance: Theory of mind and children's desire to perform. *Child Development*, *86*(2), 651–658. Available from http://dx.doi.org/10.1111/cdev.12314.

Chappell, K. (2007). The dilemmas of teaching for creativity: Insights from expert specialist dance teachers. *Thinking Skills and Creativity*, *2*, 39–56. Available from http://dx.doi.org/10.1016/j.tsc.2007.01.001.

Charles, R. E., & Runco, M. A. (2000–2001). Developmental trends in the evaluative and divergent thinking of children. *Creativity Research Journal*, *13*(3–4), 417–437.

Chartrand, D., & Chatfield, S. J. (2005). A critical review of the prevalence of secondary amenorrhea in ballet dancers. *Journal of Dance Medicine and Science*, *9*(3–4), 74–80.

Charyton, C., & Snelbecker, G. E. (2007). Engineers' and musicians' choices of self-descriptive adjectives as potential indicators of creativity by gender and domain. *Psychology of Aesthetics, Creativity, and the Arts*, *1*(2), 91–99. Available from http://dx.doi.org/10.1037/1931-3896.1.2.91.

Chatard, A., Pyszczynski, T., Arndt, J., Selimbegovic, L., Konan, P. N., & van der Linden, M. (2012). Extent trauma exposure and PTSD symptom severity as predictors of anxiety-buffer functioning. *Psychological Trauma: Theory, Research, Practice, and Policy*, *4*(1), 47–55. Available from http://dx.doi.org/10.1037/a0021085.

Chavez-Eakle, R. A. (2007). From incubation to insight: Working memory and the role of the cerebellum. *Creativity Research Journal*, *19*, 31–34.

Chavez-Eakle, R. A., Eakle, A. J., & Cruz-Fuentes, C. (2012). The multiple relations between creativity and personality. *Creativity Research Journal*, *24*(1), 76–82. Available from http://dx.doi.org/10.1080/10400419.2012.64923.3.

Chavez-Eakle, R. A., Lara, M. C., & Cruz, C. (2006). Personality: A possible bridge between creativity and psychopathology. *Creativity Research Journal*, *18*(1), 27–38.

Cheek, J. M., & Buss, A. H. (1981). Shyness and sociability. *Journal of Personality and Social Psychology*, *41*(2), 330–339.

Chimenti, R. L., van Dillen, L. R., Prather, H., Hunt, D., Cheimenti, P. C., & Khoo-Summers, L. (2013). Underutilization of worker's compensation insurance among professional orchestral musicians. *Medical Problems of Performing Artists*, *28*(1), 54–60.

Chirico, A., Serino, S., Cipresso, P., Gaggioli, A., & Riva, G. (2015). When music "flows". State and trait in musical performance, composition and listening: A systematic review. *Frontiers in Psychology*, *6*, 906. Available from http://dx.doi.org/10.3389/fpsyg.2015.00906.

Choi, J.-B., Hong, S., Nelesen, R., Bardwell, W. A., Natarajan, L., Schubert, C., & Dimsdale, J. E. (2006). Age and ethnicity differences in short-term heart-rate variability. *Psychosomatic Medicine*, *68*, 421–426. Available from http://dx.doi.org/10.1097/01.psy.0000221378.09239.6a.

Christensen, J. F., & Calvo-Merino, B. (2013). Dance as a subject for empirical aesthetics. *Psychology of Aesthetics, Creativity, and the Arts*, *7*(1), 76–88. Available from http://dx.doi.org/10.1037/a0031827.

Chua, J. (2015). The role of social support in dance talent development. *Journal for the Education of the Gifted*, *38*(2), 169–195. Available from http://dx.doi.org/10.1177/0162353215578281.

Chui, P. H., & Deldin, P. J. (2007). Neural evidence for enhanced error detection in major depressive disorder. *American Journal of Psychiatry*, *164*(4), 608–616.

Cinzia, D. D., & Vittorio, G. (2009). Neuroaesthetics: A review. *Current Opinion in Neurobiology*, *19*, 682–687. Available from http://dx.doi.org/10.1016/j.conb.2009.09.001.

Cizek, E., Kelly, P., Kress, K., & Mattfeldt-Beman, M. (2016). Factors affecting healthful eating among touring popular musicians and singers. *Medical Problems of Performing Artists*, *31*(2), 63–68. Available from http://dx.doi.org/10.21091/mppa.2016.2013.

Clabaugh, A., & Morling, B. (2004). Stereotype accuracy of ballet and modern dancers. *Journal of Social Psychology*, *144*(1), 31–48.

Clark, T., Gupta, A., & Ho, C. H. (2014). Developing a dancer wellness program employing developmental evaluation. *Frontiers in Psychology*, *5*(731), 1–14. Available from http://dx.doi.org/10.3389/f.psyg.2014.00731.

Clark, T., Holmes, P., & Redding, E. (2011). Investigating the physiological demands of musical performance. In A. Williamon, D. Edwards, & L. Bartel (Eds.), *Proceedings of the international symposium on performance science* (pp. 137–142). Utrecht, The Netherlands: European Association of Concervatoires (AEC).

Clark, T., Williamon, A., & Redding, E. (2013). The value of health screening in music schools and conservatories. *Clinical Rheumatology, 32*, 497–500. Available from http://dx.doi.org/10.1007/s0067-013-2203-9.

Claxton, A. F., Pannells, T. C., & Rhoads, P. A. (2005). Developmental trends in the creativity of school-age children. *Creativity Research Journal, 17*(4), 327–335.

Cloitre, M., Courtois, C. A., Charuvastra, A., Caapezza, R., Stolbach, B. C., & Green, B. L. (2011). Treatment of complex PTSD: Results of the ISTSS expert clinician survey on best practices. *Journal of Traumatic Stress, 24*(6), 615–627.

Cloninger, C. R. (1994). Temperament and personality. *Current Opinion in Neurobiology, 4*, 266–273.

Cloninger, C. R., Svrakic, D. M., & Pyzybeck, T. R. (1993). A psychobiological model of temperament and character. *Archives of General Psychiatry, 50*(12), 975–990.

Clyesdale, G. (2006). Creativity and competition: The Beatles. *Creativity Research Journal, 18*(2), 129–139.

Coakley, J. (2009). From the outside in: Burnout as an organization. *Journal of Intercollegiate Sports, 2*, 35–41.

Cohen, G. (2006). Research on creativity and aging: The positive impact of the arts on health and illness. *Generations, 1*, 7–15.

Cohen, J. L., Gupta, P. K., Lichstein, E., & Chadda, K. D. (1980). The heart of a dancer: Noninvasive cardiac evaluation of professional ballet dancers. *Journal of Cardiology, 45*, 959–965.

Cohen, S. J. (Ed.), (1974). *Dance as a theatre art.* Princeton, NY: A Dance Horizons Book.

Cole, T., & Krich Chinoy, H. (Eds.), (1976). *Directors on directing: A source book of the modern theatre.* New York: Macmillan Publishing Comp.

Coleridge, S. T. (1817/1976). Fancy and imagination. In A. Rothenberg, & C. R. Hausman (Eds.), *The creativity question* (pp. 61–62). Durham, NC: Duke University Press.

Collet, C., Vernet-Maury, E., Delhomme, G., & Dittmar, A. (1997). Autonomic nervous system response patterns specificity to basic emotions. *Journal of the Autonomic Nervous System, 62*, 45–57.

Collins, M. A., & Amabile, T. M. (1999). Motivation and creativity. In R. J. Sternberg (Ed.), *Handbook of creativity* (pp. 297–312). New York: Cambridge University Press.

Collishaw, S., Pickles, A., Messer, J., Rutter, M., Shearer, C., & Maughan, B. (2007). Resilience to adult psychopathology following childhood maltreatment: Evidence from a community sample. *Child Abuse and Neglect, 31*, 211–229. Available from http://dx.doi.org/10.1016/j.chiabu.2007.02.004.

Conde, V., Altenmuller, E., Villringer, A., & Ragert, P. (2012). Task-irrelevant auditory feedback facilitates motor performance in musicians. *Frontiers in Psychology, 3*(146), 1–10. Available from http://dx.doi.org/10.3389/f.psyg.2012.00146.

Conroy, D. E., Poczwardowski, A., & Henschen, K. P. (2001). Evaluative criteria and consequences associated with failure and success for elite athletes and performing artists. *Journal of Applied Sport Psychology, 13*(3), 300–322. Available from http://dx.doi.org/10.1080/103132001753144428.

Constantini, N., Arieli, R., Chodick, G., & Dubnov-Raz, G. (2010). High prevalence of Vitamin D insufficiency in athletes and dancers. *Clinical Journal of Sport Medicine, 20* (5), 368–371. Available from http://dx.doi.org/10.1097/JSM.0b013e3181120712.

Cooper, R. B., & Jayatilaka, B. (2006). Group creativity: The effects of extrinsic, intrinsic, and obligation motivations. *Creativity Research Journal*, *18*(2), 153–172.

Copeland, C. T. (2016). Take some time to feel this over: Relations between mood responses, indecision, and creativity. *Creativity Research Journal*, *28*(1), 11–15. Available from http://dx.doi.org/10.1080/10400419.2016.1125247.

Copland, A. (1957). *What to listen for in music*. New York: New American Library.

Corbetta, M., Patel, G., & Shulman, G. L. (2008). The reorienting system of the human brain: From environment to Theory of Mind. *Neuron Review*, *58*, 306–324.

Corlu, M., Maes, P.-J., Muller, C., Kochman, K., & Leman, M. (2015). The impact of cognitive load on operatic singers' timing performance. *Frontiers in Psychology*, *6*(429), 1–10. Available from http://dx.doi.org/10.3389/fpsyg.2015.00429.

Corrigall, K. A., Schellenberg, E. G., & Misura, N. M. (2013). Music training, cognition, and personality. *Frontiers in Psychology*, *4*(e222), 1–10. Available from http://dx.doi.org/10.3389/fpsyg.2013.00222.

Corriveau, K. H., Harris, P. L., Meins, E., Fernhough, C., Arnott, B., Elliot, L., ... de Rosnay, M. (2009). Young children's trust in their mother's claims: Longitudinal links with attachment security in infancy. *Child Development*, *80*(3), 750–761. Available from http://dx.doi.org/10.111/j.1467-8624.2009.01295.x.

Costa, M. S. S., Ferreira, A. S., Orsini, M., Silva, E. B., & Felicio, L. R. (2016). Characteristics and prevalence of musculoskeletal injury in professional and non-professional ballet dancers. *Brazilian Journal of Physical Therapy*, *20*(2), 166–175. Available from http://dx.doi.org/10.1590/bjpt-rbf.2014.0142.

Cote, S., & Morgan, L. M. (2002). A longitudinal analysis of the association between emotion regulation, job satisfaction, and intentions to quit. *Journal of Organizational Behavior*, *23*, 947–962. Available from http://dx.doi.org/10.1002/job.174.

Coubard, O. A., Duretz, S., Lefebrve, V., Lapalus, P., & Ferrufino, L. (2011). Practice of contemporary dance improves cognitive flexibility in aging. *Frontiers in Aging Neuroscience*, *3*(13), 1–12. Available from http://dx.doi.org/10.3389/fnagi.2011.000013.

Couillandre, A., Lewton-Brain, P., & Portero, P. (2008). Exploring the effects of kinesiological awareness and mental imagery on movement intention in the performance of demi-plie. *Journal of Dance Medicine and Science*, *12*(3), 91–98.

Coulson, S. (2012). Collaborating in a competitive world: Musicians' working lives and understandings of entrepreneurship. *Work, Employment and Society*, *26*(2), 246–261. Available from http://dx.doi.org/10.1177/0950017011432919.

Coutts, R., Gilleard, W., Hennessy, M., Silk, A., Williams, G., & Weatherby, R. P. (2006). Development and assessment of an incremental fatigue protocol for contemporary dance. *Medical Problems of Performing Artists*, *21*, 65–70.

Covington, M. V., & Mueller, K. J. (2001). Intrinsic versus extrinsic motivation: An approach/avoidance reformulation. *Educational Psychology Review*, *13*(2), 157–176. Available from http://dx.doi.org/10.1040-726X/01/0600-0157$19.50/0.

Cowgill, J. G. (2009). Breathing for singers: A comparative analysis of body types and breathing tendencies. *Journal of Singing*, *66*(2), 141–147.

Cowley, D. (2015). Treating insomnia can decrease suicide risk. *New England Journal of Medicine Journal Watch*, *21*(4), 28.

Cox, W. J., & Kenardy, J. (1993). Performance anxiety, social phobia, and setting effects in instrumental music students. *Journal of Anxiety Disorders*, *7*, 49–60.

Crabtree, J., & Green, M. J. (2016). Creative cognition and psychosis vulnerability: What's the difference. *Creativity Research Journal*, *28*(1), 24–32. Available from http://dx.doi.org/10.1080/10400419.2015.1030305.

Craig, A. D. (2004). Human feelings: Why are some more aware than others? *Trends in Cognitive Sciences*, *8*(6), 239–241. Available from http://dx.doi.org/10.1016/tics.2004.04.004.

Creasey, G. (2002). Psychological distress in college-aged women: Links with unresolved/preoccupied attachment status and the mediating role of negative mood regulation expectations. *Attachment and Human Development*, *4*, 261–277. Available from http://dx.doi.org/10.1080/14616730210167249.

Crespo, M., & Fernandez-Lansac, V. (2015). Memory and narrative of traumatic events: A literature review. *Psychological Trauma: Theory, Research, Practice, and Policy*, *8*(2), 149–156. Available from http://dx.doi.org/10.1037tra000041.

Crisafulli, A., Scott, A. C., Wensel, R., Davos, C. H., Francis, D. P., Pagliaro, P., . . . Piepoli, M. F. (2003). Muscle metaboreflex-induced increases in stroke volume. *Medicine and Science in Sports and Exercise*, *35*(2), 221–228. Available from http://dx.doi.org/10.1249/01.MSS.0000048639.02548.24.

Critchley, H. D. (2005). Neural mechanisms of autonomic, affective, and cognitive integration. *Journal of Comparative Neurology*, *493*, 154–166. Available from http://dx.doi.org/10.1002/cne.20749.

Critchley, H. D. (2012). How emotions are shaped by bodily states. *Emotion Review*, *4*(2), 163–168. Available from http://dx.doi.org/10.1177/1754073911430132.

Critchley, H. D., Wiens, S., Rotshtein, P., Ohman, A., & Dolan, R. (2004). Neural systems supporting interoception awareness. *Nature Neuroscience*, *7*(2), 189–195.

Crnivec, R. (2004). Assessment of health risks in musicians of the Slovene Philharmonic orchestra, Ljubljana, Slovenia. *Medical Problems of Performing Artists*, *19*(3), 140–145.

Cross, E. S., & Ticini, L. F. (2012). Neuroaesthetics and beyond: New horizons in applying the science of the brain to the art of dance. *Phenomenology and Cognitive Science*, *11*, 5–16. Available from http://dx.doi.org/10.1007/s11097-010-9190-y.

Crozier, W. R. (2003). Individual differences in artistic achievement: A within family case study. *Creativity Research Journal*, *15*(4), 311–319.

Crusius, T. (1991). *A teacher's introduction to philosophical hermeneutics*. Urbana, IL: National Council of Teachers of English.

Crust, L., & Swann, C. (2013). The relationship between mental toughness and dispositional flow. *European Journal of Sport Science*, *13*(2), 215–220. Available from http://dx.doi.org/10.1080/17461391.2011.635698.

Cruz, C. A. (2014). Contemporary circus dramaturgy: An interview with Louis Patrick Leroux. *Theatre Topics*, *24*(3), 269–273. Available from http://dx.doi.org/10.1353/tt.2014.0029.

Csikszentmihalyi, M. (1990). *Flow: The psychology of optimal experience*. New York: Harper & Row.

Csikszentmihalyi, M. (1996). *Creativity: Flow and the psychology of discovery and invention*. New York: Harper Collins.

Csikszentmihalyi, M. (1999). Implications of a systems perspective for the study of creativity. In R. J. Sternberg (Ed.), *Handbook of creativity* (pp. 313–335). New York: Cambridge University Press.

Culbertson, R. (1995). Embodied memory, transcendence, and telling: Recounting trauma, re-establishing the self. *New Literary History*, *26*(1), 169–195.

Cumberbatch, B. Retrieved from www.goodreads.com/quotes/tag/acting.

Cumming, J., & Duda, J. (2012). Profiles of perfectionism, body-related concerns, and indicators of psychological health in vocational dance students: An investigation of the 2×2 model of perfectionism. *Psychology of Sport and Exercise*, *13*(6), 729–738.

Cunningham, M. Retrieved from www.goodreads.com/quotes/tag/dancing.

Currier, J. M., Holland, J. M., & Neimeyer, R. A. (2006). Sense-making, grief, and the experience of violent loss: Toward a meditational model. *Death Studies*, *30*, 403–428.

Cutting, J. E., DeLong, J. E., & Brunick, K. L. (2011). Visual activity in Hollywood film: 1935 to 2005 and beyond. *Psychology of Aesthetics, Creativity, and the Arts*, *5*(2), 115–125. Available from http://dx.doi.org/10.1037/a0020995.

Dalebroux, A., Goldstein, T. R., & Winner, E. (2008). Short-term mood repair through art-making: Positive emotion is more effective than venting. *Motivation and Emotion*, *32* (4), 288–295. Available from http://dx.doi.org/10.1007/s11031-008-9105-1.

Damasio, A. (1999). *The feeling of what happens: Body and emotion in the making of consciousness*. New York: Harcourt Brace and Company.

Danese, A., & McEwen, B. S. (2012). Adverse childhood experience, allostasis, allostatic load, and age-related disease. *Physiology and Behavior*, *106*, 29–39. Available from http://dx.doi.org/10.1016/j.physbeh.2011.08.019.

Daprati, E., Iosa, M., & Haggard, P. (2009). A dance to the music of time: Aesthetically-relevant changes in body posture in performing art. *PLoS ONE*, *4*(3), e5023. Available from http://dx.doi.org/10.137/journal.pone.0005023.

Davidson, J. W., & Borthwick, S. J. (2002). Family dynamics and family scripts: A case study of musical development. *Psychology of Music*, *30*, 121–136.

Davila, E. P., Florez, H., Fleming, L. E., Lee, D. J., Goodman, E., LeBlanc, W. G., ... Clarke, T. (2010). Prevalence of the metabolic syndrome among U.S. workers. *Diabetes Care*, *33*, 2390–2395.

Davis, C. G., Wohl, M. J. A., & Verberg, N. (2007). Profiles of posttraumatic growth following an unjust loss. *Death Studies*, *31*, 603–712.

Davis, M. A. (2009). Understanding the relationship between mood and creativity: A meta-analysis. *Organizational Behavior and Human Decision Processing*, *108*, 25–38. Available from http://dx.doi.org/10.1016/j.obhdp.2008.04.001.

Davison, K. K., Earnest, M. B., & Birch, L. L. (2002). Participation in aesthetic sports and girls' weight concerns at ages 5 and 7 years. *International Journal of Eating Disorders*, *31*, 312–317.

Day, H., Koutedakis, Y., & Wyon, M. (2011). Hypermobility and dance: A review. *International Journal of Sports Medicine*, *32*, 485–489. Available from http://dx.doi.org/10.1055/s-0031-1273690.

de Assis, M. A., de Mello, M. F., Scorza, F. A., Cadrobbi, M. P., Schooedl, A. F., da Silva, S. G., ... Arida, R. M. (2008). Evaluation of physical activity habits in patients with posttraumatic stress disorder. *Clinical Science*, *64*, 473–478. Available from http://dx.doi.org/10.1590/S1807-59322008000400010.

De Bellis, M. D., Keshavan, M. S., Shifflett, H., Iyengar, S., Beers, S. R., Hall, J., & Moritz, G. (2002). Brain structures in pediatric maltreatment-related posttraumatic stress disorder: A sociodemographically matched study. *Biological Psychiatry*, *52*(11), 1066–1078. Available from http://dx.doi.org/10.1016/S00006-3223(02)01459-2.

Decety, J., Grezes, J., Costes, N., Perani, D., Jeannerod, M., Procyk, E., ... Fazio, F. (1997). Brain activity during observation of actions: Influence of action content and subject's strategy. *Brain*, *120*, 1763–1777. Available from http://dx.doi.org/10.1093/brain/120.10.1763.

de Chumaceiro, C. L. D. (2004). Serendipity and pseudoserendipity in career paths of successful women: Orchestra conductors. *Creativity Research Journal*, *16*(2 & 3), 345–356.

De Dreu, C. K., Baas, M., & Nijstad, B. A. (2008). Hedonic tone and activation level in the mood-creativity link: Toward a dual pathway of creativity model. *Journal of Personality and Social Psychology*, *94*(5), 739–756. Available from http://dx.doi.org/10.1037/0022-3514.94.5.739.

de Gelder, B. (2006). Towards the neurobiology of emotional body language. *Nature Reviews*, *7*, 242–249.

De Greef, M., van Wijck, R., Reynders, K., Toussaint, J., & Hesseling, R. (2003). Impact of the Groningen Exercise Therapy for symphony orchestra musicians program on perceived playing-related musculoskeletal disorders of professional musicians. *Medical Problems of Performing Artists*, *18*, 156–160.

de Haas, G. J. F. (2013). Synchronicity in the performing arts: Oscar Wilde's nightmare? *Clinical Rheumatology*, *32*, 493–495. Available from http://dx.doi.org/10.1007/s10067-013-2200-z.

de Jonge, J., Spoon, E., Sonnentag, S., Dormann, C., & van den Tooren, M. (2012). "Take a break?!" Off-job recovery, job demands, and job resources as predictors of health, active learning and creativity. *European Journal of Work and Organizational Psychology*, *21* (3), 321–348. Available from http://dx.doi.org/10.1080/1359432X.2011.576009.

de Jonge, L., Moreira, E. A. M., Martin, C. K., & Ravussin, E. (2010). Impact of 6-month caloric restriction on autonomic nervous system activity in healthy, overweight, individuals. *Obesity*, *18*, 414–416. Available from http://dx.doi.org/10.1038/oby.2009.408.

de Manzano, O. (2010). *Biological mechanisms in creativity and flow (Doctoral dissertation)*. Stockholm: Dept Women's and Children's Health. Karolinska Instituetet.

de Manzano, O., Cervenka, S., Jucaite, A., Hellenas, O., Farde, L., & Ullen, F. (2013). Individual differences in the proneness to have flow experiences are linked to dopamine D2 receptor availability in the dorsal striatum. *Neuroimage*, *67*, 1–6. Available from http://dx.doi.org/10.1016/j.neuroimage.2012.10.072.

de Manzano, O., Cervenka, S., Karabanov, A., Farde, L., & Ullen, F. (2010). Thinking outside a less intact box: Thalamic dopamine D2 receptor densities are negatively related to psychometric creativity in healthy individuals. *PLoS ONE*, *5*(5), e10670. Available from http://dx.doi.org/10.1371/journal.pone.0010670.

de Manzano, O., Theorell, T., Harmat, L., & Ullen, F. (2010). The psychophysiology of flow during piano playing. *Emotion*, *10*, 301–311. Available from http://dx.doi.org/10.1037/a0018432.

de Manzano, O., & Ullen, F. (2012a). Activation and connectivity patterns of the presupplementary and dorsal premotor areas during free improvisation of melodies and rhythms. *Neuroimage*, *63*, 271–280. Available from http://dx.doi.org/10.1016/j.neuroimage.2012.06.024.

de Manzano, O., & Ullen, F. (2012b). Goal-independent mechanisms for free response generation: Creative and pseudo-random performance share neural substrates. *Neuroimage*, *59*, 772–780. Available from http://dx.doi.org/10.1016/jneuroimage.2011.07.016.

De Marco, T. C., Taylor, C. L., & Friedman, R. S. (2015). Reinvestigating the effect of interpersonal sadness on mood-congruency in music preference. *Psychology of Aesthetics, Creativity, and the Arts*, *9*(1), 81–90. Available from http://dx.doi.org/10.1037/a0038691.

De Petrillo, L., & Winner, E. (2005). Does art improve mood? A test of a key assumption underlying art therapy? *Art Therapy*, *22*(4), 205–212. Available from http://dx.doi.org/10.1080/07421656.2005.10129521.

DePrince, A. P., Huntjens, R. J. C., & Dorahy, M. J. (2015). Alienation appraisals distinguish adults diagnosed with DID from PTSD. *Psychology Trauma: Theory, Research, Practice, and Policy*, *7*(6), 578–582. Available from http://dx.doi.org/10.1037/tra000069.

Derryberry, D., & Reed, M. A. (1994). Temperament and attention: Orienting toward and away from positive and negative signals. *Journal of Personality and Social Psychology*, *66*(6), 1128–1139.

Dersh, J., Gatchel, R. J., Polatin, P., & Mayer, T. (2002). Prevalence of psychiatric disorders in patients with chronic work-related musculoskeletal pain disability. *Journal of Occupational and Environmental Medicine*, *44*(5), 459–468.

Dewey, J. (1910). *How we think*. New York: Health.

Dick, F., Lee, H. L., Nusbaum, H., & Price, C. J. (2011). Auditory-motor expertise alters "speech selectivity" in professional musicians and actors. *Cerebral Cortex*, *21*, 938–948. Available from http://dx.doi.org/10.1093/cercor/bhq166.

Dick, R. W., Berning, J. R., Dawson, W., Ginsburg, R. D., Miller, C., & Shybut, G. T. (2013). Athletes and the arts—the role of sports medicine in the performing arts. *Current Sports Medicine Reports*, *12*(6), 397–403. Available from http://dx.doi.org/10.1249/jsri.0000000000000009.

Diedrich, J., Benedek, M., Jauk, E., & Neubauer, A. C. (2015). Are creative ideas novel and useful? *Psychology of Creativity, Aesthetics, and the Arts*, *9*(1), 35–40. Available from http://dx.doi.org/10.1037/a0038688.

Dietrich, A. (2004). Neurocognitive mechanisms underlying the experience of flow. *Consciousness and Cognition*, *13*, 746–761. Available from http://dx.doi.org/10.1016/j.concog.2004.07.002.

Dietrich, A. (2007). Who's afraid of a cognitive neuroscience of creativity? *Methods*, *42*, 22–27. Available from http://dx.doi.org/10.1016/j.ymeth.2006.12.009.

Dietrich, A. (2008). Imaging the imagination: The trouble with motor imagery. *Methods*, *45*, 319–324. Available from http://dx.doi.org/10.1016/j.ymeth.2008.04.004.

Dietrich, A., & Kanso, R. (2010). A review of EEG, ERP, and neuroimaging studies of creativity and insight. *Psychological Bulletin*, *136*(5), 822–848. Available from http://dx.doi.org/10.1037/a0019749.

Dijksterhuis, A., & Meurs, T. (2006). Where creativity resides: The generative power of unconscious thought. *Consciousness and Cognition*, *15*, 135–146. Available from http://dx.doi.org/10.1016/j.concog.2005.04.007.

Dijkstra, K., Kaschak, M. P., & Zwaan, R. A. (2007). Body posture facilitates retrieval of autobiographical memories. *Cognition*, *102*(1), 139–149. Available from http://dx.doi.org/10.1016/j.cognition.2005.12.009.

Di Martino, A., Scheres, A., Marguiles, D. S., Kelly, A. M. C., Uddin, L. Q., Shehzad, Z., ... Milham, M. P. (2008). Functional connectivity of human striatum: A resting state fMRI study. *Cerebral Cortex*, *18*, 2735–2747. Available from http://dx.doi.org/10.1093/cercor/bhn041.

Dishman, R. K., Motl, R. W., Sallis, J. F., Dunn, A. L., Birnbaum, A. S., Welk, G. J., ... Jobe, J. B. (2005). Self-management strategies mediate self-efficacy and physical activity. *American Journal of Preventive Medicine*, *29*(1), 10–18. Available from http://dx.doi.org/10.1016/j.amepre.2005.03.012.

Dobson, M. C. (2011). Insecurity, professional sociability, and alcohol: Young freelance musicians' perspectives on work and life in the music profession. *Psychology of Music*, *39*(2), 240–260. Available from http://dx.doi.org/10.1177/0305735610373562.

Doinita, N. E., & Maria, N. D. (2015). Attachment and parenting styles. *Procedia-Social and Behavioral Sciences*, *203*, 199–204. Available from http://dx.doi.org/10.1016/j.sbspro.2015.08.282.

Domene, P. A., & Easton, C. (2014). Combined triaxial accelerometry and heart rate telemetry for the physiological characterization of Latin dance in non-professional adults. *Journal of Dance Medicine and Science*, *18*(1), 29–36. Available from http://dx.doi.org/10.12678/1089-313X.18.1.29.

Dommerholt, J. (2009). Performing arts medicine: Instrumentalist musicians: Part I—General considerations. *Journal of Bodywork and Movement Therapies*, *13*, 311–319. Available from http://dx.doi.org/10.1016/j.jbmt.2009.02.003.

Dommerholt, J. (2010a). Performing arts medicine: Instrumentalist musicians: Part II—Examination. *Journal of Bodywork and Movement Therapies*, *14*, 65–72. Available from http://dx.doi.org/10.1016/j.jbmt.2009.02.004.

Dommerholt, J. (2010b). Performing arts medicine: Instrumentalist musicians: Part III—Case histories. *Journal of Bodywork and Movement Therapies*, *14*, 127–138. Available from http://dx.doi.org/10.1016/j.jbmt.2009.02.005.

Dong, M., Giles, W. H., Felitti, V. J., Dube, S. R., Williams, J. E., Chapman, D. P., & Anda, R. F. (2004). Insights into causal pathways for ischemic heart disease: Adverse childhood experiences study. *Circulation*, *110*, 1761–1766.

Donnelly, J. Retrieved from www.goodreads.com/quotes/tag/acting.

Dotti, A., Fioravanti, M., Balotta, M., Tozzi, F., Cannella, C., & Lassari, R. (2002). Eating behavior of ballet dancers. *Eating Weight Disorder*, *7*, 60–67.

Drapeau, C. W., & DeBrule, D. S. (2013). The relationship of hypomania, creativity and suicidal ideation in undergraduates. *Creativity Research Journal*, *25*(1), 75–79. Available from http://dx.doi.org/10.1080/10400419.2013.752231.

Drus, M., Kozbelt, A., & Hughes, R. R. (2014). Creativity, psychopathology, and emotion processing: A liberal response bias for remembering negative information is associated with higher creativity. *Creativity Research Journal*, *26*(3), 251–262. Available from http://dx.doi.org/10.1080/10400419.2014.929400.

Dubrovsky, S. (2015a). Feeling threatened? Get some rest. *New England Journal of Medicine Journal Watch*, *21*(9), 68.

Dubrowsky, S. (2015b). How can a mother's childhood experiences be transmitted to her children? *New England Journal of Medicine Journal Watch*, *21*(12), 94.

Ducher, G., Kukuljan, S., Hill, B., Garnham, A. P., Nowson, C. A., & Kimlin, M. G. (2011). Vitamin D status and musculoskeletal health in adolescent male ballet dancers: A pilot study. *Journal of Dance Medicine and Science*, *15*(3), 99–107.

Dunn, L. W., Corn, A. L., & Morelock, M. J. (2004). The relationship between scores on the ICMIC and selected talent domains: An investigation with gifted adolescents. *Gifted Child Quarterly*, *48*(2), 133–142.

Dupuy, O., Bherer, L., Audiffren, M., & Bosquet, L. (2013). Night and postexercise cardiac autonomic control in functional overreaching. *Applied Physiology and Nutritional Metabolism*, *38*, 200–208. Available from http://dx.doi.org/10.1139/apnm.2012.0203.

Durante, J. F., Durante, C. L., & Furiasse, J. G. (2002). *The mitral valve prolapse syndrome/dysautonomia survivor guide*. Oakland, CA: New Harbinger Publications, Inc.

Dutton, S. E. (2001). Urban youth development—Broadway style: Using theatre and group work as vehicles for positive youth development. *Social Work and Groups*, *23*(4), 39–59.

Dyce, J. A., & O'Conner, B. P. (1994). The personality of popular musicians. *Psychology of Music*, *22*, 168–173.

Dykas, M. J., Woodhouse, S. S., Jones, J. D., & Cassidy, J. (2014). Attachment-related biases in adolescents' memory. *Child Development*, *85*(6), 2185–2201. Available from http://dx.doi.org/10.1111/cdev.12268.

Eccles, J. A., Beacher, F. D. C., Gray, M. A., Jones, C. L., Minati, L., Harrison, N. A., & Critchley, H. D. (2012). Brain structure and joint hypermobility: Relevance to the expression of psychiatric symptoms. *British Journal of Psychiatry*, *200*, 508–509. Available from http://dx.doi.org/10.1192/bjp.bp.111.092460.

Eccles, J. A., Owens, A. P., Mathias, C. J., Umeda, S., & Critchley, H. D. (2015). Neurovisceral phenotypes in the expression of psychiatric symptoms. *Frontiers in Neuroscience*, *9*(4), 1–13. Available from http://dx.doi.org/10.3389/fnins.2015.00004.

Edwards, J.A. Retrieved from www.goodreads.com/quotes/tag/acting.

Egner, T., & Gruzelier, J. H. (2003). Ecological validity of neurofeedback: Modulation of slow wave EEG enhances musical performance. *Neuroreport*, *14*(9), 1221–1224.

Egner, T., & Hirsch, J. (2005). The neural correlates and functional integration of cognitive control in a Stroop task. *Neuroimage*, *24*, 539–547. Available from http://dx.doi.org/10.1016/j.neuroimage.2004.09.007.

Ehlert, U., Gaab, J., & Heinrichs, M. (2001). Psychoneuroendocrinology contributions to the etiology of depression, posttraumatic stress disorder, and stress-related bodily disorders: The role of the hypothalamus-pituitary-adrenal axis. *Biological Psychology*, *57*, 141–152.

Ehn, E. (2007). A space for truth: Meditations on theatre and the Rawandan Genocide. *American Theatre*, *24*(3), 34–73.

Eisenman, R. (2001). Creativity, risk-taking, sex differences and birth order. *Journal of Evolutionary Psychology*, *28*(3), 189.

Ekegren, C. L., Quested, R., & Brodrick, A. (2014). Injuries in pre-professional ballet dancers: Incidence, characteristics and consequences. *Journal of Science and Medicine in Sport*, *17*, 271–275. Available from http://dx.doi.org/10.1016/j.sams.2013.07.013.

Ellamil, M., Dobson, C., Beeman, M., & Christoff, K. (2012). Evaluative and generative modes of thought during the creative process. *Neuroimage*, *59*, 1783–1794. Available from http://dx.doi.org/10.1016/j.neuroimage.2011.08.008.

Elliot, A. J., & Covington, M. V. (2001). Approach and avoidance motivation. *Educational Psychology Review*, *13*(2), 73–92. Available from http://dx.doi.org/10.1040-726X/01/0600-0073$19.50/0.

Elliot, A. J., & Reis, H. T. (2003). Attachment and exploration in adulthood. *Journal of Personality and Social Psychology*, *85*(2), 317–331. Available from http://dx.doi.org/10.1037/0022-3514.85.2.317.

Elliot, A. J., & Thrash, T. M. (2001). Achievement goals and the hierarchical model of achievement motivation. *Education Psychology Review*, *13*(2), 139–156. Available from http://dx.doi.org/10.1040-726X/01/0600-0139$19.50/0.

Elliot, A. J., & Thrash, T. M. (2002). Approach-avoidance motivation in personality: Approach and avoidance temperaments and goals. *Journal of Personality and Social Psychology*, *82* (5), 804–818. Available from http://dx.doi.org/10.1037//0022-3514.82.5.804.

Ellis, R. J., & Thayer, J. F. (2010). Music and autonomic nervous system (dys)function. *Music Perception: An Interdisciplinary Journal*, *27*(4), 317–326. Available from http://dx.doi.org/10.1525/mp.2010.27.4.317.

Enders, L., Spector, J. T., Altenmuller, E., Schmidt, A., Kelin, C., & Jabusch, H.-C. (2011). Musician's dystonia and comorbid anxiety: Two sides of one coin? *Movement Disorders*, *26*(3), 539–542. Available from http://dx.doi.org/10.1002/mds.23607.

Endler, N. S., & Parker, J. D. A. (1990). *Coping inventory for stressful situations: Manual* (2nd ed.). Toronto: Multi-Health Systems.

Engelhard, I. M., van den Hout, M. A., & Lommen, M. J. J. (2009). Individuals high in neuroticism are not more reactive to adverse events. *Personality and Individual Differences*, *47*, 697–700. Available from http://dx.doi.org/10.1016/j.paid.2009.05.031.

Enman, N. M., Sabban, E. L., McGonigle, P., & Van Bockstaele, E. J. (2015). Targeting the neuropeptide Y system in stress-related psychiatric disorders. *Neurobiology of Stress*, *1*, 33–43. Available from http://dx.doi.org/10.1016/j.ynstr.2014.09.007.

Ericsson, K. A. (2007). Deliberate practice and the modifiability of body and mind: Toward a science of the structure and acquisition of expert and elite performance. *International Journal of Sport Psychology*, *38*(1), 4–34.

Ericsson, K. A. (2013). Training history, deliberate practice and elite sports performance: An analysis in response to Tucker an Collins review—what makes champions? *British Journal of Sports Medicine*, *47*, 533–535.

Ericsson, K. A. (2014). Why expert performance is special and cannot be extrapolated from studies of performance in the general population: A response to criticism. *Intelligence*, *45*, 81–103. Available from http://dx.doi.org/10.1016/j.intell.2013.12.001.

Ericsson, K. A., Roring, R. W., & Nandagopal, K. (2007). Giftedness and evidence for reproducibly superior performance: An account based on the expert performance framework. *High Ability Studies*, *18*, 3–56.

Erikson, E. H. (1959/1980). *Identity and the life cycle*. New York: W. W. Norton & Company.

Essex, M. J., Boyce, W. T., Hertzman, C., Lam, L. L., Armstrong, J. M., Neumann, S. M. A., & Kobor, M. S. (2013). Epigenetic vestiges of early developmental adversity: Childhood stress exposure and DNA methylation in adolescence. *Child Development*, *84* (1), 58–75. Available from http://dx.doi.org/10.111/j.1467-8624.2011.01641.x.

Etzel, J. A., Johnsen, E. L., Dickerson, J., Tranel, D., & Adolps, R. (2006). Cardiovascular and respiratory response during musical mood induction. *International Journal of Psychophysiology*, *61*, 57–69. Available from http://dx.doi.org/10.1016/j.ijpsycho.2005.10.025.

Eusanio, J., Thomson, P., & Jaque, S. V. (2014). Perfectionism, shame, and self-concept in dancers: A mediation analysis. *Journal of Dance Medicine and Science*, *13*(3), 106–114.

Evans, G. W., & Kim, P. (2013). Childhood poverty, chronic stress, self-regulation, and coping. *Child Development Perspectives*, *7*(1), 43–48. Available from http://dx.doi.org/10.1111/cdep.12013.

Eysenck, H. J. (1997). Creativity and personality. In M. Runco (Ed.), *The creativity research handbook* (Vol. 1, pp. 41–66). Cresskill NJ: Hampton Press.

Eysenck, H. J., & Eysenck, S. B. G. (1991). *The Eysenck Personality Questionnaire—Revised*. Sevenoaks: Hodder & Stoughton.

Fancourt, D., Ockelford, A., & Belai, A. (2014). The psychoneuroimmunological effects of music: A systematic review and a new model. *Brain, Behavior, and Immunity*, *36*, 15–26. Available from http://dx.doi.org/10.1016/j.bbi.2013.10.014.

Farina, B., Speranza, A. M., Imperatori, C., Quintiliani, M. I., & Marca, G. D. (2015). Change in heart rate variability after adult attachment interview in dissociative patients. *Journal of Trauma and Dissociation*, *16*, 170–180. Available from http://dx.doi.org/10.1080/15299732.2014.975309.

Farrell, M. P. (2001). *Collaborative circles: Friendship dynamics and creative work*. Chicago: University of Chicago Press.

Fearon, D. D., Copeland, D., & Saxon, T. F. (2013). The relationship between parenting styles and creativity in a sample of Jamaican children. *Creativity Research Journal*, *25* (1), 119–128. Available from http://dx.doi.org/10.1080/10400419.2013.752287.

Fehm, L., & Schmidt, K. (2006). Performance anxiety in gifted adolescent musicians. *Journal of Anxiety Disorders*, *20*(1), 98–109. Available from http://dx.doi.org/10.1016/janxdis.2004.11.011.

Feist, G. J. (1998). A meta-analysis of personality in scientific and artistic creativity. *Personality and Social Psychology Review*, *2*(4), 290–309.

Feist, G. J. (1999). The influence of personality on artistic and scientific creativity. In R. J. Sternberg (Ed.), *Handbook of creativity* (pp. 273–296). Cambridge, UK: Cambridge University Press.

Feist, G. J., & Barron, F. X. (2003). Predicting creativity from early to late adulthood: Intellect, potential, and personality. *Journal of Research in Personality*, *37*, 62–88. Available from http://dx.doi.org/10.1016/S0092-6566(02)00526-6.

Felitti, V. J., & Anda, R. F. (2010). The relationship of adverse childhood experiences to adult medical disease, psychiatric disorders and sexual behavior: Implications for healthcare. In R. A. Lanius, E. Vermetten, & C. Pain (Eds.), *The impact of early life trauma on health and disease: The hidden epidemic* (pp. 77–87). Cambridge: Cambridge University Press.

Felman, S. (1995). Education and crisis, or the vicissitudes of teaching. In C. Caruth (Ed.), *Trauma: Exploration in memory*. Baltimore, MD: The John Hopkins University Press.

Felman, S., & Laub, D. (1992). *Testimony: Crises of witnessing in literature, psychoanalysis, and history*. New York: Routledge.

Ferguson, S., Moere, A.M., & Cabrera, D. (2005). Seeing sound: Real-time sound visualization in visual feedback loops used for training musicians. *Proceedings of the ninth international conference on information visualization, IV'05*, 97–103. Available from http://dx.doi.org/10.1109/iv.2005.114.

Fink, A., Benedeck, M., Grabner, R. H., Staudt, B., & Neubauer, A. C. (2007). Creativity meets neuroscience: Experimental tasks for the neuroscientific study of creative thinking. *Methods*, *42*, 68–76. Available from http://dx.doi.org/10.1016/j.ymet.2006.12.001.

Fink, A., Graif, B., & Neubauer, A. C. (2009). Brain correlates underlying creative thinking: EEG alpha activity in professional vs. novice dancers. *Neuroimage*, *46*(3), 854–862. Available from http://dx.doi.org/10.1016/j.neuroimage.2009.02.036.

Fink, A., Schwab, D., & Papousek, I. (2011). Sensitivity of EEG upper alpha activity to cognitive and affective creativity interventions. *International Journal of Psychophysiology*, *82*, 233–239. Available from http://dx.doi.org/10.1016/j.ijpsycho.2011.09.003.

Fink, A., Slamar-Halbedi, M., Unterrainer, H. F., & Weiss, E. M. (2012). Creativity: Genius, madness, or a combination of both? *Psychology of Aesthetics, Creativity, and the Arts*, *6* (1), 11–18. Available from http://dx.doi.org/10.1037/a0024874.

Fink, A., Weber, B., Koschutnig, K., Benedek, M., Reishofer, G., Ebner, F., . . . Weiss, E. M. (2014). Creativity and schizotypy from the neuroscience perspective. *Cognition, Affect, Behavior and Neuroscience*, *14*, 378–387. Available from http://dx.doi.org/10.3758/s13415-013-0210-6.

Fink, A., & Woschnjak, S. (2011). Creativity and personality in professional dancers. *Personality and Individual Differences*, *51*, 754–758. Available from http://dx.doi.org/10.1016/j.paid.2011.06.024.

Fink, B., Weege, B., Flugge, J., Roder, S., Neave, N., & McCarty, K. (2012). Men's personality and women's perception of their dance quality. *Personality and Individual Differences*, *52*, 232–235. Available from http://dx.doi.org/10.1016/j.paid.2011.10.008.

Finke, R. A. (1996). Imagery, creativity, and emergence. *Consciousness and Cognition*, *5*, 381–393.

Fiorelli, J. A., & Russ, S. W. (2012). Pretend play, coping, and subjective well-being in children: A follow up study. *American Journal of Play*, *5*(1), 81–103.

Firth, L., Alderson-Day, B., Woods, N., & Fernyhough, C. (2015). Imaginary companions in childhood: Relations to imagination skills and autobiographical memory in adults. *Creativity Research Journal*, *27*(4), 308–313. Available from http://dx.doi.org/10.1080/10400419.2015.1087240.

Fisher, J. E., Mohanty, A., Herrington, J. D., Koven, N. S., Miller, G. A., & Heller, W. (2004). Neuropsychological evidence for dimensional schizotypy: Implications for creativity and psychopathology. *Journal of Research in Personality, 38*(1), 24–31. Available from http://dx.doi.org/10.1016/j.jrp.2003.09.014.

Fitts, M. (2007). "Drop it like its hot": Culture industry laborers and their perspectives on rap music video production. *Meridians: Feminism, Race, Transnationalism, 8*(1), 211–235.

Fjellman-Wiklund, A., & Chesky, K. (2006). Musculoskeletal and general health problems of acoustic guitar, electric guitar, electric bass, and banjo players. *Medical Problems of Performing Artists, 21*(4), 169–176.

Fjellman-Wiklund, A., Grip, H., Andersson, H., Karlsson, J. S., & Sundelin, G. (2004). EMG trapezius muscle activity pattern in string players: Part II—Influences of basic body awareness therapy on the violin playing technique. *International Journal of Industrial Ergonomics, 33*, 357–367. Available from http://dx.doi.org/10.1016/j.ergon.2003.10.008.

Fjellman-Wiklund, A., Grip, H., Karlsson, J. S., & Sundelin, G. (2004). EMG trapezius muscle activity pattern in string players: Part 1—Is there variability in the playing technique? *International Journal of Industrial Ergonomics, 33*, 347–356. Available from http://dx.doi.org/10.1016/j.ergon.2003.10.007.

Flaherty, A. W. (2011). Brain illness and creativity: Mechanisms and treatment risks. *Canadian Journal of Psychiatry, 56*(3), 132–143. Available from http://dx.doi.org/10.1177/070674371105600303.

Fleith, D. D.-S., Rodrigues, M. A. M., Viana, M. C. A., & Cerqueira, T. C. S. (2000). The creation process of Brazilian musicians. *First Quarter, 34*(1), 61–75.

Fleming, A. S., Kraemer, G. W., Gonzalez, A., Lovic, V., Rees, S., & Melo, A. (2002). Mothering begets mothering: The transmission of behavior and its neurobiology across generations. *Pharmacology Biochemistry, and Behavior, 73*, 61–75.

Fletcher, P. C., Happe, F., Frith, U., Baker, S. C., Dolan, R. J., Frackowiak, R. S. J., & Frith, C. D. (1995). Other minds in the brain: A functional imagining study of "theory of mind" in story comprehension. *Cognition, 57*, 109–128.

Floras, J. S. (2009). Sympathetic nervous system activation in human heart failure: Clinical implications of an updated model. *Journal of American College of Cardiology, 54*(5), 375–385. Available from http://dx.doi.org/10.1016/j.jass.2009.03.061.

Foilb, A. R., & Christianson, J. P. (2016). Serotonin 2C receptor antagonist improves fear discrimination and subsequent safety signal recall. *Progress in Neuropsychopharmacology and Biological Psychiatry, 65*, 78–84. Available from http://dx.doi.org/10.1016/j.pnpbp.2015.08017.

Foley, E. C., & Bird, H. A. (2013). Hypermobility in dance: Assets, no liability. *Clinical Rheumatology, 32*, 455–461. Available from http://dx.doi.org/10.1007/s10067-013-2191-9.

Fonagy, P., & Target, M. (1997). Attachment and reflective function: Their role in self-organization. *Development and Psychopathology, 9*, 679–700.

Foreman-Wernet, L., & Dervin, B. (2011). Cultural experience in context: Sense-making the arts. *The Journal of Arts Management, Law, and Society, 41*, 1–37. Available from http://dx.doi.org/10.1080/10632921.2011.545725.

Forgeard, M. J. (2013). Perceiving benefits after adversity: The relationship between self-reported posttraumatic growth and creativity. *Psychology of Aesthetics, Creativity, and the Arts, 7*(3), 245–264. Available from http://dx.doi.org/10.1037/a0031223.

Fornia, G. L., & Frame, M. W. (2001). The social and emotional needs of gifted children: Implications for family counseling. *The Family Journal, 9*(4), 384–390. Available from http://dx.doi.org/10.1177/1066480701094005.

Fourie, M. M., Rauch, H. L., Morgan, B. E., Ellis, G. F. R., Jordaan, E. R., & Thomas, K. G. F. (2011). Guilt and pride are heartfelt, but not equally so. *Psychophysiology, 48*, 888–899. Available from http://dx.doi.org/10.1111/j.1469-8986.2010.01157x.

Fox, M.J. Retrieved from www.Betterworld.net.

Foxman, I., & Brugel, B. (2006). Musician health and safety: Preventing playing-related musculoskeletal disorders. *Aaohn Journal, 54*(7), 309–316.

Frade, C. (2005). New modes of business organization and precarious employment: Towards the recommendation of labour? *Journal of European Social Policy, 15*(2), 107–121. Available from http://dx.doi.org/10.1177/0958928705051509.

Franco, R. A., & Andrus, J. G. (2007). Common diagnoses and treatments in professional voice users. *Otolaryngologic Clinics of North America, 40*, 1025–1061. Available from http://dx.doi.org/10.1016/j.otc.2007.05.008.

Frank, C., Kobesova, A., & Kolar, P. (2013). Dynamic neuromuscular stabilization and sports rehabilitation. *The International Journal of Sports Physical Therapy, 8*(1), 62–73.

Franklin, M. S., Moore, K. S., Yip, C.-Y., Jonides, J., Rattray, K., & Moher, J. (2008). The effects of musical training on verbal memory. *Psychology of Music, 36*(3), 1–13. Available from http://dx.doi.org/10.1177/0305735607086044.

Fraser, D. (2010). Creativity, mood disorders and the aesthetic. *Gifted Education International, 27*, 84–96.

Freed, S., & D'Andrea, W. (2015). Autonomic arousal and emotion in victims of interpersonal violence: Shame proneness but not anxiety predicts vagal tone. *Journal of Trauma and Dissociation, 16*, 367–383. Available from http://dx.doi.org/10.1080/15299732.2015.1004771.

Freedberg, D., & Gallese, V. (2007). Motion, emotion and empathy in esthetic experience. *Trends in Cognitive Sciences, 11*(5), 197–203. Available from http://dx.doi.org/10.1016/j.tics.2007.02.003.

Freeman, W. J. (2000). *How brains make up their minds*. New York: Columbia University Press.

Freud, S. (1908/1985). Creative writers and daydreaming. In A. Dickson (Ed.), *Art and literature: The Pelican Freud library* (Vol. 14, pp. 130–141). New York: Penguin Books.

Freud, S. (1914/1985). The Moses of Michelangelo. In A. Dickson (Ed.), *Art and literature: The Pelican Freud library* (Vol. 14, pp. 251–282). New York: Penguin Books.

Freud, S. (1942/1985). Psychopathic characters on the stage. In A. Dickson (Ed.), *Art and literature: The Pelican Freud library* (Vol. 14, pp. 120–127). New York: Penguin Books.

Friedman, B. H. (2010). Feelings and the body: The Jamesian perspective on autonomic specificity of emotion. *Biological Psychology, 84*, 383–393. Available from http://dx.doi.org/10.1016/j.biopsycho.2009.10.006.

Friesen, K. J., Rozenek, R., Clippinger, K., Gunter, K., Russo, A. C., & Sklar, S. E. (2011). Bone mineral density and body composition of collegiate modern dancers. *Journal of Dance Medicine and Science, 15*(1), 31–36.

Fritz, B. S., & Avsec, A. (2007). The experience of flow and subjective well-being of music students. *Horizons of Psychology, 16*(2), 5–17.

Frye, N. (1963). *The educated imagination*. Toronto, Canada: Canadian Broadcasting Organization.

Fryer, M. (2012). Some key issues in creativity research and evaluation as seen from a psychological perspective. *Creativity Research Journal, 24*(1), 21–28. Available from http://dx.doi.org/10.1080/10400419.2012.649236.

Fuhrmann, T. L., Brayer, A., Andrus, N., & McIntosh, S. (2010). Injury prevention for modern dancers: A pilot study of an educational intervention. *Journal of Community Health, 35*, 527–533. Available from http://dx.doi.org/10.1007/s10900-010-9223-z.

Fujioka, T., Fujioka, A., Tan, N., Chowdhury, I., Mouri, H., Sakata, Y., & Nakamura, S. (2001). Mild prenatal stress enhances learning performance in the non-adopted rat offspring. *Neuroscience*, *103*(2), 301–307.

Furnham, A., & Bachtiar, V. (2008). Personality and intelligence as predictors of creativity. *Personality and Individual Differences*, *45*, 613–617. Available from http://dx.doi.org/10.1016/j.paid.2008.06.023.

Gaab, N., Gaser, C., Zaehle, T., Jancke, L., & Schlaug, G. (2003). Functional anatomy of pitch memory—and fMRI study with sparse temporal sampling. *Neuroimage*, *19*, 1417–1426. Available from http://dx.doi.org/10.1016/S1053-8119(03)00224-6.

Gajewski, A. K., & Poznanska, A. (2008). Mortality of top athletes, actors, and clergy in Poland: 1924–2000 follow-up study of the long term effect of physical activity. *European Journal of Epidemiology*, *23*, 335–340. Available from http://dx.doi.org/10.1007/s10654-008-9237-3.

Gallegos, A. M., Lytle, M. C., Moynihan, J. A., & Talbot, N. L. (2015). Mindfulness-based stress reduction to enhance psychological functioning and improve inflammatory biomarkers in trauma-exposed women: A pilot study. *Psychological Trauma: Theory, Research, Practice, and Policy*, *7*(6), 525–532. Available from http://dx.doi.org/10.1037/tra000053.

Gambichler, T., Boms, S., & Freitag, M. (2004). Contact dermatitis and other skin conditions in instrumental musicians. *Biomed Central Dermatology*, *4*(3), 1–12.

Gamboa, J. M., Hagins, M., & Manal, T. J. (2005). An analysis to define the clinical practice of physical therapy for performing artists. *Journal of Dance Medicine and Science*, *9*(2), 41–55.

Gamboa, J. M., Roberts, L. A., Maring, J., & Fergus, A. (2008). Injury patterns in elite preprofessional ballet dancers and the utility of screening programs to identify risk characteristics. *Journal of Orthopaedic and Sports Physical Therapy*, *38*(3), 126–136.

Garcia-Campayo, J., Asso, E., & Alda, M. (2011). Joint hypermobility and anxiety: The state of the art. *Current Psychiatric Reports*, *13*, 18–25. Available from http://dx.doi.org/10.1007/s11920-010-0164-0.

Garcia-Dantas, A., & Quested, E. (2015). The effect of manipulated and accurate assessment feedback on the self-efficacy of dance students. *Journal of Dance Medicine and Science*, *19*(1), 22–30. Available from http://dx.doi.org/10.12678/1089-313X.19.1.22.

Gardner, H. (1983). *Frames of the mind: The theory of multiple intelligences*. New York: Basic Books.

Gardner, H. (1993). *Creating minds*. New York: Basic Books.

Garrido, S., & Schubert, E. (2011). Individual differences in the enjoyment of negative emotion in music: A literature review and experiment. *Music Perception: An Interdisciplinary Journal*, *28*(3), 279–296. Available from http://dx.doi.org/10.1525/mp.2011.28.3.279.

Gaztambide-Fernandez, R. A., Saifer, A., & Desai, C. (2013). "Talent" and the misrecognition of social advantage in specialized arts education. *Roeper Review*, *35*, 124–135. Available from http://dx.doi.org/10.1080/02783193.2013.766964.

Geeves, A. M., McIlwain, D. J. F., & Sutton, J. (2016). Seeing yellow: "Connection" and routine in professional musicians' experience of music performance. *Psychology of Music*, *44*(2), 183–201. Available from http://dx.doi.org/10.1177/0305735614560841.

Geisler, F. C., & Kubiak, T. (2009). Heart rate variability predicts self-control in goal pursuit. *European Journal of Personality*, *23*, 623–633. Available from http://dx.doi.org/10.1002/per.727.

Gembris, H., & Heye, A. (2014). Growing older in a symphony orchestra: The development of the age-related self-concept and the self-estimated performance of professional musicians in a lifespan perspective. *Music Scientiae*, *18*(4), 371–391. Available from http://dx.doi.org/10.1177/1029864914548912.

George, J. M., & Zhou, J. (2002). Understanding when bad moods foster creativity and good ones don't: The role of context and clarity of feelings. *Journal of Applied Psychology*, *87*(4), 687–697. Available from http://dx.doi.org/10.1037//0021-9010.87.4.687.

Geyer, H. L., & Bressman, S. B. (2006). The diagnosis of dystonia. *Lancet Neurology*, *5*, 780–790.

Ghacibeh, G. A., Shenker, J. I., Shenal, B., Uthman, B. M., & Heilman, K. M. (2006). Effect of vagus nerve stimulation on creativity and cognitive flexibility. *Epilepsy and Behavior*, *8*, 720–725. Available from http://dx.doi.org/10.1016/j.yebeh.2006.03.008.

Gibbs, R. W. (2003). Embodied experience and linguistic meaning. *Brain and Language*, *84*, 1–15.

Gibson, C., Folley, B. S., & Park, S. (2009). Enhanced divergent thinking and creativity in musicians: A behavioral and near-infrared spectroscopy study. *Brain and Cognition*, *69*, 162–169. Available from http://dx.doi.org/10.1016/j.bandc.2008.07.009.

Gibson, C., & Mumford, M. D. (2013). Evaluation, criticism, and creativity: Criticism content and effects on creative problem solving. *Psychology of Aesthetics, Creativity, and the Arts*, *7*(4), 314–331. Available from http://dx.doi.org/10.1037/a0032616.

Giesbrecht, T., Geraerts, E., & Merckelbach, H. (2007). Dissociation, memory commission errors, and heightened autonomic reactivity. *Psychiatry Research*, *150*(3), 277–285. Available from http://dx.doi.org/10.1016/j.psychres.2006.04.016.

Giesbrecht, T., & Merckelbach, H. (2006). Dreaming to reduce fantasy?—Fantasy proneness, dissociation, and subjective sleep experiences. *Personality and Individual Differences*, *41*(4), 697–706. Available from http://dx.doi.org/10.1016/j.paid.2006.02.015.

Gillespie, W., & Myors, B. (2000). Personality of rock musicians. *Psychology of Music*, *28*, 154–165.

Gillie, B. L., & Thayer, J. F. (2014). Individual differences in resting heart rate variability and cognitive control in posttraumatic stress disorder. *Frontiers in Psychology*, *5*(758), 1–7. Available from http://dx.doi.org/10.3389/fpsych.2014.00758.

Glass, J. M., Lyden, A. K., Petzke, F., Stein, P., Whalen, G., Ambrose, K., ... Clauw, D. (2004). The effect of brief exercise cessation on pain, fatigue, and mood symptom development in healthy, fit individuals. *Journal of Psychosomatic Research*, *57*(4), 391–398.

Glaveanu, V. P. (2010). Creativity as cultural participation. *Journal of Theory of Social Behavior*, *41*, 48–67.

Glimcher, P. W. (2011). Understanding dopamine and reinforcement learning: The dopamine reward prediction error hypothesis. *Proceedings of the National Academy of Science*, *108*(3), 15647–15654. Available from http://dx.doi.org/10.1073/pnas.1014269108.

Gluck, J., Ernst, R., & Unger, F. (2002). How creative define creativity: Definitions reflect different types of creativity. *Creativity Research Journal*, *14*(1), 55–67.

Godin, M. E., & Thomson, P. (2011). Theatre. In M. A. Runco, & S. R. Pritzker (Eds.), *Encyclopaedia of creativity* (2nd ed., Vol. 2, pp. 465–472). London: Elsevier Publishing Company.

Goldsmith, H. H., Buss, A. H., Plomin, R., Rothbart, M. K., Thomas, A., Chess, S., ... McCall, R. B. (1987). Roundtable: What is temperament? Four approaches. *Child Development*, *58*(2), 505–529.

Goldstein, A. L., Flett, G. L., Wekerle, C., & Wall, A.-M. (2009). Personality, child maltreatment, and substance use: Examining correlates of deliberate self-harm among university students. *Canadian Journal of Behavioral Science*, *41*(4), 241–251. Available from http://dx.doi.org/10.1037/a0014847.

Goldstein, D. S., Robertson, D., Esler, M., Straus, S. E., & Eisenhofer, G. (2002). Dysautonomias: Clinical disorders of the autonomic nervous system. *Annual Internal Medicine, 137*(9), 753–763.

Goldstein, T. R. (2009). Psychological perspectives on acting. *Psychology of Aesthetics, Creativity, and the Arts, 3*(1), 6–9. Available from http://dx.doi.org/10.1037/a0014644.

Goldstein, T. R. (2015). Predictors of engagement in and transfer from acting training. *Psychology of Aesthetics, Creativity, and the Arts, 9*(3), 266–273. Available from http://dx.doi.org/10.1037/a0039106.

Goldstein, T. R., Tamir, M., & Winner, E. (2013). Expressive suppression and acting class. *Psychology of Aesthetics, Creativity, and the Arts, 7*(2), 191–196. Available from http://dx.doi.org/10.1037/a0030209.

Goldstein, T. R., & Winner, E. (2009). Living in alternative inner worlds: Early signs of acting talent. *Creativity Research Journal, 21*(1), 117–124. Available from http://dx.doi.org/10.1080/10400410802633749.

Goldstein, T. R., & Winner, E. (2010–2011). Engagement in role play, pretense, and acting classes predict advanced theory of mind skill in middle childhood. *Imagination, Cognition and Personality, 30*(3), 249–258. Available from http://dx.doi.org/10.2190/IC.30.3.c.

Goldstein, T. R., & Winner, E. (2012). Enhancing empathy and theory of mind. *Journal of Cognition and Development, 13*(1), 19–37. Available from http://dx.doi.org/10.1080/15248372.2011.573514.

Goldstein, T. R., Wu, K., & Winner, E. (2009). Actors are skilled in theory of mind but not empathy. *Imagination, Cognition and Personality, 29*(2), 115–133. Available from http://dx.doi.org/10.2190/IC.29.2.c.

Golland, Y., Keissar, K., & Levit-Binnun, N. (2014). Studying the dynamics of autonomic activity during emotional experience. *Psychophysiology, 51*(11), 1101–1111. Available from http://dx.doi.org/10.1111/psyp.12261.

Golomer, E., Mbongo, F., Toussaint, Y., Cadiou, M., & Israel, I. (2010). Right hemisphere in visual regulation of complex equilibrium: The female ballet dancers' experience. *Neurological Research, 32*(4), 409–415. Available from http://dx.doi.org/10.1179/174313209X382476.

Gomez, P., & Danuser, B. (2004). Affective and physiological responses to environmental noises and music. *International Journal of Psychophysiology, 53*, 91–103. Available from http://dx.doi.org/10.1016/j.ijpsycho.2004.02.002.

Goncalo, J. A., & Staw, B. M. (2006). Individualism-collectivism and group creativity. *Organizational Behavior and Human Decision Processes, 100*(1), 96–109. Available from http://dx.doi.org/10.1016/j.obhdp.2005.11.003.

Goncy, E. A., & Waehler, C. A. (2006). An empirical investigation of creativity and musical experience. *Psychology of Music, 34*(3), 307–321. Available from http://dx.doi.org/10.1177/0305735606064839.

Goodwin, F. K., & Jamison, K. R. (2007). *Manic-depressive illness: Bipolar disorders and recurrent depression*. New York: Oxford University Press.

Gopnik, A., & Walker, C. M. (2013). Considering counterfactuals: The relationship between causal learning and pretend play. *American Journal of Play, 6*(1), 15–28.

Gordon, J., & Gridley, M. C. (2013). Musical preference as a function of stimulus complexity of piano jazz. *Creativity Research Journal, 25*(1), 143–146. Available from http://dx.doi.org/10.1080/10400419.2013.752303.

Gow, K., Lang, T., & Chant, D. (2004). Fantasy proneness, paranormal beliefs and personality features in out-of-body experiences. *Contemporary Hypnosis, 21*(3), 107–125.

Grabner, R. H., Fink, A., & Neubauer, A. C. (2007). Brain correlates of self-rated originality of ideas: Evidence from event-related power and phase-locking changes in the EEG. *Behavioral Neuroscience*, *121*(1), 224–230. Available from http://dx.doi.org/10.1037/0735-7044.121.1.224.

Graham, M. (1991). *Blood memory: An autobiography*. New York: Doubleday.

Graham, M. Retrieved from www.goodreads.com/quotes/tag/dance.

Graham, T. (2007). After the gifts are opened, what's next? *Herald Democrat, Section: Religion*, January 5.

Grape, C., Sandgren, M., Hansson, L.-O., Ericson, M., & Theorell, T. (2003). Does singing promote well-being? An empirical study of professional and amateur singers during a singing lesson. *Integrative Physiological and Behavioral Science*, *38*(1), 65–74.

Gratz, K. L., & Chapman, A. L. (2007). The role of emotional responding and childhood maltreatment in the development and maintenance of deliberate self-harm in male undergraduates. *Psychology of Men and Masculinity*, *8*(1), 1–14. Available from http://dx.doi.org/10.1037/1524-9220.8.1.1.

Gratz, K. L., & Roemer, L. (2004). Multidimensional assessment of emotion regulation and dysregulation: Development, factor structure, and initial validation of the difficulties in emotion regulation scale. *Journal of Psychopathology and Behavioral Assessment*, *26*(1), 41–54.

Greengross, G., Martin, R. A., & Miller, G. (2012). Personality traits, intelligence, humor styles, and humor production ability of professional stand-up comedians compared to college students. *Psychology of Aesthetics, Creativity, and the Arts*, *6*(1), 74–82. Available from http://dx.doi.org/10.1037/a0025774.

Greengross, G., & Miller, G. F. (2009). The Big Five personality traits of professional comedians compared to amateur comedians, comedy writers, and college students. *Personality and Individual Differences*, *47*(2), 79–83. Available from http://dx.doi.org/10.1016/j.paid.2009.01.045.

Gregory, J., & Embrey, D. G. (2009). Reducing the effects of profound catastrophic trauma for former child soldiers: Companion recovery model. *Traumatology*, *15*(1), 52–62. Available from http://dx.doi.org/10.1177/1534765608323442.

Grewe, O., Nagel, F., Kopiez, R., & Altenmuller, E. (2007). Listening to music as a re-creative process: Physiological, psychological, and psychoacoustical correlates of chills and strong emotions. *Music Perception*, *24*(3), 297–314. Available from http://dx.doi.org/10.1525/mp.2007.24.3.297.

Griskevicius, V., Cialdini, R. B., & Kenrick, D. T. (2006). Peacocks, Picasso, and parental investment: The effects of romantic motives on creativity. *Journal of Personality and Social Psychology*, *91*(1), 63–76. Available from http://dx.doi.org/10.1037/0022.3514.91.1.63.

Gross, J. J., & Thompson, R. A. (2007). Emotion regulation: Conceptual foundations. In J. J. Gross (Ed.), *Handbook of emotion regulation* (pp. 3–46). New York: Guilford Press.

Grossman, F. K., Sorsoli, L., & Kia-Keating, M. (2006). A gale force wind: Meaning making by male survivors of childhood sexual abuse. *American Journal of Orthopsychiatry*, *76*(4), 434–443. Available from http://dx.doi.org/10.1037/0002-9432.76.4.434.

Grove, J. R., Main, L. C., & Sharp, L. (2013). Stressors, recovery processes, and the manifestations of training distress in dance. *Journal of Dance Medicine and Science*, *17*(2), 70–78.

Gruzelier, J. (2009). The theory of alpha/theta neurofeedback, creative performance enhancement, long distance functional connectivity and psychological integration. *Cognitive Process*, *10*(Suppl. 1), S101–S109. Available from http://dx.doi.org/10.1007/s10339-008-0248-5.

Gruzelier, J. (2014). EEG-neurofeedback for optimizing performance. II: Creativity, the performing arts and ecological validity. *Neuroscience and Biobehavioral Reviews*, *44*, 142–158. Available from http://dx.doi.org/10.1016/j.neurbiorev.2013.11.004.

Gruzelier, J., Inoue, A., Smart, R., Steed, A., & Steffert, T. (2010). Acting performance and flow state enhanced with sensory-motor rhythm neurofeedback comparing ecologically valid immersive VR and training screen scenarios. *Neuroscience Letters*, *480*, 112–116. Available from http://dx.doi.org/10.1016/j.neulet.2010.06.019.

Gruzelier, J. H., Holmes, P., Hirst, L., Bulpin, K., Rahman, S., van Run, C., & Leach, J. (2014). Replication of elite music performance enhancement following alpha/theta neurofeedback and application to novice performers and improvisation with SMR benefits. *Biological Psychology*, *95*, 96–107. Available from http://dx.doi.org/10.1016/j.biopsycho.2013.11.001.

Gruzelier, J. H., Thompson, T., Redding, E., Brandt, R., & Steffert, T. (2014). Application of alpha/theta neurofeedback and heart rate variability training to young contemporary dancers: State anxiety and creativity. *International Journal of Psychophysiology*, *93*, 105–111. Available from http://dx.doi.org/10.1016/j.ipsycho.2013.05.004.

Guidetti, L., Gallota, M. C., Emerenziani, P., & Baldari, C. (2007). Exercise intensities during a ballet lesson in female adolescents with different technical ability. *International Journal of Sports Medicine*, *28*, 736–742. Available from http://dx.doi.org/10.1055/s-2007-964909.

Gunnar, M. R., & Adam, E. K. (2012). The hypothalamic-pituitary-adrenocortical system and emotion: Current wisdom and future directions. *Monograms of the Society for Research in Child Development*, *77*(2), 109–119.

Gunty, A. L., Frazier, P. A., Tennen, H., Tomich, P., Tashiro, T., & Park, C. (2011). Moderators of the relation between perceived and actual posttraumatic growth. *Psychological Trauma: Theory, Research, Practice, and Policy*, *3*(1), 61–66. Available from http://dx.doi.org/10.1037/a0020485.

Gupta, A. (2009). Vygotskian perspectives on using dramatic play to enhance children's development and balance creativity with structure in the early childhood classroom. *Early Childhood Development and Care*, *179*(8), 1041–1054. Available from http://dx.doi.org/10.1080/03004430701731654.

Guptill, C. (2008). Musicians' health: Applying the ICF framework in research. *Disability and Rehabilitation*, *30*(12–13). Available from http://dx.doi.org/10.1080/09638280701800517.

Guptill, C. (2011). The lived experience of working as a musician with an injury. *Work*, *40* (3), 269–280. Available from http://dx.doi.org/10.3233/WOR-2011-1230.

Guptill, C. (2012). Injured professional musicians and the complex relationship between occupation and health. *Journal of Occupational Science*, *19*(3), 258–270. Available from http://dx.doi.org/10.1080/14427591.2012.670901.

Gusewell, A., & Ruch, W. (2014). Are musicians particularly sensitive to beauty and goodness? *Psychology of Aesthetics, Creativity, and the Arts*, *8*(1), 96–103. Available from http://dx.doi.org/10.1037/a0035217.

Gute, D., & Gute, G. (2015). *How creativity works in the brain*. Washington, DC: National Endowment for the Arts.

Gute, G., Gute, D. S., Nakamura, J., & Csikszentmihalyi, M. (2008). The early lives of highly creative persons: The influence of the complex family. *Creativity Research Journal*, *20* (4), 343–357. Available from http://dx.doi.org/10.1080/10400410802391207.

Hackney, M. E., Kantorovich, S., & Earhart, G. M. (2007). A study on the effects of Argentine Tango as a form of partnered dance for those with Parkinson Disease and the healthy elderly. *American Journal of Dance Therapy*, *29*(2), 109–127. Available from http://dx.doi.org/10.1007/s10465-007-9039-2.

Hadida, A. L. (2013). Institutions, assets combinations, and film performance: A US–France comparison. *Psychology of Aesthetics, Creativity, and the Arts*, *7*(2), 155–170. Available from http://dx.doi.org/10.1037/a0030166.

Hagen, E., & Bryant, G. (2003). Music and dance as a coalition signaling system. *Human Nature*, *14*(1), 21–51.

Hagen, U., & Frankel, H. (1973). *Respect for acting*. New York: MacMillan Publishing Company, Inc.

Hagins, M. (2013). The use of stabilization exercises and movement reeducation to manage pain and improve function in a dancer with focal degenerative joint disease of the spine. *Journal of Dance Medicine and Science*, *15*(3), 136–142.

Hagood, T. K., & Kahlich, L. C. (2007). Research in choreography. In L. Bresler (Ed.), *International handbook of research in arts education* (Vol. 16, pp. 517–531). The Netherlands: Springer.

Hahnengress, M. L., & Boning, D. (2010). Cardiopulmonary changes during clarinet playing. *European Journal of Applied Physiology*, *110*, 1199–1208. Available from http://dx.doi.org/10.1007/s00421-010-1576-6.

Hakim, A. J., & Grahame, R. (2004). Non-musculoskeletal symptoms in joint hypermobility syndrome: Indirect evidence for autonomic dysfunction. *Rheumatology*, *43*(9), 1194–1202. Available from http://dx.doi.org/10.1093/rheumatology/keh286.

Hakim, A. J., & Sahota, A. (2006). Joint hypermobility and skin elastic: The hereditary disorders of connective tissue. *Clinics in Dermatology*, *24*, 521–533.

Hall, H. K., & Hill, A. P. (2012). Perfectionism, dysfunctional achievement striving and burnout in aspiring athletes: The motivational implications for performing artists. *Theatre, Dance and Performance Training*, *3*(2), 216–228. Available from http://dx.doi.org/10.1080/194432927.2012.693534.

Hall, P. A., Elias, L. J., Fong, G. T., Harrison, A. H., Borowsky, R., & Sarty, G. E. (2008). A social neuroscience perspective on physical activity. *Journal of Sport and Exercise Psychology*, *30*, 432–449.

Hallam, S. (2002). Musical motivation: Towards a model synthesizing the research. *Music Education Research*, *4*(2), 225–244. Available from http://dx.doi.org/10.1080/1461380022000011939.

Haller, C. S., & Courvoisier, D. S. (2010). Personality and thinking style in different creative domains. *Psychology of Aesthetics, Creativity, and the Arts*, *4*(3), 149–160. Available from http://dx.doi.org/10.1037/a0017084.

Hallett, M. G., & Hoffman, B. (2014). Performing under pressure: Cultivating the peak performance mindset for workplace excellence. *Counselling Psychology Journal: Practice and Research*, *66*(3), 212–230. Available from http://dx.doi.org/10.1037/cpb000009.

Halpern, A. R. (2012). Dynamic aspects of musical imagery. *Annals of the New York Academy of Sciences*, *1252*, 200–205. Available from http://dx.doi.org/10.1111/j.1749-6632.2011.06442.x.

Hambrick, D. Z., Oswald, F. L., Altmann, E. M., Meinz, E. J., Gobet, F., & Campitelli, G. (2014). Deliberate practice: Is that all it takes to become an expert? *Intelligence*, *45*, 34–45. Available from http://dx.doi.org/10.1016/j.intell.2013.04.001.

Hamerman, D. (1999). Toward an understanding of frailty. *Annals of Internal Medicine*, *130*, 945–950.

Hamilton, L. H., & Robson, B. (2006). Performing arts consultation: Developing expertise in this domain. *Professional Psychology: Research and Practice*, *37*(3), 254–259. Available from http://dx.doi.org/10.1037/0735-7028.37.3.254.

Hamilton, L. H., Solomon, R., & Solomon, J. (2006). A proposal for standardized psychological screening for dancers. *Journal of Dance Medicine and Science*, *10*(1 & 2), 40–45.

Hammond, R. A. (2009). *Respecting babies: A new look at Magda Gerber's RIE approach*. Washington, DC: Zero to Three.

Hammond, T., Gialloreto, C., Kubas, H., & Davis, H. (2013). The prevalence of failure-based depression among elite athletes. *Clinical Journal of Sport Medicine*, *23*(4), 273–277. Available from http://dx.doi.org/10.1097/JSM.0b013e318287b870.

Hancox, J. E., Quested, E., & Duda, J. L. (2015). Suitability of the perceived motivational climate in Sport Questionnaire-2 for dance research: A think aloud approach. *Journal of Dance Medicine and Science*, *19*(4), 149–162. Available from http://dx.doi.org/10.12678/1089.313x.19.4.149.

Hannah, M. T., Domino, G., Hanson, R., & Hannah, W. (1994). Acting and personality change: The measurement of change in self-perceived personality characteristics during the actor's character development process. *Journal of Research in Personality*, *28*, 277–286.

Hannon, E. E., & Trainor, L. J. (2007). Music acquisition: Effects of enculturation and formal training on development. *Trends in Cognitive Sciences*, *11*(11), 466–472. Available from http://dx.doi.org/10.1016/j.tics.2007.08.008.

Hanrahan, C., & Vergeer, I. (2000–2001). Multiple uses of mental imagery by professional modern dancers. *Imagination, Cognition and Personality*, *20*(3), 231–255.

Harari, D., Bakermans-Kranenburg, M. J., de Kloet, C. S., Geuze, E., Vermetten, E., Westenberg, H. G. M., & van IJzendoorn, M. H. (2009). Attachment representations in Dutch veterans with and without deployment-related PTSD. *Attachment and Human Development*, *11*(6), 515–536. Available from http://dx.doi.org/10.1080/14616730903282480.

Hargreaves, D., & Tiggemann, M. (2002). The effect of television commercials on mood and body dissatisfaction: The role of appearance-schema activation. *Journal of Social and Clinical Psychology*, *21*(3), 287–308.

Hargreaves, D., & Tiggemann, M. (2003). Longer-term implications for responsiveness to "thin-ideal" television: Support for a cumulative hypothesis of body image disturbance? *European Eating Disorder Review*, *11*(6), 465–477. Available from http://dx.doi.org/10.1002/erv.v11:6/issuetoc.

Hargreaves, D., & Tiggemann, M. (2009). Muscular ideal media images and men's body image: Social comparison processing and individual vulnerability. *Psychology of Men and Masculinity*, *10*(2), 109–119. Available from http://dx.doi.org/10.1037/a0014691.

Hargreaves, D. J. (2012). Musical imagination: Perception and production, beauty and creativity. *Psychology of Music*, *40*(5), 539–557. Available from http://dx.doi.org/10.1077/0305735612444893.

Harrington, D. M., & Anderson, A. M. (1981). Creativity, masculinity, femininity, and three models of psychological androgyny. *Journal of Personality and Social Psychology*, *41* (4), 744–757. Available from http://dx.doi.org/10.1037/0022-3514.41.4.744.

Harris, D. A. (2001). Using β-blockers to control stage fright: A dancer's dilemma. *Medical Problems of Performing Artists*, *16*(2), 72–76.

Harris, D. A. (2007). Dance/movement therapy approaches to fostering resilience and recovery among African adolescent torture survivors. *Torture Violence*, *17*(2), 134–155.

Harris, R., & de Jong, B. M. (2014). Cerebral activations related to audition-driven performance imagery in professional musicians. *PLoS ONE*, *9*(4), e93681. Available from http://dx.doi.org/10.1371/journal.pone.0093681.

Harrison, A., Tchanturia, K., & Treasure, J. (2010). Attentional bias, emotion recognition, and emotion regulation: State of trait? *Biological Psychiatry*, *68*, 755–761. Available from http://dx.doi.org/10.1016/j.biopsych.2010.04.037.

Hass, R. W. (2014). Domain-specific exemplars affect implicit theories of creativity. *Psychology of Aesthetic, Creativity, and the Arts, 8*(1), 44–52. Available from http://dx.doi.org/10.1037/a0035368.

Hass, R. W., & Weisberg, R. W. (2009). Career development in two seminal American songwriters: A test of the equal odds rule. *Creativity Research Journal, 21*(2–3), 183–190. Available from http://dx.doi.org/10.1080/10400410902855275.

Hass, R. W., & Weisberg, R. W. (2015). Revisiting the 10-year rule for composers from the Great American Songbook: On the validity of two measures of creative production. *Psychology of Aesthetics, Creativity, and the Arts, 9*(4), 471–479. Available from http://dx.doi.org/10.1037.aca0000021.

Hassandra, M., Goudas, M., & Chroni, S. (2003). Examining factors associated with intrinsic motivation in physical education: A qualitative approach. *Psychology of Sport and Exercise, 4*, 211–233. Available from http://dx.doi.org/10.1016/S1469-0292(02)00006-7.

Haught, C. (2015). The role of constraints in creative sentence production. *Creativity Research Journal, 27*(2), 160–166. Available from http://dx.doi.org/10.1080/10400419.2015.1030308.

Hauschildt, M., Peters, J. V., Moritz, S., & Jelinek, L. (2011). Heart rate variability in response to affective scenes in posttraumatic stress disorder. *Biological Psychology, 88*, 215–222. Available from http://dx.doi.org/10.1016/j.biopsycho.2011.08.004.

Hautala, A. J., Kiviniemi, A. M., & Tulppo, M. P. (2009). Individual responses to aerobic exercise: The role of the autonomic nervous system. *Neuroscience and Biobehavioral Reviews, 33*, 107–115. Available from http://dx.doi.org/10.1016/j.neurobiorev.2008.04.009.

Haven, T. J. (2009). "That part of the body is just gone": Understanding and responding to dissociation and physical health. *Journal of Trauma and Dissociation, 10*, 204–218. Available from http://dx.doi.org/10.1080/15299730802624569.

Haworth-Hoeppner, S. (2000). The critical shapes of body image: The role of culture and family in the production of eating disorders. *Journal of Marriage and Family, 62*, 212–227.

Hawthorne, M. (2015). Study shines a spotlight on darkness with the entertainment industry. *Victoria News*. Retrieved from www.theage.com.au/victoria/study-shines-a-spotlight-shines-on-darkness-within-the-entertainment-industry-20150909-gjinxp#ixzz3lRf4KBKe.

Hays, K. F. (2002). The enhancement of performance excellence among performing artists. *Journal of Applied Sport Psychology, 14*, 299–312. Available from http://dx.doi.org/10.1080/10413200290103572.

Hefferon, K. M., & Ollis, S. (2006). "Just clicks": An interpretive phenomenological analysis of professional dancers' experience of flow. *Research in Dance Education, 7*(2), 141–159. Available from http://dx.doi.org/10.1080/14647890601029527.

Heide, F. J., Porter, N., & Saito, P. K. (2012). Do you hear the people sing? Musical theatre and attitude change. *Psychology of Aesthetics, Creativity, and the Arts, 6*(3), 224–230. Available from http://dx.doi.org/10.1037/a0027574.

Heiland, T. L., Murray, D. S., & Edley, P. P. (2008). Body image of dancers in Los Angeles: The cult of slenderness and media influence among dance students. *Research in Dance Education, 9*(3), 257–275. Available from http://dx.doi.org/10.1080/14647890802386932.

Heim, C., Plotsky, P. M., & Nemeroff, C. B. (2004). Importance of studying the contributions of early adverse experience to neurobiological findings in depression. *Neuropsychopharmacology, 29*, 641–648. Available from http://dx.doi.org/10.1038/sj.npp.1300397.

Hein, G. (2014). Empathy and resilience in a connected world. In M. Kent, M. C. Davis, & J. W. Reich (Eds.), *The resilience handbook: Approaches to stress and trauma* (pp. 144–155). New York: Routledge.

Heitkemper, M., Jarrett, M., Cain, K. C., Burr, R., Levy, R. L., Feld, A., & Hertig, V. (2001). Autonomic nervous system function in women with irritable bowel syndrome. *Digestive Diseases and Sciences*, *46*(6), 1276–1284.

Helson, R. (1999). A longitudinal study of creative personality in women. *Creativity Research Journal*, *12*(2), 89–101.

Heming, M. J. E. (2004). Occupational injuries suffered by classical musicians through overuse. *Clinical Chiropractic*, *7*(2), 55–66. Available from http://dx.doi.org/10.1016/j.clch.2004.02.008.

Henke, K. (2010). A model for memory systems based on processing modes rather than consciousness. *Nature Review Neuroscience*, *11*, 523–532. Available from http://dx.doi.org/10.1038/nrn2850.

Hennessy, B. A., & Amabile, T. M. (2010). Creativity. *Annual Review of Psychology*, *61*, 569–598. Available from http://dx.doi.org/10.1146/annurev.psych.093008.100416.

Henry, B. L., Minassian, A., Paulus, M. P., Geyer, M. A., & Perry, W. (2010). Heart rate variability in bipolar mania and schizophrenia. *Journal of Psychiatry Research*, *44*(3), 168–176. Available from http://dx.doi.org/10.1016/j.psychires.2009.07.011.

Heredia, L., Hinkamp, D., Brodsky, M., & Llapur, C. (2014). Playing-related problems among musicians of the Orquesta Buena Vista Social Club and supporting bands. *Medical Problems of Performing Artists*, *29*(2), 80–85.

Herer, B. (2000). The longevity and causes of death of jazz musicians, 1990–1999. *Medical Problems of Performing Artists*, *15*(3), 119–122.

Herholz, S. C., & Zatorre, R. J. (2012). Musical training as a framework for brain plasticity: Behavior, function, and structure. *Neuron*, *76*(8), 486–502. Available from http://dx.doi.org/10.1016/j.neuron.2012.10.011.

Hernandez, B. M. (2012). Addressing occupational stress in dancers. *Journal of Physical Education, Recreation and Dance*, *83*(6), 3–50. Available from http://dx.doi.org/10.1080/07303084.2012.10598784.

Hesse, E., & Main, M. (1999). Second-generation effects of unresolved trauma as observed in non-maltreating parents: Dissociated, frightening and threatening parental behavior. *Psychoanalytic Inquiry*, *19*, 481–540.

Hesse, E., & Main, M. (2000). Disorganized infant, child, and adult attachment: Collapse in behavioral and attentional strategies. *Journal of the American Psychoanalytic Association*, *48*, 1097–1127.

Hesse, E., & Main, M. (2006). Frightened, threatening, and dissociative parental behavior in low-risk samples: Description, discussion, and interpretations. *Development & Psychopathology*, *18*(2), 309–343.

Hetzler, E. T. (2008). Actors and emotion in performance. *Studies in Theatre and Performance*, *28*(1), 59–78. Available from http://dx.doi.org/10.1386/stap.28.1.59/3.

Hewitt, P. L., & Flett, G. L. (2004). *Multidimensional perfectionism scale technical manual*. Toronto: Multi-Health Systems.

Hiddleston, T. Retrieved from www.goodreads.com/quotes/tag/acting.

Hildebrandt, H., Nubling, M., & Candia, V. (2012). Increment of fatigue, depression, and stage fright during the first year of high-level education in music students. *Medical Problems of Performing Artists*, *27*(1), 43–48.

Hilscher, M. C., Cupchik, G. C., & Leonard, G. (2008). Melodrama and film noir on today's big screen: How modern audiences experience yesterday's classics. *Psychology of*

Aesthetics, Creativity, and the Arts, *2*(4), 203–212. Available from http://dx.doi.org/10.1037/a0012700.

Hincapie, C. A., Morton, E. J., & Cassidy, J. D. (2008). Musculoskeletal injuries and pain in dancers: A systematic review. *Archives of Physical Medicine and Rehabilitation*, *89*, 1819–1829. Available from http://dx.doi.org/10.1016/j.apmr.2008.02.020.

Hiraldo, P. (2010). The role of critical race theory in higher education. *The Vermont Connection*, *31*, 53–59.

Hiroa, K., & Kobayashi, R. (2013). Health-related quality of life and sense of coherence among the unemployed with autotelic, average, and non-autotelic personalities: A cross-sectional survey in Hiroshima, Japan. *PLoS ONE*, *8*(9), e73915. Available from http://dx.doi.org/10.1371/journal.pone.0073915.

Hoch, A. Z., Papanek, P., Szabo, A., Widlansky, M. E., & Gutterman, D. D. (2011). Folic acid supplementation improves vascular function in professional dancers with endothelial dysfunction. *Journal of Physical Medicine and Rehabilitation*, *3*, 1005–1012. Available from http://dx.doi.org/10.1016/j.pmrj.2011.02.014.

Hodge, A. (Ed.), (2000). *Twentieth century actor training*. New York: Routledge.

Hodges, D. A. (2014). Music as an agent of resilience. In M. Kent, M. C. Davis, & J. W. Reich (Eds.), *The resilience handbook: Approaches to stress and trauma* (pp. 100–112). New York: Routledge.

Hodgkins, C. W., Kennedy, J. G., & O'Loughlin, P. F. (2008). Tendon injuries in dance. *Clinics in Sports Medicine*, *27*, 279–288. Available from http://dx.doi.org/10.1016/j.csm.2007.12.003.

Hoeckner, B., Wyatt, E. W., Decety, J., & Nusbaum, H. (2011). Film music influences how viewers relate to movie characters. *Psychology of Aesthetics, Creativity, and the Arts*, *5* (2), 146–153. Available from http://dx.doi.org/10.1037/a0021544.

Hoffman, J., & Russ, S. (2012). Pretend play, creativity, and emotion regulation in children. *Psychology of Aesthetics, Creativity, and the Arts*, *6*(2), 175–184. Available from http://dx.doi.org/10.1037/a0026299.

Hoffman, S. L., & Hanrahan, S. J. (2012). Mental skills for musicians: Managing music performance anxiety and enhancing performance. *Sport, Exercise, and Performance Psychology*, *1*(1), 17–28. Available from http://dx.doi.org/10.1037/a0025409.

Holm-Hadulla, R. M., & Bertolino, A. (2014). Creativity, alcohol and drug abuse: The pop icon Jim Morrison. *Psychopathology*, *47*, 167–173. Available from http://dx.doi.org/10.1159/000354617.

Holt, R. R. (2002). Quantitative research on the primary process. Method and findings. *Journal of the American Psychoanalytic Association*, *50*, 457–482.

Hopper, D. M., Grisbrook, T. L., Newnham, P. J., & Edwards, D. J. (2014). The effects of vestibular stimulation and fatigue on postural control in classical ballet dancers. *Journal of Dance Medicine and Science*, *18*(2), 67–73. Available from http://dx.doi.org/10.12678/1089-313X.18.2.67.

Horan, R. (2009). The neuropsychological connection between creativity and mediation. *Creativity Research Journal*, *21*(2–3), 199–222. Available from http://dx.doi.org/10.1080/10400410902858691.

Howard, F. (2008). Managing stress or enhancing wellbeing? Positive psychology's contributions to clinical supervision. *Australian Psychologist*, *43*(2), 105–113. Available from http://dx.doi.org/10.1080/00050060801978647.

Hudson, B. (2002). The effects of the Alexander Technique on the respiratory system of the singer/actor: Part I. *Journal of Singing*, *59*(1), 9–17.

Hunter, P. G., Schellenberg, E. G., & Schimmack, U. S. (2010). Feelings and perceptions of happiness and sadness induced by music: Similarities, differences, and mixed emotions. *Psychology of Aesthetics, Creativity, and the Arts*, *4*(1), 47–56. Available from http://dx.doi.org/10.1037/a0016873.

Hunter, S. T., Bedell, K. E., & Mumford, M. D. (2007). Climate for creativity: A quantitative review. *Creativity Research Journal*, *19*(1), 69–90.

Hutchinson, C. U., Sachs-Ericsson, N. J., & Ericsson, K. A. (2013). Generalizable aspects of the development of expertise in ballet across countries and cultures: A perspective from the expert performance approach. *High Ability Studies*, *24*(1), 21–47. Available from http://dx.doi.org/10.1080/13598139.2013.780966.

Hutt, K., & Redding, E. (2014). The effect of an eyes-closed dance-specific training program on dynamic balance in elite pre-professional ballet dancers: A randomized controlled pilot study. *Journal of Dance Medicine and Science*, *18*(1), 3–11. Available from http://dx.doi.org/10.12678/1089-313X.18.1.3.

Hyde, K. L., Lerch, J., Norton, A., Forgeard, M., Winner, E., Evans, A. C., & Schlaug, G. (2009). Musical training shapes structural brain development. *Journal of Neuroscience*, *29*(10), 3019–3025. Available from http://dx.doi.org/10.1523/JNEURSCI.5118.08.2009.

Iacoboni, M. (2009). Neurobiology of imitation. *Current Opinions in Neurobiology*, *19*, 661–665. Available from http://dx.doi.org/10.1016/j.conb.2009.09.008.

Iltis, P. W. (2003). Ventilation, carbon dioxide drive, and dyspnea associated with French horn playing: A pilot study. *Medical Problems of Performing Artists*, *18*, 47–51.

Inesta, C., Terrados, N., Garcia, D., & Perez, J. A. (2008). Heart rate in professional musicians. *Journal of Occupational Medicine and Toxicology*, *3*(16), 1–11. Available from http://dx.doi.org/10.1186/1745-6673-3-16.

Irwin, H. J. (1990). Fantasy proneness and paranormal beliefs. *Psychological Reports*, *66*, 655–658.

Isen, A. M., & Reeve, J. (2005). The influence of positive affect on intrinsic and extrinsic motivation: Facilitating enjoyment of play, responsible work behavior, and self-control. *Motivation and Emotion*, *29*(4), 297–325. Available from http://dx.doi.org/10.1007/s11031-006-9019-8.

Ishiimura, I., & Kodama, M. (2009). Flow experience in everyday activities of Japanese college students: Autotelic people and time management. *Japanese Psychological Research*, *51*(1), 47–54. Available from http://dx.doi.org/10.1111/j.1468-5884.2009.00387.x.

Ivcevic, Z., & Brackett, M. A. (2015). Predicting creativity: Interactive effects of openness to experience and emotion regulation ability. *Psychology of Aesthetics, Creativity, and the Arts*, *9*(4), 480–487. Available from http://dx.doi.org/10.1037/a0039826.

Ivcevic, Z., & Mayer, J. D. (2006–2007). Creative types and personality. *Imagination, Cognition and Personality*, *26*(1–2), 65–86.

Ivcevic, Z., Mayer, J. D., & Brackett, M. A. (2007). Emotional intelligence and emotional creativity. *Journal of Personality*, *75*(2), 199–236. Available from http://dx.doi.org/10.1111/j.1467-6494.2007.00437.x.

Iwanaga, M., Kobayashi, A., & Kawasaki, C. (2005). Heart rate variability with repetitive exposure to music. *Biological Psychology*, *70*, 61–66. Available from http://dx.doi.org/10.1016/j.biopsycho.2004.11.015.

Izaks, G., & Westendorp, R. G. J. (2003). Ill or just old? Towards a conceptual framework of the relation between ageing and disease. *Biomed Central: Geriatrics*, *3*(7), 1–6.

Jabusch, H.-C., & Altenmuller, E. (2004). Anxiety as an aggravating factor during onset of focal dystonia in musicians. *Medical Problems of Performing Artists*, *19*(2), 75–81.

Jabusch, H.-C., Muller, S. V., & Altenmuller, E. (2004). Anxiety in musicians with focal dystonia and those with chronic pain. *Movement Disorders*, *19*(10), 1169–1175. Available from http://dx.doi.org/10.1002/mds.20110.

Jackson, P. L., & Decety, J. (2004). Motor cognition: A new paradigm to study self-other interactions. *Current Opinion in Neurobiology*, *14*, 259–263.

Jackson, S. A., & Eklund, R. C. (2004). *The flow scales manual*. Morgantown, WV: Fitness Information Technology, Inc.

Jacobs, A., & Sadie, S. (1985). *The limelight book of opera*. New York: Limelight Editions.

Jacobs, C. L., Hincapie, C. A., & Cassidy, J. D. (2012). Musculoskeletal injuries and pain in dancers: A systematic review update. *Journal of Dance Medicine and Science*, *16*(2), 74–84.

Jaeger, J., Lindblom, K. M., Parker-Guilbert, K., & Zoellner, L. A. (2014). Trauma narratives: It's what you say, not how you say it. *Psychological Trauma: Theory, Research, Practice, and Policy*, *6*(5), 473–481. Available from http://dx.doi.org/10.1037/a0035239.

Jakobson, L. S., Cuddy, L. L., & Kilgour, A. R. (2003). Time tagging: A key to musician's superior memory. *Music Perception*, *20*(3), 307–313.

Jamison, J. Retrieved from www.goodreads.com/quotes/tag/dancing.

Jankovic, J., & Ashoori, A. (2008). Movement disorders in musicians. *Movement Disorders*, *23*(14), 1957–1965. Available from http://dx.doi.org/10.1002/mds.22255.

Jansch, C., Harmer, C., & Cooper, M. J. (2009). Emotional processing in women with anorexia nervosa and in healthy volunteers. *Eating Behavior*, *10*, 184–191. Available from http://dx.doi.org/10.1016/j.eatbeh.2009.06.001.

Janyacharoen, T., Phusririt, C., Angkapattamakul, S., Hurst, C. P., & Sawanyawisuth, K. (2015). Cardiopulmonary effects of traditional Thai dance on menopausal women: A randomized controlled trial. *Journal of Physical Therapy Science*, *27*, 2569–2572.

Jaque, S. V., Karamanukyan, I. H., & Thomson, P. (2015). A psychophysiological study of orchestra conductors. *Medical Problems in Performing Artists*, *30*(4), 189–196.

Jarvin, L., & Subotnik, R. F. (2010). Wisdom from conservatory faculty: Insights on success in classical music performance. *Roeper Review*, *32*, 78–87. Available from http://dx.doi.org/10.1080/02783191003587868.

Jeffers, C. S. (2009). On empathy: The mirror neuron system and art education. *International Journal of Education and the Arts*, *10*(15), 1–17.

Jerome, E. M., & Liss, M. (2005). Relationship between sensory processing style, adult attachment, and coping. *Personality and Individual Differences*, *38*, 1341–1352. Available from http://dx.doi.org/10.1016/j.paid.2004.08.016.

Johnson, R. A. (1986). *Inner work: Using dreams and active imagination for personal growth*. San Francisco: Harper-Collins Publishers.

Johnson, S. L., Tharp, J. A., & Holmes, M. K. (2015). Understanding creativity in bipolar I disorder. *Psychology of Aesthetics, Creativity, and the Arts*, *9*(3), 319–327. Available from http://dx.doi.org/10.1037/a0038852.

Johnston, M. A. (1999). Influences of adult attachment exploration. *Psychological Reports*, *84*, 31–34.

Jola, C., Davis, A., & Haggard, P. (2011). Proprioceptive integration and body representation: Insights into dancers' expertise. *Experimental Brain Research*, *213*(2–3), 257–265. Available from http://dx.doi.org/10.1007/s00221-011-2743-7.

Jones, M. E., Roy, M. M., & Verkuilen, J. (2014). The relationship between reflective rumination and musical ability. *Psychology of Aesthetics, Creativity, and the Arts*, *8*(2), 219–226. Available from http://dx.doi.org/10.1037/a0035634.

Julmi, C., & Scherm, E. (2015). The domain-specificity of creativity: Insights from new phenomenology. *Creativity Research Journal*, *27*(2), 151–159. Available from http://dx.doi.org/10.1080/10400419.2015.1030310.

Jung, W. H., Kang, D.-H., Kim, E., Shin, K. S., Jang, J. H., & Kwon, J. S. (2013). Abnormal corticostriatal-limbic functional connectivity in obsessive-compulsive disorder during reward processing and resting state. *Neuroimage: Clinical*, *3*, 27–38. Available from http://dx.doi.org/10.1016/j.nicl.2013.06.013.

Juslin, P. N., & Laukka, P. (2003). Communication of emotions in vocal expression and music performance: Different channels, same code? *Psychological Bulletin*, *129*(5), 770–814. Available from http://dx.doi.org/10.1037/0033-2909.129.5.770.

Kaess, M., Parzer, P., Mattern, M., Plener, P. L., Bifulco, A., Resch, F., & Brunner, R. (2013). Adverse childhood experiences and their impact on frequency, severity, and the individual function of nonsuicidal self-injury in youth. *Psychiatry Research*, *206*, 265–272. Available from http://dx.doi.org/10.1016/j.psychres.2012.10.012.

Kagan, J., & Snidman, N. (2004). *The long shadow of temperament*. Cambridge, MA: Harvard University Press.

Kahari, K., Zachau, G., Eklof, M., & Moller, C. (2004). The influence of music and stress on musicians' hearing. *Journal of Sound and Vibration*, *277*, 627–631. Available from http://dx.doi.org/10.1016/j.jsv.2004.03.025.

Kahari, K., Zachau, G., Eklof, M., Sandsjo, L., & Moller, C. (2003). Assessment of hearing and hearing disorders in rock/jazz musicians. *International Journal of Audiology*, *42*, 279–288.

Kalliopuska, M. (1989). Empathy, self-esteem and creativity among junior ballet dancers. *Perceptual and Motor Skills*, *69*, 1227–1234.

Kaneko, Y., Lianzo, S., & Dawson, W. J. (2005). Pain as an incapacitating factor in symphony orchestra musicians in Sao Paulo, Brazil. *Medical Problems of Performing Artists*, *20*(4), 168–174.

Kantor-Martynuska, J., & Horabik, J. (2015). Granularity of emotional responses to music: The effect of musical expertise. *Psychology of Aesthetics, Creativity, and the Arts*, *9*(3), 235–247. Available from http://dx.doi.org/10.1037/a0039107.

Kaplan, E. W. (2005). Going the distance: Trauma, social rupture, and the work of repair. *Theater Topics*, *15*(2), 171–183.

Karabanov, A., Cervenka, S., de Manzano, O., Forssberg, H., Farde, L., & Ullen, F. (2010). Dopamine D2 receptor density in the limbic striatum is related to implicit but not explicit movement sequence learning. *Proceedings of the National Academy of Sciences*, *107*(16), 7574–7579. Available from http://dx.doi.org/10.1073/pnas.0911805107.

Karageorghis, C. I., Smith, D. L., & Priest, D.-L. (2012). Effects of voice enhancement technology and relaxing music on the frequency of imagery among break dancers. *Journal of Dance Medicine and Science*, *16*(1), 8–16.

Karasek, R., & Theorell, T. (1990). *Healthy work: Stress productivity and the reconstruction of working life*. Oxford: Basic Books.

Karavasilis, L., Doyle, A. B., & Markiewicz, D. (2003). Association between parenting style and attachment to mother in middle childhood and adolescence. *International Journal of Behavioral Development*, *27*(2), 153–164. Available from http://dx.doi.org/10.1080/01650250244000155.

Karpati, F. J., Giacosa, C., Foster, N. E. V., Penhune, V. B., & Hyde, K. L. (2016). Sensorimotor integration is enhanced in dancers and musicians. *Experimental Brain Research*, *234*, 893–903. Available from http://dx.doi.org/10.1007/s00221-015-4524-1.

Kattenstroth, J.-C., Kalish, T., Holt, S., Tegenthoff, M., & Dinse, H. R. (2013). Six months of dance intervention enhances postural, sensorimotor, and cognitive performance in elderly without affecting cardio-respiratory functions. *Frontiers in Aging Neuroscience*, *5*(5), 1–16. Available from http://dx.doi.org/10.3389/fnagi.2013.00005.

Kattenstroth, J.-C., Kolankowska, I., Kalisch, T., & Dinse, H. R. (2010). Superior sensory, motor, and cognitive performance in elderly individuals with multi-year dancing activities. *Frontiers in Aging Neuroscience*, *2*(31), 1–9. Available from http://dx.doi.org/10.3389/fnagi2010.00031.

Katz, L. F., & Rigterink, T. (2012). Domestic violence and emotion socialization. *Monographs of the Society for Research in Child Development*, *77*(2), 52–60.

Kaufman, J. C. (2005). The door that leads into madness: Eastern European poets and mental illness. *Creativity Research Journal*, *17*(1), 99–103.

Kaufman, J. C., & Baer, J. (2012). Beyond new and appropriate: Who decides what is creative? *Creativity Research Journal*, *24*(1), 83–91. Available from http://dx.doi.org/10.1080/10400419.2012.649237.

Kaufman, J. C., & Beghetto, R. A. (2013). Do people recognize the four Cs?: Examining layperson conceptions of creativity. *Psychology of Aesthetics, Creativity, and the Arts*, *7* (3), 229–236. Available from http://dx.doi.org/10.1037/a0033295.

Kaufman, J. C., Bromley, M. L., & Cole, J. C. (2006–2007). Insane, poetic, lovable: Creativity and endorsement of the "mad-genius" stereotype. *Imagination, Cognition and Personality*, *26*(1–2), 149–161.

Kaufman, J. C., Cropely, D. H., Baer, J., Reiter-Palmon, R., & Sinnett, S. (2013). Furious activity vs. understanding: How much expertise is needed to evaluate creative work? *Psychology of Aesthetics, Creativity, and the Arts*, *7*(4), 332–340. Available from http://dx.doi.org/10.1037/a0034809.

Kaufman, J. C., Pumaccahua, T. T., & Holt, R. E. (2013). Personality and creativity in realistic, investigative, artistic, social and enterprising college majors. *Personality and Individual Differences*, *54*, 913–917. Available from http://dx.doi.org/10.1016/j.paid.2013.01.013.

Kaufman, J. C., & Sexton, J. D. (2006). Why doesn't the writing cure help poets? *Review of General Psychology*, *10*(3), 268–282. Available from http://dx.doi.org/10.1037/1089-2680.10.3.268.

Kaufman-Cohen, Y., & Ratzon, N. Z. (2011). Correlation between risk factors and musculoskeletal disorders among classical musicians. *Occupational Medicine*, *61*, 90–95. Available from http://dx.doi.org/10.1093/occmed/kqq196.

Kaufmann, G. (2003). Expanding the mood-creativity equation. *Creativity Research Journal*, *15*(2–3), 131–135.

Kaufmann, G., & Vosberg, S. K. (1997). "Paradoxical" mood effects on creative problem-solving. *Cognition and Emotion*, *11*(2), 151–170.

Kaufmann, G., & Vosberg, S. K. (2002). The effects on early and late idea production. *Creativity Research Journal*, *14*(3–4), 317–330.

Kava, K. S., Larson, C. A., Stiller, C. H., & Maher, S. F. (2010). Trunk endurance exercise and the effect on instrumental performance: A preliminary study comparing Pilates exercise and a trunk and proximal upper extremity endurance exercise program. *Music Performance Research*, *3*(1), 1–30.

Kawabata, M., & Mallett, C. J. (2011). Flow experience in physical activity: Examination of the internal structure of flow from a process-related perspective. *Motivation and Emotion*, *35*, 393–402. Available from http://dx.doi.org/10.1007/s11031-011-9221-1.

Keeler, J. R., Roth, E. A., Neuser, B. L., Spitsbergen, J. M., Waters, D. J. M., & Vianney, J.-M. (2015). The neurochemistry and social flow of singing: Bonding and oxytocin. *Frontiers in Human Neuroscience*, *9*, 518. Available from http://dx.doi.org/10.3389/fnhum.2015.00518.

Keller, J., & Bless, H. (2008). Flow and regulatory compatibility: An experimental approach to the flow model of intrinsic motivation. *Personality and Social Psychology Bulletin*, *34*(2), 196–209. Available from http://dx.doi.org/10.1177/0146167207310026.

Keller, P. E. (2012). Mental imagery in music performance: Underlying mechanisms and potential benefits. *Annals of New York Academy of Sciences*, *1252*, 206–213. Available from http://dx.doi.org/10.1111/j.1749-6632.2011.06439.x.

Kelly, G. (1996). Mechanical overload and skeletal muscle fiber hyperplasia: A meta-analysis. *Journal of Applied Physiology*, *81*(4), 1584–1588.

Kelman, B. B. (2000). Occupational hazards in female ballet dancers: Advocate for a forgotten population. *AAOHN Journal*, *48*(9), 430–434.

Keltner, D., & Haidt, J. (2003). Approaching awe, a moral, spiritual, and aesthetic emotion. *Cognition and Emotion*, *17*(2), 297–314. Available from http://dx.doi.org/10.1080/02699930244000318.

Kendall-Tackett, K. (2002). The health effects of childhood abuse: Four pathways by which abuse can influence health. *Child Abuse and Neglect*, *26*, 715–729.

Kendall-Tackett, K., & Klest, B. (2009). Causal mechanisms and multidirectional pathways between trauma, dissociation, and health. *Journal of Trauma and Dissociation*, *10*, 129–134. Available from http://dx.doi.org/10.1080/15299730802624510.

Kenny, D., & Ackermann, B. (2015). Performance-related musculoskeletal pain, depression and music performance anxiety in professional orchestral musicians: A population study. *Psychology of Music*, *43*(1), 43–60. Available from http://dx.doi.org/10.1177/0305735613493953.

Kenny, D., Driscoll, T., & Ackermann, B. (2014). Psychological well-being in professional orchestral musicians in Australia: A descriptive population study. *Psychology of Music*, *42*(2), 210–232. Available from http://dx.doi.org/10.1177/0305735612463950.

Kenny, D. T. (2005). A systematic review of treatments for music performance anxiety. *Anxiety, Stress & Coping: An International Journal*, *18*(3), 183–208. Available from http://dx.doi.org/10.1080/10615800500167258.

Kenny, D. T., & Ackermann, B. (2013). Depression and music performance anxiety are associated with severity of performance related musculoskeletal pain in professional orchestral musicians. *International Symposium on Performance Science*, 187–192.

Kenny, D. T., Arthey, S., & Abbass, A. (2014). Intensive short-term dynamic psychotherapy for severe music performance anxiety. *Medical Problems in Performing Artists*, *29*(1), 3–7.

Kenny, D. T., & Asher, A. (2016). Life expectancy and cause of death in popular musicians: Is the popular musician lifestyle the road to ruin? *Medical Problems of Performing Artists*, *31*(1), 37–44.

Kenny, D. T., Davis, P., & Oates, J. (2004). Music performance anxiety and occupational stress amongst opera chorus artists and their relationship with state and trait anxiety and perfectionism. *Anxiety Disorders*, *18*, 757–777. Available from http://dx.doi.org/10.1016/j.janxdis.2004.09.004.

Kenny, D. T., & Osborne, M. S. (2006). Music performance anxiety: New insights from young musicians. *Advances in Cognitive Psychology*, *2*(2–3), 103–112.

Kerns, K. A., Abraham, M. M., Schlegelmilch, A., & Morgan, T. A. (2007). Mother-child attachment in later middle childhood: Assessment approaches and associations with mood and emotion regulation. *Attachment and Human Development*, *9*(1), 33–53. Available from http://dx.doi.org/10.1080/14616730601151441.

Kessler, R. C., McLaughlin, K. A., Green, J. G., Gruber, M. J., Sampson, N. A., Zaslavsky, A. M., ... Williams, D. R. (2010). Childhood adversities and adult psychopathology in the WHO World Mental Health surveys. *British Journal of Psychiatry*, *197*, 378–385. Available from http://dx.doi.org/10.1192/bjp.bp.110.080499.

Keysers, C., & Gazzola, V. (2009). Expanding the mirror: Vicarious activity for actions, emotions and sensations. *Current Opinion in Neurobiology*, *19*, 666–671. Available from http://dx.doi.org/10.1016/j.conb.2009.10.006.

Khalfa, S., Roy, M., Rainville, P., Bella, S. D., & Peretz, I. (2008). Role of tempo entrainment in psychophysiological differentiation of happy and sad music? *International Journal of Psychophysiology*, *68*, 17–26. Available from http://dx.doi.org/10.1016/j.ijpsycho.2007.12.001.

Kharkhurin, A. V. (2014). Creativity. 4 in 1: Four-criterion construct of creativity. *Creativity Research Journal*, *26*(3), 338–352. Available from http://dx.doi.org/10.1080/10400419.2014.929424.

Kibler, J. L. (2009). Posttraumatic stress and cardiovascular disease risk. *Journal of Trauma and Dissociation*, *10*, 135–150. Available from http://dx.doi.org/10.1080/15299730802624577.

Kilic, E. Z., Kilic, C., & Yilmaz, S. (2008). Is anxiety sensitivity a predictor of PTSD in children and adolescents? *Journal of Psychosomatic Research*, *65*, 81–86. Available from http://dx.doi.org/10.1016/j.psychores.2008.02.013.

Kilpatrick, D. G., Resnick, H. S., Milanak, M., Miller, M. W., Keyes, K. M., & Friedman, M. J. (2013). National estimates of exposure to traumatic events and PTSD prevalence using DSM-IV and DSM-5 criteria. *Journal of Traumatic Stress*, *26*, 537–547. Available from http://dx.doi.org/10.1002/jts.21848.

Kirschenbaum, R. J. (1998). The creativity classification system: An assessment theory. *Roeper Review*, *21*(1), 20–26.

Kiviniemi, A. M., Tulppo, M. P., Eskelinen, J. J., Savolainen, A. M., Kapanen, J., Heinonen, I. H. A., ... Kalliokoski, K. K. (2015). Autonomic function predicts fitness response to short-term high-intensity interval training. *International Journal of Sports Medicine*, *36*, 915–921. Available from http://dx.doi.org/10.1055/s-0035-1549854.

Kizkin, S., Karlidag, R., Ozcan, C., & Ozisik, H. I. (2006). Reduced P50 auditory sensory gating response in professional musicians. *Brain and Cognition*, *61*, 249–254. Available from http://dx.doi.org/10.1016/j.bandc.2006.01.006.

Klasen, M., Weber, R., Kirchner, T. T. J., Mathiak, K. A., & Mathiak, K. (2012). Neural contributions to flow experiences during video game playing. *Social Cognition and Affective Neuroscience*, *7*, 485–495. Available from http://dx.doi.org/10.1093/scan/nsr021.

Kleber, B., Birbaumer, N., Veit, R., Trevorrow, T., & Lotze, M. (2007). Overt and imagined singing of an Italian aria. *Neuroimage*, *36*(3), 889–900.

Kleber, B., Veit, R., Birbaumer, N., Gruzelier, J., & Lotze, M. (2010). The brain of opera singers: Experience-dependent changes in functional activation. *Cerebral Cortex*, *20*, 1144–1152. Available from http://dx.doi.org/10.1093/cercor/bhp177.

Kleber, B., Zeitouni, A. G., Friberg, A., & Zatorre, R. J. (2013). Experience-dependent modulation of feedback integration during singing: Role of the right anterior insula. *Journal of Neuroscience*, *33*(14), 6070–6080. Available from http://dx.doi.org/10.1523/JNERUORSCI.4418-12.2013.

Klein, A. M., & Johns, M. M. (2007). Vocal emergencies. *Otolaryngological Clinics of North America*, *40*, 1063–1080. Available from http://dx.doi.org/10.1016/j.0tc.2007.05.009.

Kline, J. B., Krauss, J. R., Maher, S. F., & Qu, X. (2013). Core strength training using a combination of home exercises and a dynamic sling system for the management of low back pain in pre-professional ballet dancers: A case series. *Journal of Dance Medicine and Science*, *17*(1), 24–33. Available from http://dx.doi.org/10.12678/1089-313X.17.1.24.

Klinger, E., Henning, V. R., & Janssen, J. M. (2009). Fantasy-proneness dimensionalized: Dissociative component is related to psychopathology, daydreaming as such is not. *Journal of Research in Personality*, *43*, 506–510. Available from http://dx.doi.org/10.1016/j.jrp.2008.12017.

Knezevic, M., & Ovsenik, M. (2002). Can creativity in conditions of war trauma be a danger to personal development? *Human Relations*, *55*(9), 1139–1153.

Kochanska, G., & Kim, S. (2013). Early attachment organization with both parents and future behavior problems: Infancy to middle childhood. *Child Development*, *84*(1), 283–296. Available from http://dx.doi.org/10.1111/j.1467-8624.2012.01852.x.

Koechlin, E., & Summerfield, C. (2007). An information theoretical approach to prefrontal executive function. *Trends in Cognitive Sciences*, *11*(6), 229–235. Available from http://dx.doi.org/10.1016/j.tics.2007.04.005.

Koeneke, S., Lutz, K., Wustenberg, T., & Jancke, L. (2004). Long-term training affects cerebellar processing in skilled keyboard players. *Neuroreport*, *15*(8), 1279–1282. Available from http://dx.doi.org/10.1097/01.wnr.0000127463.10147e7.

Kogan, N. (2002). Careers in the performing arts: A psychological perspective. *Creativity Research Journal*, *14*(1), 1–16. Available from http://dx.doi.org/10.1207/S15326934CRJ1401_1.

Kogan, N., & Kangas, B. L. (2006). Careers in the dramatic arts: Comparing genetic and interactional perspectives. *Empirical Studies in the Arts*, *24*(1), 43–54.

Kok, L. M., Vlieland, T. P. M. V., Fiocco, M., Kaptein, A. A., & Nelissen, R. G. H. H. (2013). Musicians' illness perceptions of musculoskeletal complaints. *Clinical Rheumatology*, *32*, 487–492. Available from http://dx.doi.org/10.1007/s10067-013-2199-1.

Koutedakis, Y. (2000). "Burnout" in dance: The physiological viewpoint. *Journal of Dance Medicine and Science*, *4*(4), 122–127.

Koutedakis, Y., & Jamurtas, A. (2004). The dancer as a performing athlete: Physiological considerations. *Sports Medicine*, *34*(10), 651–661.

Koutedakis, Y., Hukam, H., Metsios, G., Nevill, A., Giakas, G., Jamurtas, A., & Myszkewycz, L. (2007). The effects of three months of aerobic and strength training on selected performance- and fitness-related parameters in modern dance students. *Journal of Strength and Conditioning Research*, *21*(3), 808–812.

Kowal, J., & Fortier, M. S. (1999). Motivational determinants of flow: Contributions from self-determination theory. *The Journal of Social Psychology*, *139*(3), 355–368. Available from http://dx.doi.org/10.1080/00224549909598391.

Krampe, R. T., & Ericsson, K. A. (1996). Maintaining excellence: Deliberate practice and elite performance in young and older pianists. *Journal of Experimental Psychology: General*, *125*(4), 331–359. Available from http://dx.doi.org/10.1037/0096-3445.125.4.331.

Krasnow, D., Mainwaring, L., & Kerr, G. (1999). Injury, stress, and perfectionism in young dancers and gymnasts. *Journal of Dance Medicine & Science*, *3*(2), 51–58.

Krasnow, D. H., & Wilmerding, M. V. (2015). *Motor learning and control for dance: Principles and practices for performers and teachers*. Champaign, IL: Human Kinetics.

Kraus, N., & Chandrasekaran, B. (2010). Music training for the development of auditory skills. *Nature Reviews*, *11*, 599–605.

Krause, V., Schnizler, A., & Pollok, B. (2010). Functional network interactions during sensorimotor synchronization in musicians and non-musicians. *Neuroimage*, *52*, 245–251. Available from http://dx.doi.org/10.1016/j.neuroimage.2010.03.081.

Kreibig, S. D. (2010). Autonomic nervous system activity in emotion: A review. *Biological Psychology*, *84*, 394–421. Available from http://dx.doi.org/10.1016/j.biopsycho.2010.03.010.

Kreibig, S. D., Gendolla, G. H. E., & Scherer, K. R. (2012). Goal relevance and goal conduciveness appraisals lead to differential autonomic reactivity in emotional responding to performance feedback. *Biological Psychology*, *91*, 365–375. Available from http://dx.doi.org/10.1016/j.biopsycho.2012.08.007.

Kring, A. M., Smith, D. A., & Neale, J. M. (1994). Individual differences in dispositional expressiveness: Development and validation of the Emotional Expressivity Scale. *Journal of Personality and Social Psychology*, *66*(5), 934–949.

Kristensen, T. (2004). The physical context of creativity. *Creativity and Innovation Management*, *13*(2), 89–96.

Kucharska-Pietura, K., Nikolaou, V., Masiak, M., & Treasure, J. (2004). The recognition of emotion in the faces and voice of anorexia nervosa. *International Journal of Eating Disorders*, *35*(1), 42–47. Available from http://dx.doi.org/10.1002/eat.10219.

Kuhlman, K. R., Vargas, I., Geiss, E. G., & Lopez-Duran, N. L. (2015). Age of trauma onset and HPA axis dysregulation among trauma-exposed youth. *Journal of Traumatic Stress*, *28*, 572–579. Available from http://dx.doi.org/10.1002/jts.22054.

Kuhn, D. (2002). The effects of active and passive participation in musical activity on the immune system as measured by salivary immunoglobulin A (SlgA). *Journal of Music Therapy*, *39*(1), 30–39. Available from http://dx.doi.org/10.1093/jmt/39.1.30.

Kushner, R. (2003). Understanding the links between performing artists and audiences. *Journal of Arts Management, Law, and Society*, *33*(2), 114–126. Available from http://dx.doi.org/10.1080/10632920309596570.

Kusserow, M., Candia, V., Amft, O., Hildebrandt, H., Folkers, G., & Troster, G. (2012). Monitoring stage fright outside the laboratory: An example in a professional musician using wearable sensors. *Medical Problems of Performing Artists*, *27*(1), 21–30.

Kyaga, S., Landen, M., Boman, M., Hultman, C. M., Langstrom, N., & Lichtenstein, P. (2013). Mental illness, suicide, and creativity: 40-year prospective total population study. *Journal of Psychiatric Research*, *47*, 83–90. Available from http://dx.doi.org/10.1016/j.jpsychires.2012.09.010.

Kyaga, S., Lichtenstein, P., Boman, M., Hultman, C., Langstrom, N., & Landen, M. (2011). Creativity and mental disorder: Family study of 300,000 people with severe mental disorder. *The British Journal of Psychiatry*, *199*, 373–379. Available from http://dx.doi.org/10.1192/bjp.bp.110.085316.

Lacaille, N., Koestner, R., & Gaudreau, P. (2007). On the value of intrinsic rather than traditional achievement goals for performing artists: A short-term prospective study. *International Journal of Music Education*, *25*(3), 245–257.

Lack, S. A., Kumar, V. K., & Arevalo, S. (2003). Fantasy proneness, creativity capacity, and styles of creativity. *Perceptual and Motor Skills*, *96*, 19–24.

Lading, O., & Schellenberg, E. G. (2012). Liking unfamiliar music: Effects of felt emotion and individual differences. *Psychology of Aesthetics, Creativity, and the Arts*, *6*(2), 146–154. Available from http://dx.doi.org/10.1037/a0024671.

Laible, D., Panfile, T., & Makariev, D. (2008). The quality and frequency of mother-toddler conflict: Links with attachment and temperament. *Child Development*, *79*(2), 426–443.

Lamont, A. (2012). Emotion, engagement and meaning in strong experiences of music performance. *Psychology of Music*, *40*(5), 574–594. Available from http://dx.doi.org/10.1177/0305735612448510.

Lane, A., Hewston, R., Redding, E., & Whyte, G. P. (2003). Mood changes following modern-dance classes. *Social Behavior and Personality*, *31*(5), 453–460. Available from http://dx.doi.org/10.2224/sbp.2003.31.5.453.

Lane, A. M., Whyte, G. P., Terry, P. C., & Nevill, A. M. (2005). Mood, self-set goals and examination performance: The moderating effect of depressed mood. *Personality and Individual Differences*, *39*(1), 143–153. Available from http://dx.doi.org/10.1016/j.paid.2004.12.015.

Langendorfer, F. (2008). Personality differences among orchestral instrumental groups: Just a stereotype? *Personality and Individual Differences*, *44*, 608–618. Available from http://dx.doi.org/10.1016/j.paid.2007.09.027.

Langendorfer, F., Hodapp, V., Kreutz, G., & Bongard, S. (2006). Personality and performance anxiety among professional orchestral musicians. *Journal of Individual Differences*, *27*(3), 162–171. Available from http://dx.doi.org/10.1027/1614-0001.27.3.162.

Lankford, D. E., Bennion, T. W., King, J., Hessing, N., Lee, L., & Heil, D. P. (2014). The energy expenditure of recreational ballroom dance. *International Journal of Exercise Science*, *7*(3), 228–235.

Lantz, J., & Raiz, L. (2003). Play and art in existential trauma therapy with children and their parents. *Contemporary Family Therapy*, *25*(2), 165–177.

Laposa, J. M., & Alden, L. E. (2008). The effect of pre-existing vulnerability factors on a laboratory analogue trauma experience. *Journal of Behavior Therapy and Experimental Psychiatry*, *39*, 424–435. Available from http://dx.doi.org/10.1016/j.jbtep.2007.11.002.

Lau, S., & Cheung, P. C. (2010). Developmental trends of creativity: What twists of turn do boys and girls take at different grades? *Creativity Research Journal*, *22*(3), 329–336. Available from http://dx.doi.org/10.1080/10400419.2010.503543.

Laugharne, J., Lilee, A., & Janca, A. (2010). Role of psychological trauma in the cause and treatment of anxiety and depressive disorders. *Current Opinion in Psychiatry*, *23*, 25–29. Available from http://dx.doi.org/10.1097/YCO.0b013e3283345dc5.

Leahy, T., Pretty, G., & Tenenbaum, G. (2002). Prevalence of sexual abuse in organised competitive sport in Australia. *Journal of Sexual Aggression: An International, Interdisciplinary Forum for Research, Theory and Practice*, *8*(2), 16–36. Available from http://dx.doi.org/10.1080/13552600208413337.

Leaver, R., Harris, E. C., & Palmer, K. T. (2011). Musculoskeletal pain in elite professional musicians from British symphonies. *Occupational Medicine*, *61*(8), 549–555. Available from http://dx.doi.org/10.1093/occmed/kqr129.

LeBlanc, A., Jin, Y. C., Oberr, M., & Sivola, C. (1997). Effects of audience on music performance anxiety. *Journal of Research in Music Education*, *45*, 480–496.

LeBouthillier, D. M., McMillan, K. A., Thibodeau, M. A., & Asmundson, G. J. (2015). Types and number of traumas associated with suicidal ideation and suicide attempts in PTSD: Findings from a U.S. nationally representative sample. *Journal of Traumatic Stress*, *28*, 183–190. Available from http://dx.doi.org/10.1002/jts.22010.

LeBoutillier, N., & Marks, D. F. (2003). Mental imagery and creativity: A meta-analytic review study. *British Journal of Psychology*, *94*, 29–44.

Ledoux, E., Cloutier, E., Ouellet, F., Gagnon, I., Thuilier, C., & Ross, J. (2009). *Occupational risks in the performing arts: An exploratory study (Report R-607)*. Montreal, QC: IRSST.

Lee, E. A. D., Bissett, J. K., Carter, M. A., Cowan, P. A., Pyne, J. M., Speck, P. M., ... Tolley, E. A. (2013). Preliminary findings of the relationship of lower heart rate variability with military sexual trauma and presumed posttraumatic stress disorder. *Journal of Traumatic Stress*, *26*, 249–256. Available from http://dx.doi.org/10.1002/jts.21797.

Lee, H.-S., Park, H. Y., Yoon, J. O., Kim, J. S., Chun, J. M., Aminata, I. W., … Jeon, I.-H. (2013). Musicians' medicine: Musculoskeletal problems in string players. *Clinics in Orthopedic Surgery*, *5*, 155–160. Available from http://dx.doi.org/10.4055/cios.2013.5.3.155.

Lee, S.-H., Carey, S., Dubey, R., & Matz, R. (2012). Intervention program in college instrumental musicians, with kinematics analysis of cello and flute playing: A combined program of yogic breathing and muscle strengthening-flexibility exercises. *Medical Problems of Performing Artists*, *27*(2), 85–94.

Leijnse, J. N. A. L., Hallett, M., & Sonneveld, G. J. (2015). A multifactorial conceptual model of peripheral neuromusculoskeletal predisposing factors in task-specific focal hand dystonia in musicians: Etiologic and therapeutic implications. *Biological Cybernetics*, *109*, 109–123. Available from http://dx.doi.org/10.1007/s00422-014-0631-5.

Lemons, G. (2011). Diverse perspectives of creativity testing: Controversial issues when used for inclusion into gifted programs. *Journal for the Education of the Gifted*, *34*(5), 742–772. Available from http://dx.doi.org/10.1177/0162353211417221.

Lench, H. C., Levine, L. J., & Roe, E. (2010). Trait anxiety and achievement goals as predictors of self-reported health in dancers. *Journal of Dance Medicine and Science*, *14*(4), 163–170.

Lese, A.-C. (2014). The contribution of biomechanics and of tai chi exercises to the psychological and development of training actors. *Procedia—Social and Behavioral Sciences*, *149*, 495–502. Available from http://dx.doi.org/10.1016/j.sbspro.2014.08.297.

Leslie, D., & Rantisi, N. M. (2011). Creativity and place in the evolution of a cultural industry: The case of Cirque du Soleil. *Urban Studies*, *48*(9), 1771–17787. Available from http://dx.doi.org/10.1177/0042098010377475.

Leube, D. T., Knoblich, G., Erb, M., Grodd, W., Bartels, M., & Kircher, T. T. J. (2003). The neural correlates of perceiving one's own movements. *Neuroimage*, *20*, 2084–2090.

Leudar, I., & Costall, A. (2004). On the persistence of the "problem of other minds" in psychology: Chomsky, Grice and Theory of Mind. *Theory and Psychology*, *14*(5), 601–621. Available from http://dx.doi.org/10.1177/0959354304046175.

Leudar, I., Costall, A., & Francis, D. (2004). Theory of mind: A critical assessment. *Theory and Psychology*, *14*(5), 571–578. Available from http://dx.doi.org/10.1177/0959354304046173.

Levenson, R. W., Ekman, P., & Friesen, W. V. (1990). Voluntary facial action generates emotion-specific autonomic nervous system activity. *Psychophysiology*, *27*(4), 363–384.

Levin, C. L. (2006). *Flow and motivation in male ballet dancers (Doctoral dissertation)*. The Wright Institute. Retrieved from ProQuest Dissertations and Theses. http://gateway.proquest.com/openurl:pqdiss:3231334.

Levin, R., & Fireman, G. (2001–2002). The relation of fantasy-proneness, psychological absorption and imaginative involvement to nightmare prevalence and nightmare distress. *Imagination, Cognition and Personality*, *21*(2), 111–129.

Levin, R., Simeon, D., & Guralnick, O. (2004). Role of fantasy proneness, imaginative involvement, and psychological absorption in depersonalization disorder. *Journal of Nervous and Mental Disease*, *192*(1), 69–71. Available from http://dx.doi.org/10.1097/01.nmd.0000106003.46153.54.

Levitin, D. J., & Rogers, S. E. (2005). Absolute pitch: Perception, coding, and controversies. *Trends in Cognitive Sciences*, *9*(1), 26–33. Available from http://dx.doi.org/10.1016/j.tics.2004.11.007.

Lewis, M. D., & Granic, I. (Eds.), (2000). *Emotion, development, and self-organization: Dynamic systems approaches to emotional development*. New York: Cambridge University Press.

Li, W., Yang, W., Li, W., Li, Y., Wei, D., Li, H., ... Zhang, Q. (2015). Brain structure and resting-state functional connectivity in university professors with high academic achievement. *Creativity Research Journal*, *27*(2), 139–150. Available from http://dx.doi.org/10.1080/10400419.2015.1030311.

Licht, C. M. M., Vreeburg, S. A., Dortland, A. K. B. R., Giltay, E. J., Hoogendijk, W. J. G., DeRijk, R. H., ... Penninx, B. W. J. H. (2010). Increased sympathetic and decreased parasympathetic activity rather than changes in hypothalamic-pituitary-adrenal axis activity is associated with metabolic abnormalities. *Journal of Clinical Endocrinology and Metabolism*, *95*(5), 2458–2466.

Lieberman, H. R., Falco, C. M., & Slade, S. S. (2002). Carbohydrate administration during a day of sustained aerobic activity improves vigilance, as assessed by a novel ambulatory monitoring device, and mood. *American Journal of Clinical Nutrition*, *76*, 120–127.

Liebschutz, J., Saltz, R., Brower, V., Keane, T. M., Lloyd-Travaglini, C., Averbuch, T., & Samet, J. H. (2007). PTSD in urban primary care: High prevalence and low physician recognition. *Journal of Internal Medicine*, *22*, 719–726. Available from http://dx.doi.org/10.1007/s11606-007-0161-0.

Liederbach, M., Dilgen, F. E., & Rose, D. J. (2008). Incidence of anterior cruciate ligament injuries among elite ballet and modern dancers. *American Journal of Sports Medicine*, *36*(9), 1779–1788. Available from http://dx.doi.org/10.1177/0363546508323644.

Liederbach, M., Hagins, M., Gamboa, J. M., & Welsh, T. M. (2012). Assessing and reporting dancer capacities, risk factors, and injuries: Recommendations from the IADMS standard measures consensus initiative. *Journal of Dance Medicine and Science*, *16*(4), 139–153.

Liederbach, M., Kremenic, I. J., Orishimo, K. F., Pappas, E., & Hagins, M. (2014). Comparison of landing biomechanics between male and female dancers and athletes. Part 2: Influence of fatigue and implications for anterior cruciate ligament injury. *American Journal of Sports Medicine*, *42*(5), 1089–1095. Available from http://dx.doi.org/10.1177/0363546514524525.

Liederbach, M., Schanfein, L., & Kremenic, I. J. (2013). What is known about the effect of fatigue on injury occurrence among dancers? *Journal of Dance Medicine and Science*, *17*(3), 101–108. Available from http://dx.doi.org/10.12678/1089-313X.17.3.101.

Liiv, H., Jurimae, T., Maestu, J., Purge, P., Hannus, A., & Jurimae, J. (2012). Physiological characteristics of elite dancers of different dance styles. *European Journal of Sport Science*, *14*(S10), S429–S436. Available from http://dx.doi.org/10.1080/17461391.2012.711861.

Lill, A. (2014). From local to global: The evolution of musical play in secondary schools. *International Journal of Play*, *3*(3), 251–266. Available from http://dx.doi.org/10.1080/21594937.2014.976036.

Lille, D. (2010). *Equipoise: The life and work of Alfredo Corvino*. New York: Dance Movement Press.

Lim, S., & Smith, J. (2008). The structural relationships of parenting style, creativity personality, and loneliness. *Creativity Research Journal*, *20*(4), 412–419. Available from http://dx.doi.org/10.1080/10400410802391868.

Limont, W., Dreszer-Drogorob, J., Bedynksa, S., Sliwinska, K., & Jastrzebska, D. (2014). "Old wine in new bottles"? Relationships between overexcitabilities, the Big Five personality traits and giftedness in adolescents. *Personality and Individual Differences*, *69*, 199–204. Available from http://dx.doi.org/10.1016/j.paid.2014.06.003.

Lin, C.-F., Lee, I.-J., Liao, J.-H., Wu, H.-W., & Su, F.-C. (2011). Comparison of postural stability between injured and uninjured ballet dancers. *American Journal of Sports Medicine*, *39*(6), 1324–1331. Available from http://dx.doi.org/10.1177/0363546510393943.

Lin, J.-S., Chang, W.-S., & Liang, C. (2015). The imagination constructs with science students: Interplay among social climate, intrinsic motivation, and personality traits. *Imagination, Cognition, and Personality*, *34*(4), 340–359. Available from http://dx.doi.org/10.1177/0276236615572589.

Lind, M. J., Aggen, S. H., Kendler, K. S., York, T. P., & Amstadter, A. B. (2016). An epidemiologic study of childhood sexual abuse and adult sleep disturbances. *Psychological Trauma: Theory, Research, Practice, and Policy*, *8*(2), 198–205. Available from http://dx.doi.org/10.1037/tra000080.

Lindell, A. K. (2011). Lateral thinkers are not so laterally minded: Hemispheric asymmetry, interaction, and creativity. *Laterality*, *16*(4), 479–498. Available from http://dx.doi.org/10.1080/1357650X.2010.497813.

Lindmark, S., Lonn, L., Wiklund, U., Tufvesson, M., Olsson, T., & Eriksson, J. W. (2005). Dysregulation of the autonomic nervous system can be a link between visceral adiposity and insulin resistance. *Obesity Research*, *13*(4), 717–728.

Lindt, A. K. (2011). Lateral thinkers are not so laterally minded: Hemispheric asymmetry, interaction, and creativity. *Laterality*, *16*(4), 479–498. Available from http://dx.doi.org/10.1080/1357650X.2010.497813.

Linley, P. A. (2003). Positive adaptation to trauma: Wisdom as both process and outcome. *Journal of Traumatic Stress*, *16*(6), 601–610.

Liu, D., Liao, H., & Loi, R. (2012). The dark side of leadership: A three-level investigation of the cascading effect of abusive supervision on employee creativity. *Academy Management Journal*, *55*(5), 1187–1212. Available from http://dx.doi.org/10.5465/amj.2010.0400.

Llinas, R. R. (2001). *I of the vortex: From neurons to self.* Cambridge, MA: The MIT Press.

Loggia, M. L., Modli, J. S., & Bushnell, M. C. (2008). Empathy hurts: Compassion for another increases both sensory and affective components of pain perception. *Pain*, *136* (1–2), 168–176. Available from http://dx.doi.org/10.1016/j.pain.2007.07.017.

Lolich, M., Vazquez, G. H., Zapata, S., & Akiskal, K. K. (2015). Affective temperaments in tango dancers. *Journal of Affective Disorders*, *173*, 27–30. Available from http://dx.doi.org/10.1016/j.jad.2014.10.018.

Lord, V. M., Cave, P., Hume, V. J., Flude, E. J., Evans, A., Kelly, J. L., ... Hopkinson, N. S. (2010). Singing teaching as a therapy for chronic respiratory disease: A randomized controlled trial and qualitative evaluation. *BioMed Central Pulmonary Medicine*, *10*(41), 1–7. Available from http://dx.doi.org/10.1186/1471-2466-10-41.

Loring, P. A. (2007). The most resilient show on earth: The circus as a model for viewing identity, change, and chaos. *Ecology and Society*, *12*(1), 9. Available from http://www.ecologyandsociety.org/vol12/iss1/art/.

Lotze, M., Montoya, P., Erb, M., Hulsmann, E., Flor, H., Klose, U., ... Grodd, W. (1999). Activation of cortical and cerebellar motor areas during executed and imagined hand movements: An fMRI study. *Journal of Cognitive Neuroscience*, *11*(5), 491–501.

Lotze, M., Scheler, G., Tan, H.-R. M., Braun, C., & Birbaumer, N. (2003). The musician's brain: Functional imaging of amateurs and professionals during performance and imagery. *Neuroimage*, *20*, 1817–1829. Available from http://dx.doi.org/10.1016/j.neuroimage.2003.07.018.

Lubart, T. L. (2001). Models of the creative process: Past, present and future. *Creativity Research Journal*, *13*(3–4), 295–308. Available from http://dx.doi.org/10.1207/S15326934CRJ1334_07.

Lubet, A. (2002). Disability studies and performing arts medicine. *Medical Problems of Performing Artists*, *17*(2), 59–62.

Luby, J. L., Barch, D. M., Belden, A., Gaffrey, M. S., Tillman, B., Babb, C., . . . Botteron, K. N. (2012). Maternal support in early childhood predicts hippocampal volumes at school age. *Proceedings From the National Academy of Science*, *109*(8), 2854–2859. Available from http://dx.doi.org/10.1073/pnas.1118003109.

Ludwig, A. M. (1998). Method and madness in the arts and sciences. *Creativity Research Journal*, *11*(2), 93–101.

Luh, D.-B., & Lu, C.-C. (2012). From cognitive style to creativity achievement: The mediating role of passion. *Psychology of Aesthetics, Creativity, and the Arts*, *6*(3), 282–288. Available from http://dx.doi.org/10.1037/a0026868.

Lukaschek, K., Kruse, J., Emeny, R. T., Lacruz, M. E., von Eisenhart Rothe, A., & Ladwig, K.-H. (2013). Lifetime traumatic experiences and their impact on PTSD: A general population study. *Social Psychiatry and Psychiatric Epidemiology*, *48*, 525–532. Available from http://dx.doi.org/10.1007/a00127-012-0585-7.

Luke, A. C., Kinney, S. A., D'Hemecourt, P. A., Baum, J., Owen, M., & Micheli, L. J. (2002). Determinants of injuries in young dancers. *Medical Problems of Performing Artists*, *17*, 105–112.

Luo, Z. C., Fraser, W. D., Julien, P., Deal, C. L., Audibert, F., Smith, G. N., . . . Walker, M. (2006). Tracing the origins of "fetal origins" of adult diseases: Programming by oxidative stress? *Medical Hypothesis*, *66*, 38–44.

Lussier-Ley, C., & Durand-Bush, N. (2009). Exploring the role of feel in the creative experience of modern dancers: A realistic tale. *Research in Dance Education*, *10*(3), 199–217. Available from http://dx.doi.org/10.1080/14647890993324154.

Lynn, S. J., & Rhue, J. W. (1988). Fantasy proneness: Hypnosis, developmental antecedents and psychopathology. *American Psychologist*, *43*(1), 35–40.

Ma, H.-H. (2009). The effect size of variables associated with creativity: A meta-analysis. *Creativity Research Journal*, *21*(1), 30–42. Available from http://dx.doi.org/10.1080/10400410802633400.

MacArthur, L. J. (2011). Behind closed doors: Emotional abuse in the music studio. *International Symposium on Performance Science*, 387–392.

MacDonald, R., & Wilson, G. (2005). Musical identities of professional jazz musicians: A focus group investigation. *Psychology of Music*, *33*(4), 395–417. Available from http://dx.doi.org/10.1177/0305735605056151.

MacEachen, E., Clarke, J., Franche, R.-L., & Irvin, E. (2006). Systemic review of the qualitative literature on returning to work after injury. *Scandinavian Journal of Work Environment*, *32*(4), 257–269.

Macfarlane, J. D., & Rietveld, A. B. M. (2013). Malignancies presenting at a performing arts medicine clinic. *Clinical Rheumatology*, *32*, 453–454. Available from http://dx.doi.org/10.1007/s10067-013-2197-3.

Macintyre, J., & Joy, E. (2000). Foot and ankle injuries in dance. *Clinics in Sports Medicine*, *19*(2), 351–368.

MacLean, C. R. (2011). Differences in Myers–Briggs personality types among high school band, orchestra, and choir members. *Journal of Research in Music Education*, *59*(1), 85–100. Available from http://dx.doi.org/10.1177/002242941035579.

Mageau, G. A., Vallerand, R. J., Charest, J., Salvy, S.-J., Lacaille, N., Bouffard, T., & Koestner, R. (2009). On the development of harmonious and obsessive passion: The role of autonomy support, activity specialization, and identification with the activity. *Journal of Personality*, *77*(3), 601–646. Available from http://dx.doi.org/10.1111/j.1467-6494.2009.00559.x.

Maier, M. A., Bernier, A., Pekrun, R., Zimmermann, P., Strasser, K., & Grossmann, K. E. (2005). Attachment state of mind and perceptual processing of emotional stimuli.

Attachment and Human Development, *7*(1), 67–81. Available from http://dx.doi.org/10.1080/14616730500039606.

Main, M. (2000). The organized categories of infant, child and adult attachment: Flexible vs. inflexible attention under attachment-related stress. *Journal of the American Psychoanalytic Association*, *48*(4), 1055–1096.

Main, M., Goldwyn, R., & Hesse, E. (2003). *Adult attachment scoring and classification systems*. Unpublished manuscript, University of California at Berkeley.

Main, M., Hesse, E., & Kaplan, N. (2005). Predictability of attachment behavior and representational processes at 1, 6 and 19 years of age: The Berkeley longitudinal study. In K. E. Grossman, K. Grossman, & E. Waters (Eds.), *Attachment from infancy to adulthood: Lessons from the longitudinal studies* (pp. 245–304). New York: The Guilford Press.

Mallorqui-Bague, N., Garfinkel, S. N., Engels, M., Eccles, J. A., Pailhez, G., Bulbena, A., & Critchley, H. D. (2014). Neuroimaging and psychophysiological investigation of the link between anxiety, enhanced affective reactivity and interoception in people with joint hypermobility. *Frontiers in Psychology*, *5*(1162), 1–8. Available from http://dx.doi.org/10.3389/fpsyg.2014.01162.

Malpede, K. (1996). Teaching witnessing: A class wakes to the genocide in Bosnia. *Theatre Topics*, *6*(2), 167–179.

Malpede, K. (1999). Chilean testimonies: An experiment in theatre of witness. *Journal of Contemporary Psychotherapy*, *29*, 307–316.

Maltby, J., Day, L., McCutcheon, L. E., Houran, J., & Ashe, D. (2006). Extreme celebrity worship, fantasy proneness and dissociation: Developing the measurement and understanding of celebrity worship within a clinical personality context. *Personality and Individual Differences*, *40*(2), 273–283. Available from http://dx.doi.org/10.1016/j.paid.2005.07.004.

Maltby, N., Tolin, D. F., Worhunsky, P., O'Keefe, T. M., & Kiehl, K. A. (2005). Dysfunctional action monitoring hyperactivates frontal-striatal circuits in obsessive-compulsive disorder: An event-related fMRI study. *Neuroimage*, *24*, 495–503. Available from http://dx.doi.org/10.1016/j.neuroimage.2004.08.041.

Manchester, R. A. (2006). Toward better prevention of injuries among performing artists. *Medical Problems of Performing Artists*, *21*(1), 1–2.

Mancini, A. D., & Bonanno, G. A. (2006). Resilience in the face of potential trauma: Clinical practices and illustrations. *Journal of Clinical Psychology*, *62*, 971–985. Available from http://dx.doi.org/10.1002/jclp.20283.

Manger, T. A., & Motta, R. W. (2005). The impact of an exercise program on PTSD, anxiety, and depression. *International Journal of Emergency Mental Health*, *7*(1), 49–57.

Mangset, P., Kleppe, B., & Royseng, S. (2012). Artists in an iron cage? Artists' work in performing arts institutions. *Journal of Arts Management, Law, and Society*, *42*, 156–175. Available from http://dx.doi.org/10.1080/10632921.2012.727773.

Marchant-Haycox, S. E., & Wilson, G. D. (1992). Personality and stress in performing artists. *Personality and Individual Differences*, *13*(10), 1061–1068.

Marinelli, R., Bindi, R., Marchi, S., Castellani, E., Carli, G., & Santarcangelo, E. (2012). Hypnotizability-related differences in written language. *International Journal of Clinical and Experimental Hypnosis*, *60*(1), 54–66. Available from http://dx.doi.org/10.1080/00207144.2011.622196.

Marjanovic-Shane, A., & White, E. J. (2014). When the footlights are off: A Bakhtinian interrogation of play as postupok. *International Journal of Play*, *3*(2), 119–135. Available from http://dx.doi.org/10.1080/21594937.2014.931686.

Marsalis, W. Retrieved from www.goodreads.com/quotes/tag/music.

Martens, A., Greenberg, J., Allen, J. J. B., Hayes, J., Schimel, J., & Johns, M. (2010). Self-esteem and autonomic physiology: Self-esteem levels predict cardiac vagal tone. *Journal of Research in Personality*, *44*, 573–584. Available from http://dx.doi.org/10.1016/j.jrp.2010.07.001.

Martin, C. T., Ric, A., & Hristovski, R. (2015). Creativity and emergence of specific dance movements using instructional constraint. *Psychology of Aesthetics, Creativity, and the Arts*, *9*(1), 65–74. Available from http://dx.doi.org/10.1037/a0038706.

Martin, J. J., & Cutler, K. (2002). An exploratory study of flow and motivation in theatre actors. *Journal of Applied Sport Psychology*, *14*, 344–352. Available from http://dx.doi.org/10.1080/10413200290103608.

Martin, L. L. (2001). Mood as input: A configural view of mood effect. In L. L. Martin, & G. L. Clore (Eds.), *Theories of mood and cognition: A user's guidebook* (pp. 135–157). New Jersey, NJ: Erlbaum.

Martindale, C. (2007). Recent trends in the psychological study of aesthetics, creativity, and the arts. *Empirical Studies of the Arts*, *25*(2), 121–141.

Martyn-Stevens, B. E., Brown, L. E., Beam, W. C., & Wiersma, L. D. (2012). Effects of a dance season on the physiological profile of collegiate female modern dancers. *Medicina Sportiva*, *16*(1), 1–5. Available from http://dx.doi.org/10.5004/17342260.987830.

Marusak, H. A., Etkin, A., & Thomason, M. E. (2015). Disrupted insula-based circuit organization and conflict interference in trauma-exposed youth. *Neuroimage: Clinical*, *8*, 516–525. Available from http://dx.doi.org/10.1016/j.nicl.2015.04.007.

Maslach, C., & Jackson, S. E. (1981). The measurement of experienced burnout. *Journal of Occupational Behavior*, *2*, 99–113.

Maslach, C., Schaufeli, W. B., & Leiter, M. P. (2001). Job burnout. *Annual Review of Psychology*, *52*, 397–422.

Masten, A. S. (2001). Ordinary magic: Resilience processes in development. *American Psychologist*, *56*(3), 227–238. Available from http://dx.doi.org/10.1037//0003-066X.56.3.227.

Master, S., De Biase, N., Chiari, B. M., & Laukkanen, A.-M. (2008). Acoustic and perceptual analyses of Brazilian male actors' and nonactors' voices: Long-term average spectrum and the "actor's format". *Journal of Voice*, *22*(2), 146–154. Available from http://dx.doi.org/10.1016/j.jvoice.2006.09.006.

Master, S., Guzman, M., Axocar, M. J., Munoz, D., & Bortnem, C. (2015). How do laryngeal and respiratory functions contribute to differentiate actors/actresses and untrained voices? *Journal of Voice*, *29*(3), 333–345. Available from http://dx.doi.org/10.1016/j.jvoice.2014.09.003.

Matare, J. (2009). Creativity or musical intelligence? A comparative study of improvisation/improvisation performance by European and African musicians. *Thinking Skills and Creativity*, *4*, 194–203. Available from http://dx.doi.org/10.1016/j.tsc.2009.09.005.

Mathur, S., & Parameswaran, G. (2015). Gender neutrality in play of young migrant children: An emerging trend or an outlier? *American Journal of Play*, *7*(2), 174–200.

Matinsen, O. L. (2011). The creative personality: A synthesis and development of the creative person profile. *Creativity Research Journal*, *23*(3), 185–202. Available from http://dx.doi.org/10.1080/10400419.2011.595656.

Matos, M., Pinto-Gouveia, J., & Costa, V. (2013). Understanding the importance of attachment in shame traumatic memory relation to depression: The impact of emotion regulation processes. *Clinical Psychology and Psychotherapy*, *20*, 149–165. Available from http://dx.doi.org/10.1002/cpp.786.

Mayer, J. D., Perkins, D. M., Caruso, D. R., & Salovey, P. (2001). Emotional intelligence and giftedness. *Roeper Review*, *23*(3), 131–137. Available from http://dx.doi.org/10.1080/02783190109554084.

Mayseless, N., Eran, A., & Shamay-Tsoory, S. G. (2015). Generating original ideas: The neural underpinning of originality. *Neuroimage*, *116*, 232–239. Available from http://dx.doi.org/10.1016.neuroimage.2015.05.030.

McAdams, D. P., & Olson, B. D. (2010). Personality development: Continuity and change over the life course. *Annual Review of Psychology*, *61*, 517–542. Available from http://dx.doi.org/10.1146/annurev.psych.093008.100507.

McAdams, D. P., & Pals, J. L. (2006). A new Big Five: Fundamental principles for an integrative science of personality. *American Psychologist*, *61*(3), 204–217. Available from http://dx.doi.org/10.1037/0003-066X.61.3.204.

McCabe, A., & Peterson, C. (2011–2012). Predictors of adult narrative elaboration, emotion, attachment, and gender. *Imagination, Cognition and Personality*, *31*(4), 327–344. Available from http://dx.doi.org/10.2190/IC.31.4.f.

McClary, R. (2007). Healing the psyche through music, myth, and ritual. *Psychology of Aesthetics, Creativity, and the Arts*, *1*(3), 155–159. Available from http://dx.doi.org/10.1037/1931-3896.1.3.155.

McCoy, J. M., & Evans, G. W. (2002). The potential role of the physical environment in fostering creativity. *Creativity Research Journal*, *14*(3 & 4), 409–426.

McCrae, R. R. (2007). Aesthetic chills as a universal marker of openness to experience. *Motivation and Emotion*, *31*, 5–11. Available from http://dx.doi.org/10.1007/s11031-007-9053-1.

McCrae, R. R., & Costa, P. C., Jr. (1987). Validation of the five-factor model across instruments and observers. *Journal of Personality and Social Psychology*, *52*, 81–90.

McCraty, R., Atkinson, M., Tomasino, D., & Stuppy, W. P. (2001). Analysis of twenty-four hour heart rate variability in patients with panic disorder. *Biological Psychology*, *56*, 131–150.

McCraty, R., Barrios-Choplin, B., Atkinson, M., & Tomasino, D. (1998). The effects of different types of music on mood, tension, and mental clarity. *Alternate Therapies and Health Medicine*, *4*(1), 75–84.

McDougall, S. J., Widdop, R. E., & Lawrence, A. J. (2005). Central autonomic integration of psychological stressors: Focus on cardiovascular modulation. *Autonomic Neuroscience: Basic and Clinical*, *123*, 1–11.

McEldowney, K. M., Hopper, L. S., Etlin-Stein, H., & Redding, E. (2013). Fatigue effects on quadriceps and hamstrings activation in dancers performing drop landings. *Journal of Dance Medicine and Science*, *17*(3), 109–114. Available from http://dx.doi.org/10.12678/1089-313X.17.3.109.

McEwen, B. S., Eiland, L., Hunter, R. G., & Miller, M. M. (2012). Stress and anxiety: Structural plasticity and epigenetic regulation as a consequence of stress. *Neuropharmacology*, *62*, 3–12.

McEwen, B. S., Gray, J. D., & Nasca, C. (2015). Recognizing resilience: Learning from the effects of stress on the brain. *Neurobiology of Stress*, *1*, 1–11. Available from http://dx.doi.org/10.1016/j.ynstr.2014.09.001.

McGloin, J. M., & Widom, C. S. (2001). Resilience among abused and neglected children grown up. *Development and Psychopathology*, *13*(4), 1021–1038.

McGregor, E. Retrieved from www.goodreads.com/quotes/tag/acting.

McIntyre, P. (2008). Creativity and cultural production: A study of contemporary western popular music songwriting. *Creativity Research Journal*, *20*(1), 40–52. Available from http://dx.doi.org/10.1080/10400410701841898.

McKinley, N. M., & Hyde, J. S. (1996). The objectified body consciousness scale. *Psychology of Women Quarterly*, *20*, 181–215.

McKinney, C., & Power, L. (2012). Childhood playtime, parenting, and psychopathology in emerging adults: Implications for research and play therapists. *International Journal of Play Therapy*, *21*(4), 215–231. Available from http://dx.doi.org/10.1037/a0029172.

McManus, I. C., Jonvik, H., Richards, P., & Paice, E. (2011). Vocation and avocation: Leisure activities correlate with professional engagement, but not burnout, in a cross-sectional survey of UK doctors. *BioMed Central*, *9*(100), 1–18. Available from http://dx.doi.org/10.1186/1741-7015-9-100.

McPherson, G. E. (2009). The role of parents in children's musical development. *Psychology of Music*, *37*(1), 91–110. Available from http://dx.doi.org/10.1177/0305735607086049.

McPherson, M., & Limb, C. J. (2013). Difficulties in the neuroscience of creativity: Jazz improvisation and the scientific method. *Annals of the New York Academy of Sciences*, *1303*, 80–83. Available from http://dx.doi.org/10.1111/nyas.12174.

McWilliams, L. A., Cox, B. J., & Enns, M. W. (2003). Use of coping inventory for stressful situations in a clinically depressed sample: Factor structure, personality correlates, and prediction of distress. *Journal of Clinical Psychology*, *59*(12), 1371–1385. Available from http://dx.doi.org/10.1002/jclp.10228.

Meaney, M. J. (2001). Maternal care, gene expression, and the transmission of individual differences in stress reactivity across generations. *Annual Review of Neuroscience*, *24*, 1161–1192.

Meister, I. G., Krings, T., Foltys, H., Boroojerdi, B., Uller, M., Topper, R., & Thron, A. (2004). Playing piano in the mind—an fMRI study of music imagery and performance in pianists. *Cognitive Brain Research*, *19*, 219–228. Available from http://dx.doi.org/10.1016/j.cogbrainres.2003.12.005.

Melinder, A., Baugerud, G. A., Ovenstad, K. S., & Goodman, G. S. (2013). Children's memories of removal: A test of attachment theory. *Journal of Traumatic Stress*, *26*, 125–133. Available from http://dx.doi.org/10.1002/jts.21784.

Mendaglio, S., & Tillier, W. (2006). Dabrowski's theory of positive disintegration and giftedness: overexcitability research findings. *Journal for the Education of the Gifted*, *30*(1), 68–87.

Mendlewicz, L., Linkowski, P., Bazelmans, C., & Philippot, P. (2005). Decoding emotional facial expressions in depressed and anorexic patients. *Journal of Affective Disorders*, *89*, 195–199. Available from http://dx.doi.org/10.1016/j.jad.2005.07.010.

Menon, V. (2011). Large-scale brain networks and psychopathology: A unifying triple network model. *Trends in Cognitive Sciences*, *15*(10), 483–506. Available from http://dx.doi.org/10.1016/j.tics.2011.08.003.

Menon, V., & Levitin, D. J. (2005). The rewards of music listening: Response and physiological connectivity of the mesolimbic system. *Neuroimage*, *28*, 175–184. Available from http://dx.doi.org/10.1016/j.neuroimage.2005.05.053.

Merckelbach, H. (2004). Telling a good story: Fantasy proneness and the quality of fabricated memories. *Personality and Individual Differences*, *37*(7), 1371–1382. Available from http://dx.doi.org/10.1016/j.paid.2004.01.007.

Merckelbach, H., & van de Ven, V. (2001). Another white Christmas: Fantasy proneness and reports of "hallucinatory experiences" in undergraduate students. *Journal of Behavior Therapy and Experimental Psychiatry*, *32*, 137–144.

Merriam Webster Dictionary and Thesaurus. http://www.merriam-webster.com.

Mesquita, B., & Albert, D. (2007). The cultural regulation of emotions. In J. J. Gross (Ed.), *Handbook of emotion regulation* (pp. 486–503). New York: Guilford Press.

Mesurado, B., & de Minzi, M. C. R. (2013). Child's personality and perception of parental relationship as correlates of optimal experience. *Journal of Happiness Studies, 14*, 199–214. Available from http://dx.doi.org/10.1007/s10902-012-9324-8.

Metzl, E. S. (2009). The role of creative thinking in resilience after Hurricane Katrina. *Psychology of Aesthetics, Creativity, and the Arts, 3*(2), 112–123. Available from http://dx.doi.org/10.1037/a0013479.

Micheli, L. J., Cassella, M., Faigenbaum, A. D., Southwick, H., & Ho, V. (2005). Preseason to postseason changes in body composition of professional ballet dancers. *Journal of Dance Medicine and Science, 9*(2), 56–59.

Mika, E. (2005). Theory of positive disintegration as a model of personality development for exceptional individuals. In N. L. Hafenstein, B. Kutrumbos, & J. Delisle (Eds.), *Perspectives in gifted education: Complexities of emotional development, spirituality and hope* (Vol. 3, pp. 4–32). Publication Institute for the Development of Gifted Education, Ricks Center for Gifted Children: University of Denver, CO.

Mikulincer, M., & Sheffi, E. (2000). Adult attachment style and cognitive reactions to positive affect: A test of mental categorization and creative problem solving. *Motivation and Emotion, 24*(3), 149–174.

Milevsky, A., Schlechter, M., Netter, S., & Keehn, D. (2007). Maternal and paternal parenting styles in adolescent associations with self-esteem, depression and life-satisfaction. *Journal of Child and Family Studies, 16*, 39–47. Available from http://dx.doi.org/10.1007/s10826-006-9066-5.

Miller, C. (2006). Dance medicine: Current concepts. *Physical Medicine and Rehabilitation Clinics of North America, 17*, 803–811. Available from http://dx.doi.org/10.1016/j.pmr.2006.06.005.

Miller, C., & Moa, G. (1998). Injury characteristics and outcomes at a performing arts school clinic. *Medical Problems of Performing Artists, 13*(3), 114–116.

Miller, E. M., & Cohen, L. M. (2012). Engendering talent in others: Expanding domains of giftedness and creativity. *Roeper Review, 34*, 104–113. Available from http://dx.doi.org/10.1080/02783193.2012.660684.

Miller, K. E., & Quigley, B. M. (2011). Energy drink use and substance use among musicians. *Journal of Caffeine Research, 1*(1), 67–73. Available from http://dx.doi.org/10.1089/caf.2011.0003.

Miller, K. E., & Quigley, B. M. (2012). Sensation-seeking, performance genres, and substance use among musicians. *Psychology of Music, 40*(4), 389–410. Available from http://dx.doi.org/10.1177/0305735610387776.

Milligan, K., Atkinson, L., Trehub, S. E., Benoit, D., & Poulton, L. (2003). Maternal attachment and the communication of emotion through song. *Infant Behavior and Development, 26*, 1–13.

Mills, M. J., & Fullagar, C. J. (2008). Motivation and flow: Toward an understanding of the dynamics of the relation in architecture students. *The Journal of Psychology: Interdisciplinary and Applied, 142*(5), 533–556. Available from http://dx.doi.org/10.3200/JRLP.142.5.533-556.

Milton, J., Small, S. L., & Solodkin, A. (2008). Imaging motor imagery: Methodological issues related to expertise. *Methods, 45*, 336–341. Available from http://dx.doi.org/10.1016/j.ymeth.2008.05.002.

Miskovic, V., & Schmidt, L. A. (2012). New directions in the study of individual differences in temperament: A brain-body approach to understanding fearful and fearless children. *Monographs of the Society for Research in Child Development, 77*(2), 28–38.

Mistiaen, W., Roussel, N. A., Vissers, D., Daenen, L., Truijen, S., & Nijs, J. (2012). Effects of aerobic endurance, muscle strength, and motor control exercise on physical fitness and musculoskeletal injury rate in preprofessional dancers: An uncontrolled trial. *Journal of Manipulative Physiology Therapy*, *35*, 381–389.

Mittal, C., & Griskevicius, V. (2014). Sense of control under uncertainty depends on people's childhood environment: A life history theory approach. *Journal of Personality and Social Psychology*, *107*(4), 621–637. Available from http://dx.doi.org/10.1037/a0037398.

Mittal, C., Griskevicius, V., Simpson, J. A., Sung, S., & Young, E. S. (2015). Cognitive adaptations to stressful environments: When childhood adversity enhances adult executive functioning. *Journal of Personality and Social Psychology*, *109*(4), 604–621. Available from http://dx.doi.org/10.1037/pspi000028.

Miura, A., Kudo, K., & Nakazawa, K. (2013). Action-perception coordination dynamics of whole-body rhythmic movement in stance: A comparison study of street dancers and non-dancers. *Neuroscience Letters*, *544*, 157–162. Available from http://dx.doi.org/10.1016/j.neurlet.2013.04.005.

Miura, A., Kudo, K., Ohtsuki, T., & Kanehisa, H. (2011). Coordination modes in sensorimotor synchronization of whole-body movement: A study of street dancers and non-dancers. *Human Movement Science*, *30*, 1260–1271. Available from http://dx.doi.org/10.1016/j.humov.2010.08.006.

Moffit, D. M., Russ, A. C., & Mansell, J. L. (2015). Marching band camp injury rates at the collegiate level. *Medical Problems of Performing Artists*, *30*(2), 96–99.

Moller, S. J. (2015). Imagination, playfulness, and creativity in children's play with different toys. *American Journal of Play*, *7*(3), 322–346.

Molnar-Szakacs, I., & Overy, K. (2006). Music and mirror neurons: From motion to "e"motion. *Social Cognition and Affective Neuroscience*, *1*, 235–241. Available from http://dx.doi.org/10.1093/scan/nsl029.

Monk, C. S., McClure, E. B., Nelson, E. E., Zarahn, E., Bilder, R. M., Leibenluft, E., ... Pine, D. S. (2003). Adolescent immaturity in attention-related brain engagement to emotional facial expression. *Neuroimage*, *20*, 420–428. Available from http://dx.doi.org/10.1016/S1053-8119(03)00355-0.

Monsma, E. V., & Overby, L. Y. (2004). The relationship between imagery and competitive anxiety in ballet auditions. *Journal of Dance Medicine and Science*, *8*(1), 11–18.

Montero, B. (2006). Proprioception as an aesthetic sense. *Journal of Aesthetics and Art Criticism*, *64*(2), 231–242.

Montgomery, S. S., & Robinson, M. D. (2003). What becomes of undergraduate dance majors? *Journal of Cultural Economics*, *27*, 57–71.

Montinaro, A. (2010). The music brain: Myth and science. *World Neurosurgery*, *73*(5), 442–453. Available from http://dx.doi.org/10.1016/J.WNEU.2010.02.060.

Montuori, A. (2003). The complexity of improvisation and the improvisation of complexity: Social science, art and creativity. *Human Relations*, *56*(2), 237–255.

Moore, D. W., Bhadelia, R. A., Billings, R. L., Fulwiler, C., Heilman, K. M., Rood, K. M. J., & Gansler, D. A. (2009). Hemispheric connectivity and the visual-spatial divergent-thinking component of creativity. *Brain and Cognition*, *70*, 267–272. Available from http://dx.doi.org/10.1016/j.bande.2009.02.011.

Moore, M. (2007). Golgi tendon organs: Neuroscience update with relevance to stretching and proprioception in dancers. *Journal of Dance Medicine and Science*, *11*(3), 85–92.

Moore, M., & Russ, S. W. (2008). Follow-up of a pretend play intervention: Effects on play, creativity, and emotional processes in children. *Creativity Research Journal*, *20*(4), 427–436. Available from http://dx.doi.org/10.1080/10400410802391892.

Moos, R. H., & Holahan, C. J. (2003). Dispositional and contextual perspectives on coping: Toward an integrative framework. *Journal of Clinical Psychology*, *59*(12), 1387−1403. Available from http://dx.doi.org/10.1002/jclp.10229.

Mor, S., Day, H. I., Flett, G. L., & Hewitt, P. L. (1995). Perfectionism, control, and components of performance anxiety in professional artists. *Cognitive Therapy and Research*, *19*(2), 207−225.

Morelock, M. (1996). On the nature of giftedness and talent: Imposing order on chaos. *Roeper Review*, *19*(1), 4−12. Available from http://dx.doi.org/10.1080/02783199609553774.

Morinville, A., Miranda, D., & Gaudreau, P. (2013). Music listening motivation is associated with global happiness in Canadian late adolescents. *Psychology of Aesthetics, Creativity, and the Arts*, *7*(4), 384−390. Available from http://dx.doi.org/10.1037/a0034495.

Morley, A. P., Narayanan, M., Mines, R., Molokhia, A., Baxter, S., Craig, G., ... Craig, I. (2012). AVPR1A and SLC6A4 polymorphisms in choral singers and non-musicians: A gene association study. *PLoS ONE*, *7*(2), 331763. Available from http://dx.doi.org/10.1371/journal.pone.0031763.

Morrin, N., & Redding, E. (2013). Acute effects of warm-up stretch protocols on balance, vertical jump height, and range of motion in dancers. *Journal of Dance Medicine and Science*, *17*(1), 34−40. Available from http://dx.doi.org/10.12678/1089-313X.17.1.34.

Morris, A. L., Cleary, A. M., & Still, M. L. (2008). The role of autonomic arousal in feelings of familiarity. *Consciousness and Cognition*, *17*, 1378−1385. Available from http://dx.doi.org/10.1016/j.concog.2008.04.005.

Mosing, M. A., Magnusson, P. K. E., Pedersen, N. L., Nakamura, J., Madison, G., & Ullen, F. (2012). Heritability of proneness for psychological flow experiences. *Personality and Individual Differences*, *53*, 699−704. Available from http://dx.doi.org/10.1016/j.paid.2012.05.035.

Mosing, M. A., Pedersen, N. L., Cesarini, D., Johannesson, M., Magnusson, P. K. E., Nakamura, J., ... Ullen, F. (2012). Genetic and environmental influences on the relationship between flow proneness, locus of control and behavioral inhibition. *PLoS ONE*, *7*(11), e47958. Available from http://dx.doi.org/10.1371/journal.pone.0047958.

Motta-Valencia, K. (2006). Dance-related injury. *Physical Medicine and Rehabilitation Clinics of North America*, *17*, 697−723. Available from http://dx.doi.org/10.1016/j.pmr.2006.06.001.

Mottweiler, C. M., & Taylor, M. (2014). Elaborated role play and creativity in preschool age children. *Psychology of Aesthetics, Creativity, and the Arts*, *8*(3), 277−286. Available from http://dx.doi.org/10.1037/a0036083.

Movius, H. L., & Allen, J. J. B. (2005). Cardiac vagal tone, defensiveness, and motivational style. *Biological Psychology*, *68*, 147−162. Available from http://dx.doi.org/10.1016/j.biopsycho.2004.03.019.

Mueller-Wohfahrt, H.-W., Haensel, L., Mithoefer, K., Ekstrand, J., English, B., McNally, S., ... Ueblacker, P. (2012). Terminology and classification of muscle injuries in sport: The Munich consensus statement. *British Journal of Sports Medicine*, *10*, 1−9. Available from http://dx.doi.org/10.1136/bjsports-2012-091448.

Mulpuru, S. K., Vasavada, B. C., Punukollu, G. K., & Patel, A. G. (2008). Cardiovocal syndrome: A systematic review. *Heart, Lung, and Circulation*, *17*, 1−4. Available from http://dx.doi.org/10.1016/j.hlc.2007.04.007.

Murgia, C. (2013a). Overuse, tissue fatigue, and injuries. *Journal of Dance Medicine and Science*, *17*(3), 92−100. Available from http://dx.doi.org/10.12678/1089-313X.17.3.92.

Murgia, C. (2013b). Overuse, fatigue, and injury: Neurological, psychological, physiological, and clinical aspects. *Journal of Dance Medicine and Science*, *17*(2), 51–52. Available from http://dx.doi.org/10.12678/1089-313X.17.2.51.

Murphy, A., Steele, M., Dube, S. R., Bate, J., Bonuck, K., Meissner, P., ... Steele, H. (2014). Adverse Childhood Experiences (ACEs) questionnaire and Adult Attachment Interview (AAI): Implications for parent child relationships. *Child Abuse and Neglect*, *38*, 224–233. Available from http://dx.doi.org/10.1016/j.chiabu.2013.09.004.

Musacchia, G., Sams, M., Skoe, E., & Kraus, N. (2007). Musicians have enhanced subcortical auditory and audiovisual processing of speech and music. *Proceedings of the National Academy of Science*, *104*(40), 15894–15898. Available from http://dx.doi.org/10.1073/pnas.0701498104.

Myers, L., Fleming, M., Lancman, M., Perrine, K., & Lancman, M. (2013). Stress coping strategies in patients with psychogenic non-epileptic seizures and how they relate to trauma symptoms, alexithymia, anger and mood. *Seizure*, *22*, 634–639. Available from http://dx.doi.org/10.1016/j.seizure.2013.04018.

Nadal, M., & Skov, M. (2013). Introduction to the special issue: Toward an interdisciplinary neuroaesthetics. *Psychology of Aesthetics, Creativity, and the Arts*, *7*(1), 1–12. Available from http://dx.doi.org/10.1037/a00318482.

Nagel, J. J. (2009). How to destroy creativity in music students: The need for emotional and psychological support services in music schools. *Medical Problems of Performing Artists*, *24*(1), 15–18.

Nagy, Z. (2015). The apperception of musical creativity: Performance as ritual, composition as self-realization. *Creativity Research Journal*, *27*(1), 68–75. Available from http://dx.doi.org/10.1080/1040419.2014.961784.

Nakahara, H., Furuya, S., Francis, P. R., & Kinoshita, H. (2010). Psycho-physiological responses to expressive piano performance. *International Journal of Psychophysiology*, *75*, 268–276. Available from http://dx.doi.org/10.1016/j.ijpsycho.2009.12.008.

Nakahara, H., Furuya, S., Masuko, T., Francis, P. R., & Kinoshita, H. (2011). Performing music can induce greater modulation of emotion-related psychophysiological response than listening to music. *International Journal of Psychophysiology*, *81*, 152–158. Available from http://dx.doi.org/10.1016/j.ijpsycho.2011.06.003.

Nam, D. H., Lim, J.-V., Ahn, C. M., & Choi, H.-S. (2004). Specially programmed respiratory muscle training for singers by using respiratory muscle training device (ultrabreathe). *Yonsei Medical Journal*, *45*(5), 810–817.

National Endowment for the Arts. (2013). *News*. http://arts.gov/news/2013/us-bureau-economic-analysis-and-national-endowment-arts-release-preliminary-report-impact.

National Endowment for the Arts (2015). *A decade of arts engagement: Findings from the survey of public participation in the arts, 2002–2012 (NEA Research Report No. 58)*. Washington, DC: National Endowment for the Arts.

Nawrocka, A., Mynarski, W., Powerska-Didowska, A., Grabara, M., & Garbaciak, W. (2014). Musculoskeletal pain among Polis music school students. *Medical Problems of Performing Artists*, *19*(2), 64–69.

Neal, J., & Frick-Horbury, D. (2001). The effects of parenting styles and childhood attachment patterns on intimate relationships. *Journal of Instructional Psychology*, *28*(3), 178.

Necka, E., & Hlawcz, T. (2013). Who has an artistic temperament? Relationships between creativity and temperament among artists and bank officers. *Creativity Research Journal*, *25* (2), 182–188. Available from http://dx.doi.org/10.1080/10400419.2013.783744.

Neff, K., Hsieh, Y., & Dejitterat, K. (2005). Self-compassion, achievement goals, and coping with academic failure. *Self and Identity*, *4*, 263–287.

Neihart, M. (1998). Creativity, the arts, and madness. *Roeper Review*, *21*(1), 47–50.

Neihart, M. (1999). The impact of giftedness on psychological well-being: What does the empirical literature say? *Roeper Review*, *22*(1), 10–17.

Neiminen, P. (2000). Stereotypes and self-ratings among Finnish dancers. *Research in Dance Education*, *1*(2), 133–154. Available from http://dx.doi.org/10.1080/713694262.

Nelmes, J. (2004). *An introduction to film studies* (3rd ed.). London: Routledge.

Nelson, B., & Rawlings, D. (2009). How does it feel? The development of the Experience of Creativity Questionnaire. *Creativity Research Journal*, *21*(1), 43–53. Available from http://dx.doi.org/10.1080/10400410802633442.

Nelson, B., & Rawlings, D. (2010). Relating schizotypy and personality to the phrenology of creativity. *Schizophrenia Bulletin*, *36*(2), 388–399. Available from http://dx.doi.org/10.1093/schbul/sbn098.

Nemiro, J. (1997). Interpretive artists: A qualitative exploration of the creative process of actors. *Creativity Research Journal*, *10*(2 & 3), 229–239.

Nestler, E. J., Barrot, M., DiLeone, R. J., Eisch, A. J., Gold, S. J., & Monteggia, L. M. (2002). Neurobiology of depression. *Neuron*, *34*, 13–25.

Nettle, D. (2006). Psychological profiles of professional actors. *Personality and Individual Differences*, *40*, 375–383. Available from http://dx.doi.org/10.1016/j.paid.2005.07.008.

Neuwirth, B. Retrieved from www.Brainyquote.com.

Newberg, A. B., & d'Aquili, E. G. (2000). The creative brain/the creative mind. *Zygon*, *35* (1), 53–68.

Newman, B. M., & Newman, P. R. (2015). *Development through the life: A psychosocial approach* (12th ed.). Stamford, CT: Cengage Learning.

Ng, L. C., Ahishakiye, N., Miller, D. E., & Meyerowitz, B. E. (2015). Narrative characteristics of genocide testimonies predict posttraumatic stress disorder symptoms years later. *Psychological Trauma: Theory, Research, Practice, and Policy*, *7*(3), 303–311. Available from http://dx.doi.org/10.1037/tra000024.

Nijs, J., Wouters, K., Cras, P., & Daenen, L. (2015). The role of sensorimotor incongruence in pain in professional dancers. *Motor Control*, *19*, 271–288. Available from http://dx.doi.org/10.1123/ijsnem.2013.0074.

Nitschke, J. B., Sarinopoulos, I., Oathes, D. J., Johnstone, T., Whalen, P. J., Davidson, R. J., & Kalin, N. H. (2009). Anticipatory activation in the amygdala and anterior cingulate in generalized anxiety disorder and prediction of treatment. *American Journal of Psychiatry*, *166* (3), 302–310. Available from http://dx.doi.org/10.1176/appi.ajp.07101682.

Noh, Y.-E., Morris, T., & Andersen, M. B. (2007). Psychological intervention programs for reduction of injury in ballet dancers. *Research in Sports Medicine*, *15*, 13–32. Available from http://dx.doi.org/10.1080/15438620600987064.

Noh, Y.-E., Morris, T., & Andersen, M. B. (2011). Psychosocial factors and ballet injuries. *International Journal of Sport and Exercise Psychology*, *3*(1), 79–90. Available from http://dx.doi.org/10.1080/1612197X.2005.9671759.

Noice, H., & Noice, T. (2006). What studies of actors and acting can tell us about memory and cognitive functioning. *Current Directions in Psychological Science*, *15*(1), 14–18.

Noice, T., & Noice, H. (2002). The expertise of professional actors: A review of recent research. *High Ability Studies*, *13*(1), 7–19. Available from http://dx.doi.org/10.1080/13598130220132271.

Nolet, R. (2013). Virtuoso hands. *Clinical Rheumatology*, *32*, 435–438. Available from http://dx.doi.org/10.1007/s10067-013-2186-6.

Noon, M., Hoch, A. Z., McNamara, L., & Schimke, J. (2010). Injury patterns in female Irish dancers. *American Academy of Physical Medicine and Rehabilitation*, *2*, 1030–1034. Available from http://dx.doi.org/10.1016/j.pmrj.2010.05.013.

Nordin, S. M., & Cumming, J. (2006a). The development of imagery in dance: Part I: Qualitative findings from professional dancers. *Journal of Dance Medicine and Science*, *10*(1 & 2), 28–34.

Nordin, S. M., & Cumming, J. (2006b). The development of imagery in dance: Part II: Quantitative findings from a mixed sample of dancers. *Journal of Dance Medicine and Science*, *10*(1 & 2), 21–27.

Nordin-Bates, S. M., Quested, E., Walker, I., & Redding, E. (2012). Climate change in the dance studio: Findings from the UK centers for advanced training. *Sport, Exercise, and Performance Psychology*, *1*(1), 3–16. Available from http://dx.doi.org/10.1037/a0025316.

Nordin-Bates, S. M., Walker, I., Baker, J., Garner, J., Hardy, C., Irvine, S., ... Blevins, P. (2011). Injury, imagery, and self-esteem in dance healthy minds in injured bodies. *Journal of Dance Medicine and Science*, *15*(2), 78–85.

Nordt, C., Warnke, I., Seifritz, E., & Kawohl, W. (2015). Modelling suicide and unemployment: A longitudinal analysis covering 63 countries, 2000–2011. *Lancet Psychiatry*, *2* (3), 239–245. Available from http://dx.doi.org/10.1016/s2215-0366(14)00118-7.

Nusbaum, E. C., Silvia, P. J., Beaty, R. E., Burgin, C. J., Hodges, D. A., & Kwapil, T. R. (2014). Listening between the notes: Aesthetic chills in everyday music listening. *Psychology of Aesthetics, Creativity, and the Arts*, *8*(1), 104–109. Available from http://dx.doi.org/10.1037/a0034867.

Nuwer, R. (2016). The right stuff: What psychological and physical traits separate the world's best athletes from the rest of us? *Scientific American Mind*, *27*(4), 39–44.

Nyberg, L., Eriksson, J., Larsson, A., & Marklund, P. (2006). Learning by doing versus learning by thinking: An fMRI study of motor and mental training. *Neuropsychologia*, *44*, 711–717. Available from http://dx.doi.org/10.1016/j.neuropsychologia.2005.08.006.

Obradovic, J., & Boyce, W. T. (2012). Developmental psychophysiology of emotion processes. *Monograms of the Society for Research in Child Development*, *77*(2), 120–128.

O'Bryan, E. M., McLeish, A. C., Kraemer, K. M., & Fleming, J. B. (2015). Emotion regulation difficulties and posttraumatic stress disorder symptom cluster severity among trauma-exposed college students. *Psychological Trauma: Theory, Research, Practice, and Policy*, *7*(2), 131–137. Available from http://dx.doi.org/10.1037/a0037764.

O'Connor, M.-F., Allen, J. J. B., & Kaszniak, A. W. (2005). Emotional disclosure for whom? A study of vagal tone in bereavement. *Biological Psychology*, *68*, 135–146. Available from http://dx.doi.org/10.1016/biopsycho.2004.04.003.

Oddey, A. (1994). *Devising theatre: A practical and theoretical handbook*. New York: Routledge.

O'Donnell, K., Brydon, L., Wright, C. E., & Steptoe, A. (2008). Self-esteem levels and cardiovascular and inflammatory responses to acute stress. *Brain, Behavior, and Immunity*, *22*, 1241–1247. Available from http://dx.doi.org/10.1016/j.bbi.2008.06.012.

Ohira, H., Ichikawa, N., Nomura, M., Isowa, T., Kimura, K., Kanayama, N., ... Yamada, J. (2010). Brain and autonomic associations accompanying stochastic decision-making. *Neuroimage*, *49*, 1024–1037. Available from http://dx.doi.org/10.1016/j.neuroimage.2009.07.060.

Ohnishi, T., Matsuda, H., Asada, T., Aruga, M., Hirakata, M., Nishikawa, M., ... Imabayashi, E. (2001). Functional anatomy of musical perception in musicians. *Cerebral Cortex*, *11*, 754–760.

Oldman, G. Retrieved from www.goodreads.com/quotes/tag/acting.

Olszewski-Kubilius, P., Subotnik, R. F., & Worrell, F. (2015). Antecedent and concurrent psychosocial skills that support levels of achievement within talent domains. *High Ability Studies*, *26*(2), 195–210. Available from http://dx.doi.org/10.13598139.2015.1095077.

Onder, M., Cosar, B., Oztas, M. O., & Candansayar, S. (2000). Stress and skin diseases in musicians: Evaluation of the beck depression scale, general psychologic profile (the Brief Symptom Inventory [BSI]), beck anxiety scale and stressful life events in musicians. *Biomedicine and Pharmacotherapy*, *54*, 258–262.

Ontai, L. L., & Thompson, R. A. (2008). Attachment, parent-child discourse and theory-of-mind development. *Social Development*, *17*(1), 47–60.

Oppezzo, M., & Schwartz, D. L. (2014). Give your ideas some legs: The positive effect of walking on creative thinking. *Journal of Experimental Psychology*, *40*(4), 1142–1152. Available from http://dx.doi.org/10.1037/a0036577.

Oreck, B., Baum, S., & McCartney, H. (2000). *Artistic talent development for urban youth: The promise and the challenge (No. RM00144)*. Storrs: University of Connecticut, The National Research Center on the Gifted and Talented.

Oreck, B. A., Owen, S. V., & Baum, S. M. (2003). Validity, reliability, and equity issues in an observational talent assessment process in the performing arts. *Journal for the Education of the Gifted*, *27*(1), 62–94.

Ortigo, K. M., Westen, D., DeFife, J. A., & Bradley, B. (2013). Attachment, social cognition, and posttraumatic stress symptoms in a traumatized, urban population: Evidence for the mediating role of object relations. *Journal of Traumatic Stress*, *26*, 361–368. Available from http://dx.doi.org/10.1002/jts.21815.

Osborne, M. S., Greene, D. J., & Immel, D. T. (2014). Managing performance anxiety and improving mental skills in conservatoire students through performance psychology training: A pilot study. *Psychology of Well-Being: Theory, Research and Practice*, *4*, 18.

Osipovich, D. (2006). What is a theatrical performance? *The Journal of Aesthetics and Art Criticism*, *64*(4), 461–470.

Osula, B., & Irvin, S. M. (2009). Cultural awareness in intercultural mentoring: A model for enhancing mentoring relationships. *International Journal of Leadership Studies*, *5*(1), 37–50.

Otis, C. L., Drinkwater, B., Johnson, M., Loucks, A., & Wilmore, J. (1997). The female athlete triad. *Medicine and Sports Exercise*, *29*(5), i–ix.

Otis, J. D., Keane, T. M., & Kerns, R. D. (2003). An examination of the relationship between chronic pain and posttraumatic stress disorder. *Journal of Rehabilitation Research and Development*, *40*(5), 397–406.

Overby, L. Y., Hall, C., & Haslam, I. (1997–98). A comparison of imagery used by dance teachers, figure skating coaches, and soccer coaches. *Imagination, Cognition and Personality*, *17*(4), 323–337.

Paarup, H. M., Baelum, J., Holm, J. M., Manniche, C., & Wedderkopp, N. (2011). Prevalence and consequences of musculoskeletal symptoms in symphony orchestra musicians vary by gender: A cross-sectional study. *BioMed Central Musculoskeletal Disorders*, *12*, 223.

Paarup, H. M., Baelum, J., Manniche, C., Holm, J. W., & Wedderkopp, N. (2012). Occurrence and co-existence of localized musculoskeletal symptoms and findings in work-attending orchestra musicians: An exploratory cross-sectional study. *BioMed Central*, *5*(541), 1–15.

Paget, D. (2007). Acting with facts: Actors performing the real in British theater and television since 1990. A preliminary report on a new research project. *Studies in Documentary Film*, *1*(2), 165–176. Available from http://dx.doi.org/10.1386/sdf.1.2.165/1.

Pagona, B., & Costas, M. (2008). The development of motor creativity in elementary school children and its retention. *Creativity Research Journal*, *20*(1), 72–80. Available from http://dx.doi.org/10.1080/10400410701842078.

Palmiero, M., Nakatani, C., Raver, D., Belardinelli, M. O., & van Leeuen, C. (2010). Abilities within and across visual and verbal domains: How specific is their influence on creativity? *Creativity Research Journal*, *2*(4), 369–377. Available from http://dx.doi.org/10.1080/10400419.2010.523396.

Panero, M. E., Godlstein, T. R., Rosenberg, R., Hughes, H., & Winner, E. (2016). Do actors possess traits associated with high hypnotizability? *Psychology of Aesthetics, Creativity, and the Arts*, *10*(2), 233–239. Available from http://dx.doi.org/10.1037/aca000044.

Panksepp, J. (1998). *Affective neuroscience: The foundations of human and animal emotions*. New York: Oxford Press.

Papageorgi, I., Hallam, S., & Welch, G. F. (2007). A conceptual framework for understanding musical performance anxiety. *Research Studies in Music Education*, *28*(1), 83–107.

Paparizos, A. L., Tripp, D. A., Sullivan, M. J. L., & Rubenstein, M. L. (2005). Catastrophizing and pain perception in recreational ballet dancers. *Journal of Sport Behavior*, *28*(1), 35–50.

Pappas, E., Kremenic, I., Liederbach, M., Orishimo, K. F., & Hagins, M. (2011). Time to stability differences between male and female dancers after landing from a jump on flat and inclined floors. *Clinical Journal of Sport Medicine*, *21*(4), 325–329. Available from http://dx.doi.org/10.1097/JSM.0b013e31821f5cfb.

Parasuraman, S., & Purohit, Y. S. (2000). Distress and boredom among orchestra musicians: The two faces of stress. *Journal of Occupational Health Psychology*, *5*(1), 74–83. Available from http://dx.doi.org/10.1037/1076-8998.5.1.74.

Parent, S., & Demers, G. (2011). Sexual abuse in sport: A model to prevent and protect athletes. *Child Abuse Review*, *20*(2), 120–133. Available from http://dx.doi.org/10.1002/car.1135.

Park, C. L., & Slattery, J. M. (2014). Resilience interventions with a focus on meaning and values. In M. Kent, M. C. Davis, & J. W. Reich (Eds.), *The resilience handbook: Approaches to stress and trauma* (pp. 270–282). New York: Routledge.

Park, H. R. P., Kirk, I. J., & Waldie, K. E. (2015). Neural correlates of creative thinking and schizotypy. *Neuropsychologia*, *73*, 94–107. Available from http://dx.doi.org/10.1016/j.neuropsychologia.2015.05.007.

Parra, A. (2006). "Seeing and feeling ghosts": Absorption, fantasy proneness, and healthy schizotypy as predictors of crisis apparition experiences. *The Journal of Parapsychology*, *70*(2), 357–372.

Parsons, L. M., Sergent, J., Hodges, D. A., & Fox, P. T. (2005). The brain basis of piano performance. *Neuropsychologia*, *43*, 199–215. Available from http://dx.doi.org/10.1016/j.neuropsychologia.2004.11.007.

Pata, D., Welsh, T., Bailey, J., & Range, V. (2014). Improving turnout in university dancers. *Journal of Dance Medicine and Science*, *18*(4), 169–177. Available from http://dx.doi.org/10.12678/1089-313X.18.4.169.

Patel, A. D. (2012). The OPERA hypothesis: Assumptions and clarifications. *Annals of the New York Academy of Sciences*, *1252*, 124–128. Available from http://dx.doi.org/10.1111/j.1749-6632.2011.06426.

Patston, L. L., Kirk, I. J., Rolfe, M. H. S., Corballis, M. C., & Tippett, L. J. (2007). The unusual symmetry of musicians: Musicians have equilateral interhemispheric transfer for visual information. *Neuropsychologia*, *45*, 2059–2065. Available from http://dx.doi.org/10.1016/j.neuropsychologia.2007.02.001.

Patston, L. L., & Tippett, L. J. (2011). The effect of background music on cognitive performance in musicians and nonmusicians. *Music Perception*, *29*(2), 173–183. Available from http://dx.doi.org/10.1525/MP.2011.29.2.173.

Patston, T., & Loughlan, T. (2014). Playing with performance: The use and abuse of beta-blockers in the performing arts. *Victorian Journal of Music Education*, *1*, 3–10.

Paulus, E. J., Argo, T. R., & Egge, J. A. (2013). The impact of posttraumatic stress disorder on blood pressure and heart rate in a veteran population. *Journal of Traumatic Stress*, *26*, 169–172. Available from http://dx.doi.org/10.1002/jts.21785.

Pavis, P. (2008). On faithfulness: The difficulties experienced by the text/performance couple. *Theatre Research International*, *33*(2), 117–126. Available from http://dx.doi.org/10.1017/S0307883308003635.

Payne, H. (2009). Pilot study to evaluate dance movement psychotherapy (the BodyMind Approach) in patients with medically unexplained symptoms: Participant and facilitator perceptions and a summary discussion. *Body, Movement, and Dance in Psychotherapy*, *4*(2), 77–94. Available from http://dx.doi.org/10.1080/17432970902018008.

Pearson, B. L., Russ, S. W., & Spannagel, S. A. C. (2008). Pretend play and positive psychology: Natural companions. *The Journal of Positive Psychology*, *3*(2), 110–119. Available from http://dx.doi.org/10.1080/17439760701760617.

Pedersen, C., Erleben, K., & Sporring, J. (2006). Ballet balance strategies. *Simulation Modelling Practice and Theory*, *14*, 1135–1142. Available from http://dx.doi.org/10.1016/j.simpat.2006.09.009.

Pedersen, M. E., Wilmerding, M. V., Kuhn, B. T., & Encinias-Sandoval, E. (2001). Energy requirements of the American professional flamenco dancer. *Medical Problems of Performing Artists*, *16*, 47–52.

Peifer, C., Schulz, A., Schachinger, H., Bauman, N., & Antoni, C. H. (2014). The relation of flow-experience and physiological arousal under stress—Can u shape it? *Journal of Experimental Social Psychology*, *53*, 62–69. Available from http://dx.doi.org/10.1016/j.jesp.2014.01.009.

Peleg, M., Lev-Wiesel, R., & Yaniv, D. (2014). Reconstruction of self-identity of Holocaust child survivors who participated in "Testimony Theatre." *Psychological Trauma, Theory, Research, Practice, and Policy*, *6*(4), 411–419. Available from http://dx.doi.org/10.1037/a003834.

Pelletier, M., Bouthillier, A., Levesque, J., Carrier, S., Breault, C., Paquette, V., ... Beauregard, M. (2003). Separate neural circuits for primary emotions? Brain activity during self-induced sadness and happiness in professional actors. *Neruoreport*, *14*(8), 1111–1116. Available from http://dx.doi.org/10.1097/01.wnr.0000075421.59944.69.

Peng, H.-T., Chen, W. C., Kernozek, T. W., Kim, K., & Song, C.-Y. (2015). Influences of patellofemoral pain and fatigue in female dancers during ballet jump-landing. *International Journal of Sports Medicine*, *36*, 747–753. Available from http://dx.doi.org/10.1055/s-0035-1547220.

Penniment, K. J., & Egan, S. J. (2012). Perfectionism and learning experiences in dance class as risk factors for eating disorders in dancers. *European Eating Disorders Review*, *20* (1), 13–22.

Perez, L. G., Abrams, M. P., Lopez-Martinez, A. E., & Asmundson, G. J. G. (2012). Trauma exposure and health: The role of depressive and hyperarousal symptoms. *Journal of Traumatic Stress*, *25*, 641–648. Available from http://dx.doi.org/10.1002/jts.21762.

Perez, R. M., Solana, R. S., Murillo, D. B., & Hernandez, F. J. M. (2014). Visual availability, balance performance and movement complexity in dancers. *Gait and Posture*, *40*, 556–560. Available from http://dx.doi.org/10.1016/j.gaitpost.2014.06.021.

Perez-Fabello, M. J., & Campos, A. (2007). Influence of training in artistic skills on mental imaging capacity. *Creativity Research Journal*, *19*(2–3), 227–232.

Perez-Fabello, M. J., & Campos, A. (2011). Dissociative experiences and creativity in fine arts students. *Creativity Research Journal, 23*, 38–41.

Perrone, K. M., Ksiazak, T. M., Wright, S. L., Vannatter, A., Crane, A. L., & Tanney, A. (2010). Multigenerational giftedness: Perceptions of giftedness across three generations. *Journal for the Education of the Gifted, 33*(4), 606–627.

Perry, B. D. (1999). Memories of fear: How the brain stores and retrieves physiologic states, feelings, behaviors and thoughts from traumatic events. In J. Goodwin, & R. Attias (Eds.), *Splintered reflections: Images of the body in trauma* (pp. 3–38). New York: Basic Books.

Perry-Smith, J. E., & Shalley, C. E. (2003). The social side of creativity: A static and dynamic social network perspective. *The Academy of Management Review, 28*(1), 89–106.

Petrie, K. J., Pennebaker, J. W., & Sivertsen, B. (2008). Things we said today: A linguistic analysis of the Beatles. *Psychology of Aesthetics, Creativity, and the Arts, 2*(4), 197–202. Available from http://dx.doi.org/10.1037/a0013117.

Petsche, H., Kaplan, S., von Stein, A., & Filz, O. (1997). The possible meaning of the upper and lower frequency ranges for cognitive and creative tasks. *International Journal of Psychophysiology, 26*, 77–97.

Phillips, R. J., & Strachan, I. C. (2016). Breaking up is hard to do: The resilience of the rock group as an organizational form for creating music. *Journal of Cultural Economics, 40* (1), 29–74. Available from http://dx.doi.org/10.1007/s10824-014-9226-1.

Phillips, T., & Street, J. (2015). Copyright and musicians at the digital margins. *Media, Culture and Society, 37*(3), 342–358. Available from http://dx.doi.org/10.1177/0163443714567018.

Phillips-Silver, J., & Trainor, L. J. (2007). Hearing what the body feels: Auditory encoding of rhythmic movement. *Cognition, 105*, 533–546. Available from http://dx.doi.org/10.1016/j.cognition.2006.11.006.

Piaget, J. (1951). *Play, dreams, and imitation in childhood.* New York: W. W. Norton & Company.

Piechowski, M. M., & Cunningham, K. (1985). Patterns of overexcitability in a group of artists. *Journal of Creative Behavior, 19*(3), 153–174. Available from http://dx.doi.org/10.1002/j.2162-6057.1985.tb00655.x.

Piechowski, M.M., Silverman, L., Cunningham, K., & Falk, R.F. (1982). A comparison of intellectually gifted and artists on five dimensions of mental functioning. *American Educational Research Association Annual Meeting*, New York.

Pietrek, C., Elbert, T., Weierstall, R., Muller, O., & Rockstroh, B. (2013). Childhood adversities in relation to psychiatric disorders. *Psychiatry Research, 206*, 103–110. Available from http://dx.doi.org/10.1016/j.psychres.2012.11.003.

Piffer, D., & Hur, Y.-M. (2014). Heritability of creative achievement. *Creativity Research Journal, 26*(2), 151–157. Available from http://dx.doi.org/10.1080/10400419.2014.901068.

Piirto, J. (1995). Deeper and broader: The pyramid of talent development in the context of giftedness construct. *The Education Forum, 59*(4), 363–370. Available from http://dx.doi.org/10.1080/00131729509335068.

Piirto, J. (1998). *Understanding those who create* (2nd ed.). Scottsdale, AZ: Gifted Potential Press, Inc.

Piirto, J., & Fraas, J. (2012). A mixed-methods comparison of vocational and identified-gifted high school students on the overexcitability questionnaire. *Journal for the Education of the Gifted, 35*(1), 3–34. Available from http://dx.doi.org/10.1177/0162353211433792.

Pilger, A., Haslacher, H., Ponocny-Seliger, E., Perkmann, T., Bohm, K., Budinsky, A., ... Winkler, R. (2014). Affective and inflammatory responses among orchestra musicians in performance situation. *Brain, Behavior, and Immunity*, *37*, 23–29. Available from http://dx.doi.org/10.1016/j.bbi.2013.10.018.

Pinel, J. P. J. (2000). *Biopsychology*. Toronto: Allyn & Bacon.

Pinho, A. L., de Manzano, O., Fransson, P., Eriksson, H., & Ullen, F. (2014). Connecting to create: Expertise in musical improvisation is associated with increased functional connectivity between premotor and prefrontal areas. *Journal of Neuroscience*, *34*(18), 6156–6163. Available from http://dx.doi.org/10.1523/JNEUROSCI.4769.13.2014.

Pirves, D., Augustine, G. J., Fitzpatrick, D., LaMantia, A.-S., McNamara, J. O., & Williams, S. M. (2001). *Neuroscience*. Sunderland, MA: Sinauer Associates, Inc.

Platek, S. M., Thomson, J. W., & Gallup, G. G. (2004). Cross-modal self-recognition: The role of visual, auditory, and olfactory primes. *Consciousness and Cognition*, *13*, 197–210.

Plato (c338 BCE/1976). Inspiration. In A. Rothenberg, & C. R. Hausman (Eds.), *The creativity question* (pp. 31–33). Durham, NC: Duke University Press.

Platt, M., & Freyd, J. (2012). Trauma and negative underlying assumptions in feelings of shame: An exploratory study. *Psychological Trauma: Theory, Research, Practice, and Policy*, *4*(4), 370–387. Available from http://dx.doi.org/10.1037/a0024253.

Poczwardowski, A., & Conroy, D. E. (2002). Coping responses to failure and success among elite athletes and performing artists. *Journal of Applied Sport Psychology*, *14*, 313–329. Available from http://dx.doi.org/10.1080/10413200290103581.

Policastro, E., & Gardner, H. (2005). From case studies to robust generalizations: An approach to the study of creativity. In R. J. Sternberg (Ed.), *Handbook of creativity* (pp. 213–225). New York: Cambridge University Press.

Pollich, F. E., Paterson, H. M., Bruderlin, A., & Sanford, A. J. (2001). Perceiving affect from arm movement. *Cognition*, *82*, B51–B61.

Porges, S. W. (2001). The polyvagal theory: Phylogenetic substrates of a social nervous system. *International Journal of Psychophysiology*, *42*, 29–52.

Porges, S. W. (2004). Neuroception: A subconscious system for detecting threats and safety. *Zero to Three*, 19–24, May.

Posey, E. (2002). Dance education in dance schools in the private sector: Meeting the demands of the marketplace. *Journal of Dance Education*, *2*(2), 43–49.

Power, R. A., Steinberg, S., Bjornsdottir, G., Rietveld, C., Abdellaoui, A., Nivard, M. M., ... Stefansson, K. (2015). Polygenic risk scores for schizophrenia and bipolar disorder predict creativity. *Nature Neuroscience*, *18*, 953–955. Available from http://dx.doi.org/10.1038/nn.4040.

Powers, M. B., Medina, J. L., Burns, S., Kauffman, B. K., Monfils, M., Asmundson, G. J. G., ... Smits, J. A. J. (2015). Exercise augmentation of exposure therapy for PTSD: Rationale and pilot efficacy data. *Cognitive Behavior Therapy*, *44*(4), 314–327. Available from http://dx.doi.org/10.1080/16506073.2015.1012740.

Pozo-Municio, C. (2007). Genitourinary conditions in young dancers: Relationship between urinary incontinence and foot flexibility. *Journal of Dance Medicine and Science*, *11*(2), 49–59.

Prabhu, V., Sutton, C., & Sauser, W. (2008). Creativity and certain personality traits: Understanding the mediating effects of intrinsic motivation. *Creativity Research Journal*, *20*(1), 53–66. Available from http://dx.doi.org/10.1080/10400410701841955.

Pradham, M., & Aujla, I. (2014). The relationship between passion and the psychological well-being of professional dancers. *Journal of Dance Medicine and Science*, *18*(1), 37–44. Available from http://dx.doi.org/10.12678/1089-313X.18.1.37.

Preece, S. B. (2011). Performing arts entrepreneurship: Toward a research agenda. *The Journal of Arts Management, Law, and Society*, *41*, 103–120. Available from http://dx.doi.org/10.1080/10632921.2011.573445.

Press, C. M., & Warburton, E. C. (2007). Creativity research in dance. In L. Bresler (Ed.), *International handbook of research in arts education* (Vol. 16, pp. 1273–1290). The Netherlands: Springer.

Pressing, J. (1988). Improvisation methods and models. In J. A. Sloboda (Ed.), *Generative processes in music: The psychology of performance, improvisation, and composition* (pp. 129–1780). Oxford: Clarendon Press.

Preti, A., & Miotto, P. (1999). Suicide among eminent artists. *Psychological Reports*, *84*, 291–301.

Preti, A., & Vellante, M. (2007). Creativity and psychopathology: Higher rates of psychosis proneness and nonright-handedness among creative artists compared to same age and gender peers. *Journal of Nervous and Mental Disease*, *195*(10), 837–845. Available from http://dx.doi.org/10.1097/NMD.06013e3181568180.

Preti, A., de Biasi, F., & Miotto, P. (2001). Musical creativity and suicide. *Psychological Reports*, *89*, 719–727. Available from http://dx.doi.org/10.2466/pr0.2001.89.3.719.

Price, B. R., & Pettijohn, T. F. (2006). The effect of ballet dance attire on body and self-perceptions of female dancers. *Social Behavior and Personality*, *34*(8), 991–998. Available from http://dx.doi.org/10.2224/sbp.2006.34.8.991.

Prisk, V. R., O'Loughlin, P. F., & Kennedy, J. G. (2008). Forefoot injuries in dancers. *Clinics in Sports Medicine*, *27*, 305–320. Available from http://dx.doi.org/10.1016/j.csm.2007.12.005.

Purves, D., Augustine, G. J., Fitzpatrick, D., Katz, L. C., LaMantia, A.-S., McNamara, J. O., & Williams, S. M. (2001). *Neuroscience* (2nd ed). Sunderland, MA: Sinauer Associates, Inc.

Puccini, G. Retrieved from www.goodreads.com/quotes/tag/music.

Putnam, K. T., Harris, W. W., & Putnam, F. W. (2013). Synergistic childhood adversities and complex adult psychopathology. *Journal of Traumatic Stress*, *26*, 435–442. Available from http://dx.doi.org/10.1002/jts.21833.

Quentzel, S., & Loewy, J. (2010). An integrative bio-psycho-musical assessment model for the treatment of musicians: Part I—A continuum of support. *Music Medicine*, *2*(2), 117–120.

Quested, E., & Duda, J. L. (2009). Perceptions of the motivational climate, need satisfaction, and indices of well- and ill-being among hip hop dancers. *Journal of Dance Medicine and Science*, *13*(1), 10–19.

Quested, E., & Duda, J. L. (2011). Antecedents of burnout among elite dancers: A longitudinal test of basic needs. *Psychology of Sport and Exercise*, *12*, 159–167. Available from http://dx.doi.org/10.1016/j.psysport.2010.09.003.

Quested, E., Duda, J. L., Ntoumanis, N., & Maxwell, J. P. (2013). Daily fluctuations in the affective states of dancers: A cross-situational test of basic needs theory. *Psychology of Sport and Exercise*, *14*, 586–595. Available from http://dx.doi.org/10.1016/j.psychsport.2013.02.006.

Radtke, T., Kriemler, S., Eser, P., Saner, H., & Wilhelm, M. (2013). Physical activity intensity and surrogate markers for cardiovascular health in adolescents. *European Journal of Applied Physiology*, *113*(5), 1213–1222. Available from http://dx.doi.org/10.1007/s00421-012-2542-2.

Raeburn, S. D. (2000). Psychological issues and treatment strategies in popular musicians: A review, Part 2. *Medical Problems of Performing Artists*, *15*(1), 6–16.

Raeburn, S. D. (2007). The ring of fire: Shame, fame, and rock 'n' roll. *Medical Problems in Performing Artists*, *22*, 3–9.

Raeburn, S. D., Hipple, J., Delaney, W., & Chesky, K. (2003). Surveying popular musicians' health status using convenience samples. *Medical Problems of Performing Artists*, *18* (3), 113–119.

Rafferty, S. (2010). Considerations for integrating fitness into dance training. *Journal of Dance Medicine and Science*, *14*(2), 45–49.

Raglan, G. B., & Schulkin, J. (2014). Introduction to allostasis and allostatic load. In M. Kent, M. C. Davis, & J. W. Reich (Eds.), *The resilience handbook: Approaches to stress and trauma* (pp. 44–52). New York: Routledge.

Raikkonen, K., Matthews, K. A., & Salomon, K. (2003). Hostility predicts metabolic syndrome risk factors in children and adolescents. *Health Psychology*, *22*, 279–286.

Rainville, P., Bechara, A., Naqvi, N., & Damasio, A. R. (2006). Basic emotions are associated with distinct patterns of cardiovascular activity. *International Journal of Psychophysiology*, *61*, 5–18. Available from http://dx.doi.org/10.1016/j.ijpsycho.2005.10.024.

Raisz, L. (2005). Pathogenesis of osteoporosis: Concepts, conflicts, and prospects. *Journal of Clinical Investigation*, *115*(12), 18–25.

Ramkumar, P. N., Farber, J., Arnouk, J., Varner, K. E., & McCulloch, P. C. (2016). Injuries in a professional ballet company: A ten-year retrospective study. *Journal of Dance Medicine and Science*, *20*(1), 30–37. Available from http://dx.doi.org/10.12678/1089-313X.20.1.30.

Rantisi, N. M., & Leslie, D. (2015). Circus in action: Exploring the role of a translation zone in the Cirque du Soleil's creative practices. *Economic Geography*, *91*(2), 147–164. Available from http://dx.doi.org/10.111/ecge.12082.

Rasmussen, A., & Rasmussen, P. (2015). Conceptions of student talent in the context of talent development. *International Journal of Qualitative Studies in Education*, *28*(4), 476–495. Available from http://dx.doi.org/10.1080/09518398.2014.916013.

Raston, S. M. (1997). A study of young artists: The development of artistic talent and creativity. *Creativity Research Journal*, *10*(2 & 3), 175–192.

Rauschenberger, S., & Lynn, S. J. (2002–2003). Fantasy-proneness, negative affect and psychopathology. *Imagination, Cognition and Personality*, *22*(3), 239–255.

Ravaldi, C., Vannacci, A., Bolognesi, E., Mancini, S., Faravelli, C., & Ricca, V. (2006). Gender role, eating disorder symptoms, and body image concern in ballet dancers. *Journal of Psychosomatic Research*, *61*, 529–535. Available from http://dx.doi.org/10.1016/j.jpsychorres.2006.04.016.

Ravaldi, C., Vannacci, A., Zucchi, T., Mannucci, E., Cabras, P. L., Boldrini, M., ... Ricca, V. (2003). Eating disorders and body image disturbances among ballet dancers, gymnasium users and body builders. *Psychopathology*, *36*, 247–254. Available from http://dx.doi.org/10.1159/000073450.

Raymond, D. M., Romeo, J. H., & Kumke, K. V. (2012). A pilot study of occupational injury and illness experienced by classical musicians. *Workplace Health and Safety*, *60*(1), 19–24.

Razunmnikova, O. M., Kovtun, L. T., & Krivoshchekov, S. G. (2013). Cardiorespiratory response to hypoxia in persons with different levels of creativity. *Human Physiology*, *39* (4), 105–111.

Reason, M., Kay, R., Kauppi, J.-P., Tohka, J., Jola, C., Reynolds, D., ... Pollick, F. E. (2016). Spectators' aesthetic experience of sound and movement in dance performance: A transdisciplinary investigation. *Psychology of Aesthetics, Creativity, and the Arts*, *10* (1), 42–55. Available from http://dx.doi.org/10.1037/a0040032.

Redding, E., Weller, P., Ehrenberg, S., Irvine, S., Quin, E., & Rafferty, S. (2009). The development of a high intensity dance performance fitness test. *Journal of Dance Medicine and Science*, *13*(1), 3–9.

Reddy, M. K., Seligowski, A. V., Rabenhorst, M. M., & Orcutt, H. K. (2015). Predictors of expressive writing content and posttraumatic stress following a mass shooting. *Psychological Trauma: Theory, Research, Practice, and Policy*, *7*(3), 286–294. Available from http://dx.doi.org/10.1037/a0037918.

Reiter-Palmon, R., Illies, M. Y., Buboltz, C., Cross, L. K., & Nimps, T. (2009). Creativity and domain specificity: The effect of task type on multiple indexes of creative problem-solving. *Psychology of Creativity, Aesthetics, and the Arts*, *3*(2), 73–80. Available from http://dx.doi.org/10.1037/a0013410.

Renfrow, P. J., & Gosling, S. D. (2003). The do re mi's of everyday life: The structure and personality correlates of music preferences. *Journal of Personality and Social Psychology*, *84*, 1236–1256. Available from http://dx.doi.org/10.1037/0022-3514.84.6.1236.

Renzulli, J. S. (1978). What makes giftedness? Reexamining a definition. *Phi Delta Kappan*, *60*(180–184), 261.

Renzulli, J. S. (2002). Emerging conceptions of giftedness: Building a bridge to the new century. *Exceptionality*, *10*(2), 67–75.

Renzulli, J. S. (2005). The three-ring definition of giftedness: A developmental model for promoting creative productivity. In R. J. Sternberg, & J. E. Davidson (Eds.), *Conceptions of giftedness* (2nd ed., pp. 246–279). New York: Cambridge University Press.

Reul, J. M. H. M., Collins, A., Aliba, R., Mifsud, K. R., Carter, S. D., Gutierrez-Mecinas, M., . . . Linthorst, A. C. E. (2015). Glucocorticoids, epigenetic control and stress resilience. *Neurobiology of Stress*, *1*, 44–59. Available from http://dx.doi.org/10.1016/ynstr.2014.10.001.

Reyes, C. J., & Asbrand, J. P. (2005). A longitudinal study assessing trauma symptoms in sexually abused children engaged in play therapy. *International Journal of Play Therapy*, *14*(2), 25–47. Available from http://dx.doi.org/10.1037/h0088901.

Rhodes, M. (1961/1987). An analysis of creativity. In S. G. Isaksen (Ed.), *Frontiers of creativity research: Beyond the basics* (pp. 216–222). Buffalo, NY: Bearly. (Original work published, 1961).

Rhue, J. W., & Lynn, S. J. (1987). Fantasy proneness: Developmental antecedents. *Journal of Personality*, *55*(1), 121–137.

Rich, J. D., & Weisberg, R. W. (2004). Creating all in the family: A case study in creative thinking. *Creativity Research Journal*, *16*(2 & 3), 247–259.

Richards, D. (1992). *A director's method for film and television*. Boston: Focal Press.

Richter, B., Lohle, E., Knapp, B., Weikert, M., Schlomicher-Their, J., & Verdolini, K. (2002). Harmful substances on the opera stage: Possible negative effects on singers' respiratory tracts. *Journal of Voice*, *16*(1), 72–80. Available from http://dx.doi.org/10.1016/S0892-1997(02)00074-7.

Richter, M., & Knappe, K. (2014). Mood impact on effort-related cardiovascular reactivity depends on activity context: Evidence from a task with an unfixed performance standard. *International Journal of Psychophysiology*, *93*, 227–234. Available from http://dx.doi.org/10.1016/j.ijpsycho.2014.05.002.

Rickard, I. J., & Lummaa, V. (2007). The predictive adaptive response and metabolic syndrome: Challenges for the hypothesis. *Trends in Endocrinology and Metabolism*, *18*(3), 94–99.

Rickert, D. L. L., Barrett, M. S., & Ackerman, B. J. (2014a). Injury and the orchestral environment, Part II: Organisational culture, behavioral norms, and attitudes to injury. *Medical Problems of Performing Artists*, *29*(2), 94–101.

Rickert, D. L. L., Barrett, M. S., & Ackerman, B. J. (2014b). Injury and the orchestral environment, Part I: The role of work organization and psychosocial factors in injury risk. *Medical Problems of Performing Artists*, *28*(4), 219–229.

Rickman, A. Retrieved from www.goodreads.com/quotes/tag/acting.

Rietveld, A. B. M. (2013). Dancers' and musicians' injuries. *Clinical Rheumatology*, *32*, 425–434. Available from http://dx.doi.org/10.1007/s10067-013-2184-8.

Rietveld, A. B. M., & Leijnse, J. N. A. L. (2013). Focal hand dystonia in musicians: A synopsis. *Clinical Rheumatology*, *32*, 481–486. Available from http://dx.doi.org/10.1007/s10067-013-2196-4.

Rietveld, A. B. M., Macfarlane, J. D., & de Haas, G. J. F. (2013). Some thoughts on the prevention of complaints in musicians and dancers. *Clinical Rheumatology*, *32*, 449–452. Available from http://dx.doi.org/10.1007/s10067-013-2195-5.

Rip, B., Fortin, S., & Vallerand, R. J. (2006). The relationship between passion and injury in dance students. *Journal of Dance Medicine and Science*, *10*(1 & 2), 14–20.

Risner, D. (2007). Rehearsing masculinity: Challenging the "boy code" in dance education. *Research in Dance Education*, *8*(2), 139–153. Available from http://dx.doi.org/10.1080/14617890701706107.

Riva, G. (2014). Out of my real body: Cognitive neuroscience meets eating disorders. *Frontiers in Human Neuroscience*, *8*(236), 1–20. Available from http://dx.doi.org/10.3389/fnhum.2014.002.36.

Rizzolatti, G., & Craighero, L. (2004). The mirror-neuron system. *Annual Review of Neuroscience*, *27*, 169–192. Available from http://dx.doi.org/10.1146/annurev.neuro.27.070203.144230.

Roazen, P. (1976). *Erik H. Erikson: The power and limits of a vision*. New York: Collier Macmillan Publishers.

Robb, A., & Davies, M. (2015). "Being inside the story": A phenomenology of onstage experience and the implications of flow. *About Performance*, *13*, 45–67.

Rocamora, C. (2008). Albee sizes up the dark vast: An interview. *American Theatre*, *25*(1), 30–121.

Rocha, S. F., Marocolo, M., Correa, E. N. V., Morato, G. S. G., & da Mota, R. (2014). Physical activity helps to control music performance anxiety. *Medical Problems of Performing Artists*, *29*(2), 111–112.

Rochman, D., & Diamond, G. M. (2008). From unresolved anger to sadness: Identifying physiological correlates. *Journal of Counseling Psychology*, *55*(1), 96–105. Available from http://dx.doi.org/10.1037/0022-0167.55.1.96.

Rodgers, M. W. M., O'Modhrain, S., & Craig, C. M. (2013). Temporal guidance of musicians' performance movement is an acquired skill. *Experimental Brain Research*, *226*, 221–230. Available from http://dx.doi.org/10.1007/s00221-013-3427-2.

Rodrigues-Krause, J., Cunha, G. D. S., Alberton, C. L., Follmer, B., Krause, M., & Reischak-Oliveira, A. (2014). Oxygen consumption and heart rate responses to isolated ballet exercise sets. *Journal of Dance Medicine and Science*, *18*(3), 99–105. Available from http://dx.doi.org/10.12678/1089-313X.18.3.99.

Rodrigues-Krause, J., Krause, M., Cunha, G. D. S., Perin, D., Martins, J. B., Alberton, C. L., ... Reischak-Oliveira, A. (2014). Ballet dancers cardiorespiratory, oxidative and muscle damage responses to classes and rehearsals. *European Journal of Sport Science*, *14*(3), 199–208. Available from http://dx.doi.org/10.1080/17461391.2013.777796.

Rodrigues-Krause, J., Krause, M., & Reischak-Oliveira, A. (2015). Cardiorespiratory considerations in dance: From classes to performance. *Journal of Dance Medicine and Science*, *19*(3), 91–102. Available from http://dx.doi.org/10.12678/1089-313X.19.3.91.

Romens, S. E., McDonald, J., Svaren, J., & Pollak, S. D. (2015). Associations between early life stress and gene methylation in children. *Child Development*, *86*(1), 303–309. Available from http://dx.doi.org/10.1111/cdev.12270.

Romeo, R. D. (2010). Adolescence: A central event in shaping stress reactivity. *Developmental Psychobiology*, *52*, 244–253. Available from http://dx.doi.org/10.1002/dev.20437.

Roncaglia, I. (2006). Retirement as a career transition in ballet dancers. *International Journal of Educational and Vocational Guidance*, *6*, 181–193. Available from http://dx.doi.org/10.1007/s10775-0006-9106-0.

Ronkko, R., Heliovaara, M., Malmivaara, Roine, R., Seitsalo, S., & Sanio, P. (2007). Musculoskeletal pain disability and quality of life among retired dancers. *Journal of Dance Medicine and Science*, *11*(4), 105–109.

Rooks, C., Veledar, E., Goldberg, J., Votaw, J., Shah, A., Bremner, J. D., & Vaccarino, V. (2015). Long-term consequences of early trauma on coronary heart disease: Role of familial factors. *Journal of Traumatic Stress*, *28*, 456–459. Available from http://dx.doi.org/10.1002/jts.22044.

Root-Bernstein, M., & Root-Bernstein, R. (2006). Imaginary worldplay in childhood and maturity and its impact on adult creativity. *Creativity Research Journal*, *18*(4), 405–425.

Rosenvinge, J. H., & Pettersen, G. (2015). Epidemiology of eating disorders part II: An update with a special reference to the DSM-5. *Advances in Eating Disorders: Theory, Research and Practice*, *3*(2), 198–220. Available from http://dx.doi.org/10.1080/21662630.2014.940549.

Roset-Liobet, J., Rosines-Cubellis, D., & Salo-Orfila, J. M. (2000). Identification of risk factors for musicians in Catalonia (Spain). *Medical Problems of Performing Artists*, *15*(4), 167–174.

Ross, S. R., & Keiser, H. N. (2014). Autotelic personality through a five-factor lens: Individual differences in flow-propensity. *Personality and Individual Differences*, *59*, 3–8. Available from http://dx.doi.org/10.106/j.paid.2013.09.029.

Rossiter, K., Kontos, P., Colantonio, A., Gilbert, J., Gray, J., & Keightley, M. (2008). Staging data: Theatre as a tool for analysis and knowledge transfer in health research. *Social Science and Medicine*, *66*(1), 130–145. Available from http://dx.doi.org/10.1016/j.socsimed.2007.07.021.

Rostan, S. M., Pariser, D., & Gruber, H. E. (2002). A cross-cultural study of the development of artistic talent, creativity and giftedness. *High Ability Studies*, *13*, 125–155. Available from http://dx.doi.org/10.1080/1359813022000048789.

Rothbart, M. K. (2007). Temperament, development, and personality. *Current Directions in Psychological Science*, *16*(4), 207–212.

Rothbart, M. K., Ahadi, S. A., & Evans, D. E. (2000). Temperament and personality: Origins and outcomes. *Journal of Personality and Social Psychology*, *78*(1), 122–135. Available from http://dx.doi.org/10.1037//0022.3514.78.1.122.

Rothbart, M. K., Ahadi, S. A., Hershey, K. L., & Fisher, P. (2001). Investigations of temperament at three to seven years: The children's behavior questionnaire. *Child Development*, *72*(5), 1394–1408.

Rothbart, M. K., Ellis, L. K., & Posner, M. I. (2004). Temperament and self-regulation. In R. F. Baumeister, & K. D. Vohs (Eds.), *Handbook of self-regulation: Research, theory, and application* (pp. 357–370). New York: The Guilford Press.

Rothbart, M. K., & Sheese, B. E. (2007). Temperament and emotion regulation. In J. J. Gross (Ed.), *Handbook of emotion regulation* (pp. 331–350). New York: Guilford Press.

Rothenberg, A., & Hausman, C. R. (Eds.), (1976). *The creativity question*. Durham, NC: Duke University Press.

Roussel, N., De Kooning, M., Schutt, A., Mottram, S., Truijen, S., Nijs, J., & Daenen, L. (2013). Motor control and low back pain in dancers. *International Journal of Sports Medicine, 34*, 138–143. Available from http://dx.doi.org/10.1055/s-0032-1321722.

Roy, N., Ryker, K. S., & Bless, D. M. (2000). Vocal violence in actors: An investigation into its acoustic consequences and the effects of hygienic laryngeal release training. *Journal of Voice, 14*(2), 215–230. Available from http://dx.doi.org/10.1016/S0892-1997(00)80029-6.

Roy-Byrne, P. (2013a). Biomarkers of suicidality. *New England Journal of Medicine, Journal Watch, 19*(1), 79.

Roy-Byrne, P. (2013b). Depression, inflammation, and exercise: An evolving story. *New England Journal of Medicine, Journal Watch, 19*(12), 94.

Roy-Byrne, P. (2015a). Unemployment linked to increased suicide risk worldwide. *New England Journal of Medicine, Journal Watch, 21*(4), 28–29.

Roy-Byrne, P. (2015b). How suicide attempts by parents may lead to increased risk for suicide attempts by their children. *New England Journal of Medicine, Journal Watch, 21*(3), 17–18.

Rubini, E. C., Souza, A. C., Mello, M. L., Bacurau, R. F. P., Cabral, L. F., & Farinatti, P. T. V. (2011). Immediate effect of static and proprioceptive neuromuscular facilitation stretching on hip adductor flexibility in female ballet dancers. *Journal of Dance Medicine and Science, 15*(4), 177–181.

Ruiz, M. H., Strubing, F., Jabusch, H.-C., & Altenmuller, E. (2011). EEG oscillatory patterns are associated with error prediction during music performance and are altered in musician's dystonia. *Neuroimage, 55*, 1791–1803. Available from http://dx.doi.org/10.1016/j.neuroimage.2010.12.050.

Runco, M. A. (2004). Creativity. *Annual Review of Psychology, 55*, 657–687.

Runco, M. A., & Jaeger, G. J. (2012). The standard definition of creativity. *Creativity Research Journal, 24*(1), 92–96. Available from http://dx.doi.org/10.1080/10400419.2012.650092.

Russ, S. W. (2003). Play and creativity: Developmental issues. *Scandinavian Journal of Educational Research, 47*(3), 291–303. Available from http://dx.doi.org/10.1080/0031383032000079254.

Russ, S. W. (2007). Pretend play: A resource for children who are coping with stress and managing anxiety. *New York State Psychologist, 9*(4), 13–17.

Russ, S. W., & Wallace, C. E. (2013). Pretend play and creative processes. *American Journal of Play, 6*(1), 136–148.

Russell, J. A. (2013). Preventing dance injuries: Current perspectives. *Journal of Sports Medicine, 4*, 199–210.

Ruthsatz, J., Detterman, D., Griscom, W. S., & Cirullo, B. A. (2008). Becoming an expert in the musical domain: It takes more than just practice. *Intelligence, 26*, 330–338. Available from http://dx.doi.org/10.1016/j.intell.2007.08.003.

Rutter, I. M. (2006). Implications of resilience concepts for scientific understanding. *Annals of the New York Academy of Science, 1094*, 1–12. Available from http://dx.doi.org/10.1196/1nn11s.1376.002.

Rutter, L. A., Weatherill, R. P., Krill, S. C., Orazem, R., & Taft, C. T. (2013). Posttraumatic stress disorder symptoms, depressive symptoms, exercise, and health in college students. *Psychological Trauma: Theory, Research, Practice, and Policy, 5*(1), 56–61. Available from http://dx.doi.org/10.1037/a0021996.

Ryan, C. (2009). An investigation into the choral singer's experience of music performance anxiety. *Journal of Research in Music Education, 57*(2), 108–126. Available from http://dx.doi.org/10.1177/0022429209336132.

Ryan, C., Wapnick, J., Lacalle, N., & Darrow, A.-A. (2006). The effects of various physical characteristics of high-level performers on adjudicators' performance ratings. *Psychology of Music, 34*(4), 559–572. Available from http://dx.doi.org/10.1177/0305735606608106.

Saarikallio, S., Luck, G., Burger, B., Thompson, M., & Toivianinen, P. (2013). Dance moves reflect current affective state illustrative of approach-avoidance motivation. *Psychology of Aesthetics, Creativity, and the Arts, 7*(3), 296–305. Available from http://dx.doi.org/10.1037/a0032589.

Sachs, C. (1937/1965). *World history of the dance*. New York: W.W. Norton & Company, Inc.

Sachs-Ericsson, N., Cromer, K., Hernandez, A., & Kendall-Tackett, K. (2009). A review of childhood abuse, health, and pain-related problems: The role of psychiatric disorders and current life stress. *Journal of Trauma and Dissociation, 10*, 170–188. Available from http://dx.doi.org/10.1080/15299730802624585.

Sadler, M. E., & Miller, C. J. (2010). Performance anxiety: A longitudinal study of the roles of personality and experience in musicians. *Social Psychological and Personality Science, 1*(3), 280–287. Available from http://dx.doi.org/10.1177/1948550610370492.

Sagar, S. S., & Stoeber, J. (2009). Perfectionism, fear of failure, and affective response to success and failure: The central role of fear of experiencing shame and embarrassment. *Journal of Sport and Exercise Psychology, 31*(5), 602–627.

Sakai, N. (2002). Hand pain attributed to overuse among professional pianists: A study of 200 cases. *Medical Problems of Performing Artists, 17*(4), 178–180.

Salimpoor, V. N., Benvoy, M., Longo, G., Cooperstock, J. R., & Zatorre, R. J. (2009). The rewarding aspects of music listening are related to degree of emotional arousal. *PLoS ONE, 4*, e7487. Available from http://dx.doi.org/10.137/journal.pone.0007487.

Salimpoor, V. N., & Zatorre, R. J. (2013). Neural interactions that give rise to musical pleasure. *Psychology of Aesthetics, Creativity, and the Arts, 7*(1), 62–75. Available from http://dx.doi.org/10.1037/a0031819.

Sambo, M. (2013). Physical therapy rehabilitation strategies for dancers: A qualitative study. *Journal of Dance Medicine and Science, 17*(1), 11–17. Available from http://dx.doi.org/10.12678/1089-313X.17.1.11.

Sanchez-Bernardos, M. L., & Avia, M. D. (2004). Personality correlates of fantasy proneness among adolescents. *Personality and Individual Differences, 37*, 1969–1979.

Sanchez-Bernardos, M. L., & Avia, M. D. (2006). The relationship between fantasy proneness and schizotypy in adolescents. *Journal of Nervous and Mental Disease, 194*(6), 411–414. Available from http://dx.doi.org/10.1097/01.nmd.0000222406.16820.13.

Sanchez-Bernardos, M. L., Lloreda, M. J. H., Avia, M. D., & Bragado-Alvarez, C. (2015). Fantasy proneness and personality profiles. *Imagination, Cognition and Personality, 34*(4), 327–339. Available from http://dx.doi.org/10.1177/0276236615572584.

Sanchez-Ruiz, M. J., Perez-Gonzalez, J. C., & Petrides, K. V. (2010). Trait emotional intelligence profiles of students from different university faculties. *Australian Journal of Psychology, 62*(1), 51–57. Available from http://dx.doi.org/10.1080/00049530903312907.

Sanders, S. G. (2008). Dancing through pregnancy: Activity guidelines for professional and recreational dancers. *Journal of Dance Medicine and Science, 12*(1), 17–22.

Sandgren, M. (2002). Voice, soma, and psyche: A qualitative and quantitative study of opera singers. *Medical Problems of Performing Artists, 17*(1), 11–21.

Sandgren, M. (2009). Health anxiety instead of performance anxiety among opera singers. *Proceedings of the 7th Trinidad Conference of European Society for the Cognitive Sciences of Music.*

Santiago, P. O., Ursano, R. J., Gray, C. L., Pynoos, R. S., Spiegel, D., Lewis-Fernandez, R., ... Fullerton, C. S. (2013). A systematic review of PTSD prevalence and trajectories in DSM-5 defined trauma exposed populations: Intentional and non-intentional traumatic events. *PLoS ONE, 8*(4), e59236. Available from http://dx.doi.org/10.1371/journal.pone.0059236.

Sapolsky, R. M. (2007). Stress, stress-related disease, and emotional regulation. In J. J. Gross (Ed.), *Handbook of emotion regulation* (pp. 606–615). New York: Guilford Press.

Sartin, J. S. (2010). Contagious rhythm: Infectious disease of 20th century musicians. *Clinical Medicine and Research, 8*(2), 106–113. Available from http://dx.doi.org/10.3121/cmr.2010.882.

Sataloff, R. T., Rosen, D. C., & Levy, S. (1999). Medical treatment of performance anxiety: A comprehensive approach. *Medical Problems of Performing Artists, 14*(3), 122–126.

Sawyer, R. K. (2000). Improvisational cultures: Collaborative emergence and creativity in improvisation. *Mind, Culture, and Activity, 7*(3), 180–185. Available from http://dx.doi.org/10.1207/S15327884MCA0703_05.

Sawyer, R. K. (2006a). *Explaining creativity: The science of human innovation.* New York: Oxford University Press.

Sawyer, R. K. (2006b). Group creativity: Musical performance and collaboration. *Psychology of Music, 34*(2), 148–165. Available from http://dx.doi.org/10.11770305735606061850.

Sawyer, R. K. (2011). The cognitive neuroscience of creativity: A critical review. *Creativity Research Journal, 23*(2), 137–154. Available from http://dx.doi.org/10.1080/10400419.2011.571191.

Sawyer, R. K., & DeZutter, S. (2009). Distributed creativity: How collective creations emerge from collaboration. *Psychology of Aesthetics, Creativity, and the Arts, 3*(2), 81–92. Available from http://dx.doi.org/10.1037/a0013282.

Saxbe, D. E., Margolin, G., Shapiro, L. A. S., & Baucom, B. R. (2012). Does dampened physiological reactivity protect youth in aggressive family environments. *Child Development, 83*(3), 821–830. Available from http://dx.doi.org/10.1111/j.1467-8624.2012.01752.x.

Saxe, R., & Kanwisher, N. (2003). People thinking about people: The role of the temporo-parietal junction in "theory of mind." *Neuroimage, 19*, 1835–1842. Available from http://dx.doi.org/10.1016/S1053-8119(03)00230-1.

Scharf, M., Mayseless, O., & Kivenson-Baron, I. (2004). Adolescents' attachment representations and developmental tasks in emerging adulthood. *Developmental Psychology, 40*(3), 430–444. Available from http://dx.doi.org/10.1037/0012.1649.40.3.430.

Schaufeli, W. B., Leiter, M. P., & Maslach, C. (2009). Burnout: 35 years of research and practice. *Career Development International, 14*(3), 204–220. Available from http://dx.doi.org/10.1108/13620430910966406.

Schellenberg, E. G., & von Scheve, C. (2012). Emotional cues in American popular music: Five decades of the top 40. *Psychology of Aesthetics, Creativity, and the Arts, 6*(3), 196–203. Available from http://dx.doi.org/10.1037/a0028024.

Scheper, M. C., de Vries, D. E., Juul-Kristensen, B., Nollet, F., & Engelbert, R. H. H. (2014). The functional consequences of Generalized Joint Hypermobility: A cross-sectional study. *BMC Musculoskeletal Disorders, 15*, e243. Available from http://dx.doi.org/10.1186/1471-2474-15-243.

Scheper, M. C., de Vries, J. E., de Vos, R., Verbunt, J., Nollet, F., & Engelbert, R. H. H. (2013). Generalized joint hypermobility in professional dancers: A sign of talent or vulnerability. *Rheumatology*, *52*(4), 651–658. Available from http://dx.doi.org/10.1093/rheumatology/kes220.

Schiller, W. (2005). Children's perceptions of live arts performances: A longitudinal study. *Early Child Development and Care*, *176*(6), 543–552. Available from http://dx.doi.org/10.1080/0300443050013141 1.

Schilling, M. (2005). A "small-world" network model of cognitive insight. *Creativity Research Journal*, *17*(2–3), 131–154.

Schimmenti, A. (2012). Unveiling the hidden self: Developmental trauma and pathological shame. *Psychodynamic Practice*, *18*(2), 195–211. Available from http://dx.doi.org/10.1080/14753634.2012.664873.

Schlaug, G., Forgeard, M., Zhu, L., Norton, A., Norton, A., & Winner, E. (2009). Training-induced neuroplasticity in young children. *Annals of New York Academy of Science*, *1169*, 205–208. Available from http://dx.doi.org/10.1111/j.1749-6632.2009.04842.x.

Schlaug, G., Norton, A., Overy, K., & Winner, E. (2005). Effects of music training on the child's brain and cognitive development. *Annals of New York Academy of Science*, *1060*, 219–230. Available from http://dx.doi.org/10.1196/annals.1360.015.

Schlesinger, J. (2009). Creative mythconceptions: A closer look at the evidence for the "Mad Genius" hypothesis. *Psychology of Aesthetics, Creativity, and the Arts*, *3*(2), 62–72. Available from http://dx.doi.org/10.1037/a0013975.

Schmahl, C. G., Elzinga, B. M., & Bremner, J. D. (2002). Individual differences in psychophysiological reactivity in adults with childhood abuse. *Clinical Psychology and Psychotherapy*, *9*(4), 271–276. Available from http://dx.doi.org/10.1002/cpp.325.

Schmeichel, B. J., & Baumeister, R. F. (2004). Self-regulatory strength. In R. F. Baumeister, & K. D. Vohs (Eds.), *Handbook of self-regulation: Research, theory, and application* (pp. 84–98). New York: The Guilford Press.

Schmidt, A., Jabusch, H.-C., Altenmuller, E., Enders, L., Saunders-Pullman, R., Bressman, S. B., ... Hagenah, J. (2011). Phenotypic spectrum of musician's dystonia: A task-specific disorder? *Movement Disorders*, *26*(3), 546–549. Available from http://dx.doi.org/10.1002/mds.23526.

Schmit, J. M., Regis, D. I., & Riley, M. A. (2005). Dynamic patterns of postural sway in ballet dancers and track athletes. *Experimental Brain Research*, *163*, 370–378. Available from http://dx.doi.org/10.1007/s00221-004-2185-6.

Schneider, E., & Chesky, K. (2011). Social support and performance anxiety of college music students. *Medical Problems of Performing Artists*, *26*(3), 157–163.

Schoeb, V., & Zosso, A. (2012). "You cannot perform music without taking care of your body": A qualitative study on musicians' representation of body and health. *Medical Problems of Performing Artists*, *27*(3), 129–136.

Schoenleber, M., & Berenbaum, H. (2012). Shame regulation in personality pathology. *Journal of Abnormal Psychology*, *121*(2), 433–446. Available from http://dx.doi.org/10.1037/a0025281.

Schoenleber, M., Sippel, L. M., Jakupcak, M., & Tull, M. T. (2015). Role of trait shame in the association between posttraumatic stress and aggression among men with a history of interpersonal trauma. *Psychological Trauma: Theory, Research, Practice, and Policy*, *7*(1), 43–49. Available from http://dx.doi.org/10.1037/a0037434.

Schore, A. N. (1994). *Affect regulation and the origin of the self: The neurobiology of emotional development*. Hillsdale, NJ: L. Erlbaum Associates.

Schore, A. N. (2003). *Affect regulation and the repair of the self.* New York: W.W. Norton & Company.

Schuele, S. U., & Lederman, R. J. (2004). Occupational disorders in instrumental musicians. *Medical Problems of Performing Artists, 19*(3), 123–128.

Schuengel, C., Bakermans-Kranenberg, M. J., & van Ijzendoorn, M. H. (1999). Frightening maternal behavior linking unresolved loss and disorganized infant attachment. *Journal of Consulting and Clinical Psychology, 67*, 54–63.

Schuldberg, D. (2000–2001). Creativity and psychopathology: Categories, dimensions, and dynamics. *Creativity Research Journal, 13*(1), 105–110.

Scialom, M., Goncalves, A., & Padovani, C. R. (2005). Work and injuries in dancers: Survey of a professional dance company in Brazil. *Medical Problems of Performing Artists, 21*, 29–33.

Scibinetti, P., Tocci, N., & Pesce, C. (2011). Motor creativity and creative thinking in children: The diverging role of inhibition. *Creativity Research Journal, 23*(3), 262–272. Available from http://dx.doi.org/10.1080/10400419.2011.595993.

Scoffier-Meriaux, S., Falzon, C., Lewton-Brain, P., Filaire, E., & d'Arripe-Longueville, F. (2015). Big five personality traits and eating attitudes in intensely training dancers: The mediating role of internalized thinness norms. *Journal of Sports Science and Medicine, 14*, 627–633.

Sears, S. F., Urizar, G. G., & Evans, G. D. (2000). Examining a stress-coping model of burnout and depression in extension agents. *Journal of Occupational Health Psychology, 5* (1), 56–62. Available from http://dx.doi.org/10.1037/1076-8998.5.1.56.

Seeley, W. W., Menon, V., Schatzberg, A. F., Keller, J., Glover, G. H., Kenna, H., ... Greicius, M. D. (2007). Dissociable intrinsic connectivity networks for salience processing and executive control. *Journal of Neuroscience, 27*, 2349–2356.

Seery, M. D., Holman, E. A., & Silver, R. C. (2010). Whatever does not kill us: Cumulative lifetime adversity, vulnerability and resilience. *Journal of Personality and Social Psychology, 99*(6), 1025–1041. Available from http://dx.doi.org/10.1037/a0021344.

Segerstrom, S. C., & Nes, L. S. (2007). Heart rate variability reflects self-regulatory strength, effort, and fatigue. *Psychological Science, 18*(3), 275–281. Available from http://dx.doi.org/10.1111/j.1467-9280.2007.01888.x.

Selby, E. C., Shaw, E. J., & Houtz, J. C. (2005). The creative personality. *Gifted Child Quarterly, 49*(4), 300–314.

Seligman, M. E. (1975/1992). *Helplessness.* San Francisco, CA: Freeman.

Seligman, M. E. P., Steen, T. A., Park, N., & Peterson, C. (2005). Positive psychology progress: Empirical validation of interventions. *American Psychologist, 60*(5), 410–421. Available from http://dx.doi.org/10.1037//0003-066X.60.5.410.

Seton, M. (2010). The ethics of embodied actor training and habitual vulnerability. *International Journal of Ethics in Theatre and Performance, 1*(1), 5–20. Available from http://dx.doi.org/10.1386/peet.1.1.5_1.

Sevdalis, V., & Keller, P. E. (2011). Captured by motion: Dance, action understanding, and social cognition. *Brain and Cognition, 77*, 231–236. Available from http://dx.doi.org/10.1016/j.bandc.2011.08.005.

Sevdalis, V., & Raab, M. (2014). Empathy in sports, exercise, and the performing arts. *Psychology of Sport and Exercise, 15*, 173–179. Available from http://dx.doi.org/10.1016/j.psychsport.2013.10.013.

Shah, S., Weiss, D. S., & Burchette, R. J. (2012). Injuries in professional modern dancers: Incidence, risk factors, and management. *Journal of Dance Medicine and Science, 16* (1), 17–25.

Shahaeian, A., Peterson, C. C., Slaughter, V., & Wellman, H. M. (2011). Culture and the sequence of steps in theory of mind development. *Developmental Psychology*, *47*(5), 1239–1247. Available from http://dx.doi.org/10.1037/a0023899.

Shakespeare (1600/1987). *The tragedy of Hamlet, Prince of Denmark*. New York: New American Library.

Sheets-Johnstone, M. (1999). Emotion and movement: A beginning empirical-phenomenological analysis of their relationship. In R. Nunez, & W. J. Freeman (Eds.), *Reclaiming cognition: The primacy of action, intention, and emotion* (pp. 259–277). Bowling Green, OH: Imprint Academic.

Sherman, A. J., Mayall, E., & Tasker, S. L. (2014). Can a prescribed turnout conditioning program reduce the differential between passive and active turnout in pre-professional dancers? *Journal of Dance Medicine and Science*, *18*(4), 159–168. Available from http://dx.doi.org/10.12678/1089-313X.18.4.159.

Sherwood, L. (2010). *Human physiology from cells to systems* (7th ed.). Pacific Grove, CA: Brooks/Cole.

Shimamura, A. P., Cohn-Sheehy, B. I., & Shimamura, T. A. (2014). Perceiving movement across film edits: A psychocinematic analysis. *Psychology of Aesthetics, Creativity, and the Arts*, *8*(1), 77–80. Available from http://dx.doi.org/10.1037/a0034595.

Shin, H.-W., Kang, S. K., Hallett, M., & Sohn, Y. H. (2012). Reduced surround inhibition in musicians. *Experimental Brain Research*, *219*, 403–408. Available from http://dx.doi.org/10.1007/s00221-012-3102-z.

Shirtcliff, E. A., Granger, D. A., Booth, A., & Johnson, D. (2005). Low salivary cortisol levels and externalizing behavior problems in youth. *Development and Psychopathology*, *17*, 167–184. Available from http://dx.doi.org/10.1017/S0954579405050091.

Shrira, A., Palgi, Y., Ben-Ezra, M., & Shmotkin, D. (2011). Transgenerational effects of trauma in midlife: Evidence for resilience and vulnerability in offspring of holocaust survivors. *Psychological Trauma: Theory, Research, Practice, and Policy*, *3*(4), 394–402. Available from http://dx.doi.org/10.1037/a0020608.

Shuter-Dyson, R. (2000). Profiling music students: Personality and religiosity. *Psychology of Music*, *28*, 190–196.

Siedler, R. D., Noll, D. C., & Thiers, G. (2004). Feedforward and feedback processes in motor control. *Neuroimage*, *22*, 1775–1783. Available from http://dx.doi.org/10.1016/j.neuroimage.2004.05.003.

Siegel, D. (2012). *The developing mind: How relationships and the brain interact to shape who we are* (2nd ed.). New York: Guilford Press.

Sills, B. Retrieved from www.goodreads.com/quotes/tag/music.

Silvestrini, N., & Gendolla, G. H. E. (2009). The joint effect of mood, task valence, and task difficulty on effort-related cardiovascular response and facial EMG. *International Journal of Psychophysiology*, *73*, 226–234. Available from http://dx.doi.org/10.1016/j.ijpsycho.2009.03.004.

Silvia, P. J., Fayn, K., Nusbaum, E. C., & Beaty, R. E. (2015). Openness to experience and awe response to nature and music: Personality and profound aesthetic experiences. *Psychology of Aesthetics, Creativity, and the Arts*, *9*(4), 376–384. Available from http://dx.doi.org/10.1037/aca000028.

Silvia, P. J., Kaufman, J. C., & Pretz, J. E. (2009). Is creativity domain-specific? Latent class models of creative accomplishments and creative self-descriptions. *Psychology of Aesthetics, Creativity, and the Arts*, *3*(3), 139–148. Available from http://dx.doi.org/10.1037/a0014940.

Silvia, P. J., & Kimbrel, N. A. (2010). A dimensional analysis of creativity and mental illness: Do anxiety and depression symptoms predict creative cognition, creative accomplishments, and creative self-concepts? *Psychology of Aesthetics, Creativity, and the Arts*, *4*(1), 2–10. Available from http://dx.doi.org/10.1037/a001694.

Silvia, P. J., Thomas, K. S., Nusbaum, E. C., Beaty, R. E., & Hodges, D. A. (2016). How does music training predict cognitive abilities? A bifactor approach to musical expertise and intelligence. *Psychology of Aesthetics, Creativity, and the Arts*, *10*(2), 184–190. Available from http://dx.doi.org/10.1037/aca0000058.

Simeon, D., Riggio-Rosen, A., Guralnik, O., Knutelska, M., & Nelson, D. (2003). Depersonalization disorder: Dissociation and affect. *Journal of Trauma and Dissociation*, *4*(4), 63–76. Available from http://dx.doi.org/10.1300/J229v04n04_05.

Simeonova, D. I., Chang, K. D., Strong, C., & Ketter, T. A. (2005). Creativity in familial bipolar disorder. *Journal of Psychiatric Research*, *39*, 623–631. Available from http://dx.doi.org/10.1016/j.jpsychires.2005.01.005.

Simeons, V. L., & Tervaniemi, M. (2013). Musician–instrument relationship as a candidate index for professional well-being in musicians. *Psychology of Aesthetics, Creativity, and the Arts*, *7*(2), 171–180. Available from http://dx.doi.org/10.1037/a0030164.

Simon, L. (2006). Managing creative projects: An empirical synthesis of activities. *International Journal of Project Management*, *24*, 116–126. Available from http://dx.doi.org/10.1016/j.ijproman.2005.09.002.

Simonton, D. K. (1999a). *Origins of genius: Darwinian perspectives on creativity*. New York: Oxford University Press.

Simonton, D. K. (1999b). Creativity from a historiometric perspective. In R. J. Sternberg (Ed.), *Handbook of creativity* (pp. 116–133). New York: Cambridge University Press.

Simonton, D. K. (2000). Creativity: Cognitive, personal, developmental, and social aspects. *American Psychologist*, *55*(1), 151–158. Available from http://dx.doi.org/10.1037//003.066X.55.1.151.

Simonton, D. K. (2004). Film awards as indicators of cinematic creativity and achievement: A quantitative comparison of the Oscars and six alternatives. *Creativity Research Journal*, *16*(2 & 3), 163–172.

Simonton, D. K. (2005). Film as art versus film as business: Differential correlates of screenplay characteristics. *Empirical Studies of the Arts*, *23*(2), 93–117.

Simonton, D. K. (2007a). Film music: Are award-winning scores and songs heard in successful motion pictures? *Psychology of Aesthetics, Creativity, and the Arts*, *1*(2), 53–60. Available from http://dx.doi.org/10.1037/1931-3896.1.2.53.

Simonton, D. K. (2007b). Cinema composers: Career trajectories for creative productivity in film music. *Psychology of Aesthetics, Creativity, and the Arts*, *1*(3), 160–169. Available from http://dx.doi.org/10.1037/1931-3896.1.3.160.

Simonton, D. K. (2014). More method in the mad-genius controversy: A historiometric study of 204 historic creators. *Psychology of Aesthetics, Creativity, and the Arts*, *8*(1), 53–61. Available from http://dx.doi.org/10.1037/a0035367.

Simor, P., Csoka, S. Y., & Bodizs, R. (2010). Nightmares and bad dreams in patients with borderline personality disorder: Fantasy as a coping skill. *European Journal of Psychiatry*, *24* (1), 28–37. Available from http://dx.doi.org/10.4321/S0213-61632010000100004.

Singer, J. L., & Singer, D. G. (2015). Professional paths over six decades researching and practicing play. *International Journal of Play*, *4*(2), 190–202. Available from http://dx.doi.org/10.1080/21594937.2015.1060570.

Slade, J. M., Landers, D. M., & Martin, P. E. (2002). Muscular activity during real and imagined movements: A test of inflow explanations. *Journal of Sport and Exercise Psychology*, *24*, 151–167.

Sledjeski, E. M., & Delhanty, D. L. (2012). Prior peritraumatic dissociative experiences affect autonomic reactivity during trauma recall. *Journal of Trauma and Dissociation*, *13*, 32–50. Available from http://dx.doi.org/10.1080/15299732.2011.608628.

Smith, D. J. (2003). A framework for understanding the training process leading to elite performance. *Sports Medicine*, *33*(15), 1103–1126.

Smitt, M. S., & Bird, H. A. (2013). Measuring and enhancing proprioception in musicians and dancers. *Clinical Rheumatology*, *32*, 469–473. Available from http://dx.doi.org/10.1007/s10067-013-2193-7.

Smol, E., & Fredyk, A. (2012). Supplementary low-intensity aerobic training improves aerobic capacity and does not affect psychomotor performance in professional female ballet dancers. *Journal of Human Kinetics*, *31*, 79–87. Available from http://dx.doi.org/10.2478/v10078-012-0008-6.

Snowden, P. L., & Christian, L. G. (1999). Parenting the young gifted child: Supportive behaviors. *Roeper Review*, *21*(3), 215–221.

Somer, E. (2001). An ethnographic study of former dissociative identity disordered patients. *Imagination, Cognition and Personality*, *20*(4), 315–345. Available from http://dx.doi.org/10.2190/Y9BT-0EUC-P8CQ-UGYG.

Somer, E. (2002). Maladaptive daydreaming: A qualitative inquiry. *Journal of Contemporary Psychotherapy*, *32*(2–3), 197–212. Available from http://dx.doi.org/10.1023/A:1020597026919.

Somer, E., Lehrfeld, J., Bigelsen, J., & Jopp, D. S. (2016). Development and validation of the Maladaptive Daydreaming Scale (MDS). *Consciousness and Cognition*, *39*, 77–91. Available from http://dx.doi.org/10.1016/j.concog.2015.12.001.

Sonnby-Borgstrom, M., & Jonsson, P. (2004). Dismissing-avoidant pattern of attachment and mimicry reactions at different levels of information processing. *Scandinavian Journal of Psychology*, *45*, 103–113.

Sousa, M., Carvalho, P., Moreira, P., & Teixeria, V. H. (2013). Nutrition and nutritional issues for dancers. *Medical Problems of Performing Artists*, *28*(3), 119–123.

Sowden, P. T., Clements, L., Redlich, C., & Lewis, C. (2015). Improvisation facilitates divergent thinking and creativity: Realizing a benefit of primary school arts education. *Psychology of Aesthetics, Creativity, and the Arts*, *9*(2), 128–138. Available from http://dx.doi.org/10.1037/aca000018.

Spahn, C., Burger, T., Hildebrandt, H., & Seidenglanz, K. (2005). Health locus of control and preventive behavior among students of music. *Psychology of Music*, *33*(3), 256–268. Available from http://dx.doi.org/10.1177030573560505373 3.

Spahn, C., Echternach, M., Zander, M. F., Voltmer, E., & Richter, B. (2010). Music performance anxiety in opera singers. *Logopedics Phoniatrics Vocology*, *35*(4), 175–182. Available from http://dx.doi.org/10.3109/14015431003720600.

Spahn, C., Ell, N., & Seidenglanz, K. (2001). Psychosomatic findings in musician patients at a department of hand surgery. *Medical Problems of Performing Artists*, *16*, 144–151.

Spahn, C., Nusseck, M., & Zander, M. (2014). Long-term analysis of health status and preventive behavior in music students across an entire university program. *Medical Problems of Performing Artists*, *29*(1), 8–15.

Spear, L. P. (2000). The adolescent brain and age-related behavioral manifestations. *Neuroscience and Biobehavioral Reviews*, *24*, 417–463.

Spector, J. T., & Brandfonbrener, A. G. (2007). Methods of evaluation of musician's dystonia: Critique of measurement tools. *Movement Disorders*, *22*(3), 309–312. Available from http://dx.doi.org/10.1002/mds.21214.

Spence, C. (2001). Prevalence rates for medical problems among flautists: A comparison of the UNT-Musician Health Survey and the Flute Health Survey. *Medical Problems of Performing Artists*, *16*, 99–101.

Springer, K. W., Sheridan, J., Kuo, D., & Carnes, M. (2003). The long-term health outcomes of childhood abuse: An overview and a call to action. *Journal of General Internal Medicine*, *18*, 864–870.

Srivastava, S., Childers, M. E., Baek, J. H., Strong, C. M., Hill, S. J., Warsett, K. S., ... Ketter, T. A. (2010). Toward interaction of affective and cognitive contributors to creativity in bipolar disorders: A controlled study. *Journal of Affective Disorders*, *125*, 27–34. Available from http://dx.doi.org/10.1016/j.jad.2009.12.018.

Sroufe, L. A. (1985). Attachment classifications from the perspective of infant-caregiver relationships and infant temperament. *Child Development*, *56*, 1–14.

Sroufe, L. A. (2005). Attachment and development: A prospective, longitudinal study from birth to adulthood. *Attachment and Human Development*, *7*(4), 349–367. Available from http://dx.doi.org/10.1080/14616730500365928.

Sroufe, L. A., Egeland, B., Carlson, E., & Collins, W. A. (2005). *The development of the person: The Minnesota study of risk and adaptation from birth to adulthood*. New York: Guilford Press.

Stahl, J. (2016). *Being misty* (pp. 28–31). Dance Magazine, May.

Stams, G.-J. J. M., Juffer, F., & van IJzendoorn, M. H. (2002). Maternal sensitivity, infant attachment, and temperament in early childhood predict adjustment in middle childhood: The case of adopted children and their biologically unrelated parents. *Developmental Psychology*, *38* (5), 806–821. Available from http://dx.doi.org/10.1037//0012-1649.38.5.806.

Standley, J. M. (2008). Aesthetic responses and earliest memories: The difference between music and literature. *Creativity Research Journal*, *20*(4), 437–444. Available from http://dx.doi.org/10.1080/10400410802278891.

Stanislavski, K. (1924/1963). *An actor prepares*. New York: Routledge.

Stanislavski, K. Retrieved from www.goodreads.com/quotes/tag/acting.

Steele, H., Steele, M., & Croft, C. (2008). Early attachment predicts emotion recognition at 6 and 11 years old. *Attachment and Human Development*, *10*(4), 379–393. Available from http://dx.doi.org/10.1080/146730802461409.

Stein, M. I. (1953). Creativity and culture. *Journal of Psychology*, *36*, 31–32.

Steinberg, H., Sykes, E. A., Moss, T., Lowery, S., LeBoutillier, N., & Dewey, A. (1997). Exercise enhances creativity independently of mood. *British Journal of Sports Medicine*, *31*, 240–245.

Steinberg, N., Hershkovitz, I., Peleg, S., Dar, G., Masharawi, Y., Zeev, A., & Sieve-Ner, I. (2013). Morphological characteristics of the young scoliotic dancer. *Physical Therapy in Sport*, *14*, 213–220. Available from http://dx.doi.org/10.1016/j.ptsp.2012.07.003.

Steinberg, N., Siev-Ner, I., Peleg, S., Dar, G., Masharawi, Y., Zeev, A., & Hershkovitz, I. (2013). Injuries in female dancers aged 8 to 16 years. *Journal of Athletic Training*, *48* (1), 118–123. Available from http://dx.doi.org/10.4085/1062-6050-48.1.106.

Stern, R. M., & Lewis, N. L. (1968). Ability of actors to control their GSRS and express emotions. *Psychophysiology*, *4*(3), 294–299.

Sternbach, D. J. Retrieved from www.gmu/centers/artswellness/overuseinjury.

Sternberg, R. J. (Ed.), (1999). *Handbook of creativity*. Cambridge, UK: Cambridge University Press.

Sternberg, R. J. (2001). What is the common thread of creativity?: Its dialectical relation to intelligence and wisdom. *American Psychologist*, *56*(4), 360–362. Available from http://dx.doi.org/10.1037//0003-066X.56.4.360.

Sternberg, R. J. (2005). Creativity or creativities? *International Journal of Human-Computer Studies*, *63*, 370–382. Available from http://dx.doi.org/10.1016/j.ijhcs.2005.04.003.

Sternberg, R. J. (2006). The nature of creativity. *Creativity Research Journal*, *18*(1), 87–98.

Sternberg, R. J. (2012). The assessment of creativity: An investment-based approach. *Creativity Research Journal*, *24*(1), 3–12. Available from http://dx.doi.org/10.1080/10400419.2012.652925.

Sternberg, R. J., & Lubart, T. I. (1991). An investment theory of creativity and its development. *Human Development*, *34*(1), 1–31.

Stevens, C., Malloch, S., McKechnie, S., & Steven, N. (2003). Choreographic cognition: The time-course and phenomenology of creating a dance. *Pragmatics and Cognition*, *11*(2), 297–326. Available from http://dx.doi.org/10.1075/pc.11.2.06ste.

Stevens, C., & McKechnie, S. (2005). Thinking in action: Thought made visible in contemporary dance. *Cognitive Processing*, *6*(4), 243–252. Available from http://dx.doi.org/10.1007/s10339-005-0014-x.

Stewart, L., Henson, R., Kampe, K., Walsh, V., Turner, R., & Frith, U. (2003). Brain changes after learning to read and play music. *Neuroimage*, *20*, 71–83. Available from http://dx.doi.org/10.1016/S1053-8119(03)00248-9.

Stinson, S. W. (2004). My body/myself: Lessons from dance education. In L. Bresler (Ed.), *Knowing bodies, moving minds: Towards embodied teaching and learning*. London: Kluwer Academic.

St-Louis, A. C., & Vallerand, R. J. (2015). A successful creative process: The role of passion and emotions. *Creativity Research Journal*, *27*(2), 175–187. Available from http://dx.doi.org/10.1080/10400419.2015.1030314.

Stoeber, J., Harris, R. A., & Moon, P. S. (2007). Perfectionism and the experience of pride, shame, and guilt: Comparing healthy perfectionists, unhealthy perfectionists, and non-perfectionists. *Personality and Individual Differences*, *43*(1), 131–141. Available from http://dx.doi.org/10.1016/j.paid.2006.11.012.

Stoeber, J., & Otto, K. (2006). Positive conceptions of perfectionism: Approaches, evidence, challenges. *Personality and Social Psychology Review*, *10*(4), 295–319. Available from http://dx.doi.org/10.1207/s15327957pspr1004_2.

Stovall-McClough, K. C., & Cloitre, M. (2006). Unresolved attachment, PTSD, and dissociation in women with childhood abuse histories. *Journal of Counseling and Clinical Psychology*, *74*(2), 219–228.

Strait, D. L., Kraus, N., Skoe, E., & Ashley, R. (2009). Musical experience and neural efficiency—effects of training on subcortical processing of vocal expressions of emotion. *European Journal of Neuroscience*, *29*, 661–668. Available from http://dx.doi.org/10.1111/j.1460-9568.2009.06617.x.

Strait, D. L., Parberry-Clark, A., Hittner, E., & Kraus, N. (2012). Musical training during early childhood enhances the neural encoding of speech in noise. *Brain and Language*, *123*, 191–201. Available from http://dx.doi.org/10.1016/j.bandl.2012.09.001.

Stremikis, B. A. (2002). The personal characteristics and environmental circumstances of successful women musicians. *Creativity Research Journal*, *14*(1), 85–92. Available from http://dx.doi.org/10.1207/S15326934CRJ1401_7.

Stright, A. D., Gallagher, K. C., & Kelley, K. (2008). Infant temperament moderates relations between maternal parenting in early childhood and children's adjustment in first grade. *Child Development*, *79*(1), 186–200.

Strohle, A. (2009). Physical activity, exercise, depression and anxiety disorders. *Biological Psychiatry*, *116*, 777–784. Available from http://dx.doi.org/10.1007/s00702-008-0092-x.

Strong, C. M., Nowakowska, C., Santosa, C. M., Wang, P. W., Kraemer, H. C., & Ketter, T. A. (2007). Temperament-creativity relationships in mood disorder patients, healthy controls and highly creative individuals. *Journal of Affective Disorders*, *100*, 41–48. Available from http://dx.doi.org/10.1016/j/jad.2006.10.015.

Stuckey, M. I., Tordi, N., Mourot, L., Gurr, L. J., Rakobowchuk, M., Millar, P. J., ... Kamath, M. V. (2012). Autonomic recovery following sprint interval exercise. *Scandinavian Journal of Medical Science and Sports*, *22*, 756–763. Available from http://dx.doi.org/10.111/j.1600-0838.2011.01320.x.

Studer, R. K., Danuser, B., Wild, P., Hildebrandt, H., & Gomez, P. (2014). Psychophysiological activation during preparation, performance, and recovery in high- and low-anxious music students. *Applied Psychophysiology and Biofeedback*, *39*, 45–57. Available from http://dx.doi.org/10.1007/s10484-014-9240-2.

Sturm, I., Baak, J., Storek, B., Traore, A., & Thuss-Patience, P. (2014). Effect of dance on cancer-related fatigue and quality of life. *Support Care Cancer*, *22*, 2241–2249. Available from http://dx.doi.org/10.1007/s00520-014-2181-8.

Sturman, D. A., & Moghaddam, B. (2011). The neurobiology of adolescence: Changes in brain architecture, functional dynamics, and behavioral tendencies. *Neuroscience and Biobehavioral Reviews*, *35*, 1704–1712. Available from http://dx.doi.org/10.1016/j.neurobiorev.2011.04.003.

Subotnik, R. F., Olszewski-Kubilius, P., & Worrell, F. C. (2011). Rethinking giftedness and gifted education: A proposed direction forward based on psychological science. *Psychological Science in the Public Interest*, *12*(1), 3–54. Available from http://dx.doi.org/10.1177/1529100611418056.

Sue, D. W., Capodilupo, C. M., Torino, G. C., Bucceri, J. M., Holder, A. M. B., Nadal, K. L., & Esquiin, M. (2007). Racial microaggresions in everyday life: Implications for clinical practice. *American Psychologist*, *62*(4), 271–286. Available from http://dx.doi.org/10.1037/003-066x.62.4.271.

Sullivan, P., & McCarthy, J. (2009). An experimental account of the psychology of art. *Psychology of Aesthetics, Creativity, and the Arts*, *3*(3), 181–187. Available from http://dx.doi.org/10.1037/a0014292.

Sundberg, J., & Thomasson, M. (2001). Consistency of inhalatory breathing patterns in professional operatic singers. *Journal of Voice*, *15*(3), 373–383.

Suzuki, T. (1986). *The way to acting: The theatre writings of Tadashi Suzuki*. New York: Theatre Communications Group.

Swami, V., Malpass, F., Harvard, D., Benford, K., Costescu, A., Sofitiki, A., & Taylor, D. (2013). Metalheads: The influence of personality and individual differences on preference for heavy metal. *Psychology of Aesthetics, Creativity, and the Arts*, *7*(4), 377–383. Available from http://dx.doi.org/10.1037/a0034493.

Swart, I. (2014). Overcoming adversity: Trauma in the lives of music performers and composers. *Psychology of Music*, *42*(3), 386–402. Available from http://dx.doi.org/10.1177/0305735613475371.

Swart, I. (2015). Ego boundaries and self-esteem: Two elusive facets of the psyche of performing musicians. *Psychology of Music*, *29*, 1–19. Available from http://dx.doi.org/10.1177/0305735615590283.

Swart, I., van Niekerk, C., & Hartman, W. (2010). Trauma-related dissociation as a factor affecting musicians' memory for music: Some possible solutions. *Australian Journal of Music Education*, 2, 117–134.

Symons, D. K., & Clark, S. E. (2000). A longitudinal study of mother-child relationships and theory of mind in the preschool period. *Social Development*, *9*, 3–23.

Szyf, M., & Bick, J. (2013). DNA methylation: A mechanism for embedding early life experience in the genome. *Child Development*, *84*(1), 49–57. Available from http://dx.doi.org/10.1111/j.1467-8624.2012.01793.x.

Tagore, R. (1928). *Fireflies*. New York: Collier Books.

Tal-Or, N., & Papirman, Y. (2007). The fundamental attribution error in attributing fictional figures' characters to the actors. *Media Psychology*, *9*, 331–345.

Tan, F. B., & Chou, J. B. (2011). Dimensions of autotelic personality and their effects on perceived playfulness in the context of mobile information and entertainment services. *Australian Journal of Information Systems*, *17*(1).

Tannenbaum, A. (1992). Early signs of giftedness: Research and commentary. In P. Klein, & A. Tannenbaum (Eds.), *To be young and gifted* (pp. 104–133). Norwood, NJ: Ablex Publishing Corp.

Tantia, J. F. (2012). Authentic movement and the autonomic nervous system: A preliminary invesgiation. *American Journal of Dance Therapy*, *34*, 53–73. Available from http://dx.doi.org/10.1007/s10465-012-9131-0.

Taylor, J., & Estanol, E. (2015). *Dance psychology for artistic and performance excellence*. Champaign, IL: Human Kinetics.

Taylor, L. D. (1997). MMPI-2 and ballet majors. *Personality and Individual Differences*, *22* (4), 521–526.

Taylor, M., & Mottweiler, C. M. (2002). Imaginary companions: Pretending they are real but knowing they are not. *American Journal of Play*, *1*(1), 47–54.

Taylor, M., Mottweiler, C. M., Naylor, E. R., & Levernier, J. G. (2015). Imaginary worlds in middle childhood: A qualitative study of two pairs of coordinated paracosms. *Creativity Research Journal*, *27*(2), 167–174. Available from http://dx.doi.org/10.1080/10400419.2015.1030318.

Taylor, S., & Cox, B. J. (1998). An expanded anxiety sensitivity index: Evidence for a hierarchic structure in a clinical sample. *Journal of Anxiety Disorders*, *12*(5), 463–483.

Taylor, S. E., Pham, L. B., Rivkin, I. D., & Armor, D. A. (1998). Harnessing the imagination: Mental simulation, self-regulation, and coping. *American Psychologist*, *53*(4), 429–439.

Teicher, M., Anderson, C. M., & Polcari, A. (2012). Childhood maltreatment is associated with reduced volume in the hippocampal subfields CA3, dentate gyrus and subiculum. *Proceedings of the National Academy of Science*, *109*(9), E563–E572. Available from http://dx.doi.org/10.1073/pnas.1115396109.

Teicher, M., Tomoda, A., & Andersen, S. L. (2006). Neurobiological consequences of early stress and childhood maltreatment: Are results from human and animal studies comparable? *Annals of New York Academy of Science*, *1071*, 313–323. Available from http://dx.doi.org/10.1196/annals.1364.024.

Tenenbaum, G., Edmonds, W. A., & Eccles, D. W. (2008). Emotions, coping strategies, and performance: A conceptual framework for defining affect-related performance zones. *Military Psychology*, *20*(Suppl. 1), S11–S37. Available from http://dx.doi.org/10.1080/08995600701804772.

Teng, C.-I. (2011). Who are likely to experience flow? Impact to temperament and character on flow. *Personality and Individual Differences*, *50*, 863–868. Available from http://dx.doi.org/10.1016/j.paid.2011.01.012.

Tentolouris, N., Argyrakopoulou, & Katsilambros, N. (2008). Perturbed autonomic nervous system function in metabolic syndrome. *Neuromolecular Medicine*, *10*, 169–178. Available from http://dx.doi.org/10.1007/s12017-008-8022-5.

Tharp, T. (2003). *The creative habit: Learn and use it for life: A practical guide*. New York: Simon & Schuster.

Tharp, T. Retrieved from www.goodreads.com/quotes/tag/dancing.

Thayer, J. F., Ahs, F., Fredrikson, M., Sollers, J. J., III, & Wager, D. D. (2012). A meta-analysis of heart rate variability and neuroimaging studies: Implications for heart rate variability as a marker of stress and health. *Neuroscience and Biobehavioral Reviews*, *36*, 747–756. Available from http://dx.doi.org/10.1016/j.neubiorev.2011.11.009.

Thayer, J. F., & Sternberg, E. (2006). Beyond heart rate variability: Vagal regulation of allostatic systems. *Annals of the New York Academy of Science*, *1088*, 361–372. Available from http://dx.doi.org/10.1196/annals.1366.014.

Thom, T., Haase, N., Rosamond, W., Howard, V. J., Rumsfeld, J., Manolio, T., ... American Heart Association Statistics Committee and Stroke Statistics Subcommittee (2006). Heart disease and stroke statistics—2006 update: A report from the American Heart Association Statistics Committee and Stroke Statistics Subcommittee. *Circulation*, *1113* (6), e85–151, *14* -

Thomas, A., & Chess, S. (1977). *Temperament and development*. New York: Brunner/Mazel.

Thomas, J. J., Keel, P. K., & Heatherton, T. F. (2005). Disordered eating attitudes and behaviors in ballet students: Examination of environmental and individual risk factors. *International Journal of Eating Disorders*, *38*(2), 263–268.

Thomas, J. J., Keel, P. K., & Heatherton, T. F. (2011). Disordered eating and injuries among adolescent ballet dancers. *Eating Weight Disorder*, *16*, e216–e222.

Thomas, J. P., & Nettelbeck, T. (2014). Performance anxiety in adolescent musicians. *Psychology of Music*, *42*(4), 624–634. Available from http://dx.doi.org/10.1177/030573561345151.

Thomas, K. M., & Duke, M. (2007). Depressed writing: Cognitive distortions in the works of depressed and nondepressed poets and writers. *Psychology of Aesthetics, Creativity, and the Arts*, *1*(4), 204–218. Available from http://dx.doi.org/10.1037/1931-3896.1.4.204.

Thomasson, M., & Sundberg, J. (1999). Consistency of phonatory breathing patterns in professional operatic singers. *Journal of Voice*, *13*(4), 529–541.

Thompson, C. M., Ruthmann, S. A., Anttila, E., & Doan, W. J. (2014). The influence of body mapping on student musicians' performance experiences. *International Journal of Education and the Arts*, *15*(7), 1–28.

Thomson, P. (2004). The impact of trauma on the embryo and fetus: An application of the diathesis-stress model and the neurovulnerability-neurotoxicity model. *Journal of Prenatal and Perinatal Psychology and Health*, *19*(1), 9–63.

Thomson, P. (2007). Down will come baby: Prenatal development, primitive defences and gestational dysregulation. *Journal of Trauma and Dissociation*, *8*(3), 85–114.

Thomson, P. (2010). Loss and disorganization from an attachment perspective. *Death Studies*, *34*(10), 893–914. Available from http://dx.doi.org/10.1080/07481181003765410.

Thomson, P. (2011). Dance and creativity. In M. A. Runco, & S. R. Pritzker (Eds.), *Encyclopaedia of creativity* (2nd ed., Vol. 1, pp. 343–350). London: Elsevier Publishing Company.

Thomson, P. (2012a). Shared experiences: Prenatal relational model and group process. In I. N. H. Harwood, & M. Pines (Eds.), *Self experiences in group: Intersubjective and self psychology pathways to human understanding* (2nd ed., pp. 11–33). New York: Routledge Publishers.

Thomson, P. (2012b). Movement and mindedness: Dance as a group process. In I. N. H. Harwood, & M. Pines (Eds.), *Self experiences in group: Intersubjective and self psychology pathways to human understanding* (2nd ed., pp. 213–234). New York: Routledge Publishers.

Thomson, P., & Godin, M. E. (2011). Acting. In M. A. Runco, & S. R. Pritzker (Eds.), *Encyclopaedia of creativity* (2nd ed., Vol. 1, pp. 1–8), London: Elsevier Publishing Company.

Thomson, P., & Jaque, S. V. (2011–2012). Anxiety and the influences of flow, trauma, and fantasy experiences on dancers. *Cognition, Imagination and Personality, 32*(2), 165–178.

Thomson, P., & Jaque, S. V. (2011a). Testimonial theatre-making: Establishing or dissociating the self. *Psychology of Aesthetics, Creativity, and the Arts, 5*(3), 229–236.

Thomson, P., & Jaque, S. V. (2011b). Psychophysiological study: Ambulatory measure of the ANS in performing artists. In A. Williamon, D. Edwards, & L. Bartel (Eds.), *Proceedings of the international symposium on performance science* (pp. 149–154). Utrecht, The Netherlands: European Association of Concervatoires (AEC).

Thomson, P., & Jaque, S. V. (2012a). Dissociation and the adult attachment interview in artists. *Attachment and Human Development, 14*(2), 145–160.

Thomson, P., & Jaque, S. V. (2012b). Holding a mirror up to nature: Psychological vulnerability in actors. *Psychology of Aesthetics, Creativity, and the Arts, 6*(4), 361–369.

Thomson, P., & Jaque, S. V. (2012c). Dancing with the Muses: Dissociation and flow. *Journal of Trauma and Dissociation, 13*(4), 478–489.

Thomson, P., & Jaque, S. V. (2013). Exposing shame in dancers and athletes: Shame, trauma, and dissociation in a non-clinical population. *Journal of Trauma and Dissociation, 141*, 1–16. Available from http://dx.doi.org/10.1080/15299732.2012.757714.

Thomson, P., & Jaque, S. V. (2014). Unresolved mourning, supernatural beliefs and dissociation: A mediation analysis. *Attachment and Human Development, 16*(5), 499–514.

Thomson, P., & Jaque, S. V. (2015a). Shame and fantasy in athletes and dancers. *Cognition, Imagination and Personality, 34*(3), 291–305.

Thomson, P., & Jaque, S. V. (2015b). Posttraumatic, stress disorder and psychopathology in dancers. *Medical Problems in Performing Artists, 30*(3), 157–162.

Thomson, P., & Jaque, S. V. (2016a). Overexcitability and optimal flow in talented dancers, singers and athletes. *Roeper Review, 38*(1), 32–39. Available from http://dx.doi.org/10.1080/02783193.2015.1112865.

Thomson, P., & Jaque, S. V. (2016b). Overexcitability: A psychological comparison between dancers, opera singers and athletes. *Roeper Review, 38*(2), 84–92. Available from http://dx.doi.org/10.1080/02783193.2016.1150373.

Thomson, P., & Jaque, S. V. (2016c). Exquisite moments: Achieving optimal flow in three different activity-based groups regardless of early childhood adversity. *American Journal of Play, 8*(3), 346–362.

Thomson, P., & Jaque, S. V. (2016d). Visiting the muses: Creativity, coping and PTSD in talented dancers and athletes. *American Journal of Play, 8*(3), 363–378.

Thomson, P., Jaque, S. V., & Baltz, A. (2017). Intensive opera training program effects: A psychological investigation. *International Journal of Music Education, 35*(1), 1–11. Available from http://dx.doi.org/10.1177/0255761416667467.

Thomson, P., Keehn, E. B., & Gumpel, T. P. (2009). Generators and interpreters in a performing arts population: Dissociation, trauma, fantasy proneness, and affective states. *Creativity Research Journal, 21*(1), 72–91. Available from http://dx.doi.org/10.1080/10400410802633533.

Thomson, P., Kibarska, L., & Jacque, S. V. (2011). Comparison of dissociative experiences between rhythmic gymnasts and female dancers. *International Journal of Sport and Exercise Psychology*, *9*(3), 238–250.

Throsby, D., & Zednick, A. (2011). Multiple job-holding and artistic careers: Some empirical evidence. *Cultural Trends*, *20*(1), 9–24. Available from http://dx.doi.org/10.1080/09548963.2011.540809.

Thullier, F., & Moufti, H. (2004). Multi-joint coordination in ballet dancers. *Neuroscience Letters*, *369*, 80–84. Available from http://dx.doi.org/10.1016/j.neulet.2004.08.011.

Thyssen, H. H., Clevin, L., Olesen, S., & Lose, G. (2002). Urinary incontinence in elite female athletes and dancers. *International Urogynecology Journal Pelvic Floor Dysfunction*, *13*(1), 15–17.

Tietze, R. L. (2006). Jazz and the American identity. *Psychology of Aesthetics, Creativity, and the Arts*, *5*(1), 33–41. Available from http://dx.doi.org/10.1037/1931-3896.S.1.33.

Tietze, R. L. (2008). Jazz and American identity: Case study of a college course. *Psychology of Aesthetics, Creativity, and the Arts*, *2*(4), 245–255. Available from http://dx.doi.org/10.1037/a0010843.

Tiggemann, M., & Slater, A. (2001). A test of objectification theory in former dancers and non-dancers. *Psychology of Women Quarterly*, *25*, 57–64. Available from http://dx.doi.org/10.1111/1471-6402.00007.

Timmermans, B., Vanderwegen, J., & De Bodt, M. S. (2005). Outcome of vocal hygiene in singers. *Current Opinion in Otolaryngology and Head and Neck Surgery*, *13*(3), 138–142.

To, M. L., Fisher, C., Ashkanasy, N. M., & Rowe, P. A. (2012). Within-person relationships between mood and creativity. *Journal of Applied Psychology*, *97*(3), 599–612. Available from http://dx.doi.org/10.1037/a0026097.

Togari, A., & Arai, M. (2008). Pharmacological topics of bone metabolism: The physiological function of the sympathetic nervous system in modulating bone resorption. *Journal of Pharmacological Sciences*, *106*, 542–546. Available from http://dx.doi.org/10.1254/jphs.FM0070227.

Toledo, S. D., Akuthota, V., Drake, D. F., Nadler, S. F., & Chou, L. H. (2004). Sports and perforing arts medicine, 6: Issues relating to dance. *Archives of Physical Medicine and Rehabilitation*, *85*(S1), S75–S78. Available from http://dx.doi.org/10.1053/j.apmr.2003.12.004.

Tolson, G. H., & Cuyjet, M. J. (2007). Jazz and substance abuse: Road to creative genius or pathway to premature death. *International Journal of Law and Psychiatry*, *30*, 530–538. Available from http://dx.doi.org/10.1016/j.ijlp.2007.09.004.

Tomlinson, M., Cooper, P., & Murray, L. (2005). The mother-infant relationship and infant attachment in a South African Peri-Urban settlement. *Child Development*, *76*(5), 1044–1054.

Torrents, C., Castaner, M., Reverter, F., Morey, G., & Joffre, T. (2015). Dance teachers' aesthetic perception of kinematic parameters. *Psychology of Aesthetics, Creativity, and the Arts*, *9*(4), 442–450. Available from http://dx.doi.org/10.1037/a0039757.

Tosevski, D. L., Milovancevic, M. P., & Gajic, S. D. (2010). Personality and psychopathology of university students. *Current Opinion in Psychiatry*, *23*, 48–52. Available from http://dx.doi.org/10.1097/YCO.0b013e32833d625.

Trapnell, P. D., & Campbell, J. D. (1999). Private self-consciousness and the five-factor model of personality: Distinguishing rumination from reflection. *Journal of Personality and Social Psychology*, *76*, 284–304. Available from http://dx.doi.org/10.1037/0022-3514.76.2.284.

Trepanier, S.-G., Fernet, C., Austin, S., Forest, J., & Vallerand, S. J. (2014). Linking job demands and resources to burnout and work engagement: Does passion underlie these differential relationships? *Motivation and Emotion, 38*(3), 353–366.

Triposkiadis, F., Karayannis, G., Giamouzis, G., Skoularigis, J., Louridas, G., & Butler, J. (2009). The sympathetic nervous system in heart failure. *Journal of American College of Cardiology, 54*(19), 1747–1762. Available from http://dx.doi.org/10.1016/j.jacc.2009.05.015.

Tsakiris, M., Tajadura-Jimenez, A., & Costantini, M. (2011). Just a heartbeat away from one's body: Interoception sensitivity predicts malleability of body-representations. *Proceedings of the Royal Society Bulletin, 278*, 2470–2476. Available from http://dx.doi.org/10.1098/rspb.2010.2547.

Tsekhmistrenko, T. A., & Vasil'eva, V. A. (2001). Structural transformations of the associative cortex as the morphological base of development of human cognitive functions from birth to 20 years of age. *Human Physiology, 27*(5), 544–550.

Tseng, M. M.-C., Fang, D., Lee, M.-B., Chie, W.-C., Liu, J.-P., & Chen, W. J. (2007). Two-phase survey of eating disorders in gifted dance and non-dance high-school students in Taiwan. *Psychological Medicine, 37*, 1085–1096. Available from http://dx.doi.org/10.1017/S0033291707000323.

Tsigos, C., & Chrousos, G. P. (2002). Hypothalamic-pituitary-adrenal axis, neuroendocrine factors and stress. *Journal of Psychosomatic Research, 53*, 865–871.

Tugade, M. M., & Fredrickson, B. L. (2004). Resilient individuals use positive emotions to bounce back from negative emotional experiences. *Journal of Personality and Social Psychology, 86*(2), 320–333. Available from http://dx.doi.org/101037/0022-3514.86.2.320.

Turiano, N. A., Spiro, A., & Mroczek, D. K. (2012). Openness to experience and mortality in men: Analysis of trait and facets. *Journal of Aging and Health, 24*(4), 654–672. Available from http://dx.doi.org/10.1177/0898264311431303.

Turner, B. S., & Wainwright, S. P. (2003). Corps de ballet: The case of the injured ballet dancer. *Sociology of Health and Illness, 25*(4), 269–288.

Turner, J. E., & Schallert, D. L. (2001). Expectancy–value relationships of shame reactions and shame resiliency. *Journal of Educational Psychology, 93*(2), 320–329. Available from http://dx.doi.org/10.1037/0022-0663.93.2.320.

Turner-Stokes, L., & Reid, K. (1999). Three-dimensional motion analysis of upper limb movement in the bowing arm of string-playing musicians. *Clinical Biomechanics, 14*, 426–433.

Twitchett, E., Angioi, M., Koutedakis, Y., & Wyon, M. (2009). Video analysis of classical ballet performance. *Journal of Dance Medicine and Science, 13*(4), 124–128.

Twitchett, E., Angioi, M., Koutedakis, Y., & Wyon, M. (2010). The demands of a working day among female professional ballet dancers. *Journal of Dance Medicine and Science, 14*(4), 127–132.

Twitchett, E., Broderick, A., Nevill, A. M., Koutedakis, Y., Angioi, M., & Wyon, M. (2010). Does physical fitness affect injury occurrence and time loss due to injury in elit vocational ballet students? *Journal of Dance Medicine and Science, 14*(1), 26–31.

Twitchett, E., Nevill, A., Angioi, M., Koutedakis, Y., & Wyon, M. (2011). Development, validity, and reliability of a ballet-specific aerobic fitness test. *Journal of Dance Medicine and Science, 15*(3), 123–127.

Ullen, F., de Manzano, O., Almeida, R., Magnusson, P. K., Pedersen, N. L., Nakamura, J., ... Madison, G. (2012). Proneness for psychological flow in everyday life: Associations with personality and intelligence. *Personality and Individual Differences, 52*, 167–172. Available from http://dx.doi.org/10.1016/j.paid.2011.10.003.

Ulrich, M., Keller, J., Hoening, K., Waller, C., & Gron, G. (2014). Neural correlates of experimentally induced flow experiences. *Neuroimage*, *86*, 194–2002. Available from http://dx.doi.org/10.106/j.neuroimage.2013.08.019.

University of California (2014). *Performing arts safety manual*. Office of the President, Regents of California.

Unwin, M. M., Kenny, D. T., & Davis, P. J. (2002). The effects of group singing on mood. *Psychology of Music*, *30*, 175–185.

Upton, A.-A. (2011). Real people as actors—actors as real people. *Studies in Theatre and Performance*, *31*(2), 209–222. Available from http://dx.doi.org/10.1386/stap.31.2.209_1.

Urban, K. K. (1991). On the development of creativity in children. *Creativity Research Journal*, *4*(2), 177–191.

U. S. Department of Labor, Bureau of Labor Statistics. (2015). *Employer-reported workplace injuries and illnesses—2014* (USDL-15-2086). Retrived from http://www.bls.gov/iif/osh-sum.htm.

Uzzi, B., & Spiro, J. (2005). Collaboration and creativity: The small world problem. *American Journal of Sociology*, *111*(2), 447–504.

Vaag, J., Bjorgaard, J. H., & Bjerkeset, O. (2016). Use of psychotherapy and psychotropic medication among Norweigian musicians compared to the general workforce. *Psychology of Music*, *21*, 1–15. Available from http://dx.doi.org/10.1177/0305735616637132.

Vaag, J., Giaever, F., & Bjerkeset, O. (2014). Specific demands and resources in the career of the Norwegian freelance musician. *Arts and Health*, *6*(3), 205–222. Available from http://dx.doi.org/10.1080/17533015.2013.863789.

Vallerand, R. J., Mageau, G. A., Elliot, A. J., Dumais, A., Demers, M.-A., & Rousseau, F. (2008). Passion and performance attainment in sport. *Psychology of Sport and Exercise*, *9*, 373–392. Available from http://dx.doi.org/10.1016/j.psychsport.2007.05.003.

Van Delinder, J. (2005). Taylorism, managerial control strategies, and the ballets of Balanchine and Stravinsky. *American Behavioral Scientist*, *48*(1), 1439–1452. Available from http://dx.doi.org/10.1177/0002764205277189.

van der Kolk, B. A. (2006). Clinical implications of neuroscience research in PTSD. *Annuals of the New York Academy of Science*, *1071*, 277–293. Available from http://dx.doi.org/10.1196/annals.1364.022.

van der Kolk, B. A. (2014). *The body keeps the score: Brain, mind, and body in the healing of trauma*. New York: Viking.

van der Meer, L., Groenewold, N. A., Nolen, W. A., Pijnenborg, M., & Aleman, A. (2011). Inhibit yourself and understand the other: Neural basis of distinct processes underlying Theory of Mind. *Neuroimage*, *56*, 2364–2374. Available from http://dx.doi.org/10.1016/j.neuroimage.2011.03.053.

van Dyck, E., Maes, P.-J., Hargreaves, J., Lesaffre, M., & Leman, M. (2013). Expressing induced emotions through free dance movement. *Journal of Nonverbal Behavior*, *37*, 175–190. Available from http://dx.doi.org/10.1007/s10919-013-0153-1.

van Dyck, E., Vansteenkiste, Lenoir, M., Lesaffre, M., & Leman, M. (2014). Recognizing induced emotions of happiness and sadness from dance movement. *PLoS ONE*, *9*(2), e89773. Available from http://dx.doi.org/10.1371/journal.pone.0089773.

van Kemenade, J. F. L. M., van Son, M. J. M., & van Heesch, N. C. A. (1995). Performance anxiety among professional musicians in symphonic orchestras: A self-report study. *Psychological Reports*, *77*, 555–562.

van Ryzin, M. J., Leve, L. D., Neiderhiser, J. M., Shaw, D. S., Natsuaki, M. N., & Reiss, D. (2015). Genetic influences can protect against unresponsive parenting in the prediction of child social competence. *Child Development*, *86*, 667–680. Available from http://dx.doi.org/10.1111/cdev.12335.

Van Weelden, K. (2002). Relationships between perceptions of conducting effectiveness and ensemble performance. *Journal of Research and Music Education*, *50*(2), 165–176.

Vellante, M., Zucca, G., Preti, A., Sisti, D., Rocchi, M. B. L., Akiskal, K. K., & Akiskal, H. S. (2011). Creativity and affective temperaments in con-clinical professional artists: An empirical psychometric investigation. *Journal of Affective Disorders*, *135*, 28–36. Available from http://dx.doi.org/10.1016/j/jad.2011.06.062.

Vellers, H. L., Irwin, C., & Lightfoot, J. T. (2015). Heart rate response of professional musicians when playing music. *Medical Problems of Performing Artists*, *30*(2), 100–105.

Verducci, S. (2000). A moral method? Thoughts on cultivating empathy through method acting. *Journal of Moral Education*, *29*(1), 87–99. Available from http://dx.doi.org/10.1080/030572400102952.

Verghese, J. (2006). Cognitive and mobility profile of older social dancers. *Journal of American Geriatric Society*, *54*, 1241–1244. Available from http://dx.doi.org/10.1111/j.1532-5415.2006.00808.x.

Verhaeghen, P., Joorman, J., & Aikman, S. N. (2014). Creativity, mood, and the examined life: Self-reflective rumination boosts creativity, brooding breeds dysphoria. *Psychology of Aesthetics, Creativity, and the Arts*, *8*(2), 211–218. Available from http://dx.doi.org/10.1037/a0035594.

Verhaeghen, P., Khan, R., & Joorman, J. (2005). Why we sing the blues: The relation between self-reflective rumination, mood, and creativity. *Emotion*, *5*, 226–232. Available from http://dx.doi.org/10.1037/1528-3542.2.226.

Vessey, W. B., & Mumford, M. D. (2012). Heuristics as a basis for assessing creative potential: Measures, methods, and contingencies. *Creativity Research Journal*, *24*(1), 41–54. Available from http://dx.doi.org/10.1080/10400419.2012.652928.

Vickhoff, B., Malmgren, H., Astrom, R., Nyberg, G., Ekstrom, S.-R., Engwall, M., ... Jornsten, R. (2013). Music structure determines heart rate variability of singers. *Frontiers in Psychology*, *4*(334), 1–16. Available from http://dx.doi.org/10.3389/f.psyg.2013.00334.

Vogeley, K., Bussfeld, P., Newen, A., Herrmann, S., Happe, F., Falkai, P., ... Zilles, K. (2001). Mind reading: Neural mechanisms of theory of mind and self perspective. *Neuroimage*, *14*, 170–181. Available from http://dx.doi.org/10.1006/nimg.2001.0789.

Vohs, K. D., & Baumeister, R. F. (2004). Understanding self-regulation. In R. F. Baumeister, & K. D. Vohs (Eds.), *Handbook of self-regulation: Research, theory, and application* (pp. 1–9). New York: The Guilford Press.

von Stumm, S., Chung, A., & Furnham, A. (2011). Creative ability, creative ideation and latent classes of creative achievement: What is the role of personality? *Psychology, Aesthetics, Creativity, and the Arts*, *5*(2), 107–114. Available from http://dx.doi.org/10.1037/a0020499.

Vosberg, S. K. (1998). The effects of positive and negative mood in divergent-thinking performance. *Creativity Research Journal*, *11*(2), 165–172.

Vuoskoski, J. K., & Eerola, T. (2012). Can sad music really make you sad? Indirect measures of affective states induced by music and autobiographical memories. *Psychology of Aesthetics, Creativity, and the Arts*, *6*(3), 204–213. Available from http://dx.doi.org/10.1037/a0026937.

Vusst, P., Gebauer, L., Hansen, N. C., Jorgensen, S. R., Moller, A., & Linnet, J. (2010). Personality influences career choices: Sensation seeking in professional musicians. *Music Education Research*, *12*(2), 219–230. Available from http://dx.doi.org/10.1080/14613801003746584.

Vuust, P., Pallesen, K. J., Bailey, C., can Zuijen, T. L., Gjedde, A., Roepstorff, A., & Ostergaard, L. (2005). To musicians, the message is in the meter: Pre-attentive neuronal responses to incongruent rhythm are left-lateralized in musicians. *Neuroimage*, *24*, 560–564. Available from http://dx.doi.org/10.1016/j.neuroimage.2004.08.039.

Wainwright, S. P., Williams, C., & Turner, B. S. (2005). Fractured identities: Injury and the balletic body. *Health*, *9*(1), 49–66. Available from http://dx.doi.org/10.1177/1363459305048097.

Wainwright, S. P., & Turner, B. S. (2006). "Just crumbling to bits"? An exploration of the body, aging, injury and career in classical ballet dancers. *Sociology*, *40*(2), 237–255. Available from http://dx.doi.org/10.1177/0038038506062031.

Walker, I. J., & Nordin-Bates, S. M. (2010). Performance anxiety experiences of professional ballet dancers: The importance of control. *Journal of Dance Medicine and Science*, *14* (4), 133–145.

Wallace, C. E., & Russ, S. W. (2015). Pretend play, divergent thinking, and math achievement in girls: A longitudinal study. *Psychology of Aesthetics, Creativity, and the Arts*, *9* (3), 296–305. Available from http://dx.doi.org/10.1037/a0039006.

Wallace, K. (1994). Female pelvic floor functions, dysfunctions, and behavioral approaches to treatment. *Clinical Sports Medicine*, *13*(2), 459–481.

Wallas, G. (1926). *The art of thought*. New York: Harcourt Brace.

Wallert, J., & Madison, G. (2014). Recovery after aerobic exercise is manipulated by tempo change in a rhythmic sound pattern, as indicated by autonomic reaction on heart functioning. *Frontiers in Human Neuroscience*, *8*(738), 1–13. Available from http://dx.doi.org/10.3389/fnhum.2014.00738.

Walter, J. S. (2009). Sound exposure levels experienced by university wind band members. *Medical Problems of Performing Artists*, *24*(2), 63–70.

Walzak, P., McCabe, P., Madill, C., & Sheard, C. (2008). Acoustic changes in student actors voices after 12 months of training. *Journal of Voice*, *22*(3), 300–313. Available from http://dx.doi.org/10.1016/j.jvoice.2006.10.006.

Wangh, S. (2000). *An acrobat of the heart: A physical approach to acting inspired by the work of Jerzy Grotowski*. New York: Vantage Books.

Wanke, E. M., Arendt, M., Mill, H., Koch, F., Davenport, J., Fischer, A., & Groneberg, D. A. (2014). The theatrical stage as accident site in professional dance. *Medical Problems of Performing Artists*, *29*(1), 32–36.

Wanke, E. M., Arendt, M., Mill, H., Koch, F., Wanke, A., & Groneberg, D. A. (2014). Traumatic injuries in revue dancers. *Journal of Dance Medicine and Science*, *18*(1), 22–28. Available from http://dx.doi.org/10.12678/1089-313X.18.1.22.

Wanke, E. M., Mill, H., Wanke, A., Davenport, J., & Koch, F. (2012). Dance floors as injury risk: Analysis and evaluation of acute injuries caused by dance floors in professional dance with regard to preventative aspects. *Medical Problems of Performing Artists*, *27* (3), 137–142.

Ward, S. A. (2008). The voice of dance education: Health and the power of dance. *Journal of Physical Education, Recreation and Dance*, *79*(4), 33–36. Available from http://dx.doi.org/10.1080/07303084.2008.10598161.

Wasley, D., Taylor, A., Backx, K., & Williamon, A. (2012). Influence of fitness and physical activity on cardiovascular reactivity to musical performance. *Work*, *41*, 27–32. Available from http://dx.doi.org/10.3233/WOR-20121240.

Wassiliwizky, E., Wagner, V., Jacobsen, T., & Menninghaus, W. (2015). Art-elicited chills indicate states of being moved. *Psychology of Aesthetics, Creativity, and the Arts*, *9*(4), 405–416. Available from http://dx.doi.org/10.1037/aca000023.

Watanabe, D., Savion-Lemieux, T., & Penhune, V. B. (2007). The effect of early musical training on adult motor performance: Evidence for a sensitive period in motor learning. *Experimental Brain Research, 176*, 332–340. Available from http://dx.doi.org/10.1007/s00221-006-0619-z.

Waters, E., Merrick, S., Treboux, D., Crowell, J., & Albersheim, L. (2000). Attachment security infancy and early adulthood: A twenty-year longitudinal study. *Child Development, 71*(3), 684–689.

Watson, D. E., Nordin-Bates, S. M., & Chappell, K. A. (2012). Facilitating and nurturing creativity in pre-vocational dancers: Findings from the UK centers for advanced training. *Research in Dance Education, 13*(2), 153–173. Available from http://dx.doi.org/10.1080/14647893.2012.694415.

Watson, E. (2007). Who or what creates? A conceptual framework for social creativity. *Human Resource Development Review, 6*(4), 419–441. Available from http://dx.doi.org/10.1177/1534484307308255.

Watt, C., Watson, S., & Wilson, L. (2007). Cognitive and psychological mediators of anxiety: Evidence from a study of paranormal belief and perceived childhood control. *Personality and Individual Differences, 42*(2), 335–343. Available from http://dx.doi.org/10.1016/j.paid.2006.07.015.

Weber, A., & Jaekel-Reinhart, A. (2000). Burnout syndrome: A disease of modern societies? *Occupational Medicine, 50*(7), 512–517.

Weege, B., Lange, B. P., & Fink, B. (2012). Women's visual attention to variation in men's dance quality. *Personality and Individual Differences, 53*, 236–240. Available from http://dx.doi.org/10.1016/j.paid.2012.03.011.

Weine, S. M. (1996). The witnessing imagination: Social trauma, creative artists, and witnessing professionals. *Literature and Medicine, 15*(2), 167–182.

Weinfield, N. S., Whalley, G. J., & Egeland, B. (2004). Continuity, discontinuity, and coherence in attachment from infancy to late adolescence: Sequelae of organization and disorganization. *Attachment and Human Development, 6*(1), 73–97. Available from http://dx.doi.org/10.1080/14616730310001659566.

Weinstein, A. A., Deuster, P. A., & Kop, W. J. (2007). Heart rate variability as a predictor of negative mood symptoms induced by exercise withdrawal. *Medicine and Science in Sports and Exercise, 39*(4), 735–741. Available from http://dx.doi.org/10.1249/mss.0b013e31802f590c.

Weisberg, R. B., Bruce, S. E., Machan, J. T., Kessler, R. C., Culpepper, L., & Keller, M. B. (2002). Nonpsychiatric illness among primary care patients with trauma histories and posttraumatic stress disorder. *Psychiatric Services, 53*(7), 848–854.

Weisberg, R. W. (2006). *Creativity: Understanding innovation in problem solving, science, invention and the arts*. New Jersey: Wiley and Sons, Inc.

Weisberg, R. W. (2015). On the usefulness of "value" in the definition of creativity. *Creativity Research Journal, 27*(2), 111–124. Available from http://dx.doi.org/10.1080/10400419.2015.1030320.

Wellisch, M., & Brown, J. (2013). Many faces of a gifted personality: Characteristics along a complex gifted spectrum. *Talent Development and Excellence, 5*(2), 43–58.

Wellman, H. M., Fang, F., & Peterson, C. C. (2011). Sequential progressions of theory-of-mind scale: Longitudinal perspectives. *Child Development, 82*(3), 780–792. Available from http://dx.doi.org/10.1111/j.1467-8624.2011.01583.x.

Wells, R., Outhred, T., Heathers, J. A. J., Quintana, D. S., & Kemp, A. H. (2012). Matter over mind: A randomized controlled trial of single-session biofeedback training on performance anxiety and heart rate variability in musicians. *PLoS ONE, 7*(10), e46597. Available from http://dx.doi.org/10.1371/journal.pone.0046597.

Wheeler, B. B. (2003). The institutionalization of an American avant-garde: Performance art as democratic culture, 1970–2000. *Sociological Perspectives*, *46*(4), 491–512. Available from http://dx.doi.org/10.1525/sop.2003.46.4.491.

Whipple, N., Bernier, A., & Mageau, G. A. (2011). A dimensional approach to maternal attachment state of mind: Relations to maternal sensitivity and maternal autonomy support. *Developmental Psychology*, *47*(2), 396–403. Available from http://dx.doi.org/10.1037/a0021310.

Whitman, R. D., Holocomb, E., & Zanes, J. (2010). Hemispheric collaboration in creative subjects: Cross-hemisphere priming in a lexical decision task. *Creativity Research Journal*, *22*(2), 109–118. Available from http://dx.doi.org/10.1080/10400419.2010.481480.

Wichi, R. B., de Angelis, K., Jones, L., & Irigoyen, M. C. (2009). A brief review of chronic exercise intervention to prvent autonomic nervous system changes during the aging process. *Clinics*, *64*(3), 253–258. Available from http://dx.doi.org/10.1590/S1807-59322009000300017.

Wigman, M. (1966). *Mary wigman: The language of dance*. Middletown, CT: Wesleyan University Press.

Wilhelm, K., Kovess, V., Rios-Seidel, C., & Finch, A. (2004). Work and mental health. *Social Psychiatry and Psychiatric Epidimiology*, *39*, 866–873. Available from http://dx.doi.org/10.1007/s00127-004-0869-7.

Wilke, C., Priebus, J., Biallas, B., & Frobose, I. (2011). Motor activity as a way of prevening musculoskeletal problems in string musicians. *Medical Problems of Performing Artists*, *26*(1), 24–29.

Wilkie, I. (2015). "Too many actors and too few jobs": A case for curriculum extension in UL vocational actor training. *London Review of Education*, *13*(1), 31–42.

Williams, C., & Ratel, S. (2009). *Human muscle fatigue*. New York: Routledge.

Wills, G. I. (2003). Forty lives in the bebop business: Mental health in a group of eminent jazz musicians. *British Journal of Psychiatry*, *183*, 255–259.

Wilson, M., & Deckert, J. L. (2009). A screening program for dancers administered by dancers. *Journal of Dance Medicine and Science*, *13*(3), 67–72.

Wilson, S. C., & Barber, T. X. (1983). The fantasy-prone personality: Implications for understanding imagery, hypnosis and parapsychological phenomena. In A. A. Sheikh (Ed.), *Imagery: Current theory, research and application* (pp. 340–387). New York: John Wiley & Sons.

Winner, E. (2000). The origins and ends of giftedness. *American Psychologist*, *55*(1), 159–169. Available from http://dx.doi.org/10.1037/0003-066X.55.1.159.

Wittmann, B. C., Bunzek, N., Dolan, R. J., & Duzel, E. (2007). Anticipation of novelty recruits reward system and hippocampus while promoting recollection. *Neuroimage*, *38*, 194–202. Available from http://dx.doi.org/10.1016/j.neuroimage.2007.06.038.

Wolfradt, U., & Engelmann, S. (2003). Depersonalization, fantasies, and coping behavior in clinical context. *Journal of Clinical Psychology*, *59*(10), 1117–1124. Available from http://dx.doi.org/10.1002/jclp.10204.

Wolfradt, U., Hempel, S., & Miles, J. N. V. (2003). Perceived parenting styles, depersonalization, anxiety and coping behavior in adolescents. *Personality and Individual Differences*, *34*(3), 521–532. Available from http://dx.doi.org/10.1016/S1053-8119(03)00248-9.

Wolkenstein, L., Schonenberg, M., Schirm, E., & Hautzinger, M. (2011). I can see what you feel, but I can't deal with it: Impaired theory of mind in depression. *Journal of Affective Disorders*, *132*, 104–111. Available from http://dx.doi.org/10.1016/j.ad.2011.02010.

Wollner, C. (2012). Is empathy related to perception of emotional expression in music? A multimodal time-series analysis. *Psychology of Aesthetics, Creativity, and the Arts*, *6*(3), 214–223. Available from http://dx.doi.org/10.1037/a0027392.

Woo, J.-M., & Postoiache, T. T. (2008). The impact of work environment on mood disorders and suicide: Evidence and implications. *International Journal of Disability and Human Development*, *7*(2), 185–200. Available from http://dx.doi.org/10.1515/IJDHD.2008.7.2.185.

Woodward, J., & Sikes, P. L. (2015). The creative thinking ability of musicians and nonmusicians. *Psychology of Aesthetics, Creativity, and the Arts*, *9*(10), 75–80. Available from http://dx.doi.org/10.1037/a0038177.

Woody, R. H. (1999). The musician's personality. *Creativity Research Journal*, *12*(4), 241–250.

Wright, M. O., Crawford, E., & Del Castillo, D. (2009). Childhood emotional maltreatment and later psychological distress among college students: The mediating role of maladaptive schemas. *Child Abuse and Neglect*, *33*, 59–68. Available from http://dx.doi.org/10.1016/j.chiabu.2008.12.007.

Wyon, M. (2005). Cardiorespiratory training for dancers. *Journal of Dance Medicine and Science*, *9*(1), 7–12.

Wyon, M. (2010). Preparing to perform periodization and dance. *Journal of Dance Medicine and Science*, *14*(2), 67–72.

Wyon, M. A., & Koutedakis, Y. (2013). Muscular fatigue: Considerations for dance. *Journal of Dance Medicine and Science*, *17*(2), 63–69. Available from http://dx.doi.org/10.12678/1089-313X.17.2.63.

Wyon, M. A., & Redding, E. (2005). Physiological monitoring of cardiorespiratory adaptations during rehearsal and performance of contemporary dance. *Journal of Strength and Conditioning Research*, *19*(3), 611–614.

Wyon, M. A., Deighan, M. A., Nevill, A. M., Doherty, M., Morrison, S. L., Allen, N., ... George, S. (2007). The cardiorespiratory, anthropmetric, and performance characteristics of an international/national touring ballet company. *Journal of Strength and Conditioning Research*, *21*(2), 389–393.

Yeragani, V. K., Sobolewski, E., Kay, J., Jampala, V. C., & Igel, G. (1997). Effect of age on long-term heart rate variability. *Cardiovascular Research*, *35*, 35–42.

Yoshie, M., Kudo, K., & Ohtsuki, T. (2008). Effects of psychological stress on state anxiety, electromyographic activity, and arpeggio performance in pianists. *Medical Problems of Performing Artists*, *23*, 120–132.

Yoshie, M., Kudo, K., Murakoshi, T., & Ohtsuki, T. (2009). Music performance anxiety in skilled pianists: Effects of social-evaluative performance situation on subjective autonomic, and electromyographic reactions. *Experimental Brain Research*, *199*, 117–126. Available from http://dx.doi.org/10.1007/s00221-009-1979-y.

Young, I. T., Iglewicz, A., Glorioso, D., Lanouette, N., Seay, K., Ilapakurti, M., & Zisook, S. (2013). Sicide bereavement and complicated grief. *Dialogues in Clinical Neuroscience*, *14*(2), 177–186.

Young, K. S., Parsons, C. E., Stein, A., & Kringelbach, M. L. (2012). Interpreting infant vocal distress: The ameliorative effect of musical training in depression. *Emotion*, *12*(6), 1200–1205. Available from http://dx.doi.org/10.1037/a0028705.

Young, L. N., Winner, E., & Cordes, S. (2013). Heightened incidence of depressive symptoms in adolescents involved in the arts. *Psychology of Aesthetics, Creativity, and the Arts*, *7*(2), 197–202. Available from http://dx.doi.org/10.1037/a0030468.

Yu, B.-H., Nelesen, R., & Ziegler, M. G. (2001). Mood states and impedance cardiography-derived hemodynamics. *Annals of Behavioral Medicine*, *23*(1), 21–25.

Yu, X.-L., Zhang, C., & Zhang, J.-B. (2014). Causal interactions between the cerebral cortex and the autonomic nervous system. *Science China Life Science*, *57*(5), 532–538. Available from http://dx.doi.org/10.1007/s11427-014-4627-0.

Zabelina, D. L., & Robinson, M. D. (2010a). Creativity as flexible cognitive control. *Creativity Research Journal*, *4*(3), 136–143. Available from http://dx.doi.org/10.1037/a0017379.

Zabelina, D. L., & Robinson, M. D. (2010b). Don't be so hard on yourself: Self-compassion facilitates creative originality among self-judgmental individuals. *Creativity Research Journal*, *22*(3), 288–293. Available from http://dx.doi.org/10.1080/10400419.2010.503538.

Zabellina, D. L., Felps, D., & Blanton, H. (2013). The motivational influence of self-guides on creative pursuits. *Psychology of Aesthetics, Creativity, and the Arts*, *7*(3), 112–118. <http://dx.doi.org/10.1037/a0030464>

Zachariou, A., & Whitebread, D. (2015). Musical play and self-regulation: Does musical play allow for the emergence of self-regulatory behaviors? *International Journal of Play*, *4*(2), 116–135. Available from http://dx.doi.org/10.1080/21594937.2015.1060572.

Zajac, K., & Kobak, R. (2009). Caregiver unresolved loss and abuse and child behavior problems: Intergenerational effects in a high risk sample. *Development and Psychopathology*, *21*, 173–187.

Zakowski, S. G., Harris, C., Krueger, N., Laubmeier, K. K., Garrett, S., Flanigan, R., & Johnson, P. (2003). Social barriers to emotional expression and their relations to distress in male and female cancer patients. *British Journal of Health Psychology*, *8*, 271–286.

Zander, M. F., Voltmer, E., & Spahn, C. (2010). Health promotion and prevention in higher music education: Results of a longitudinal study. *Medical Problems of Performing Artists*, *25*(2), 54–65.

Zatorre, R. J., & Halpern, A. R. (2005). Mental concerts: Musical imagery and auditory cortex. *Neuron*, *47*, 9–12. Available from http://dx.doi.org/10.1016//j.neuron.2005.06.013.

Zausner, T. (1998). When walls become doorways: Creativity, chaos theory, and physical illness. *Creativity Research Journal*, *11*(1), 21–28. Available from http://dx.doi.org/10.1207/s15326934crj1101_3.

Zaza, C., Charles, C., & Muszynski, A. (1998). The meaning of playing-related musculoskeletal disorders to classical musicians. *Social Science and Medicine*, *47*(12), 2013–2023.

Zhang, L., & Sternberg, R. J. (2011). Revisiting the investment theory of creativity. *Creativity Research Journal*, *23*(3), 229–238. Available from http://dx.doi.org/10.1080/10400419.2011.595974.

Zhang, W., Zhang, Q., Yu, B., & Zhao, L. (2015). Knowledge map of creativity research based on keywords network and co-word analysis, 1992–2011. *Quality and Quantity*, *49*, 1023–1038. Available from http://dx.doi.org/10.1007/s11135-014-0032-9.

Zinn, M., McCain, C., & Zinn, M. (2000). Musical performance anxiety and the high-risk model of threat perception. *Medical Problems of Performing Artists*, *15*, 65–71.

Zoellner, T., & Maercker, A. (2006). Posttraumatic growth in clinical psychology—A critical review and introduction of a two component model. *Clinical Psychology Review*, *26*, 626–653. Available from http://dx.doi.org/10.1016/j.cpr.2006.01.008.

Zuskin, E., Schachter, E. N., Kolcic, I., Polasek, O., Mustajbegovic, J., & Arumugam, U. (2005). Health problems in musicians: A review. *Acta Dermatovenereal Croat*, *13*(4), 247–251.

Websites

www.afm.org
https://www.sagaftra.org
www.dansetrack.com

Index

Note: Page numbers followed by "*f*," "*t*," and "*b*" refer to figures, tables, and boxes, respectively.

D

E

R

S

T

9780128040515